002

59 Springer Series in Chemical Physics
Edited by P. Toennis

Springer Series in Chemical Physics

Editors: Vitalii I. Goldanskii   Fritz P. Schäfer   J. Peter Toennies

Managing Editor: H. K. V. Lotsch

Henning Bockhorn (Ed.)

Soot Formation in Combustion

Mechanisms and Models

With 287 Figures

Springer-Verlag
Berlin Heidelberg New York
London Paris Tokyo
Hong Kong Barcelona
Budapest

Professor Dr. Henning Bockhorn

Fachbereich Chemie, Universität Kaiserslautern
Erwin-Schrödinger-Strasse, Gebäude 54
D-67663 Kaiserslautern, Germany

Series Editors

Professor Dr. Fritz Peter Schäfer

Max-Planck-Institut
für Biophysikalische Chemie
D-37077 Göttingen-Nikolausberg, Germany

Professor Vitalii I. Goldanskii

Institute of Chemical Physics
Academy of Sciences
Ulitsa Kossigyna 4
Moscow, 117334, Russia

Professor Dr. J. Peter Toennies

Max-Planck-Institut
für Strömungsforschung
Bunsenstrasse 10
D-37073 Göttingen, Germany

Managing Editor: Dr. Helmut K. V. Lotsch
Springer-Verlag, Tiergartenstrasse 17,
D-69121 Heidelberg, Germany

ISBN 3-540-58398-X Springer-Verlag Berlin Heidelberg New York

CIP data applied for

Typesetting: Camera ready by author
SPIN 10423771   54/3140 - 5 4 3 2 1 0 - Printed on acid-free paper

Preface

The formation of soot during the incomplete combustion of hydrocarbons remains one of the least solved problems of combustion. Despite essential progress in understanding single phenomena in recent years, no comprehensive theory or models that predict the most important phenomena over a wide range of boundary conditions are currently available.

The necessity of scientific research and discussion for the problem of soot formation in combustion of hydrocarbons is documented – in addition to colloquia devoted to soot formation during regularly occurring national and international meetings on combustion – in symposia or workshops that pick up special aspects of soot formation. In the last decade the series of workshops started with a symposium held in 1980 in the General Motors Research Laboratories, Warren, MI, "Particulate Carbon: Formation During Combustion" (cf. D.C. Siegla, G.W. Smith (Eds), Particulate Carbon - Formation During Combustion, Plenum Press, New York 1981). The series was continued in 1981 with a workshop about "Soot in Combustion Systems" organized by J. Lahaye and G. Prado in Le Bischenberg, France (cf. J. Lahaye, G. Prado (Eds), Soot in Combustion Systems and its Toxic Properties, Plenum Press, New York 1983). In 1989 the discussion of soot formation was reopened with a round table discussion organised by H. Jander and H. Gg. Wagner in Göttingen, Germany (cf. H. Jander, H. Gg. Wagner, Soot Formation in Combustion - An International Round Table Discussion, Nachrichten der Akademie der Wissenschaften in Göttingen, II Mathematisch-Physikalische Klasse, Jahrgang 1989, Nr. 3, Vandenhoeck und Ruprecht, Göttingen 1990). The main feature of this round table discussion was that several aspects of soot formation were discussed according to different hypotheses that were emphasized and prepared by the participants. This structure guaranteed a vivid discussion of the items presented and the recorded and edited contributions to the discussions had some impact on currently pursued approaches to the problem of soot formation.

The last workshop of this series for the time being was held from 29th September to 2nd October 1991 in Heidelberg, Germany. The aim of this workshop "Mechanism and Models of Soot Formation" was twofold: First, the most recent results from worldwide research with different approaches in the field of soot formation in combustion was presented. Second, much time was devoted to the discussion of controversies, so that the essentials of different approaches were examined in detail allowing some conclusions to be drawn and some perspectives developed for future research in this field. The organisation of the workshop "Mechanisms and Models of Soot Formation" was possible with the help and cooperation of the International Science Forum (IWH), Heidelberg. Dr. Theresa Reiter and her coworkers provided the ambience that stimulated hard scientific work and intensive discussions. Major financial support for inviting the leading scientists in the field of soot formation from various parts of the world was supplied by the Volkswagen-Stiftung. This and the personal effort of Dr. Hans Plate is gratefully acknowledged. In addition, the Commission of the European Communities and the Stiftung Universität Heidelberg contributed essential financial support and guaranteed the success of the meeting. The organisation of the workshop would not have been possible without the untiring efforts of Maria Klotz.

The present book reflects the structure and the results of the workshop. The articles from the different authors are based on their contributions presented at the workshop. These articles were prepared some time after the workshop digesting and considering the discussions during the days in Heidelberg. Furthermore, the book contains all discussions on the single contributions, the summarising discussion of each part of the workshop (the single contributions have been grouped into six parts) and a final and concluding discussion. All discussions were recorded, typed, and edited in a kind of iterative procedure in close correspondence with the authors. This sometimes demanding process finally brought about a full documentation of the strengths and weaknesses of the various most recent hypotheses and approaches to a variety of problems connected with soot formation during combustion of hydrocarbons. Though the contributions to the discussions (as well as the articles) have been edited repeatedly, the written version reflects the liveliness and frankness of the discussions that created a kind of spirit of that workshop. From this, the book is more a comprehensive discussion and critical documentation of the most recent scientific approaches to soot formation rather than merely the proceedings of a conference.

The editing of this book including the discussions of all articles required the strong cooperation of all authors, contributors to the discussions, and the publisher with the editor. This excellent cooperation is appreciated very much. Many of my coworkers have been involved in the procedure of translating the contributions of the single authors into TEX , typing the contributions to the discussions from tape, forming a first written version from this, introducing the corrections from the authors, applying the conventions of the publisher to all articles, equations, figures, tables, and other pieces of text. The unceasing help of Maria Klotz, Andreas Hornung, Ursel Gerhard, Michael Marquardt and the sustained assistance of Thomas Schäfer, Markus Kraft, Thorsten Klos, Jürgen Weichmann enabled the successful completion of this book.

Kaiserslautern, February 1994   Henning Bockhorn

Contents

Part II

Part III

Part IV

Part V

Part VI

Part VII

Soot Formation in Combustion

The short introduction preceding the seven parts of this book provides an easy access to the field for newcomers before treating the single subjects in detail.

A Short Introduction to the Problem – Structure of the Following Parts

Henning Bockhorn

Interdisziplinäres Zentrum für wissenschaftliches Rechnen,
Universität Heidelberg,
69120 Heidelberg, Fed. Rep. of Germany

Under ideal conditions the combustion of hydrocarbons leads to mainly carbon dioxide and water. Ideal conditions may be specified by stoichiometric composition of the combustible mixture, i.e. the oxygen content of the mixture everywhere is sufficient to convert the fuel completely according to the formal chemical equation

$$C_xH_y + (x + y/4)\, O_2 \longrightarrow x\, CO_2 + y/2\, H_2O\,.$$

Under these conditions a maximum of heat is released and a maximum of chemical energy is available for mechanical work.

In practical combustion devices such as industrial furnaces, gas turbines, or combustion engines conditions locally deviate from ideality. If the locally present oxygen is not sufficient to convert the fuel according to the above equation, in addition to carbon dioxide and water other products of incomplete combustion such as carbon monoxide, hydrogen, hydrocarbons and soot appear. Then the time available for the formation of soot, for the mixing between the fuel-rich fluid parcels and oxidizer, and for the oxidation of soot determines the appearance of soot in the exhaust of the combustion device. In addition to reducing of the combustion efficiency during incomplete combustion the formation of soot has attracted interest for numerous other reasons.

- The formation of soot during incomplete combustion of hydrocarbons is connected with the formation of a variety of hydrocarbons. Some of them – in particular some polynuclear aromatic hydrocarbons – are suspected to have hazardous effects on human health.
- In furnaces for industrial applications or in heat generators the intermediate formation of soot is desired to enhance the heat transfer by radiation. The soot has to be oxidized before the exhaust from these devices is released into the environment.
- Soot is a product of the chemical industry with a wide field of applications, e.g. as filler in tires or other materials, as toner in copiers, or in

printing colors, etc. The properties of soot have to be tuned according to its various applications.

The above-mentioned problems connected with soot formation during incomplete combustion of hydrocarbons or the various applications of soot demand a complete knowledge of the different processes leading from hydrocarbon molecules to soot particles.

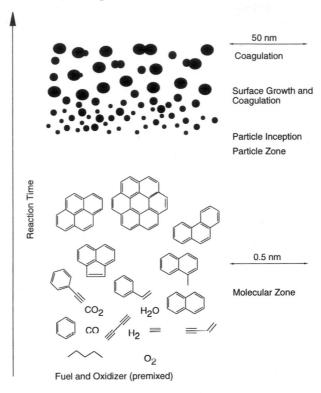

Fig. 1. A rough picture for soot formation in homogeneous mixtures (premixed flames)

It is obvious that the formation of soot, i.e. the conversion of a hydrocarbon fuel molecule containing few carbon atoms into a carbonaceous agglomerate containing some millions of carbon atoms, is an extremely complicated process. It is a kind of gaseous–solid phase transition where the solid phase exhibits no unique chemical and physical structure. Therefore, soot formation encompasses chemically and physically different processes, e.g. the formation and growth of large aromatic hydrocarbons and their transition to particles, the coagulation of primary particles to larger aggregates, and the growth of solid particles by picking up growth components from the gas phase. Figure 1 depicts a very rough picture for soot formation for a homogeneous mixture of fuel and oxidizer based on presently discussed hypotheses.

According to the picture given in Fig. 1 the hydrocarbon fuel in premixed flames is degraded during oxidation into small hydrocarbon radicals from which, under fuel-rich conditions, small hydrocarbons, particularly acetylene are formed. The latter adds hydrocarbon radicals for growth and the growing unsaturated (radicalic) hydrocarbons form aromatic rings when containing a sufficiently large number of carbon atoms. The formation of larger aromatic rings occurs mainly via the addition of acetylene. All these processes occur within molecular length scales.

The growth in the third dimension is supposed to happen by coagulation of larger aromatic structures forming primary soot particles. These primary particles quickly coagulate picking up simultaneously molecules from the gas phase for surface growth. Surface growth contributes to the major part to the final soot concentration in sooting flames while coagulation – switching the length scales to particle dimensions – determines the final size of the soot particles. The irregular aggregate structure of soot particles is also attributed to coagulation.

In non-homogeneous mixtures, e.g. diffusion flames, all chemical processes sketched roughly in Fig. 1 are overlayed by the mixing of fuel and oxidizer. The interaction of mixing and chemical reactions under turbulent conditions is an additional problem connected with the technical application of combustion.

The above-mentioned processes contribute to the formation of the bulk of soot. Many other phenomena observed during soot formation, e.g. the formation of electrically charged soot particles, the formation – charged and neutral – of fullerenes, or the formation of high molecular tarry modifications with optical properties quite different from polynuclear aromatic hydrocarbons or carbon black necessitate a refining of this picture. However, the mechanistic picture given for the formation of the bulk of soot emphasizes the ideas behind the scientific approaches to the problem of soot formation during incomplete combustion of hydrocarbons:

- Soot formation is largely determined by the chemistry that forms and converts small hydrocarbon radicals, that leads to precursor molecules from these species, and that converts the precursor molecules. Consequently, detailed information about the reactions of hydrocarbon radicals with aromatic hydrocarbons is necessary to understand the phenomena connected with soot formation. In part I of this book such information is collected, starting with an investigation of the high temperature reactions of phenyl acetylene with hydrogen atoms deriving also rate coefficients for the reactions of phenyl with acetylene. This is followed by papers about benzene and higher hydrocarbon formation during allene pyrolysis and the formation of aromatic hydrocarbons in low pressure decane and kerosene flames. The last two papers of this part are focused on larger precursor molecules: large ionic polycyclic aromatic hydrocarbons that possibly follow similar pathways to soot as neutral ones and large non-aromatic molecular structures that may contribute appreciably in an early phase to soot formation in hydrocarbon flames. These papers also contribute to the discussion of the "fine structure" of the mechanims of soot formation.

– In addition to separate elementary steps from the complex system of reactions that occur in combustion of hydrocarbons under fuel-rich conditions, a second approach is to monitor measurable properties in flames and to reproduce these properties by numerical simulation with self-consistent models. The papers contained in part II provide the experimental basis for the second approach and present models of that kind: Starting with experimental determination of radical concentrations in sooting hydrocarbon diffusion flames, which also addresses the complex interaction of mixing and chemical reactions, part II is complemented by new methods for on-line detection of large aromatic hydrocarbons in sooting flames and investigation of temperature and pressure dependence of PAH and soot concentrations over a wide range of pressures and temperatures. A detailed model for PAH and soot formation reproducing experimentally determined soot concentrations and soot particle sizes from premixed flames concludes this part.

– As mentioned above, the largest part of soot in flames is formed from surface growth reactions. Therefore, the understanding of the mass growth of soot by surface reactions is a key to the understanding of soot formation. The first paper of part III focuses on the question whether mass growth of soot or oxidation of soot particles in diffusion flames is a function of the local mixture composition or a function of the history of soot particles along their trajectories in the flames. This is followed by an investigation of temperature and concentration effects on mass growth of soot in diffusion flames where temperature and concentrations have been varied independently from other flame properties. Active site approach for surface growth reactions – a concept that comes from heterogeneous catalysis – versus recovery of surface sites by reactions with species from the gas phase is discussed extensively for different flame systems and finally experimental evidence for micro-droplets as soot particle precursors is presented.

– Practical systems and flames under different burning conditions are the challenges for any model of soot formation and oxidation. In practical systems chemical reactions interact with mixing under laminar or turbulent conditions. Part IV contains articles on soot formation, oxidation, and inhibition by additives in practical systems. Practical systems range from flames with fuel additives, counterflow partially premixed diffusion flames, shock tube devices, plug flow reactors, and diesel engines. Modelling – validated by comparison with experimental results from different flame systems – as well as experimental investigations are topics of the discussion in part IV.

- To control soot formation during combustion of hydrocarbons, all information about the single processes has to be compiled into models. Models for soot formation in laminar flames are inevitably prerequisites for models of soot formation in practical systems such as turbulent flames. The discussion of successes and uncertainties in modelling of soot formation in laminar flames is one important topic of part V of this book. Models that have been referred to several times in other parts of the book are evaluated and rated. Alternative routes, e.g. ionic mechanisms, from gaseous fuel molecules containing some carbon atoms to large soot particles containing some millions of carbon atoms are discussed, as well as other essential parts of detailed chemical models for soot formation.

- The last step towards practical combustion systems are turbulent (diffusion) flames. Part VI contains different approaches to reduce the complex chemical description of soot formation and different approaches to link the complex chemistry to a turbulent flow. Because the numerical effort for modelling turbulent flames is much larger than for one-dimensional laminar flames, models for soot formation with detailed chemical mechanisms are presently not applicable to turbulent flames, unless the large system of chemical reactions can be significantly reduced. Models for soot formation in turbulent flames presented in part VI contain a limited number of variables to describe the process of soot formation. Due to the reduction or simplification, parts of these models have to be "calibrated" for the system under consideration.

- A final discussion contained in part VII resumes the discussion of all items that have been presented during the workshop and reveals some perspectives for future work on the solution of unsolved problems and new approaches.

PAH Formation in Sooting Flames
Experimental Investigation and Kinetics

The formation of soot from incomplete combustion of hydrocarbons is largely determined by the chemistry that forms and converts small hydrocarbon radicals, that leads to precursor molecules from these species, and that converts the precursor molecules. Consequently, detailed information about the reactions of hydrocarbon radicals with aromatic hydrocarbons is necessary for developing mechanisms and models for soot formation. In part I of this book such information is collected starting with an investigation of the high temperature reactions of phenyl acetylene with hydrogen atoms deriving also rate coefficients for the reactions of phenyl with acetylene. This is followed by papers about benzene and higher hydrocarbon formation during allene pyrolysis and the formation of aromatic hydrocarbons in decane and kerosene low pressure flames. The last two papers of this part are focused on larger precursor molecules: large ionic polycyclic aromatic hydrocarbons that possibly follow similar pathways to soot as neutral ones and large non-aromatic molecular structures that may contribute appreciably in an early phase to soot formation in hydrocarbon flames. Part I - as the following parts - is completed with a concluding discussion.

Investigation of the High Temperature Reaction of Hydrogen Atoms with Phenylacetylene

Jürgen Herzler, Peter Frank

DLR, Institut für Physikalische Chemie der Verbrennung,
70569 Stuttgart, Fed. Rep. of Germany

Abstract: The reaction of $C_6H_5 - C_2H$ + H has been investigated at elevated temperatures in the postshock region behind reflected shocks. The thermal decomposition of very small amounts (1 - 3 ppm) of C_2H_5I served as a source for H-atoms. Atomic resonance absorption spectrometry (ARAS) was applied to measure simultaneously H-atom and I-atom profiles. The experiments covered the temperature range from 1190 to 1560 K at total pressures of $2.3 \cdot 10^5 \pm 0.2 \cdot 10^5$ Pa. A rate coefficient of

$$k_{1.1} = 3.0 \cdot 10^{14} T^{-0.48} e^{-700/T} \quad cm^3 mol^{-1} s^{-1}$$

is obtained for the reaction

$$C_6H_5 - C_2H \quad + \quad H \quad \longrightarrow \quad \text{Products} \tag{1.1}$$

From the available thermochemical data two product channels are plausible:

$$C_6H_5 - C_2H \quad + \quad H \quad \longrightarrow \quad C_6H_5 \quad + \quad C_2H_2, \qquad \Delta_R H_{298}^0 = 5.4 \text{ kcal mol}^{-1} \tag{1.1a}$$

$$C_6H_5 - C_2H \quad + \quad H \quad \longrightarrow \quad C_6H_4 - C_2H \quad + \quad H_2, \quad \Delta_R H_{298}^0 = 6.3 \text{ kcal mol}^{-1} \tag{1.1b}$$

The main product channel at elevated temperatures is route 1.1a. Detailed evaluation of existing thermodynamic data enables the calculation of an equilibrium constant for reaction 1.1a. From this and the rate coefficient for reaction 1.1a a rate coefficient

$$k_{1.2} = 2.7 \cdot 10^8 T^{0.7} e^{+1800/T} \quad cm^3 mol^{-1} s^{-1}$$

for the reaction of phenyl radicals with acetylene

$$C_6H_5 \quad + \quad C_2H_2 \quad \longrightarrow \quad C_6H_5 - C_2H \quad + \quad H \tag{1.2}$$

is obtained which is in good agreement with recent results from *Fahr* et al [1.1]. Reaction 1.2 is an important process in sooting flames.

1.1 Introduction

The formation of soot has attracted considerable attention due to its importance in environmental pollution and its impact on the efficiency of energy conversion processes. As the sooting tendency of a fuel in a diffusion flame is related to the specific pyrolytic mechanism of that fuel [1.2,1.3], the necessity of a comprehensive understanding of the soot chemistry is evident. Specific questions are, for example, the formation of the first aromatic ring from aliphatic species, the transition from the first to the second aromatic ring structure, and the role of aromatic radicals reacting with stable hydrocarbons.

Phenyl radicals are supposed to play an important role as intermediates in high temperature reactions leading to formation of polycyclic hydrocarbons [1.4]. Their reaction with acetylene which is present in flames as a fuel and also as an intermediate leads to phenylaceteylene (see Fig.1.1). Phenylacetylene is assumed to initiate the inception of polycyclic aromatic hydrocarbons (PAH)[1.5].

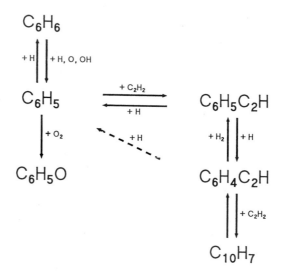

Fig. 1.1. Principal reaction pathways for phenyl formation and destruction in a benzene - air flame

In this work the rate coefficient for the reaction

$$C_6H_5 - C_2H \quad + \quad H \quad \longrightarrow \quad \text{Products} \tag{1.1}$$

is measured directly by using the shock tube technique. The thermal decomposition of C_2H_5I provides H-atoms down to temperatures of about 1150 K.

Therefore C_2H_5I was used as H-atom source. Detailed kinetic modelling of the measured species profiles brought about a rate coefficient of the reaction

$$C_6H_5 - C_2H \quad + \quad H \quad \longrightarrow \quad C_6H_5 \quad + \quad C_2H_2 \qquad (1.1a)$$

The rate coefficient for the reverse reaction

$$C_6H_5 \quad + \quad C_2H_2 \quad \longrightarrow \quad C_6H_5 - C_2H \quad + \quad H \qquad (1.2)$$

that could not be studied directly under the prevailing experimental conditions was obtained from the rate coefficient of reaction (1.1a) and the corresponding equilibrium constant.

1.2 Experimental

1.2.1 Experimental Set-up

Details of the shock tube as well as the optical setup have been described in more detail elsewhere [1.6, 1.7]. Therefore, only a short summary will be given here: The measurements were performed behind the reflected shock front close to the end flange of the shock tube. Atomic resonance absorption (ARAS) was used to monitor time dependent H- and I- atom concentrations. Microwave - excited discharges in H_2 or CH_3I diluted with He were used as light sources for the absorption measurements.

H- and I-atom concentrations were measured simultaneously recording the intensities at 121.5 nm using an oxygen spectral filter (for H-atoms) and at 164.2 nm using a 1 m vacuum-UV monochromator (for I-atoms). The transmitted intensity signals were recorded by a 2-channel digital storage oscilloscope and stored into a personal computer. A program computes an averaged signal and transforms the intensity with the aid of calibration curves into particle concentration. Because of the high sensitivity of the optical measuring technique ($[H] > 5 \cdot 10^{11} cm^{-3}$, $[I] > 1 \cdot 10^{12} cm^{-3}$) very small initial concentrations of the reactants could be used. The background absorption of phenylacetylene at 121.5 nm which was applied in surplus was measured in a separate series of experiments. No temperature dependence for the absorption cross section was found in the investigated temperature range, $\sigma_{121.5nm(C_6H_5 - C_2H)} = (10 \pm 3) \cdot 10^{-17} cm^2$.

The initial concentrations were 10 to 20 ppm $C_6H_5 - C_2H$, and 1 to 3 ppm C_2H_5I, both diluted with argon. The test gases were of high purity (Ar: 99.9999 %, $C_6H_5 - C_2H$: 97.0 %, C_2H_5I: 99.0 %).

1.2.2 Data Evaluation

The H- and I- atom concentration profiles reported here are based on calibration experiments, (see Fig.1.2).

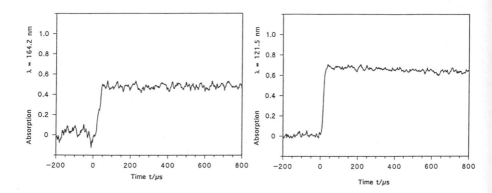

Fig. 1.2. H- and I-ARAS profiles from thermal decomposition of C_2H_5I; X_0 $_{C_2H_5J}$ = 1.5 ppm, $T_5 = 1345$ K; $p_5 = 2.12 \cdot 10^5$ Pa

For sensitive iodine-atom detection I-ARAS was exploited. To achieve an absolute calibration for iodine-atom densities known amounts of methyl-iodide, controlled by gaschromatographic analysis, were shock heated. The absorption of iodine-atoms, produced during the thermal dissociation of CH_3I was recorded. For temperatures above 1350 K the dissociation is sufficiently fast to attain constant absorption levels during the observation time interval of about 800 μs [1.7]. The calibration curve for I-ARAS was measured at total pressures of about 100 kPa. No temperature dependence in the temperature range from 1350 to 1900 K was detected.

The source of hydrogen-atoms was the thermal decomposition of ethyl iodide, followed by the fast decay of the ethyl radical at temperatures above 1100 K:

$$C_2H_5I \quad \longrightarrow \quad C_2H_5 \quad + \quad I, \qquad \Delta_R H^0_{298} \approx 56 \text{ kcal mol}^{-1} \qquad (1.3a)$$

$$C_2H_5 \quad \longrightarrow \quad C_2H_4 \quad + \quad H, \qquad \Delta_R H^0_{298} = 36.5 \text{ kcal mol}^{-1} \qquad (1.3b)$$

H-atom concentrations were calibrated using the reaction of O-atoms with H_2 to produce H-atoms in the temperature range from 1550 to 2100 K.

This procedure is suitable because the thermal dissociation of H_2 is very slow at temperatures below 2100 K. The source of O-atoms was the dissociation of N_2O which is relatively fast, so that a mixture of 2 ppm N_2O and 200 ppm H_2 is suitable to cover the above temperature range [1.6]. In addition, experiments with N_2O and H_2 concentrations of 100 ppm and 2000 ppm, respectively, were carried out to allow for calibration measurements down to temperatures of about 1250 K. From the data in ref. [1.6] an error of $\pm 20\%$ was estimated for extrapolating the rate coefficient of the N_2O decomposition to 1250 K. This results in an error of the calibration of less than 5%. No temperature dependence of the H-calibration curve was found (see Fig.1.3).

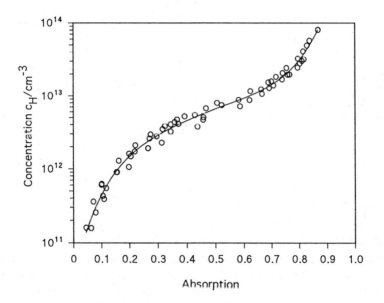

Fig. 1.3. Calibration curve for H-ARAS

Actually, the high temperature decomposition mechanism of C_2H_5I is not sufficiently clarified. A second reaction channel could be responsible for the formation and depletion of H- and I-atoms. The thermochemically most plausible reaction pathway [1.8, 1.9, 1.10]

$$C_2H_5I \quad \longrightarrow \quad C_2H_4 \quad + \quad HI, \quad \Delta_R H^0_{298} \approx 20 \, \text{kcal mol}^{-1} \quad (1.4a)$$

$$HI \quad \longrightarrow \quad H \quad + \quad I, \quad \Delta_R H^0_{298} = 71.2 \, \text{kcal mol}^{-1} \quad (1.4b)$$

can not directly bias the experimental results because the thermal dissociation of hydrogen iodide is expected to proceed relatively slow under the experimental conditions [1.11]. However, the reaction of hydrogen-atoms with hydrogen-iodide

$$\text{H} \;+\; \text{HI} \;\longrightarrow\; \text{H}_2 \;+\; \text{I} \tag{1.5}$$

which is very fast ($k_{1.5} \approx 4 \cdot 10^{13} \text{cm}^3\text{mol}^{-1}\text{s}^{-1}$ under the prevailing conditions [1.11]) could affect the apparent H- and I- atom concentration levels. Therefore a series of model calculations were carried out, varying the reaction rate ratio of $r = r_{1.4}/r_{1.3}$ from 0.1 to 1.0. In all cases with $r \geq 0.2$, no constant concentration levels of hydrogen- and iodine- atoms were obtained in contrast to the observed experimental profiles (Fig.1.2). The results of these calculations give evidence that channel 1.4 plays only a minor role in the ethyl iodide decomposition under the present experimental conditions. A more detailed investigation of the C_2H_5I-decomposition mechanism is under progress.

In all experiments the concentration of phenylacetylene which was applied in surplus did not decrease noticeably during the observation time. Besides the H-atoms no other species contribute markedly to the time dependent part of the overall absorption. Therefore, it is easy to extrapolate to the starting point of the reaction. After being corrected for the background absorption from $C_6H_5 - C_2H$ this absorption value gives a precise measure of the initial H-atom concentration. The concentration of hydrogen-atoms determined by this method agreed very well with the H-atom concentrations from the in-situ iodine-ARAS measurements. Therefore we decided to exploit this method of producing H-atoms by a thermal process for our experiments, even if there are at present some open questions concerning the kinetic model.

1.3 Experimental Results

First of all a series of experiments was carried out to investigate directly the reaction of phenyl radicals with acetylene. For this purpose mixtures of 2 to 10 ppm nitrosobenzene - the precursor molecule of phenyl radicals - and of 100 to 200 ppm acetylene were shock heated at temperatures ranging from 1150 to 1500 K. In all of these experiments relatively low H- atom concentrations where observed and the H-atom concentrations attained constant levels after about 100 to 150 μs. Figure 1.4 shows a typical measured H-atom concentration profile (squares) for one of these measurements and the corresponding calculated concentration profiles of benzene, phenylacetylene, and phenyl-radicals.

The fast conversion to a stationary H-atom concentration was observed for all experimental conditions. The reason for this fast conversion is that the H-atoms formed in the reaction

$$\text{C}_6\text{H}_5 \;+\; \text{C}_2\text{H}_2 \;\longrightarrow\; \text{C}_6\text{H}_5 - \text{C}_2\text{H} \;+\; \text{H} \tag{1.2}$$

are consumed instantaneously in the very fast recombination reaction of phenyl radicals with H-atoms to form benzene

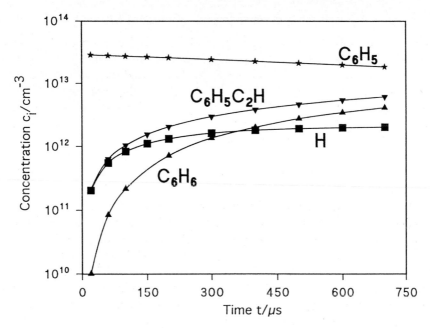

Fig. 1.4. Concentration profiles of H-atoms (measured) and of benzene, phenyl radicals and phenylacetylene (calculated); experimental conditions: $X_{0\,C_6H_5-NO}$ = 2 ppm, $X_{0\,C_2H_2}$ = 200 ppm, T_5 = 1200 K, p_5 = 2.38 · 10^5 Pa

$$C_6H_5 \quad + \quad H \quad \longrightarrow \quad C_6H_6 \tag{1.6}$$

Therefore, from the measured hydrogen atom profiles no value for the rate coefficient of reaction 1.2 could be determined.

In a next series of experiments the reverse reaction

$$C_6H_5 - C_2H \quad + H \quad \longrightarrow \quad \text{Products} \tag{1.1}$$

was studied by adding H-atoms from the thermal decomposition of ethyl iodide to phenylacetylene. In a preceeding scparate series of experiments it was validated that phenylacetylene under the present experimental conditions does not decompose in a unimolecular process at temperatures below 1570 K.

For all experiments the reactants were used in highly diluted mixtures. A typical experimental absorption profile is shown in Fig.1.5. Due to the high dilution only very few reactions contribute to the progress of the conversion. Sensitivity analysis confirmed that for the temperature interval of the present investigation only the reaction mechanism given in Table 1.1 has to be considered.

For temperatures below 1500 K the two reactions 1.1a and 1.6 are the predominating ones. Figure 1.6 shows the calculated concentration profiles of the main products and the measured hydrogen-atom concentration in

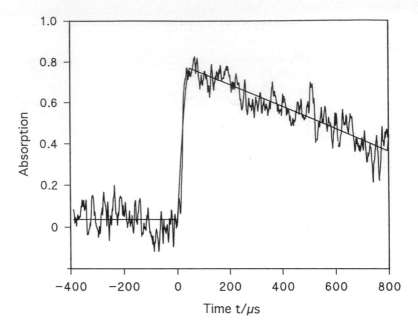

Fig. 1.5. Absorption profile at $\lambda = 121.6$ nm; experimental conditions: $X_{0\,H} = 1.4$ ppm, $X_{0\,C_6H_5-C_2H} = 15$ ppm, $T_5 = 1340$ K, $p_5 = 2.38 \cdot 10^5$ Pa

dependence on time for the relatively low temperature of about 1340 K. The calculated profiles are obtained considering only reactions 1.1a and 1.6. For a temperature range below 1500 K the experimentally measured H-atom concentrations are well predicted using this simplified mechanism.

Table 1.1. Reaction mechanism and rate coefficients

Reaction				$\log A$	n	E_a
1.1a	$C_6H_5 - C_2H + H$	$\rightarrow C_6H_5$	$+ C_2H_2$	14.4	-0.48	1.5
1.1b	$C_6H_5 - C_2H + H$	$\rightarrow C_6H_4 - C_2H$	$+ H_2$	$k_{1.1b} \leq$	$0.2 \cdot k_{1.1a}$	
1.6	C_6H_5	$+ H \rightarrow C_6H_6$		13.9		
1.7	C_6H_5	$\rightarrow l - C_6H_5$		13.7		72.5
1.8a	$l - C_6H_5$	$\rightarrow C_4H_3$	$+ C_2H_2$	62.6	-14.7	57.5
1.8b	$l - C_6H_5$	$\rightarrow C_6H_4$	$+ H$	58.4	-13.8	49.8
1.9	C_4H_3	$\rightarrow C_4H_2$	$+ H$	61.5	-13.9	61.4

Remarks: Rate coefficients are given in the form $k_i = A_i \, T^n \, e^{-E_a/RT}$; units: cm^3, mol, s, kcal;

Ref. for reaction 1.1, this work; Ref. for reactions 1.6 to 1.9, [1.12].

Rate coefficients for reaction 1.8 and 1.9 are derived from the fall-off rate expressions at $\rho \approx 1.5 \cdot 10^{-5}$ mol cm^{-3} and $\beta_c = 0.04$ according to the formalism given in ref [1.13].

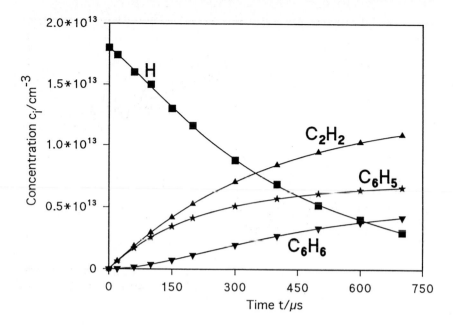

Fig. 1.6. Concentration profiles of H-atoms (measured) and of acetylene, benzene, and phenyl radicals (calculated); experimental conditions: same as in Fig.1.5

The influence of the recombination reaction 1.6 can be clearly seen at larger observation times. At large observation times more than 1/3 of the produced phenyl radicals are converted into benzene.

Figure 1.7 shows the sensitivity of a typical measured H-atom profile with respect to the rate coefficient of 1.6 (variation by a factor of 2 and 0.5, respectively). The sensitivity of the hydrogen concentration profiles with respect to the rate coefficient of reaction 1.6 is relatively low.

In contrast to this, the sensitivity of the H-atom concentration profile with respect to the rate coefficient of reaction 1.1a is very high, compare Fig. 1.8. A variation of $k_{1.1a}$ by 2 or 0.5 alters the H-atom concentration up to factors of 0.4 and 1.6, respectively.

At temperatures above 1500 K the thermal decomposition of the phenyl radicals has to be considered. This is demonstrated in Fig. 1.9. The dashed profile is calculated considering only reactions 1.1a and 1.6. The solid line which reproduces the measured data very well is calculated applying the complete reaction mechanism of Table 1.1. This mechanism comprises the unimolecular decomposition reaction of the phenyl radicals that has been investigated recently by *Braun-Unkhoff* and coworkers [1.12]. It can be clearly seen that the contributions of reactions 1.7 to 1.9 to the production of H-atoms increase with increasing reaction time.

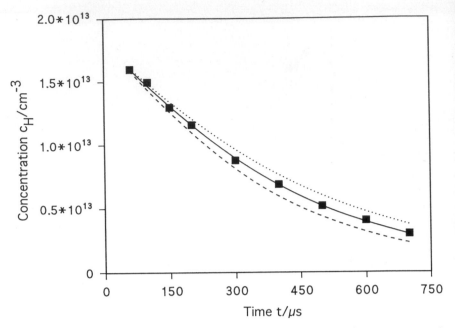

Fig. 1.7. Sensitivity of the measured H-atom concentration with respect to variations of $k_{1.6}$; experimental conditions: same as in Fig.1.5

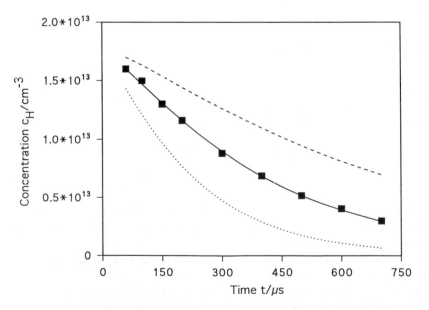

Fig. 1.8. Sensitivity of the measured H-atom concentration with respect to variations of $k_{1.1a}$; experimental conditions: same as in Fig.1.5

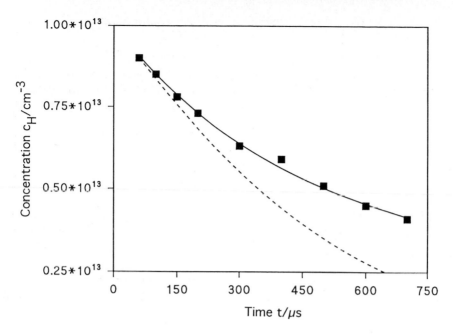

Fig. 1.9. Prediction of the H-atom concentration with different reaction mechanisms; dashed line: reaction 1.1a and 1.1b from Table 1.1; solid lines: all reactions from Table 1.1; symbols: measured H-atom concentrations; experimental conditions: $X_{0\,C_6H_5-C_2H} = 15$ ppm, $X_{0\,H} = 1$ ppm, $T_5 = 1560$ K, $p_5 = 2.27 \cdot 10^5$ Pa

1.4 Discussion of the Results

This is to our knowledge one of the first studies were under shock tube conditions the reaction of phenylacetylene with hydrogen atoms that are produced by the very fast thermal decomposition process

$$C_2H_5I \quad \longrightarrow \quad C_2H_5 \quad + \quad I \tag{1.3a}$$

$$C_2H_5 \quad \longrightarrow \quad C_2H_4 \quad + \quad H \tag{1.3b}$$

is investigated. The experimental results and computer simulations show that ethene which is also formed as a product during the thermal decomposition with $X_0 \leq 3$ ppm

  i is no additional source for H-atoms under the chosen experimental conditions

  ii reacts with negligible reaction rates with phenylacetylene and with other products resulting from the reactions given in Table 1.1.

The applied method allows the direct investigation of the over all reaction rate of the bimolecular reactions of hydrogen atoms, 1.1a and 1.1b, at elevated temperatures. From the measurements a rate coefficient of

$$k_{1.1} = 3.0 \cdot 10^{14} T^{-0.48} e^{-700/T} \quad \mathrm{cm^3 mol^{-1} s^{-1}}$$

is extracted.

For the consumption of H-atoms by phenylacetylene two possible reaction pathways have to be discussed, compare Table 1.1 (compare also the mechanism of PAH-formation proposed by *Frenklach* and coworkers [1.4]):

$$C_6H_5 - C_2H + H \quad \longrightarrow \quad C_6H_5 + C_2H_2, \qquad \varDelta_R H_{298}^0 = 5.4 \text{ kcal mol}^{-1}$$
$$(1.1a)$$

$$C_6H_5 - C_2H + H \quad \longrightarrow \quad C_6H_4 - C_2H + H_2, \quad \varDelta_R H_{298}^0 = 6.3 \text{ kcal mol}^{-1}$$
$$(1.1b)$$

In this study no product identification was possible. Therefore, from the present experimental results only the over-all rate of reactions 1.1a and 1.1b is obtained. For a temperature of 1030 K the over-all rate coefficient is $k_{1.1} = 5.4 \cdot 10^{12} \mathrm{cm^3 mol^{-1} s^{-1}}$. With the thermodynamic data given in Table 1.2 an equilibrium constant for reaction 1.1a of $K_{1.1a} = 9.3 \cdot 10^5 T^{-1.18} e^{-2500/RT}$ is calculated. Employing the results of *Fahr* [1.1] for the reverse of reaction 1.1a, $k_{1.2a} = 1.6 \cdot 10^{11} \mathrm{cm^3 mol^{-1} s^{-1}}$ at 1030 K, and the above equilibrium constant, a rate coefficient for reaction 1.1a of $k_{1.1a} = 3.7 \cdot 10^{12} \mathrm{cm^3 mol^{-1} s^{-1}}$ is obtained. For reaction channel 1.1b then a rate coefficient of $k_{1.1b} = 1.7 \cdot 10^{12} \mathrm{cm^3 mol^{-1} s^{-1}}$ at a temperature of 1030 K is derived.

The above estimations indicate that pathway 1.1a is the more favorable reaction channel. Detailed kinetic modelling of the measured species concentration profiles brings about a rate coefficent for reaction 1.1a of

$$k_{1.1a} = 2.5 \cdot 10^{14} T^{-0.48} e^{-700/T} \quad \mathrm{cm^3 mol^{-1} s^{-1}}$$

From this and the above equilibrium constant a rate coefficient for the reverse reaction 1.2

$$k_{1.2} = 2.7 \cdot 10^8 T^{0.7} e^{+1800/T} \quad \mathrm{cm^3 mol^{-1} s^{-1}}$$

is deduced.

For 1030 K this gives $k_{1.2} = 1.99 \cdot 10^{11} \quad \mathrm{cm^3 mol^{-1} s^{-1}}$ which is in good agreement with the data obtained by *Fahr* [1.1].

The heat of formation for phenyl radicals is now relatively well established. If we take uncertainties in the heat of formation of phenylacetylene of 2 kcal mol^{-1} maximum into account [1.16, 1.17], the values of $K_{1.1a}$ may change much less than by a factor of about 3. Based on these data the estimates for reaction channel 1.1a compared with reaction channel 1.1b identify reaction channel 1.1a as the dominant one. Furthermore, the rate coefficient $k_{1.2}$ that is calculated on this basis is close to that detemined experimentally by other groups.

Table 1.2. Thermodynamic data for the calculation of the equilibrium constant of reaction 1.1a

Species	H^0_{298} kcal mol^{-1}	S^0_{298} cal K^{-1} mol^{-1}	Ref.
C_6H_5	78.4 ± 0.2	69.0	1.14, 1.15
$C_6H_5 - C_2H$	75.2	76.2	1.15
$C_6H_4 - C_2H$	133.6	78.8	1.15

1.5 Acknowledgements

The authors wish to thank Prof.Th.Just for his interest in this study. We are very grateful to K.Wintergerst who performed the experimental work on ethyl iodide decomposition. We like to thank N.Ackermann and M.Kallfass for technical assistance. We are also very grateful to Dr.C.Wahl and M.Kapernaum for analyzing ethyl iodide and phenylaceteylene.

References

1.1   A. Fahr, W.G. Mallard, S.E. Stein: "Reaktions of Phenyl Radicals with Ethene, Ethyne and Benzene", in Twenty-First Symposium(International) on Combustion (The Combustion Institute, Pittsburgh 1986) p.825

1.2   S.C. Graham, J.B. Homer, J.L.J. Rosenfeld: "The Formation and Coagulation of Soot Aerosols Generated by the Pyrolysis of Aromatic Hydrocarbons", in Proceeding on Tenth International Symposium on Shock Tubes and Shock Waves, ed. by G. Kamimoto (Shock Tube Research Society, Kyoto, Japan 1975) p.621

1.3   T.S. Wang, R.A. Matula, R.C. Farmer: "Combustion Kinetics of Soot Formation from Toluene", in Eighteenth Symposium (International) on Combustion (The Combustion Institute, Pittsburgh 1981) p.1149

1.4   M. Frenklach, D.W. Clary, W.C. Gardiner Jr., S.E. Stein: "Effect of Fuel Structure on Pathways to Soot", in Twenty-First Symposium (International) on Combustion (The Combustion Institute, Pittsburgh 1986) p.1067

1.5   M. Frenklach, D.W. Clary, W.C. Gardiner Jr., S.E. Stein: "Detailed Kinetic Modeling of Soot Formation in Shock-Tube Pyrolysis of Acetylene", in Twentieth Symposium (International) on Combustion (The Combustion Institute, Pittsburgh 1984) p.887

1.6   P. Frank, Th. Just: Ber. Bunsenges. Phys. Chem. **89**, 181 (1985)

1.7   M. Braun-Unkhoff, P. Frank, Th. Just: Ber. Bunsenges. Phys. Chem. **94**, 1417 (1990)

1.8   S.W. Benson: Thermochemical Kinetics (J. Wiley, New York 1976)

1.9   J. Yuang, D.C. Conway: J. Phys. Chem. **43**, 1296 (1965)

1.10   S.W. Benson, E. O'Neal: J. Chem. Phys. **34**, 514 (1961)

1.11   D.L. Baulch, J. Duxbury, J. Grant, D.C Montague: J. Phys. Chem. Ref. Data **10**, Suppl.1, 1-1 (1981)

1.12   M. Braun-Unkhoff, P. Frank, Th. Just: "A Shock Tube Study on the Thermal Decomposition of Toluene and of the Phenyl Radical at high Temperatures ", in Twenty-Second Symposium (International) on Combustion (The Combustion Institute, Pittsburgh 1988) p. 1053

1.13   W.C. Gardiner Jr., J.Troe: "Rate Coefficients of Thermal Dissociation, Isomerization, and Recombination Reactions", in Combustion Chemistry, ed. by W.C. Gardiner Jr., (Springer-Verlag, Heidelberg, New York 1984) p.173

1.14   D.F. McMillen, D.M.Golde: Annu. Rev. Phys. Chem. **33**, 493 (1982)

1.15   M. Frenklach: private communication, data from S.E. Stein, NIST, Gaithersburgh 1991

1.16   D.W. Rogers, F.J. McLafferty: Tetrahedron **27**, 3765 (1971)

1.17   D.R. Stull, E.F. Westrum, G.C Sinke: The Thermodynamics of Organic Compounds (J. Wiley, New York 1969)

Discussion

Colket: I think that not only the rate coefficients of the reaction phenyl + acetylene are important for all these modeling efforts but also those of the reaction of phenyl + H. My question concerns the fact that you are actually measuring the rate for the reverse reaction of phenyl + acetylene. It is my experience that there is significant uncertainty not only in the heat of formation but also in the entropy for even small hydrocarbons. These uncertainties can easily explain differences by factors of two to three in the rate coefficients. Have you looked at potential uncertainties in the thermochemical data?

Frank: We have looked very carefully at the thermo-chemical data. For acetylene, H-atoms and also phenyl radicals the thermo-chemical data have uncertainties of two kcal maximum. For phenyl acetylene thermo-chemical data is well known from the work of Stull et al. In my opinion differences by factors of two to three in calculating the rate coefficients of the reaction of phenyl + acetylene from the corresponding equilibrium constant and the rate coefficient of the reverse reaction are too large to be traced back to uncertainties in the thermo-chemical data.

Homann: Can you tell us at which temperatures the thermal decomposition of phenyl is important?

Frank: Under our experimental conditions we find H atom signals which would bias the evaluation of the experimental data at temperatures above 1420 K. Below this temperature we are able to explain our experimental data solely with the reaction of phenylacetylene + H and the fast recombination reaction phenyl + H.

Benzene and Higher Hydrocarbon Formation During Allene Pyrolysis

Lisa D. Pfefferle, German Bermudez, James Boyle

Department of Chemical Engineering, Yale University, New Haven, Connecticut 06520-2159, USA

Abstract: Pyrolysis of allene in a microjet reactor at millisecond to second reaction times over a temperature range from 500 to 1700 K was used to study higher hydrocarbon growth processes from C_3 hydrocarbon species. Species detection of both stable and labile product masses was carried out using vacuum ultra violet (VUV) photoionization mass spectrometry. The first product observed with increasing temperature at ms time scales was mass 80 u followed at somewhat higher temperatures by smaller than parent-mass pyrolysis products. Low temperature benzene formation likely occurs through the molecular channel consisting of allene dimerization to a mixture of dimethylenecyclobutane isomers (DMCB) and subsequent isomerization and hydrogen loss to form benzene. Mass 79 u is observed at concentrations up to half the mass 80 u concentration prior to mass 78 u detection. The allene dimerization step is fast enough to account for the initial rate of mass 78 u production observed prior to significant allene decomposition. The 1,2-DMCB also reacts with allene to form a trimer at mass 120 u. The appearance of other high molecular weight products at masses 92 u, 116 u, 114 u and 158 u is also observed before allene decomposition occurs at greater than several percent levels. At higher temperatures and longer residence times (ms range), however, the benzene production rate shows a change in formation mechanism through a change in the apparent activation energy observed. At temperatures above 1580 K and ms time scales, mass 78 u production was consistent with formation through propargyl radical recombination. Higher hydrocarbon growth to multiple ring compounds was also observed at even longer residence times (ms range) with growth initially consistent with acetylene and possibly other small hydrocarbon addition to PAH radicals followed by reactive dimerization of small PAH compounds with the first dimers observed in the 400-600 u range.

2.1 Background

It has become increasingly apparent in the past several years that C_3 routes for soot formation in flames can be important under some if not many conditions. In several recent studies of fuel pyrolysis and combustion where both stable and labile product measurements were made (e.g. *Westmoreland* et al. [2.1], *Boyle* et al. [2.2]), C_4/C_2 growth mechanisms were determined to be not fast enough to account for observed benzene production for some of the fuels used, especially those of C_2 and C_3 hydrocarbons. *Westmoreland* et al. [2.3] have recently made a QRRK reaction pathway analysis of propargyl radical ($C_3H_3(P)$) dimerization and concluded that this route could in fact be fast enough to account for observed benzene production rates in several test flames including acetylene and ethylene. This mechanism had been proposed by *Wu* et al. [2.4] to explain benzene production in pyrolysis of allene. Benzene production from propargyl radical has also been studied by *Stein* et al. [2.5] and found to be a likely source of benzene production at flame conditions.

At low temperatures, however, the molecular channel for benzene formation can predominate [2.6, 2.7]. In early studies of the gas phase pyrolysis of allene at 800-900 K, *Bloomquist* et al. [2.6] observed dimethylenecyclobutane isomers as the primary reaction products. The 1,2-dimethylenecyclobutane (1,2-DMCB) isomer was produced at high yields by passing allene over quartz at 800 K with a 6 s mean residence time. Quantitative data was not obtained because of the high rate of polymerization of allene and its dimer. In addition, surface effects which are probably significant in this system due to the high exothermicity of the dimerization reaction were not accounted for.

In the current study mass 80 u (likely formed through allene dimerization) was the first product detected at low temperatures and residence times. At higher temperatures mass 78 u is more predominant than mass 80 u, and radical growth process for mass 78 u production become faster than the molecular channel. Two other C_3 channels have been proposed to play important roles in the formation of benzene. These are given in overall form below:

$$2 \quad C_3H_3(P) \quad \longrightarrow \quad C_6H_6 \text{ (linear)} \quad \longrightarrow \quad \text{benzene} \qquad (2.1)$$

$$C_3H_3 \quad + \quad C_3H_4(A) \quad \longrightarrow \quad C_6H_6 \text{ (benzene)} \quad + \quad H \qquad (2.2)$$

Both of these mechanisms were tested with the concentration dependence and temperature dependence observed. C_4/C_2 mechanisms were also initially tested but found to predict benzene production rates that were too low, especially as temperature increases in the 1550-1650 K range where allene decomposition becomes fast and total allene conversion was higher than 90% at 2 ms.

This study was also expanded to the pyrolysis of allene at longer (s) residence times to explore the pathway from PAH/high molecular weight hydrocarbon species to soot precursors, which is an important unresolved issue. Prior work has illustrated (e.g. *Calcote* [2.8]) that acetylene addition to condensed polyaromatic hydrocarbons alone is not fast enough to create soot masses in the ms time frame observed in flames. In addition, the H/C ratio of "soot nuclei" produced solely by the acetylene addition mechanism followed by coagulation of species of mass 1000 u and higher is too low to be consistent with the relatively high H/C ratio measurements (0.5) of young soot particles [2.9]. Analysis of dimerization of PAH [2.10, 2.11] is suggestive that this can play an important role in soot precursor growth. Some intriguing new data (*Santoro* [2.12] and *D'Alessio* et al. [2.13]) suggests formation of a soot precursor class that is more aliphatic in nature than PAH. Early work by *Homann* et al. [2.14] also found evidence for a class of reactive hydrocarbons containing more hydrogen than PAH. These factors suggest that some mechanism in addition to acetylene addition to PAH radicals is important in forming species that become the precursors for the initial soot nuclei. One possible mechanism involves reaction of large benzylic type radicals or diradicals with other stable PAH. The allene pyrolysis product profiles obtained at long (i.e. s) residence times provide evidence for reactive dimerization of small PAH species (128 to 408 u) leading to rapid growth in masses in the 408 to 662 u range and indicate possible mechanisms for such a dimerization process.

2.2 Experimental Procedure

Allene $(C_3H_4(A))$ pyrolysis was carried out in a microjet reactor source, described in detail earlier [2.15-2.17]. This is a miniature fast-flow reactor with a volume of approximately $3.2 \cdot 10^{-9}$ m^3 coupled directly to a sonic nozzle. The reactor geometry consists of an alumina multibore thermocouple insulator tube inserted into a larger alumina tube with a sapphire nozzle (5-200 mm). The inner tube is positioned to leave a reaction chamber 1 mm in length, and is sealed by fusing powdered alumina over the joints. Reactants (pure allene or allene/O_2 in this study) are introduced to the pyrolysis zone through the center-most hole in the inner alumina tube (0.4 mm inner diameter) and expanded into the reaction chamber. The reaction zone is resistively heated and temperature within the reactor zone has been calibrated using thermocouples. Thermocouples were not used continuously during experiments due to the catalytic oxidation/pyrolysis observed on the platinum/rhodium wires. Pressure within the microjet reactor was maintained at 80 kPa ± 2.67 kPa or 122 kPa ± 2.67 kPa. Under the stated operating conditions, wall reactions do not significantly affect product distributions, and collisions with the wall are much less frequent than molecule-molecule collisions. The variation in reaction product distribution with absolute pressure indicates no significant contribution of the surface until pressure is decreased by a factor of 10. A test of wall inertness was made through pyrolysis of cyclohexane in both alumina and quartz tubes; this showed no evidence of differences in the mass spectra at temperatures from 300 - 1600 K. Our analysis of this reactor (including modeling of CO/CO_2 production in combustion mode, and ion residence time studies) suggests that for short (ms) residence time operation it can be modeled as a well mixed reactor. In addition, under the low conversion reaction conditions we use in this study the differential reactor assumption can be used in the analysis of the initial reaction pathways.

Mass spectrometry (MS) has been extensively used for the detection of species in combustion environments. Unlike other techniques, e.g. laser induced fluorescence, MS is in most cases universal towards species detection: it relies upon a general rather than specific characteristic of the species being detected. In common ionization methods, however, ionization induces fragmentation due to the high excess energy (50 eV or more in standard electron impact ionization) available to be left in the ion. The spectra obtained often require intricate analysis to discern fragments from parent species in the already complicated mixtures. Single photon (VUV) ionization with energies close (within 2 eV) above the ionization potential is an alternative non-destructive method of ionization, which coupled with time of flight mass analysis has been applied by *Pfefferle* et al. [2.15], *Boyle* [2.16] and *Boyle* et al. [2.17] for the study of high temperature combustion and pyrolysis products.

The VUV photoionization mass spectrometer (VUV-MS) used for the ms residence time studies is illustrated in Figure 2.1, upper part. In this configuration the pyrolysis reactor is on axis with the time-of-flight (TOF) axis. The mass spectrometer is equipped with Wiley-McLaren type acceleration for higher resolution and an ion reflectron to compensate for initial ion energy spread, to provide a longer effective flight length (1 m) and to prevent the considerable quantities of neutral polymeric hydrocarbons produced from reaching the detector. Mass signals are displayed in real time and recorded directly onto a digital storage oscilloscope which is interfaced with a PC for data analysis. The mass resolution for the experiments described herein was measured as 325 at 78 u. Figure 2.1, lower part, shows the VUV-MS schematic for the long residence time studies where the reactor is placed at a 90° angle to the TOF axis. An Einzel lens is used in this configuration to correct for ions of unequal kinetic energy in the direction normal to the time-of-flight axis. In this configuration we also upgraded the microchannel plate detector array to Gallileo Optoelectronics HotMCP units. This allowed a 100-fold increase in the linear response region for the chevron configuration used. Linear response was obtainable from single ion counting to a 20 volt signal. In addition, detector response has not decayed over the year of operation.

VUV photons were generated by the non-linear optical mixing technique of third harmonic generation in Xe. A frequency tripled Nd-YAG laser (Quanta Ray DCR-11 or DCR 3G) operating at 10 Hz was focused into a Xenon cell with a 30 cm path length at 3.47 kPa. The signal from $C_6H_6^+$ produced by single-photon ionization of C_6H_6 from a 300 K fixed flow microjet expansion was used to monitor relative UV to VUV conversion efficiency. Optimum efficiency was found at approximately 30 mJ of energy in a 8 ns pulse at 354.6 nm, corresponding to a peak power of approximately $3.75 \cdot 10^6$ W. In our experiment the estimated absolute conversion efficiency for VUV generation using a Xe medium and a DCR-11 laser with a donut beam profile was not greater than 10^{-5}. The configuration shown in Figure 2.1, lower part, uses a Gaussian beam profile DCR 3G laser with a more uniform spatial and temporal beam profile. Multiphoton processes due to the 354.6 nm light are minimized by using differential focusing.

Ionization efficiencies at 118 nm are relatively constant ($\pm 30\%$) and high for five carbon hydrocarbons and larger, since these compounds have ionization potentials (IP) of 1 to 2 eV below the photon energy (118.2 nm = 10.49 eV). Therefore the simultaneously-obtained peak ratios are good approximations of relative concentration. However, for smaller hydrocarbons with ionization potentials lying closer to the single photon energy, absorption cross sections and ionization efficiencies vary considerably. Consequently, these species must be calibrated individually. We are developing an empirical correlation model based upon available photoelectron spectra and photon yields for a variety of small to mid-sized hydrocarbons to obtain more accurate absolute calibration. Detection efficiencies of the various molecular hydrocarbons must likewise be considered to obtain quantitative data. In the recent work of *Geno* et al. [2.18], an integrated probability of detection was calculated based upon the relative signals recovered from a variety of small peptides (m/z e = 86 to m/z e = 1059). We use this same correlation relying on the similarity of hydrocarbons and peptides (hydrocarbons with various attached functional groups). Applying these relationships to our experimental operating conditions (2 kV post-acceleration), the average probability of detecting hydrocarbon ions impacting on the microchannel plate is essentially unity for masses up through 150 u. For ions larger than approximately biphenyl, however, corrections must be applied in order to recover relative incident flux at the detector face.

Since only masses are detected, the identity of structural isomers can only be inferred from arguments of internal consistency or knowledge of kinetic mechanisms and rate constants.

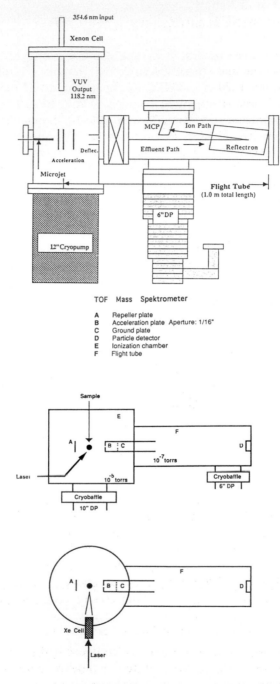

Fig. 2.1. Schematic diagram of TOF mass spectrometer: reactor outlet is along the axis of the flight tube for the ms residence time studies (upper part) and at a 90° angle to the flight tube of the mass spectrometer for the longer residence time studies (lower part)

2.3 Experimental Results and Discussion

Figure 2.2 shows the mass spectra from allene pyrolysis at increasing temperatures including 1320 K, 1360 K, 1515 K, and 1660 K at 2 ms mean residence time. These spectra are raw signal averaged data from a single run with no calibration corrections applied and the mass 40 u peak is off-scale for all except the highest temperature. As temperature was increased from 300 K to 1200 K, the first reaction product observed was mass 80 u followed at slightly higher temperatures by smaller reaction products such as mass 28 (ethylene) and mass 52 (C_4H_4).

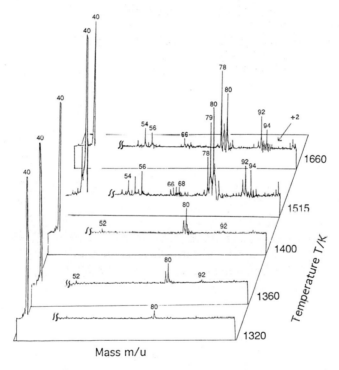

Fig. 2.2. Mass spectra taken during allene pyrolysis from microreactor at 2 ms residence time and temperatures from 1320 to 1660 K (mass 40 u is off-scale for 1320 K to 1400 K)

Flow reactor results also indicate methane and hydrogen in significant quantities, although those products were not directly measured in this experimental configuration due to their high ionization potentials (methane was indirectly measurable through its fragmentation product methyl radical). Mass 52 u (C_4H_4) is the C_4 species with the highest concentration up to 1450 K and is observed to increase coincident with ethylene. These species are observed prior to rapid allene decomposition, and are consistent with allene dimer formation to dimethylenecyclobutane (DMCB) isomers followed by decomposition through the reaction below:

$$\text{DMCB} \quad \longrightarrow \quad C_4H_4 \quad + \quad C_2H_4 \tag{2.3}$$

Starting at 1320 K a range of high mass species in addition to mass 80 u including masses 79 u, 92 u, 106 u and 120 u were observed to reach significant steady state concentrations prior to significant production of mass 78 u. In fact at these conditions, benzene production was not necessarily the rate limiting step to higher hydrocarbon production. It should also be noted that the concentration of mass 92 u increases to observable levels at low temperatures prior to significant mass 78 u production (see Fig. 2.2 and Table 2.1).

A possible low temperature pathway is the reaction of propargyl radical with vinyl acetylene:

$$C_4H_4 \quad + \quad C_3H_3 \quad \longrightarrow \quad C_7H_7 \quad \longrightarrow \quad \text{toluene} \tag{2.4}$$

The reverse of this reaction has been observed to be an important lower temperature channel in the decomposition of benzyl radical. At the low end of temperatures ($T < 1200$ K) in their toluene pyrolysis study, *Pamidimukkala* et al. [2.19] determined the major pyrolysis products to be C_7H_7 followed by C_5H_5, C_5H_3, C_4H_4 and C_4H_2. In subsequent tests with C_7H_7Cl these investigators noted that the $< C_7$ products were secondary cracking products from C_7H_7 not C_7H_8.

Mass 120 u detected at temperatures as low as 1280 K was likely formed through reaction of mass 80 u with allene as noted in early work on the thermal polymerization of allene. Mass 120 u and 160 u, which are possible allene trimers and tetramers, were not detected in our studies of ethylacetylene pyrolysis [2.15 - 17] or in our studies of methyl acetylene pyrolysis [2.15] (at temperatures and reaction times where isomerization is not equilibrated) but are present at significant levels in this study. This is another indication of allene dimer formation in this system at the low temperature reaction conditions.

A mechanism for allene dimer, trimer and tetramer formation was reported by *Weinstein* et al. [2.20] and is given below:

$$\tag{2.5}$$

$$\tag{2.6}$$

At higher temperatures and allene conversions, these species are converted through thermal decomposition and reactions with small hydrocarbon radicals.

It is instructive to compare the results from the lower temperature end (1100 - 1515 K) of this study with those of *Hidaka* et al. [2.21] for dilute allene and propyne pyrolysis behind a reflected shock wave at 2 ms residence time and temperatures ranging from 1200 to 1500 K at somewhat higher pressures (172.25 kPa - 253.31 kPa) than this study (81.06 kPa - 111.45 kPa). Even though the allene/propyne isomerization is relatively fast for these conditions, it was not equilibrated and they observed considerable differences in both product profiles and total C_3H_4 conversion between runs starting with either allene or propyne. At 1400 K for dilute (4% in Ar) propyne/allene pyrolysis they reported more than 45% total C_3H_4 conversion for allene and more than 30% conversion for propyne (both at 2 ms). The total C_3H_4 conversion reported for allene is over twice that observed in this study (for the same temperature) at higher allene concentration but about half the total pressure. During allene pyrolysis more C_6H_6 was formed than during propyne pyrolysis especially between 1350-1500 K, peaking at 1450 K and then dropping off sharply at higher temperatures. The other differences included more C_3H_6 and C_4H_4 for allene pyrolysis and considerably more C_2H_4 for allene pyrolysis, especially at the lower temperatures from 1250-1350 K. Ethylene drop off for the allene case was coincident with the fall-off in net benzene production observed. Dimethylenecyclobutane (DMCB) was inferred to be formed as the initial product at low temperature by comparison of 225 nm absorption measurements with gas chromatographic measurements of allene and propyne. This is consistent with the observations of the current study for mass 80 u appearance prior to significant allene decomposition. The greater production of C_2H_4 and C_4H_4 observed during allene pyrolysis, was likely a result of the decomposition of DMCB isomers (2.3 above) and is also consistent with the current study. Although no hydrocarbons with more than six C-atoms were reported in the study of *Hidaka* et al. [2.21], tars were noted on the tube wall at $T > 1450$ K, which is also consistent with the high mass measurements in the current study at relatively low temperatures.

An overview of the approximate concentrations of higher hydrocarbons detected at 2 ms reaction times at temperatures ranging from 1280-1580 K are given in Table 2.1. This data represents averages for several different runs at each temperature and includes normalization for estimated relative detection efficiency. Below 1450 K, high mass peaks are located at 79 u, 80 u, 92 u, 106 u and 120 u. As temperature was increased above 1450 K a broad range of hydrocarbons with 6 or more carbons were detected including mass 78 u with the largest contributions initially coming from masses 92 u, 94 u, 106 u, 116 u, 120 u, 144 u and 158 u. At 1450 K the pyrolytic decomposition of the total C_3H_4 component (allene + propyne) becomes fast, and C_4 and

Table 2.1. Relative concentrationes of selected product species during allene pyrolysis at 2 ms, 80 kPa

Molecular
Mass m/u   Temperature T/K

	1280	1320	1450	1515	1545	1580
52	**	**	4.7	4.7	7	12
54	**	**	5.9	29	23	40
56	**	**	8.2	40	56	39
66	**	**	**	3.9	4.3	6
67	**	**	4.0	3.9	3.5	3.5
78		*	3.1	23	31	47.5
79		1.5	17	30	35	43.5
80	2.4	5	25	43	50.5	52
91			2.3	2	3.5	5.5
92		**	22	35	30	
93				*	6	9.8
94			3.9	14	15	19.6
104				2.7	6.3	8.6
105				5.5	5.5	8.5
106		**	2.4	12.5	12	22
116				5.9	9.5	11.3
117				4.7	7.5	14
118				9.8	14	23
120		**	7.8	20.5	18	19
128					6.2	10
130					10.5	14
142				9	14.5	17
144			2.7	7	13	14.4
154					2	4
156				7	15	21
158			3.5	6.3	10.5	12
160				10	5	5
166					5	6
168						8.5
170					5	10
178					1	3
180					4.3	10

Remarks: *: Species present in $0 - 1$ units;

**: Species present in $1 - 2$ units;

Note: Only those hydrocarbon radical species with comparable concentration to the stable analogue were included.

C_5 radicals reach appreciable concentrations (see Table 2.1). At this point radical growth processes for higher hydrocarbon growth are expected to dominate over the molecular growth processes observed at low temperatures. Processes governing benzene production in the two growth regimes are discussed below.

2.3.1 Low Temperature Benzene Formation Mechanism

The data from both this study and that of *Hidaka* et al. [2.21] suggests that at low temperatures ($T < 1400$ K) and ms time scales the molecular channel for C_6H_6 formation through dimerization to a DMCB isomer is the important low temperature route. As noted above, *Hidaka* et al. [2.21] observed an excess of C_6H_6 from pyrolysis of allene (C_3H_4(A)) over propyne (C_3H_4(P)) at similar reaction times to this study (1-3 ms) and temperatures (1200-1500 K). At temperatures > 1250 K benzene was confirmed by GC measurements. This is consistent with our studies showing how the molecular channel can contribute to C_6H_6 production even at moderate temperatures with respect to allene/propyne isomerization. This view is supported by the observation of more C_2H_4 and C_4H_4 from the allene pyrolysis than from propyne pyrolysis even though isomerization between allene and propyne is relatively fast at these conditions but not equilibrated on the 1-3 ms time scale. As pointed out by *Kiefer* et al. [2.22], this suggests a contribution to benzene production from the molecular channel rather than through C_3H_4(A) + C_3H_3, which does not produce the correct concentration dependence or as strong a negative temperature dependence as observed.

A first step to test the viability of the possible molecular channel to benzene was to estimate rates for allene dimerization to DMCB isomers and compare the rate with the experimentally observed rate for mass 80 u production at low conversion. The thermodynamic parameters were estimated using the THERM [2.23] program and checked with experimental values where available. Equilibrium constants for the reactions 2 Allene $\rightleftharpoons 1, 2 - $ DMCB, and 2 Allene $\rightleftharpoons 1, 3 - $ DMCB were computed and found to be greater than one up to about 1400 K. At temperatures below 800 K the primary isomer predicted during early high concentration allene pyrolysis studies [2.6, 2.7] was 1,2-DMCB. At higher temperatures, however, the 1,3-isomer becomes an important product ($> 10\%$).

The rate coefficient for allene dimerization to 1,2-DMCB was estimated using transition state theory, analogy to similar reactions and limited experimental data. The dimerization rate of allene to 1,2-DMCB is pressure dependent and highly exothermic (approximately 40 kcal mol^{-1}). Table 2.2 shows the energetics of the reaction and the prediction for the dimerization rate at two temperatures using a QRRK analysis with an argon bath gas at 101.32 kPa. The activated complex for the rate limiting configuration was estimated to be approximately 15 kcal mol^{-1} above the reactants by

Gajewski et al. [2.24]. The reverse rate was calculated from equilibrium, and the rate was extrapolated to our conditions through QRRK analysis using argon as the bath gas. Experimental rate data available for dimerization to 1,2-DMCB, along with rate data of similar reactions were used to check our estimate (e.g. pyrolysis of methylenecyclobutane to allene and ethylene). The estimated rate reported on Table 2.2 for 700 and 800 K is close to the higher experimental value reported in the earlier literature [2.6, 2.7] (and references therein) shown on Table 2.3. These early measurements of allene dimerization did not account for tar formation which depleted the dimer concentration, and surface effects which accelerate the rate could also have been present. The rate coefficient determined in this study at the low temperature end (800 K) is also close to that predicted by our estimate but it should be recognized that the error bar on this estimate is likely at least ±50%, and the error bar on the experimental concentration estimate is about the same considering that other C_6H_8 isomers are likely present with varying detection efficiencies.

Table 2.2. Estimation of rate coefficients for dimerization of allene to 1,2 - DMCB

$$2(CH_2 = C = CH_2) \quad \rightarrow \quad [\ ]^* \quad \rightarrow \quad 1,2 - DMCB$$

$\Delta_F H$	$2 \cdot 45.9$	≈ 107	47.25	$kcal\ mol^{-1}$
S_{300}	116.6		74.12	$cal\ mol^{-1}K^{-1}$

$A^* \approx 1.0 \cdot 10^7\ l\ mol^{-1}s^{-1}$ at 298 K [2.24]

$E_a^* \approx 15\ kcal\ mol^{-1}$

QRRK estimate: (argon bath gas, 101.32 kPa)

$k_{dimerization} = 4.3 \cdot 10^2\ cm^3\ mol^{-1}s^{-1}$ at 700 K

$k_{dimerization} = 1.9 \cdot 10^3\ cm^3\ mol^{-1}s^{-1}$ at 800 K

Table 2.3. Estimation of rate coefficients for dimerization of allene to 1,2- DMCB from experimental data

$k_{\text{dimerization}} = 6.7 \cdot 10^3$ cm^3 mol^{-1}s^{-1} at 778 K
Data from ref. [2.6].
Further assumption: plug flow reactor.
Probable sources of error: surface reaction was possible; significant tar formation occurred.
Reaction sensitive to: pressure; surface.

$k_{\text{dimerization}} = 2.0 \cdot 10^3$ cm^3 mol^{-1}s^{-1} at 800 K
Extrapolation of low conversion data from the current experiment.

The estimates for allene dimerization rates discussed above are consistent with our experimental measurements of mass 80 u production. This does not, however, mean that early mass 78 u production observed takes place through the mass 80 u intermediate. The most probable thermal rearrangements of 1,2-DMCB, do not lead to benzene intermediates. Rupture of the carbon bond between the two methylene groups does produce species that could subsequently react to form benzene at appreciable net rates but this would be a relatively high activation energy reaction. One other possible route which is consistent with the data would be relatively fast conversion of 1,3-DCMB to methylene cyclopentene and then 1,3-cyclohexadiene, followed by formation of the cyclohexadienyl radical at mass 79 u. This mechanism would be consistent with the large mass 79 u peak observed prior to mass 78 u appearance. *Benson* et al. [2.25] proposed a multistep mechanism for conversion of 1,3-cyclohexadiene to benzene initiated by:

$$\bigcirc + \bigcirc \longrightarrow \bigcirc + \bigcirc \qquad (2.7)$$

$$\bigcirc + \bigcirc \longrightarrow \bigcirc + \bigcirc \qquad (2.8)$$

Followed by benzene formation through:

$$\bigcirc + \bigcirc \longrightarrow \bigcirc + \bigcirc \qquad (2.9)$$

$$\bigcirc \longrightarrow \bigcirc + \text{ H} \qquad (2.10)$$

This mechanism provides one explanation for the large steady state mass 79 u concentration observed in our experiments. Mass 82 u is also observed in our experiments at relatively low (1360 K) temperatures. Figure 2.3 shows relative concentrations of selected species as a function of temperature (2 ms). Mass 79 u discussed above parallels mass 80 u production until 1450 K suggesting that one is directly produced from the other as suggested above. Direct pyrolysis studies of 1,2- and 1,3-DMCB must be carried out to clarify whether a direct route to benzene from these precursors can play a role in low temperature benzene production during allene pyrolysis.

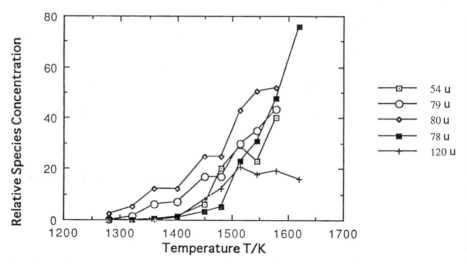

Fig. 2.3. Relative concentrations of selected higher hydrocarbon products during allene pyrolysis at 1280 - 1660 K, 2 ms

2.3.2 High Temperature Benzene Formation

At higher temperatures in the ms residence time regime, the benzene formation mechanism changes. This can be observed from the change in the temperature dependence of the benzene formation (see Fig. 2.3). The change in slope corresponds to the temperature rate at which allene conversion is also increasing rapidly from less than 1% (at 1380 K) to more than 15% (1400 K) at 2 ms and where the main conversion route changes from dimerization to decomposition as noted by the marked increase in multiplicity of product masses indicative of radical growth processes. It should be noted that at 1450 K production of species with masses higher than 78 u is significant so the actual mass 78 u production rate is greater than directly observed. An initial survey of possible mechanisms and the concentrations of possible precursors observed in this study showed that only the two C_3 channels (2.1

and 2.2) could account for the order of magnitude of mass 78 u production in the $T > 1515$ K range. Although we do not have an estimate for absolute C_3H_3 concentration in these experiments, the concentration is observed to increase sharply with temperature from 1400 to 1600 K coincident with the mass 78 u curve. A significant change in the net rate of formation of benzene and other species with masses larger than > 78 u does not occur until after C_3H_3 concentration rises significantly (i.e. mass 78 u production rate is more closely a function of C_3H_3 concentration squared than $C_3H_4(A)$ concentration times C_3H_3 concentration). Considering equilibrium for the reverse of the reaction $C_3H_4(A) + C_3H_3 \longrightarrow C_6H_6 + H$, this reaction could not produce a large positive increase in benzene production with temperature over the range from 1400 - 1600 K.

Hidaka et al. [2.21] observed more benzene formation from allene pyrolysis than from propyne pyrolysis, suggesting that the reaction $C_3H_4(A) + C_3H_3 \longrightarrow C_6H_6 + H$ could be responsible for this difference. As we have discussed above, however, the observed difference for $C_3H_4(A)$ pyrolysis is more likely due to a residual contribution to mass 78 u production from the molecular channel at temperatures below 1450 K at ms time scales.

2.3.3 Higher Hydrocarbon Formation at Long Residence Times

The allene pyrolysis experiments were repeated in the temperature range from 1100-1500 K at long (s) residence times to investigate mechanisms leading to large hydrocarbon and PAH production. Figure 2.4 illustrates the product spectra in the mass range 0-700 u for neat allene pyrolysis at 1495 K, 6 s mean residence time and 122 kPa $\pm$ 2.67 kPa. The data in these figures has been averaged over 100 laser shots but has not been normalized for efficiency of individual mass detection. At these conditions over 99 % of the allene has been converted to products with most of the carbon mass present as higher molecular weight species. This residence time/temperature snapshot is interesting because it shows the initial growth pattern of species in the mass range above 400 u. Figures 2.5-7 show expanded sections from this spectrum at several high mass ranges to highlight the different growth mechanisms observed.

Figure 2.5 shows the mass range 152-300 u at five times the magnification of Figure 2.4. This figure shows clearly that the growth to higher hydrocarbons in this mass range is consistent with the acetylene growth mechanism. The mass numbers correspond to stoichiometries of structures that are consistent with C_6 and C_5 rings in condensed configuration. The observed stoichiometries are consistent with results by *Böhm* et al. [2.26], who in detailed LC/MS studies of PAH production in this mass range in flames over a range of pressures showed no evidence of significant concentrations of PAH with side chains.

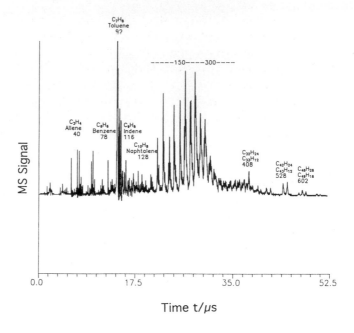

Fig. 2.4. Overview spectra of product abundances during neat allene pyrolysis at 1495 K, 122 kPa, and 6 s

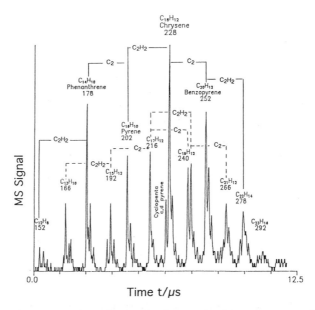

Fig. 2.5. Detail spectra of product abundances in the mass range 152 - 300 u during neat allene pyrolysis at 1495 K, 122 kPa, and 6 s

Figure 2.6 shows detail of the product distribution in the mass range 302-408 u in which peaks consistent with growth by sequential acetylene and possibly Cl additions are also present; however, more filling in of the spectrum is observed to account for a greater range of stable structures due to the different degree of ring packing arrangements possible for a given number of carbon atoms. A striking difference in the spectrum of this mass range is the approximately equal concentrations of major peaks increasing in concentration at mass 408 u, illustrating the possibility of the onset of another growth mechanism that becomes more important at higher masses. A small hydrocarbon polymerization mechanism in the absence of an equilibrium constraint would result in an approximate exponential decay of product concentrations on the high mass side of the species with the maximum concentration. This is clearly not the case in the mid to high 300 u mass range (Fig. 2.6) but is observed past the PAH maximum (at 228 u) in the mass range from 228 u to 324 u (Fig. 2.5 and 2.6). At mass 408 u a distinct change in the spectrum is observed. Mass 408 u abundance is greater than any species in the 300 u range, and fewer side peaks are present.

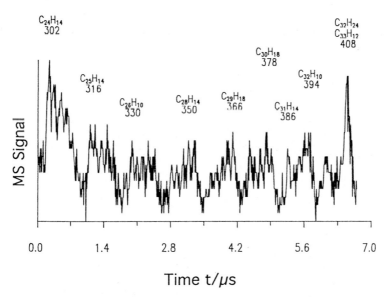

Fig. 2.6. Detail spectra of product abundances in the mass range 300 - 408 u during neat allene pyrolysis at 1495 K, 122 kPa, and 6 s

Figure 2.7 shows a ten times enlargement of the mass range 408-662 u. In this mass range significantly fewer product masses are observed. Each peak is distinct and the spacing is consistent with these species being products of dimerization of predominant peaks in the low mass PAH spectra. This is illustrated on Table 2.4 which shows dimer possibilities $A_1 + A_2 \rightleftharpoons$ Dimer $-$ H_2. Apparently, the major peaks can be predicted by reaction of masses which have isomer structures with saturated/partially saturated carbon/carbon bonds, see below for example:

Mass 154 u   Mass 180 u   Mass 190 u   Mass 204 u

All dimer combinations including a monomer mass with a possible structure of this type are noted in *italic* font on Table 2.4. These types of compounds have been detected by LC in roughly equal concentrations to condensed PAH structures during the pyrolysis and oxidative pyrolysis of a wide variety of aromatic compounds as well as n-decane [2.27]. The observed product distribution is also consistent with pre-sooting products observed in a study of soot formation from coal [2.28].

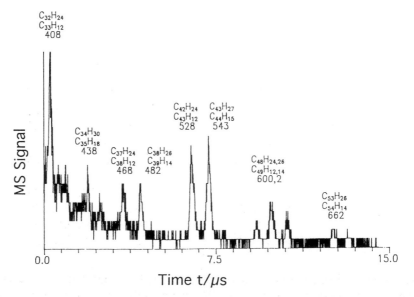

Fig. 2.7. Detail spectra of product abundances in the mass range 408 - 662 u during neat allene pyrolysis at 1495 K, 122 kPa, and 6 s

Table 2.4. Possible combinations of PAH + PAH to form high mass dimers + H_2

Monomer/u \ Dimer/u	408	420	438	468	482	527	526	528	542	544	586	600	602
78	332	344	362	392	406								
92	318	330	348	378	392								
104	306	318	336	366	380								
116		306	324										
120		302	320	350	364		408						
128			312				400	402					
135						394							
142	268			328									
152				318	332								
154	256	268		316	330		374	376	390	392			
166		256		304	318		362	364	378	380			
168	242	254		302	316		360	362	376	378			
178				292	306		350	352	366				
180	230	242			304		348	350	364	366	408		
190							338	340					
192	218	230		278	292		336	338	352		396		
194	216	228		276			334	336	350	352	394	408	
202				268			326	328	344		386	400	402
204	206	218		266			324	326	340		384		400
206		216			278		322	324	338	340		396	
216		206		254	268		312		328	330	372	386	388
218		204		252	266		310	312	326	328		384	386
226							302	304	318	320	362	376	378
228		194		242	256			302	316	318	360	374	376
230	180	192		240	254					316	358	372	374
240				230					304	306	348	362	364
242	168	180		228					302	304	346	360	362
252				218			276	278	292		336	350	352
254		168		216	230			276		292	334	348	350
256		166			228						332	346	348
266				204	218				278		322	336	338
268	142	154		202	216				276	278	320	334	336
276				194			252	254	268		312	326	328
278				192	206			252	266	268	310	324	326

One possible initiation step for reactive dimer formation that is consistent with the temperature dependence observed would be initiation through relatively high activation energy rupture of the more saturated C − C bonds in structures such as those illustrated above to form diradicals or benzylic type radicals followed by subsequent reaction with another PAH to form structures with CH_2 linkages. Early work by *Bauer* [2.29] using electron diffraction indicated the presence of single $C - C$ bonds in soot. Estimates of the formation rate of PAH diradicals or benzylic radicals and their thermodynamic stability are consistent with the observed concentration/temperature onset of dimer formation. *McMillen* et al. [2.30] have shown that arylmethyl radicals are highly stabilized (7.8 kcal mol^{-1} additional resonance stabilization energy relative to benzyl radical), and estimated the high pressure rate parameters for the unimolecular decomposition of 9-ethylanthracene to 9-anthrylmethyl radicals to be $k_\infty = 4.0 \cdot 10^{15} e^{-33383/T}$.

The high activation initiation step is also consistent with our acetylene pyrolysis data which shows no dimer formation at temperatures from 1000-1200 K even though mass 100-300 u species concentration profiles are similar to those obtained in this study at higher temperatures with dimer formation evidenced. This discounts a mechanism based on a requirement of one monomer species larger than 300 u as an alternative to the high activation energy step for reactive dimerization.

The question of whether this data can be extrapolated to examine the possible role of the reactive dimerization in an actual flame environment can be addressed roughly by comparing the growth rate of species in the 200 - 408 u range (assumed produced by acetylene addition to PAH radicals) and the 408 - 662 u range (assumed produced by the irreversible reaction of PAH species). The rates for each process are

$$r_{\text{acetylene addition}} = k_a \, [C_2H_2][PAH_1\cdot]$$

and

$$r_{\text{dimerization}} = k_d \, [PAH][PAH_2\cdot].$$

Therefore, in order to extrapolate the results of this study to flame conditions, observed dimerization rates must be normalized by the ratio of $[C_2H_2]/[PAH]_{\text{flame}}/[C_2H_2]/[PAH]_{\text{microreactor}}$. In the pyrolytic microreactor conditions the PAH concentration is significantly higher than in a flame. For the conditions of this study the ratio $[C_2H_2]/[PAH]_{\text{microreactor}}$ varies from approximately $5 - 10$. Although the flame ratio varies with conditions (fuel, local mixture fraction and temperature), the pre-sooting value for the $[C_2H_2]/[PAH]_{\text{flame}}$ ratio is estimated at approximately 100 from concentration profiles reported in the literature [2.3, 2.6]. Therefore to compare the relative importance of the dimerization reaction mechanism in the microreactor to actual flame conditions the observed rate would have to be divided by a factor of $10 - 20$. It should be noted that as discussed earlier detection sensitivity is dropping with mass in the dimerization mass range and therefore this process appears to be less dominant in Fig. 2.6 and 2.7 than it actually is.

Apparent growth rates of species in the 408-662 u range are several times as fast as those in 300 - 408 u range from the same precursor pool, with

3 − 12 times the carbons added per reaction for monomers in this mass range (carbon addition factor increases with monomer size). Therefore in this size range, the rate growth from a given monomer to twice its mass by the dimerization growth mechanism would be competitive with that by acetylene addition to PAH radicals, and this advantage would increase with monomer size. Uncertainties in this extrapolation were also evaluated. First, the PAH radicals in both mechanisms are not the same structure (phenylic type or benzylic type) and have different thermal stability. Benzyl radicals are more stable in this temperature range than phenyl radicals. The relative concentrations of these radical types, however, are also affected by their precursor concentrations and kinetic barriers for formation from the precursor. These factors are also affected by the presence of oxygen. Our preliminary results for pyrolysis of benzene and toluene in the presence of oxygen suggest that dimer formation is more, not less, predominant in the presence of oxygen.

We have shown above that a reactive dimerization growth mechanism may be competitive with other higher hydrocarbon growth mechanisms in a flame for species in the mass range above 400 u. What implications does this have for formation of soot precursors in flames? Is it more likely that species formed through this growth mechanism lead to growth of soot nuclei? Two considerations that may help answer these questions are the thermal stability of these species and their H/C ratio. Thermal stability is not directly addressable because we do not know the structure of the proposed dimers. It is clear however that species with multiple bridged or ladder structure would have high thermal stability. H/C ratio as a function of growth mechanism can be directly addressed. The possible stoichiometries of species produced from either the acetylene growth mechanism or the reactive dimerization mechanism for products in the 408-662 u mass range are illustrated on Fig. 2.5 for each major peak. It should be noted that the stoichiometry of species produced through the acetylene growth mechanism already have an H/C ratio less than that of young soot $(0.26 - 0.36)$ whereas the products postulated by a reactive dimerization mechanism have a H/C ratio consistent with a soot precursor $(0.49 - 0.75)$. Note that the highest molecular weight shown in Fig. 2.5 (662 u) would have an H/C ratio of 0.26 if grown by the acetylene growth mechanism and 0.49 if grown by a dimerization mechanism. Therefore, if formed by either growth mechanism the 662 u species would have an H/C ratio at or lower than that of young soot. In addition, even old soot has a fairly high $(0.2 - 0.25)$ H/C ratio. These observations reinforce the idea that a mechanism for soot precursor formation other than solely acetylene addition to PAH radicals must be important at least in the PAH size range above 300 u. This has been stated earlier by *Wagner* [2.9], "In order for the fuel molecules to evolve to soot, there must take place a (chemical!) condensation of species with the right hydrogen content or a (chemical!) condensation of species with a higher hydrogen content and consecutive dehydrogenation or a combination of these two pathways". We have shown that chemical dimerization can take place during the pyrolysis of allene and that this mechanism likely has relevance for flame conditions.

2.4 Conclusions

A microjet reactor coupled to a vacuum-UV photoionization time-of-flight mass spectrometer was used to study the formation of higher hydrocarbons during the pyrolysis of allene. By use of this technique, a progression of pyrolysis product spectra including hydrocarbon radicals were obtained as a function of temperature at millisecond to second reaction times. Mass 80 u was the first product species observed as temperature was increased from 300 K-1200 K at ms reaction times. This is consistent with early flow reactor studies of allene pyrolysis where allene dimerization leading to the formation of predominantly the 1,2 isomer of dimethylenecyclobutane was observed. An analysis made of a proposed mechanistic pathway for allene dimerization to DMCB isomers and possible mechanisms for subsequent conversion to benzene showed that a molecular growth mechanism is consistent with the low temperature mass 78 u observed in this and previous studies.

At higher temperatures the mass 78 u (benzene) production was observed to be consistent with formation predominantly through C_3H_3 recombination. This is observed both through the concentration dependence and the temperature at which the benzene formation rate increases dramatically.

At long (second) residence times and low temperatures (1495 K), the formation of multiple ring hydrocarbons was observed. In the mass range from 150 to 350 u growth was observed to be consistent with successive acetylene addition to PAH radicals. At higher molecular weights, however, a second competitive mechanism was observed consisting of reactive dimerization of small PAH/PAH radicals to form dimers first clearly distinguishable in the mass range above 400 u. This mechanism was estimated to be important at flame conditions and provides one explanation for the high H/C ratio observed in young soot.

References

2.1   P.R. Westmoreland, A.M. Dean, J.B. Howard, J.P. Longwell: J. Phys. Chem. **93**, 8171 (1989)
2.2   J. Boyle: "The Detection of Soot Precursors from a Microjet Reactor Using Vacuum Ultraviolet Photoionisation Mass Spectrometry"; Ph.D. Thesis, Yale University (1993)
2.3   S.D. Thomas, F. Communal, P.R. Westmoreland: Prep. Div. Fuel Chem.**36**, No. 4, 1448 (1991)

2.4   C.H. Wu, R.D. Kern: J. Phys. Chem. **91**, 6291 (1987)

2.5   S.E. Stein, J.A. Walker, M.M. Suryan, A. Fahr: "New Path to Benzene", in Twenty-Third Symposium (International) on Combustion (The Combustion Institute, Pittsburgh 1990) p. 85

2.6   A.T. Blomquist, J A. Verdol: J. Am. Chem. Soc. **78**, 109 (1956)

2.7   R.N. Meinert, C.D. Hurd: J. Am. Chem. Soc. **52**, 4540 (1930)

2.8   C. Calcote, R.J. Gill: "Comparison of the Ionic Mechanism of Soot Formation with a Free Radical Mechanism", this volume, sect. 29

2.9   H.Gg. Wagner: "Soot Formation - an Overview", in Particulate Carbon: Formation During Combustion, ed. by D.C. Siegla, G.W. Smith, (Plenum Press, New York 1983) p. 1

2.10   J.H. Miller, K.C. Smyth, W.G. Mallard: "Calculations of the Dimerization of Aromatic Hydrocarbons: Implications for Soot Formation", in Twentieth Symposium (International) on Combustion (The Combustion Institute, Pittsburgh 1984) p. 1139

2.11   J.H. Miller: "The Kinetics of Polynuclear Aromatic Hydrocarbon Agglomeration in Flames", in Twenty-Third Symposium (International) on Combustion (The Combustion Institute, Pittsburgh 1990) p. 91

2.12   R. Santoro, T.F. Richardson: "Concentration and Temperature Effects on Soot Formation in Diffusion Flames", this volume, sect. 13

2.13   A. D'Alessio, A. D'Anna, P. Minutulo: "Spectroscopic and Chemical Characterization of Soot Inception Processes in Premixed Laminar Flames at Atmospheric Pressure", this volume, sect. 5

2.14   K.-H. Homann, H.Gg. Wagner: "Some New Aspects of the Mechanism of Carbon Formation in Premixed Flames", in Eleventh Symposium (International) on Combustion (The Combustion Institute, Pittsburgh 1967) p. 371

2.15   L.D. Pfefferle, J. Boyle, J. Lobue, S. Colson: Combust. Sci. Tech. **70**, 187 (1990)

2.16   J. Boyle: "Studies of Pyrolysis and Oxidative Pyrolysis of Allene, Methyl Acetylene, Ethyl Acetylene and Butadiene Using VUV Photoionization Mass Spectrometry"; Ph.D. Thesis, Yale University, New Haven 1991

2.17   J. Boyle, L.D. Pfefferle: J. Phys. Chem. **94**, 3336 (1990)

2.18   P.W. Geno, R.D. MacFarlane: Int. J. Mass. Spectrom. Ion Processes **92**, 195 (1989)

2.19   K.M. Pamidimukkala, R.D. Kern, M.R. Patel, H.C. Wei, J.H. Kiefer: J. Phys. Chem. **91**, 2148 (1987)

2.20   B. Weinstein, A.H. Fenselau: J. Chem. Soc. **1**, 368 (1967)

2.21   Y. Hidaka, A. Namamura, A. Miyauchi, T. Shiraishi, H. Kawano: Int. J. Chem. Kin. **21**, 643 (1989)

2.22   J.H. Kiefer: " Aromatic Formation in the Decomposition of Allene and Propyne", AIChE Annual Meeting, Miami Beach, Florida, November 1-6, 1992 ; J.H. Kiefer: private communication, 1991

2.23   E. Ritter, J. Bozzelli: "THERM" Thermo Property Estimation for Radicals and Molecules, Dept. of Chemistry, New Jersey Institute of Technology

2.24   J.J. Gajewski, C.N. Shih: J. Am. Chem. Soc. **94**, 1675 (1972)

2.25   S.W. Benson, R. Shaw: J. Am. Chem. Soc. **89**, 5351 (1967)

2.26   H. Böhm, M. Bönig, C. Feldermann, H. Jander, G. Rudolph, H.Gg. Wagner: "Pressure Dependence of Formation of Soot and PAH in Sooting Flames", this volume, sect. 9
2.27   O.S.L. Bruinsma, J.A. Moulijn: Fuel Process. Technol. **18**, 213 (1988)
2.28   A.F. Sarofim, J.P. Longwell, M.J. Wornat, J. Mukherjee: "The Role of Biaryle Reactions in PAH and Soot Formation", this volume, sect. 30
2.29   S.H. Bauer: Comment to "Carbon Formation in Premixed Flammes " from U. Bonne, K. H. Homann, H. G. Wagner, in Tenth Symposium (International) on Combustion (The Combustion Institute, Pittsburgh 1965) p. 511
2.30   D.F. McMillen, P.L. Trevor, D.M. Golden: J. Am. Chem. Soc. **102**, 7400 (1980)
2.31   F.W. Lam, J.B. Howard, J.P. Longwell: "The Behavior of Polycyclic Aromatic Hydrocarbons during the Early Stages of Soot Formation", in Twenty-Second Symposium (International) on Combustion (The Combustion Institute, Pittsburgh 1988) p. 323
2.32   J.P. Longwell: "The Formation of Polycyclic Aromatic Hydrocarbons by Combustion", in Nineteenth Symposium (International) on Combustion (The Combustion Institute, Pittsburgh 1982) p. 1339
2.33   I. Glassman: "Soot Formation in Combustion Processes", in Twenty-Second Symposium (International) on Combustion (The Combustion Institute, Pittsburgh 1988) p. 295

Discussion

Santoro: What was the concentration of allene in your experiments?

Pfefferle: In most of the runs that I have shown here we used neat allene. We have also done some experiments with argon dilution but we did not observe any effect in the distribution of the product masses.

Santoro: One of the things that I was struck with in your results is that the production times for these large hydrocarbons in the temperature range from 1400 to 1500 K seem to be large compared to what we would expect from atmospheric pressure flames. The qualitative trend that we have seen seems to be very similar to what is being discussed now as a soot formation mechanism. Some of your pictures remind me of mass spectra that Professor Homann has shown recently for premixed flames; in particular, these types of shifts in the mass spectra peaks. In flames all this has to happen in a millisecond or in five milliseconds. Could you comment on this?

Pfefferle: What you have seen was the mass spectra at 1200-1500 K temperatures at 40 ms to 6 s residence time. For these conditions, species with masses in the range of 400 u appear after about 40 ms. At higher temperatures, e.g. at 1600 K which is similar to flames all this happens in a one or two milliseconds time scale.

Sarofim: I was very impressed by your results. My question is with regard to the identification of the various polyaromatics. Is that based just on mass numbers or do you have any other information?

Pfefferle: Right now most of the identification of polyaromatics is just based on mass numbers. We have a GC-MS system for backing up some of the results but we have not done that extensively yet. The other way is to tune the vacuum UV that would allow us to discriminate between single masses. But it is not a trivial thing to be able to get vacuum UV radiation at a fairly even conversion level over a certain range of frequency. We now have a technique that we can use to go from about 106 nm to 124 nm with semicontinous conversion efficiency.

Sarofim: In the distribution of high molecular weight species you have this separation by five carbon atoms. Do you have an explanation for that?

Pfefferle: I am not sure whether it is fortuitous but if you look at the pattern of possible dimers from the main peaks in the mass range from 200 u to 400 u, they all are consistent as shown on Tab. 2.4.

Glassman: As you know we did qualitative measurements in regular and inverse diffusion flames and we came to the same conclusion as you: allene forms PAH at a much faster rate than C_2 or C_4 species. Your very fundamental results are consistent with what we found in practical flames that have been investigated by sampling and chemical analysis. I think I should stress this consistency between your fundamental results and the results obtained from actual flames.

Formation of Aromatic Hydrocarbons in Decane and Kerosene Flames at Reduced Pressure

Christian Vovelle, Jean-Louis Delfau,

Marcelline Reuillon

Laboratoire de Combustion et Système Réactifs, LCSR-CNRS,
45071 Orleans Cedex 2, France

Abstract: The formation of frequently discussed soot precursors has been compared in decane and kerosene flames. A specific study on the influence of equivalence ratio on the formation of some species (acetylene, benzene, vinyl benzene and phenyl acetylene) showed that in decane flames aromatic hydrocarbons are formed from acetylene whereas in kerosene flames the main source is the aromatic part of the fuel. From this result it seems to be reasonable to represent kerosene as a mixture of decane (90%) and toluene (10%) when developing a detailed kinetic mechanism to predict benzene formation in flames burning kerosene. The mechanism was checked by comparison of computed mole fraction profiles with measured profiles of a sooting kerosene-oxygen-argon flame with an equivalence ratio of 2.2 and of a decane flame with the same equivalence ratio.

3.1 Introduction

Most studies on formation of aromatic hydrocarbons in flames have been concentrated on small fuel molecules [3.1-6]. Practical combustion systems such as automotive or airplane engines burn hydrocarbon fuels containing seven to fourteen carbon atoms. The few kinetic studies on the combustion of fuels with carbon atoms in that range have been oriented towards knock phenomena and have been conducted in a temperature range lower than 1000 K where peroxides formation dominates [3.7, 3.8]. Aromatic hydrocarbons and soot are formed at higher temperature, and in order to improve the knowledge about formation of these pollutants in practical systems there is a need of experimental and modelling studies on flames of large fuel molecules.

One fuel that is applied largely in practical combustion systems is kerosene. Because kerosene is a complex mixture with alkanes as major components, in some previous work the structure of a near sooting decane flame with an equivalence ratio of 1.9 was investigated. For this flame a kinetic model was developed to predict the mole fraction profiles of all species, including those involved in the formation of benzene, with good accuracy [3.9].

This model then was applied to kerosene flames. A direct comparison of the experimental determined structure of the above mentioned decane flame with the structure of a slightly richer (equivalence ratio 2.0) kerosene flame pointed out a quantitative agreement for the major stable and some active species but an order of magnitude difference in the maximum concentrations of benzene and others substituted aromatic hydrocarbons.

These results are in agreement with others comparative studies on the influence of aromatic fuels on PAH and soot formation. *Scully* et al. [3.10] showed that less soot was produced in a mixture of cyclohexane and benzene than with pure benzene as fuel. *Harris* et al. [3.11] had to reduce the C/O ratio of toluene/ethylene flames considerably in order to compensate the higher soot forming tendency of toluene compared with ethylene. More recently, *Lam* et al. [3.12] compared the effect of the addition of ethylene and benzene to a jet-stirred plug-flow combustor and obtained a marked increase in soot formation with benzene.

In this work a specific study is presented to identify the additional source for the formation of aromatic hydrocarbons in kerosene flames. The results show that the aromatic part of the fuel is the main contributor to the increase in the formation of aromatic hydrocarbons. This point was further specified by a detailed analysis of the structure of sooting decane and kerosene flames with an equivalence ratio of 2.2 stabilized at low pressure (6 kPa) on a flat flame burner. The experimental measured mole fraction profiles are compared with numerical predictions. Modelling of benzene formation in numerical simulation is based on acetylene addition to C_4 radicals in the decane flame and on consumption of toluene which was selected to be representative for the aromatic part in the kerosene flame.

3.2 Experimental

The low pressure ($p = 6.0$ kPa) decane and kerosene flames have been stabilized on a flat flame burner. The flame diameter was 9.5 cm. The composition of the feed gas was 8.0% fuel, 56.4% oxygen, 35.6% argon for the decane flame and 7.6% fuel, 56.8% oxygen, 35.6% argon for the kerosene flame. These conditons correspond to equivalence ratios of 2.2 for both flames where kerosene was considered as $C_{11}H_{22}$. Gas velocity at the burner exit was 24 cms^{-1}. Temperature and mole fraction profiles were measured along the centerline.

Molecular beam mass spectrometry technique was used for analyzing active and stable chemical species. Fragmentation of the fuel molecules in the electron impact ionisation source of the mass spectrometer was carefully avoided by using electron energies not higher than 1-2 eV above the ionisation potential of the respective species. This allowed the analysis of intermediate hydrocarbons formed in the flame by thermal degradation of the fuel.

The method of identification of the chemical species and calibration of the mass spectrometer have been described elsewere [3.13, 3.14].

Pt $-$ Pt 10% Rh thermocouples with wire diameters of 50 μm have been used for temperature measurements. Coating with BeO/Y$_2$O$_3$ prevented catalytic effects of the thermocouples. Heat losses due to radiation were compensated by electrical heating of the thermocouples. This was applied for thermocouple positions in the non soot forming zone of the flames. In the soot forming zone of the flames heat losses were calibrated by temperature measurements in a near sooting decane flame.

A Gilson pump for liquid chromatography was used to control the flow rate of decane or kerosene. The liquid fuel was atomized with argon in a heated small orifice and then vaporized in a heated chamber. The orifice temperature was rather critical when operating with kerosene in order to avoid fuel condensation at temperatures $T < 180°$C or polymerization at temperatures $T > 200°$C.

3.3 Results

In a previous work on modelling of acetylene flames the formation of the first aromatic ring was described by acetylene addition to C$_4$ species:

$$C_4H_5 \; + \; C_2H_2 \; \longrightarrow \; C_6H_6 \; + \; H \qquad (3.1)$$

$$C_4H_3 \; + \; C_2H_2 \; \longrightarrow \; C_6H_5 \qquad (3.2)$$

Kinetic parameters for these reactions were taken from *Westmoreland* et al. [3.15].

The reaction mechanism for rich decane flames was based on the addition of some reactions for decane consumption to the acetylene mechanism [3.9]. The quality of the predictions of the formation of aromatic hydrocarbons by this mechanism depends strongly on the accuracy of the modelling of C$_4$ species and their precursors C$_2$H$_4$ and C$_2$H$_2$.

Examination of the experimental measured mole fraction profiles in a kerosene and a decane flame showed that C$_4$ radicals and acetylene are formed in comparable amounts so that reactions 3.1 and 3.2 cannot explain the larger concentration of benzene obtained in the kerosene flame.

Therefore, a specific comparative study on the formation of benzene and two others aromatic species, phenyl acetylene and vinyl benzene, in decane and kerosene flames has been carried out.

3.3.1 Formation of Aromatic Hydrocarbons in Kerosene and Decane Flames

First of all a number of decane and kerosene flames with equivalence ratios from 1.0 to 2.5 have been investigated to qualitatively examine the differences between the two fuels and their dependence on equivalence ratios. Changes in the feed gas composition have been done keeping both overall and argon flowrates constant. Gas velocity at the burner exit was 27.5 cm s^{-1} at 298 K and 6.0 kPa for these experiments.

Figure 3.1 confirms that the maximum mole fraction of acetylene in decane flames is only slightly higher than that in kerosene flames. In both flames, a linear increase with equivalence ratio is observed for equivalence ratios above 1.6.

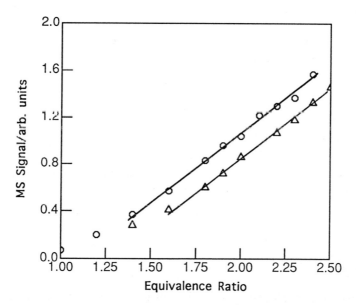

Fig. 3.1. Dependence of the maximum acetylene concentration in decane and kerosene flames in dependence on equivalence ratio; (o): decane flames; ($\triangle$): kerosene flames

The mass spectrometric measurements of benzene have been performed with an electron energy in the ion source adjusted to 13 eV. For kerosene flames a second measurement with an electron energy of 11 eV has been done to look for fragment ions formed in the ion source of the mass spectrometer. The measurements with both energies lead to a linear variation of the maximum signal with the equivalence ratio. Therefore, it can be concluded that the benzene signals are free from fragmentation effects (compare Figure 3.2). Extrapolation of the mass spectrometric signal gives a zero value for an equivalence ratio equal to 0.8.

The results from the decane flame confirm that benzene is formed in lower concentrations than in kerosene flames for flames with an equivalence ratio of 2.0 this difference is about one order of magnitude. A second difference is the variation of the maximum signal with the equivalence ratio, for the decane flame the maximum benzene concentration varies exponentially with the equivalence ratio, the exponent is higher than 1.

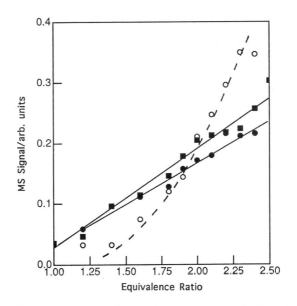

Fig. 3.2. Dependence of the maximum benzene concentration in decane and kerosene flames in dependence on equivalence ratio; ($\circ$): decane flames (signal times 10); ($\bullet$): kerosene flames 11 eV ionization energy (signal times 4); ($\blacksquare$): kerosene flames 13 eV ionization energy

The diagrams plotted in Fig. 3.2 confirm that benzene formation results from two different mechanisms in decane and kerosene flames. In the former, benzene is formed by reactions 3.1 and 3.2 so that the signal is proportional to acetylene and either C_4H_3 or C_4H_5 concentrations. Because C_4 species are formed from acetylene, benzene concentration varies with $[C_2H_2]^2$ as expected. The dashed line in Fig. 3.2 corresponds to the variation of the benzene concentration with $k\,[C_2H_2]^2$ where k adjusted to match the measured benzene concentration at an equivalence ratio of 2.4 and $[C_2H_2]$ represents the maximum signal measured for acetylene. This curve reproduces with good accuracy the experimentally determined dependence of the benzene signal on equivalence ratio. The result confirms that in the decane flames, benzene formation involves two acetylene molecules. Acetylene addition to C_4 radicals meet this condition. However, it must be mentioned that this is not the only possibility and the recombination of C_3H_3 radicals, formed by reaction of CH_2 or CH with acetylene would lead to the same relationship between the maximum benzene and acetylene signals.

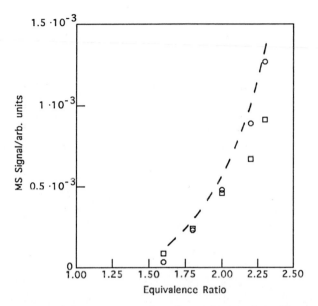

Fig. 3.3. Dependence of the maximum phenyl acetylene and vinyl benzene concentrations in decane flames in dependence on equivalence ratio; (o): phenyl acetylene; (□): vinyl benzene; dashed line: calculated phenyl acetylene signals

The linear variation of the benzene signal with the equivalence ratio in kerosene flames indicates that the aromatic components of the fuel contribute directly to the benzene formation. The procedure adopted to change the equivalence ratio of the flames (constant overall and argon flowrates) leads to the following relationship between kerosene flowrate and the equivalence ratio ϕ:

$$\dot{V}_K = (\dot{V}_o - \dot{V}_{Ar})[2\phi/(2\phi + 31)] \qquad (3.3)$$

Because 2ϕ is small compared with 31, this expression predicts a linear dependence of $\dot{V}_K$ on ϕ and, therefore, a linear dependence of the maximum benzene conzentration on the kerosene flowrate. In Equation (3.3) $\dot{V}_K$, $\dot{V}_{Ar}$, $\dot{V}_o$, represent the flowrates of kerosene, argon and the over-all flowrate, respectively.

Phenyl acetylene and vinyl benzene as well come from different sources in the flames of the two different fuels. In decane flames, these two species are

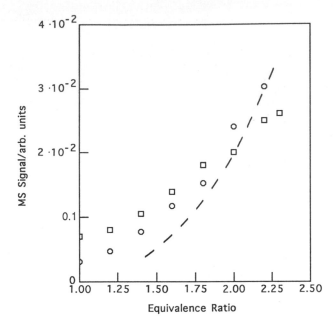

Fig. 3.4. Dependence of the maximum phenyl acetylene and vinyl benzene concentrations in kerosene flames in dependence on equivalence ratio; (o): phenyl acetylene; (□): vinyl benzene (signal times 0.4); dashed line: calculated phenyl acetylene signals

not observed for equivalence ratios lower than 1.5. A considerable increase in the signal is observed for richer flames, compare Fig. 3.3. Because it seems reasonable that these species are formed by addition of acetylene to benzene or phenyl radicals the variation of the expression $k \, [C_6H_6][C_2H_2]$ is plotted versus ϕ. Again, the value of the constant k was arbitrarily adjusted to match either the phenyl acetylene or the vinyl benzene signal at $\phi = 2.4$. The same procedure applied to the kerosene flames shows evidently that the phenyl acetylene and vinyl benzene signals do not vary with $k \, [C_6H_6][C_2H_2]$, compare Fig. 3.4. Therefore, we can conclude that phenyl acetylene and vinyl benzene measured in stoichiometric or slightly rich kerosene flames result from the aromatic components of the fuel.

3.3.2 Formation of Benzene in Kerosene Flames

The comparative study discussed in the previous section clearly shows that consumption reactions of at least one aromatic species must be added to the decane mechanism to predict the structure of rich kerosene flames. Trimethylbenzene is the main aromatic species in the kerosene that was used in this work. However, to simplify both the mechanism and to rely on existing

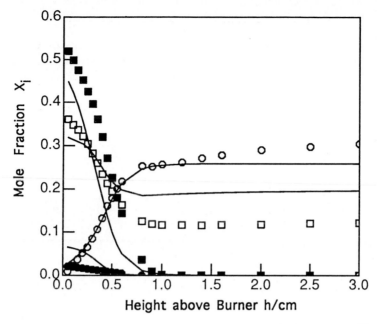

Fig. 3.5. Comparison of experimentally determined mole fraction profiles with predictions for a kerosene - O_2 - Ar flame; $\phi = 2.2$; $p = 6$ kPa; ($\bullet$): $C_{10}H_{22}$, ($\blacksquare$): O_2, ($\circ$): CO, ($\square$: Ar, Symbols: measurements, lines: predictions

kinetic data, the aromatic part of kerosene was considered to be toluene. The reactions given in Table 3.1 then were considered for its consumption.

Table 3.1. Reaction Mechanism and Rate Coefficients for toluene consumption

	Reaction						A	n	E_a
3.4	C_7H_8	+	H	$\rightarrow$	C_7H_7	+	H_2	$1.24 \cdot 10^{14}$	0.0	35.10
3.5	C_7H_7	+	H	$\rightarrow$	C_7H_8			$9.00 \cdot 10^{13}$	0.0	0.00
3.6	C_7H_8	+	H	$\rightarrow$	C_6H_6	+	CH_3	$3.50 \cdot 10^{13}$	0.0	15.50
3.7	C_7H_7			$\rightarrow$	C_3H_3	+	C_4H_4	$4.00 \cdot 10^{16}$	0.0	432.30
3.8	C_7H_7	+	O_2	$\rightarrow$	C_6H_6	+	Prod.	$1.00 \cdot 10^{13}$	0.0	0.00

Remarks: Rate coefficients are given in the form $k_i = A_i\, T^n\, e^{-E_a/RT}$; units: cm, mol, s, kJ.
Kinetic data have been taken from *Rao* et al. [3.16].

3.3.3 Modelling Kerosene Flames

The reactions given in Table 1 have been added to the mechanism validated previously for decane combustion [3.9]. This mechanism is based on a rapid consumption of the decyl radical formed by initial attack of H, O and OH on the fuel molecule. At the high temperatures prevailing in the investigated flames it was assumed that it is not necessary to distinguish between the various isomers of the $C_{10}H_{21}$ radical. The numerical simulation of the kerosene flame has been performed for a fuel composed of 10% toluene and 90% decane. The computer code from Warnatz was used with the experimental temperature profiles as input, so that the energy equation was not solved.

The thermodynamic data have been extracted from the SANDIA data base [3.17] for the species involved in the H_2, C_1 and C_2 submechanisms and the thermodynamic data from *Burcat* [3.18] for decane and toluene combustion reactions have been employed.

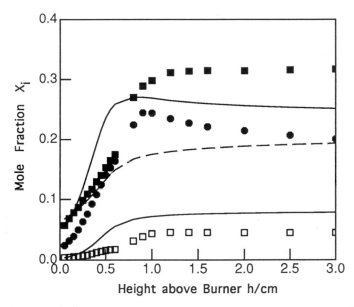

Fig. 3.6. Comparison of experimentally determined mole fraction profiles with predictions for a kerosene - O_2 - Ar flame; $\phi = 2.2$; $p = 6$ kPa; ($\bullet$): H_2O, ($\blacksquare$): H_2, ($\circ$): CO_2; Symbols: measurements, lines: predictions

In Fig. 3.5 to 3.8 the experimentally determined mole fraction profiles of the reactants, the main products, and the species involved in the formation of aromatic hydrocarbons from the alkane part of the fuel for the kerosene flame with an equivalence ratio of 2.2 are compared with predictions based on the above discussed reaction mechanism. Figure 3.9 gives a comparison of the predicted mole fraction profiles of benzene with the experimental results for a decane and a kerosene flame with the same equivalence ratio. This figure shows that in the kerosene flame benzene comes mainly from the aromatic part of the fuel. This contribution can be modelled reasonably well by adding few reactions for the consumption of one aromatic hydrocarbon to the mechanism.

The main reaction channels in the kerosene flame can be inferred from the reaction rate analysis from the numerical simulation.

Benzene is formed in the consumption of toluene very close to the burner surface. When moving downstream, the concentration of benzene is controlled by the reactions:

$$C_6H_6 \ + \ M \ \longrightarrow \ C_6H_5 \ + \ H \ + \ M \qquad (3.9)$$

$$C_6H_5 \ + \ H \ + \ M \ \longrightarrow \ C_6H_6 \ + \ M \qquad (3.10)$$

that tend to be equilibrated.

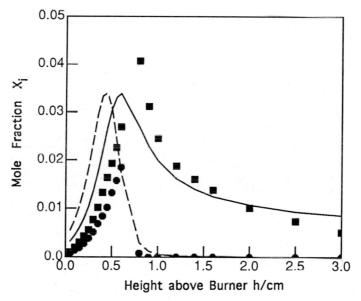

Fig. 3.7. Comparison of experimentally determined mole fraction profiles with predictions for a kerosene - O_2 - Ar flame; $\phi = 2.2$; $p = 6$ kPa; ($\bullet$): C_2H_4, ($\blacksquare$): C_2H_2; Symbols: measurements, lines: predictions

Decane consumption leads mainly to C_2H_4 and C_2H_3. The latter reacts rapidly to form C_2H_2.

The major reaction products are CO, CO_2, H_2 and H_2O. Various sources contribute to the formation of CO:

$$C_2H \ + \ O_2 \ \longrightarrow \ CO \ + \ CO \ + \ H \qquad (3.11)$$

$$CHO \ + \ M \ \longrightarrow \ CO \ + \ H \ + \ M \qquad (3.12)$$

$$C_2H_2 \ + \ O \ \longrightarrow \ CO \ + \ CH_2 \qquad (3.13)$$

$$C_2HO \ + \ H \ \longrightarrow \ CO \ + \ CH_2 \qquad (3.14)$$

$$CHO \ + \ H \ \longrightarrow \ CO \ + \ H_2 \qquad (3.15)$$

CO_2 is formed by two reactions

$$CH_2 \ + \ O_2 \ \longrightarrow \ CO_2 \ + \ H \ + \ H \qquad (3.16)$$

$$CO \ + \ OH \ \longrightarrow \ CO_2 \ + \ H \qquad (3.17)$$

in the vicinity of the burner surface, while the second reaction becomes predominant in the burned gases.

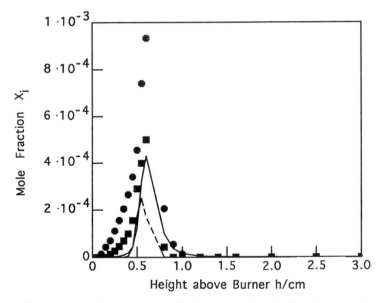

Fig. 3.8. Comparison of experimentally determined mole fraction profiles with predictions for a kerosene - O_2 - Ar flame; $\phi = 2.2$; $p = 6$ kPa; ($\bullet$): C_4H_4, ($\blacksquare$): C_4H_5; Symbols: measurements, lines: predictions

The main source for hydrogen is attack of H atoms on ethylene by:

$$C_2H_4 \ + \ H \ \longrightarrow \ H_2 \ + \ C_2H_3 \qquad (3.18)$$

The reactions of H atoms with H_2O and CH_4 are faster than this reaction, but as they tend rapidly towards equilibrium their net contribution to the formation of H_2 is small.

Very close to the burner, the reaction:

$$H_2 \ + \ OH \ \longrightarrow \ H_2O \ + \ H \qquad (3.19)$$

which is the main source of water is faster than its reverse reaction.

The kinetic parameters for the reactions 3.9 to 3.19 are taken from previous work on acetylene flames [3.9].

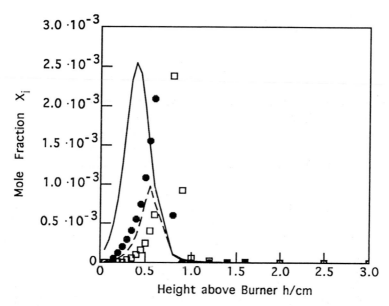

Fig. 3.9. Comparison of experimentally determined mole fraction profiles of benzene for a decane - O_2 - Ar flame and a kerosene - O_2 - Ar flame with predictions; $\phi = 2.2$; $p = 6$ kPa; (•): kerosene flame, (○): decane flame; Symbols: measurements, lines: predictions

3.4 Conclusions

In this work the formation of aromatic hydrocarbons in decane and kerosene flames was investigated. The results showed that the structures of decane and kerosene flames are similar except for benzene that is formed in larger amount in kerosene flames. A specific experimental investigation with systematic measurements of acetylene, benzene, phenyl acetylene and vinyl benzene in decane and kerosene demonstrated that the aromatic part of kerosene is the main source of aromatic hydrocarbons in kerosene flames whereas benzene is formed by the addition of acetylene to C_4 radicals in the decane flame. This difference was taken into account in developing a mechanism for modelling the kerosene flame. In addition to the decane combustion mechanism used so far only few reactions for the consumption of toluene as a representative of the aromatic part of the fuel have to be considered. This modification leads to predictions which are in reasonable agreement with the experimental mole fraction profiles in decane and kerosene flames.

References

3.1   J.A. Miller, R.E. Mitchell, M.D. Smooke, R.J. Kee: "Toward a Comprehensive Chemical Kinetic Mechanism for the Oxidation of Acetylene: Comparison of Model Predictions with Results from Flame and Shock Tube Experiments", in Nineteenth Symposium (International) on Combustion (The Combustion Institute, Pittsburgh 1988) p. 181

3.2   C.K. Westbrook, F.L. Dryer: Prog. Energy Combust. Sci. **10**, 1 (1984)

3.3   J. Warnatz: Combust. Sci. Tech. **34**, 177 (1983)

3.4   P.R. Westmoreland, J.B. Howard, J.P. Longwell: "Tests of Published Mechanisms by Comparison with Measured Laminar Flame Structure in Fuel-Rich Acetylene Combustion", in Twenty-First Symposium (International) on Combustion (The Combustion Institute, Pittsburgh 1988) p. 773

3.5   M. Frenklach, J. Warnatz: Combust. Sci. Technol. **51**, 265 (1987)

3.6   S.J. Harris, A.M. Weiner, R.J. Blint: Combust. Flame **72**, 91 (1988)

3.7   E. Axelsson, K. Brezinsky, F.L. Dryer, W.J. Pitz, C.K. Westbrook: "Chemical Kinetic Modeling of the Oxidation of Large Alkane Fuels: n-Octane and iso-Octane", in Twenty-First Symposium (International) on Combustion (The Combustion Institute, Pittsburgh 1986) p. 783

3.8   C.K. Westbrook, J. Warnatz, W.J. Pitz: "A Detailed Chemical Kinetic Reaction Mechanism for the Oxidation of iso-Octane and n-Heptane over an Extended Temperature Range and its Application to Analysis of Engine Knock", in Twenty-Second Symposium (International) on Combustion (The Combustion Institute, Pittsburgh 1986) p. 893

3.9   J.L. Delfau, M. Bouhria, M.Reuillon, O. Sanogo, R. Akrich, C. Vovelle: "Experimental and Computational Investigation of the Structure of a Sooting Decane-O_2-Ar Flame", in Twenty-Third Symposium (International) on Combustion (The Combustion Institute, Pittsburgh 1986) p. 1567

3.10   D.B. Scully, R.A. Davies: Combust. Flame **9**, 185 (1965)

3.11   S.J. Harris, A.M. Weiner: Combust. Sci. Technol. **38**, 75 (1984)

3.12   F.W. Lam, J.P. Longwell, J.B. Howard: "The Effect of Ethylene and Benzene Addition on the Formation of Polycyclic Aromatic Hydrocarbons and Soot in a Jet-Stirred/ Plug-Flow Combustor", in Twenty-Third Symposium (International) on Combustion (The Combustion Institute, Pittsburgh 1990) p. 1477

3.13   E. Bastin, J.L. Delfau, M. Reuillon, C. Vovelle, J. Warnatz: "Experimental and Computational Investigation of the Structure of a Sooting $C_2H_2 - O_2 - Ar$ Flame", in Twenty-Second Symposium (International) on Combustion (The Combustion Institute, Pittsburgh 1988) p. 313

3.14   E. Bastin: "Etude du Mécanisme de Formation des Prècurseurs des Particules de Suie dans des Flammes $C_2H_2/O_2/Ar$"; Thèse de Doctorat, Université Pierre et Marie Curie, Paris 6 (1989)

3.15 P.R. Westmoreland, A.M. Dean, J.B. Howard, J.P. Longwell: J. Phys. Chem. **93**, 8171 (1989)
3.16 V.S.Rao, G.B. Skinner: J. Phys. Chem. **88**, 4362 (1984)
3.17 J.R. Kee, M.Rupley, J.A. Miller: The Chemkin Thermodynamic Data Base, SAND87-8215-UC-4 (SANDIA Report, Livermore 1987)
3.18 A. Burcat: "Thermochemical Data for Combustion Calculations", in Combustion Chemistry, ed. by W.C. Gardiner Jr., (Springer-Verlag, Heidelberg, New York, 1984) p. 455

Discussion

Colket: Before I open the paper for discussion I want to raise the question whether what you have just shown us suggests that all of our efforts in understanding benzene formation mechanisms may be moot points. In my opinion this is not the case because obviously a lot of fuels that we have studied earlier show that direct addition of aromatic hydrocarbons increases soot formation. This is certainly consistent with your results but we have always to be careful in identifying the relative amount of aromatic hydrocarbons in the fuel with the dominant source for benzene.

Wagner: Could you give some more information about the final temperatures in your flames? Secondly, did you observe polyhedral flames on this kind of flat flame burners? It is well known that flames of higher hydrocarbons exhibit polyhedral flames on that kind of burners unless they are stabilized very close to the burner surface. Finally, could you explain what you mean by equivalence ratio? Do you have the same C/H-ratio in the decane flame as in the kerosene flame or if not how do you define the equivalence ratio?

Vovelle: The flame temperatures amounted to about 1900 K in the burnt gases of the flames of both fuels. The final flame temperatures depend more on the way of stabilisation of the flame than on the type of fuel. Even for the acetylene flames that we have investigated previously we had a final flame temperature in this range.

We did not observe polyhedral flames at the low pressures of our investigation. At higher pressures in increasing the equivalence ratio we observed polyhedral flames. They show up suddenly when increasing the equivalence ratio and we prevented them by keeping the gas velocity low.

For decane there is no problem in defining the equivalence ratio. The chemical analysis of kerosene showed that it can be represented by C_{11}, undecane. For an identical equivalence ratio there is a slight difference in the fuel content due to the aromatic hydrocarbon content. However, this aromatic hydrocarbon content is more essential for benzene formation than for the main properties of the flames.

Glassman: Going back to the way we were discussing things three years ago in Göttingen, it would be very nice to state where the position of the actual flame front is with respect to the various concentrations and where the soot formation takes place when we look at such data. Secondly, I go back to our own data on premixed flames where we had this semi-empirical correlation setting the equivalence ratio as a function of the number of $C - C$ bonds which gave some chemical and physical significance. From my conclusions - and you probably substantiated that - most of the fuels except some of the aromatic hydrocarbons break down to acetylene and then the acetylene is involved in the actual soot process not the intermediate PAH. I would like you to comment on that.

Vovelle: I agree with you. In the decane flame we have much experimental information about acetylene and acetylene is the major hydrocarbon in the burnt gases.

Peters: As far as the kinetic mechanism for benzene formation is concerned I was surprised not to see any reactions where C_3H_3 is formed in the decomposition of the fuel. Don't you think that the reaction C_3H_3 plus C_3H_3 is important for benzene formation?

Vovelle: When we started our kinetic modelling it was just the data of Westmoreland that was available and he stresses acetylene addition to C_4 for the formation of aromatic hydrocarbons. So in our first attempts we used his mechanism and his rate coefficients. After this the recombination of C_3H_3 as a main source for the formation of aromatic hydrocarbons has been discussed. However, at present some new results on the rate coefficients show that the main reaction products of C_3H_3 plus C_3H_3 are not aromatic hydrocarbons but other products. We will include all this new information in our modelling effort because in our low pressure flames we have experimentally determined the concentrations of C_3H_2, C_3H_3, C_3H_4 and others. So we can compare predictions including all possible pathways to benzene with our experimental results.

Smyth: Phil Westmoreland gave a seminar last week at NIST. Since other people have started analysing his data he is, of course, going back and re-evaluating his data on just this point in more detail. His conclusion is pretty much as you said that C_3H_3 plus C_3H_3 makes lots of mass 78 but not too much benzene. If I could sift through his comments, what he stressed was C_4H_5 plus acetylene as the predominant pathway for forming benzene in his acetylene flames.

Santoro: I noticed in the benzene profiles that in the decane flame the model very much underpredicted the early formation of benzene but matched later on in time better. In the decane plus toluene flame it did better early on in the flame and then didn't match as well later on. I was surprised by that a little bit. Is the ability of the model to make reasonable predictions at different times due to mechanistic changes between the two systems or how important is the temperature profile? It has often been pointed out that a slight shift in the temperature profile in the flame causes great changes in the species concentration.

Vovelle: Concerning the differences between the two flames you have to remember that the mechanisms are completely different. In the kerosene flame we have to model the consumption of toluene. We used the available kinetic data for the consumption of toluene but we didn't try to change the rate coefficients in our first attempts. With this model we have a good agreement early in the flame. The discrepancy later in the flame is completely due to the modelling of the toluene consumption. We consume toluene too fast in that part of the flame and so we have no more toluene to form benzene. For the decane flame the predicted benzene mole fraction profile is completely dependent on the modelling of C_2, because starting from decane we have first to form acetylene and from C_2 we form C_4 and they react to form aromatic hydrocarbons. I should mention that the discrepancy between predicted and measured mole fractions in the lower part of the flame is not very significant because we have very low concentrations and the experimentally determined mole fractions have certain experimental errors.

Growth of Large Ionic Polycyclic Aromatic Hydrocarbons in Sooting Flames

Silke Löffler, Philipp Löffler, Petra Weilmünster,

Klaus-H. Homann

Institut für Physikalische Chemie, T. H. Darmstadt,
64287 Darmstadt, Fed. Rep. of Germany

Abstract: Positive and negative ions of PAH^+ have been analysed in low-pressure premixed flames of various fuels. Mass spectra were taken through a molecular beam sampling system by a linear time-of-flight mass spectrometer and reflectron mass spectrometer. The main formation of PAH^+ occurs in the oxidation zone of the flames where their concentrations pass through a maximum. The mass increase of PAH^+ is a net result of mass growth, decomposition and oxidation. Therefore, mass growth and diminution of PAH and PAH^+ is favored by an
(a) increase in concentration of unsaturated small hydrocarbons;
(b) increase in concentration of unsaturated small hydrocarbon radicals;
(c) increase in the size of PAH or PAH^+, respectively;
(d) increase in temperature (for growth reactions which have an activation energy);
(e) increase in temperature which favors thermal decomposition;
(f) increase in concentration of oxidizing radicals such as OH;
(g) increase in the concentration of O_2;
(h) decrease in the size (or mass) and in the degree of pericondensation, increase in the number of side chains.

4.1 Introduction

There is little doubt now that large polycyclic aromatic hydrocarbons (PAH) are precursors in the homogeneous formation of soot particles from molecular reactants in flames. The role of PAH in this process has slowly become clear but is not yet fully understood. One of the main reasons is their complicated behavior in flames and the difficulty to follow their growth into the transition regime of soot formation. Most publications in this field deal with PAH having up to 7 - 8 condensed rings (300 u $\leq m \leq$ 350 u), a limit set by gas chromatrography (GC) [4.1] . In fuel-rich and sooting premixed flames, the profiles of many PAH increase within the oxidation zone to a first maximum, then decay when the temperature approaches its maximum while oxygen is still present [4.2]. In flames of aromatic fuels the decrease in the hotter part of the oxidation zone is particularly steep [4.1]. In the burned gas, when soot formation has started, PAH concentrations rise again, although much more smoothly than in the oxidation zone and without generating new small soot particles. The minimum of those PAH, which can just be separated by GC, in the zone of beginning soot formation and their re-increase in the burned gas while the rate of soot mass growth decreases, had lead to the conclusion that PAH such as acenaphtylene, pyrene, benzo(ghi)fluoranthene, coronene and others are by-products rather than intermediates in the process of soot formation [4.3].

Another difficulty in understanding the role of PAH arose from the fact that their maximum concentrations in flat flames and also in other combustion devices decrease by about one order of magnitude on the average with an increase in molecular mass of 80 to 100 u within the range accessible by GC. An application of this rule to larger PAH led to a severe mismatch with the number density of the smallest soot particles [4.4].

The difficulty to identify and to follow very large neutral PAH still exists. A new attempt to overcome this problem is given in Section 8. However, great progress has been made in studying large charged species which accompany the formation of the large neutral molecules [4.5, 6]. In a sooting hydrocarbon flame almost all of the positive charge is carried by PAH, soot particles, fullerenes and polyynes. The growth of PAH$^+$ can be followed to much larger masses into the range of small soot particles. Since fullerene ions of both signs occur together with charged PAH and soot, their role in soot formation may also be recognized [4.7].

4.2 Experimental

The experimental procedure of sampling ions from 2.67 kPa flat pre-mixed flames of hydrocarbon/oxygen via a two-stage nozzle/molecular beam sampling system and the analysis by a time-of-flight (TOF) mass spectrometer has been reported [4.5]. Recently, the linear TOF-MS was reconstructed by installing a Mamyrin-type ion reflector for better mass resolution ($(m/\Delta m)_{50\%} = 1500$) [4.6]. A similar equipment is discussed in detail in [4.17]. Acetylene, benzene, and butadiene have been used as fuels.

4.3 Nature of Polycyclic Aromatic Hydrocarbon Ions (PAH⁺)

The identification of PAH$^+$ or at least their association with certain classes of PAH$^+$ is based on reflectron TOF mass spectrometry. In addition to the atomic composition, other arguments such as thermal stability, the relation to identified neutral PAH in flames and pyrolysis, and the prevalance of peri-condensation indicate certain structures.

Beginning with idenylium, which is the smallest PAH$^+$ with more than one ring, PAH$^+$ occured with any number of C atoms (even and odd) and with varying numbers of H atoms. *Calcote* et al. [4.8] and *Michaud* et al. [4.9] who were the first to study PAH$^+$ ions in fuel-rich low-pressure flames reported exclusively molecular, i. e. non-radical PAH ions. However, higher mass resolution which allows an exact determination of the number of H atoms showed that radical PAH$^+$ prevail in many cases, particularly at higher temperature.

Even-numbered molecular PAH$^+$ with an odd number of H atoms can be regarded as protonated species which also occur as neutral molecules in the flame. For example, $C_{16}H_{11}^+$ is protonated pyrene or fluoranthene ($C_{16}H_{10}$). Odd-numbered molecular PAH$^+$ can formally be derived by loss of H$^-$ from corresponding neutral molecules which are of the types: methyl-PAH, PAH with a 5-ring containing a CH_2 group and PAH of the phenalene-typ, although the latter have not yet been detected as neutrals in flames. The ions are respectively, for example:

naphtyl-methylium   indenylium   phenalenylium

For a given formula $C_xH_y^+$ many aromatic structures are possible in principle. $C_{13}H_9^+$, for example, could occur in the following structures, if species with an unsaturated side chain are disregarded:

(4.1a)   (4.1b)   (4.1c)   (4.1d)

Although only fluorene as the neutral $C_{13}H_{10}$ with the carbon stucture (4.1c) has been detected in flames, the arguments are in favor of the phenalenylium structure (4.1d) for the ion. Whereas the relative concentration of fluorene is small, as compared to C_{12}- and C_{14}-PAH, the $C_{13}H_9^+$ is by far the most prominent ion in this mass range in C_2H_2 flames. This indicates extra thermal stability and a certain chemical inertness which can reasonably only be attributed to the totally condensed plane phenalenylium structure [4.8].

The number of H atoms for a given number of C atoms depends on the degree of peri-condensation of the aromatic ring system, on the number of 5-membered rings and on the existence and the nature of side chains. Divergence from a more circular system (peri-condensation) to an elongated or ribbon-like PAH increases the number of hydrogen atoms. Formation of 5-rings condensed with the 6-rings diminishes it. An extra H is often due to protonation.

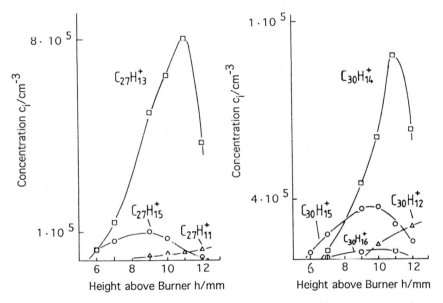

Fig. 4.1. Concentration profiles of odd-numbered and even-numbered PAH$^+$ with different hydrogen content in a sooting acetylene/oxygen flame; C/O=1.0, $v_u = 24$ cm s^{-1}, $p = 2.7$ kPa

Of the possible variations in the number of H atoms seldom more than 4 occur. The ratios of the PAH$^+$ with different H content change with temperature: Those with less H atoms are formed at higher temperature, see Fig. 4.1. This can either mean a loss of H from a protonated species to give a radical cation without change of the carbon skeleton or a PAH$^+$ with a different structure. For example, at a height of 7 mm a PAH$^+$ $C_{30}H_{15}^+$ is more abundant than other species with 30 C atoms. It is protonated $C_{30}H_{14}$. According to the principle that peri-condensation prevails in flame-PAH, the following structure is probable:

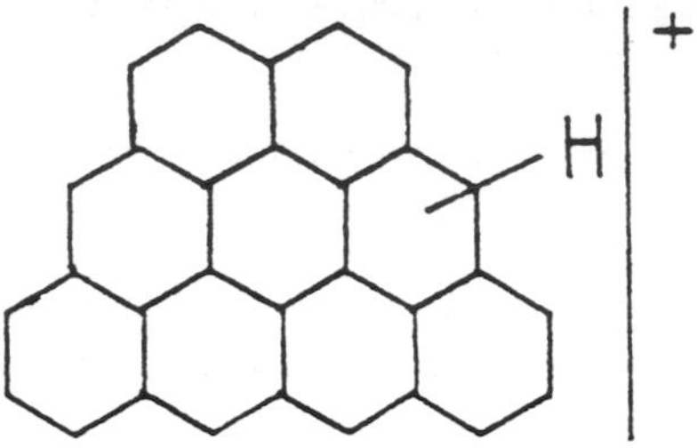

H

$+$

At greater heights in the flame, where the temperature is higher, it is outweighted by $C_{30}H_{14}^+$ which probably is the ionized PAH with the same structure. This does of course not mean that the latter is directly formed from the former. At still higher temperature, a PAH$^+$ $C_{30}H_{12}$ appears, as shown in Fig. 4.1. This can hardly be explained merely by a further loss of two H while the carbon structure is retained, since benzyne bonds have never been observed in flame-PAH. It is much more probable that this composition indicates a peri-condensed PAH with two 5-membered rings, for example:

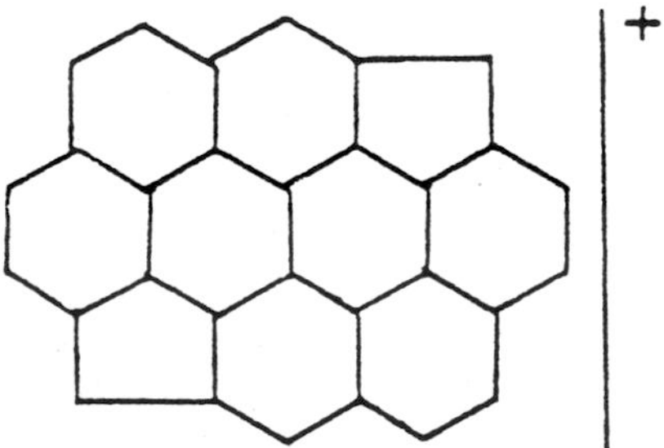

$+$

Other possible structures of hydrogen-poor PAH$^+$ must be considered, if peri-condensation with 5-rings would cause too much internal strain, particularly in small PAH. $C_{13}H_7^+$, for example, which occurs in low abundance as compared to $C_{13}H_9$ cannot be imagined as a system of three 5-rings condensed on a 6-ring. A more probable structure is that of a butadiynyl indenylium:

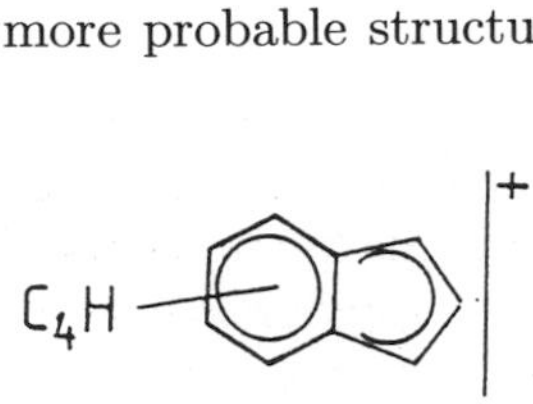

C_4H

$+$

Butadiynyl-benzene has been found in flames [4.1] and in the pyrolysis of unsaturated aliphates [4.10]. Aromates with ethynyl or polyynyl side chains are much less stable at higher temperature than mere polyynes, which is the reason why the former do not occur in high concentration. Therefore, the structure of a hydrogen-poor PAH$^+$ with a large enough ring system occuring at higher temperature is more likely to contain five-membered rings rather than ethynyl or butadiynyl side chains.

4.4 Formation and Growth of PAH$^+$

If not otherwise stated, the results refer to acetylene/oxygen flames. Some special features of PAH$^+$ in benzene flames have been reported [4.11] and will be dealt with below.

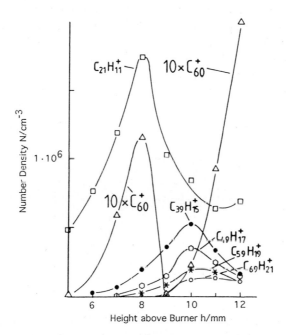

Fig. 4.2. Concentration profiles of PAH$^+$ with increasing mass and of C$_{60}^+$ (ten-fold enlarged) in the oxidation zone of a sooting acetylene/oxygen flame; burning conditions same as in Fig. 4.1

The main formation of the PAH$^+$ takes place in the oxidation zone where their concentrations go through maxima, see Fig. 4.2. After the maxima, the decrease of the PAH$^+$ with $m < 325$ u occurs almost simultaneously [4.5]. Most of this decrease of the low-mass PAH is due to thermal decomposition, probably initiated by oxidative attack. In sooting acetylene flames, PAH$^+$

with $m > 325$ u (C_{26}-PAH$^+$) continue to grow, as shown by the shift of their concentration maxima to greater heights with increasing mass. This is compatible with a growth of the PAH$^+$ by the addition of unsaturated light aliphatic hydrocarbons such as acetylene but also of butenyne, butadiyne and benzene. Which of these is the most abundant, depends on the kind of fuel. When burning acetylene, it is the hydrocarbon with the largest concentration, both in the oxidation zone and in the burned gas. With benzene as fuel, this is the major hydrocarbon in the oxidation zone, but acetylene prevails in the burned gas [4.12].

There was no sudden change in the hydrogen content of PAH$^+$ with $m > 325$ u, nor could any PAH$^+$ of extra reactivity be recognized. Up to about $C_{40}H_x^+$, odd numbered PAH$^+$ in general reach a larger concentration than even-numbered PAH$^+$. For larger PAH$^+$ this difference vanishes. The decrease in the maximum concentrations with increasing molecular mass, which is observed for smaller neutral and charged PAH [4.4], does continue to larger PAHs, but to a greatly reduced degree. There is only little change in the PAH$^+$ concentration from $C_{42}H_{16}^+$ to $C_{100}H_{24}^+$, for example, see Fig. 4.3. They reach masses of about $2.5 \cdot 10^3$ u, but a definite largest PAH$^+$ cannot be recognized.

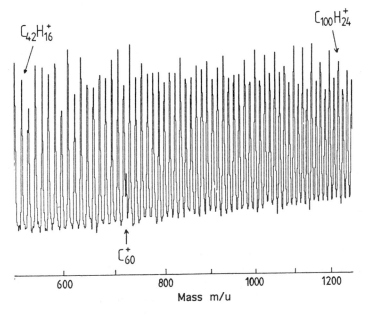

Fig. 4.3. Section of the mass spectrum of high-mass PAH$^+$ from the zone of beginning soot formation in an acetylene/oxygen flame; note the low signal height of C_{60}^+; C/O=1.0, $v_u = 24$ cm s^{-1}, $p = 2.7$ kPa, $h = 12$ mm

It is surprising that the main formation of PAH$^+$ and of neutral PAH takes place in a flame zone in which the rate of hydrocarbon oxidation increases and where the temperature increases which favors decomposition of PAH. A mass increase of PAH$^+$ therefore is a net result of mass growth, decompostion and oxidation. For ions in particular, ionisation and recombination or exchange of charges, the rates of which depend on the nature of the ion and on its size, may interfere with formation and consumption processes which take place both with ions and neutrals.

Mass growth of positively charged PAH is favored by an

(a) increase in concentration of unsaturated small hydrocarbons;

(b) increase in concentration of unsaturated small hydrocarbon radicals;

(c) increase in the size of PAH or PAH$^+$, respectively;

(d) increase in temperature (for growth reactions which have an activation energy).

A decrease in these quantitaties has the opposite effect. Diminution of PAH and PAH$^+$, on the other, hand is favored by an

(e) increase in temperature which favors thermal decomposition;

(f) increase in concentration of oxidizing radicals such as OH;

(g) increase in the concentration of O_2;

(h) decrease in the size (or mass) and in the degree of pericondensation, increase in the number of side chains.

The temperature effect is stronger on decomposition than on mass growth, since the activation energy of the former is much larger. Decomposition and oxidation take place by attacking peripheric bonds of the PAH. The relative number of these bonds increases with decreasing size and pericondensation of the PAH. Side chains are more easily split off than C-C bonds are broken in peri-condensed PAH [4.10]. With increasing size of the PAH or PAH$^+$ sticky collisions among the PAH become more frequent due to increased van der Waals-forces.

Regarding the listed quantities, they change in different and opposite ways with height in the oxidation zone. In some cases they go through maxima and decrease again. The influence of some properties, such as the concentration of O_2 and of small unsaturated aliphates and mono-aromates can better be judged from a change of the fuel and the mixture ratio. An influence of temperature can also be studied by changing the unburned gas velocity.

Judging from the formation and net growth of PAH$^+$ as a function of height in a sooting flame and with change in the mixture and in the burning conditions the items (a), (b), and (c) are the most important ones in favor of PAH$^+$ growth. Once the PAH$^+$ in sooting flames have reached masses between 300 and 400 u, their profiles indicate that their decrease in number density, after going through a maximum, is due to growth rather than to decomposition or oxidation. Continued growth of the large species takes place in spite of a simultaneous increase in temperature and a decrease in acetylene concentration, in case it is the fuel. The increase in PAH$^+$ size (opposite to item (h)) works against a decomposition favored by the temperature increase.

The overall concentration of unsaturated small hydrocarbons, such as acetylene, propyne, butenyne, benzene in the oxidation zone can be increased by burning a richer flame. In fuel-rich but non-sooting flames the PAH$^+$ reach a certain size which increases with C/O and then decompose again. Since the maximum concentration of PAH$^+$ decrease with increasing mass an observable upper limit for the size of PAH$^+$ practically depends on the sensitivity of the mass spectrometer. In a C_2H_2/O_2 flame at C/O = 0.70, $v_u = 50$ cm s^{-1}, the largest PAH$^+$ had a mass of 411 u (10 - 11 condensed rings), whereas with C/O = 0.80 a C_{33}-PAH$^+$ (17 condensed rings) was observed. In sooting flames ($0.85 \leq$ C/O ≤ 0.90), PAH$^+$ of more than $2 \cdot 10^3$ u were formed.

Items (a) through (h) can also be applied to PAH formation and growth in benzene flames. However, there are a number of differences which are mainly caused by the nature of the fuel and by the different flame structure. Phenol is an early oxidation product, from which a great number of oxo-PAH$^+$ originate which grow up to masses of about 450 u [4.11]. In a first decomposition step they form PAH$^+$ by elimination of CO, that is, by a mechanism additional to those in acetylene flames where the oxo-PAH$^+$ concentration is negligible. Consequently, the hydrogen content and therefore the structure of many PAHs in benzene flames might be different from those in acetylene flames. Some PAH$^+$ which are prominent in acetylene flames, such as $C_{13}H_9^+$, are missing. Since reflecton-TOF mass spectrometry has not yet been applied to benzene flames, details of the nature and the growth of the PAH$^+$ cannot yet be reported.

The relatively large concentrations of benzene and other monoaromatic compounds in the oxidation zone cause a rapid formation of much larger concentrations of PAH and of charged PAH than in acetylene flames. However, the growth of PAH does not extend much beyond $9 \cdot 10^2$ u, even in sooting flames. This is attributed to the flame structure in which the oxidation and the soot formation zones overlap. It causes a complete consumption of benzene and other light aromates without producing a larger concentration of acetylene and polyynes. Therefore, the growing PAH run out of low-mass hydrocarbon material. However, because of the relatively large PAH concentration, addition reactions among PAH are more frequent than in acetylene flames. As a consequence, the young soot particles are rather small but their number density is relatively large. These bimolecular reactions between PAH or PAH$^+$ and neutral PAH may also be the reason for the much larger fullerene concentration in benzene than in acetylene flames [4.14].

4.5 Formation of Charged Soot and Growth of PAH$^+$ in the Burned Gas

When in acetylene flames PAH$^+$ have reached a mass of 1500 to 2000 u, charged soot particles appear with an average mass of about $4 \cdot 10^3$ u, as shown in Fig. 4.4. The mass distribution of the PAH$^+$ does not quite smoothly develop into that of charged soot, which would be expected if only growth by addition of small hydrocarbon molecules took place. There is a minimum or shoulder in the distribution around $2 \cdot 10^3$ u. This is interpreted as being due to coagulation of PAH$^+$ of this mass range with neutral PAH of similar mass to form the first charged soot particles, simultaneously with further growth by addition of low-mass hydrocarbons. The high-mass PAH$^+$ ($m > 600$ u) soon disappear completely after the first soot particles have been formed, while lower-mass PAH$^+$ partly survive after their decrease within the oxidation zone.

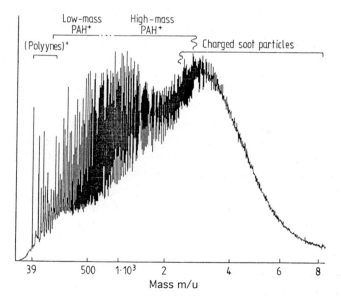

Fig. 4.4. Compressed mass spectrum of charged species in a sooting acetylene/oxygen flame at the beginning of the soot formation zone; $C/O = 1.0$, $v_u = 24 \ cm \ s^{-1}$, $p = 2.7$ kPa, $h = 12$ mm

In the burned gas of C_2H_2/O_2 flames ($h > 25$ mm, depending somewhat on the maximum flame temperature) there was a re-increase of PAH$^+$ and neutral PAH but at a much slower rate than in the oxidation zone (factor 1/50 to 1/100) [4.3, 5]. The slower rate of formation of PAH$^+$ of a certain mass range was certainly due to the greatly reduced rate of overall ion formation by chemi-ionization. However, the rate of average increase in the molecuar mass of the PAH$^+$ was also strongly diminished. Whereas in the oxidation zone the maxima in the profiles, for example, of $C_{21}H_{11}^+$ ($h = 8$ mm) and of $C_{59}H_{19}^+$ ($h = 10.5$ mm) were only 2.5 mm apart, the average PAH$^+$ mass increased only from about 225 to 250 u between 30 and 60 mm in the burned gas. The growth then ceased at about 500 u.

The concentrations of acetylene and of polyynes are lower in the burned gas than in the late oxidation zone, since some more acetylene is burned and these hydrocarbon are also consumed by the mass growth of soot particles. But this alone cannot explain the low rate of PAH$^+$ growth. The main reason seems to be the higher temperature. The temperature range in which PAH$^+$ were formed in the burned gas was 400 to 600 K higher than in the oxidation zone. The higher temperature favored decomposition of PAH$^+$ in the lower mass range. Other reasons for the low growth rate in the burned gas are the very low concentration of hydrocarbons such as butenyne, benzene and other monoaromatic compounds. Furthermore, certain hydrocarbon radicals also have a much lower concentration in the burned gas than in the oxidation zone, say at 9 mm of height. To these radicals belong C_3H_2, C_3H_3, C_2 and also aromatic radicals such as phenyl, benzyl and naphtyl [4.14]. It is therefore suggested that addition of these radicals also contributes to the growth of PAH$^+$ in the oxidation zone. Up to now, however, nothing is known about rate coefficients of ion-radical reactions.

The most impeding influence on the net PAH$^+$ growth is an increase in temperature to values where thermal decomposition becomes concurrent with or prevails over growth. This was demonstrated both by comparison of PAH$^+$ formation in the hot burned gas and the cooler parts of the oxidation zone and also by increasing the maximum flame temperature through an increase in the unburned gas velocity. The latter change has also an influence on the charged soot particles, the concentration of which is correlated with that of large PAH$^+$: The number density of soot particles is greatly reduced while their rate of growth increases. This shows that the activation energy for PAH$^+$ growth is distinctly less than that of further surface growth of charged soot particles which has been determined to be about 100 kJ mol^{-1} [4.15].

4.6 Fullerenes and Related Species as Soot Precursors?

The application of high-resolution mass spectrometry to flame ions in fuel-rich flames can also give an answer to the question whether fullerenes or so-called ico-spiral "carbon cluster" are the nuclei for the growing soot particles. These ideas have been propagated by *Kroto* et al. [4.16] both for the formation of soot in flames and of solid carbon particles in the atmosphere of carbon stars. Fullerene ions display a pre-maximum in the oxidation zone [4.13]. In the flame zone where soot formation starts they have just passed their intermediate minimum and rise again in concentration. The mass spectrum on Fig. 4.3 which was taken from this zone shows the small C_{60}^+ peak inmidst the "forest" of PAH$^+$. This demonstrates clearly that positive fullerene ions, of which C_{60}^+ always is the most prominent, do not play a role at the beginning of soot formation. Ico-spiral carbon clusters for which a nautilus-like structure is postulated should also comprise odd-numbered species. If they exist at all, their free valencies at the rim of the "opening" should be saturated with H. A careful mass spectrometric search for such particles gave no trace of them, nor could any hydrogen-containing fractions of carbon cages be detected in flames. Low concentrations of odd-numbered negative fullerene ions, for example $C_{49}H$-, $C_{51}H$-, $C_{61}H$-, $C_{75}H_2$- occur. Their masses are very near those of prominent fullerene ions (C_{50}, C_{60}, C_{74}) and they can hardly have a nautilus structure. Furthermore, their number density is by orders of magnitude lower than that of the first soot particles. In conclusion, while the growth of positive PAH$^+$ can be followed through into the range of charged soot particles, there is no indication that fullerenes or related ico-spiral carbon clusters could be precursors of soot. However, an almost opposite idea that fullerenes, in particular the large ones, are formed inside small soot particles and escape from them by evaporation, is currently being discussed [4.13].

4.7 Acknowledgements

This work was supported by the Deutsche Forschungsgemeinschaft and by the Fonds der Chemischen Industrie for which we are very much obliged.

References

4.1   H. Bockhorn, F. Fetting, H.W. Wenz: Ber. Bunsenges. Phys. Chem. **87**, 1067 (1983)
4.2   E.E. Tompkins, R. Long: "The Flux of Polycyclic Aromatic Hydrocarbons and of Insoluble Material in Pre-mixed Acetylene-Oxygen Flame", in Twelfth Symposium (International) on Combustion (The Combustion Institute, Pittsburgh 1969) p. 625
4.3   K.-H. Homann, H.Gg. Wagner: "Some New Aspects of the Mechanism of Carbon Formation in Premixed Flame", in Eleventh Symposium (International) on Combustion (The Combustion Institute, Pittsburgh 1967) p. 371
4.4   K.-H. Homann: "Formation of Large Molecules, Particulates and Ions in Premixed Hydrocarbon Flames; Progress and Unsolved Question", in Twentieth Symposium (International) on Combustion (The Combustion Institute, Pittsburgh 1984) p. 857
4.5   Ph. Gerhardt, K.-H. Homann: J. Phys. Chem. **94**, 5381 (1990)

4.6   P. Weilmünster, Th. Baum , K.-H. Homann: "High-resolution Mass Spec-
      trometry of Large PAH Ions in Sooting Ethyne/Oxygen Flames", in
      Combustion and Reaction Kinetics, 22nd International Annual Conference
      of ICT (Fraunhofer-Institut für Chemische Technologie, Karlsruhe 1991) p.
      28-1

4.7   Ph. Gerhardt, S. Löffler, K.-H. Homann: "The Formation of Polyhedral Car-
      bon Ions in Fuel-rich Acetylene and Benzene Flames", in Twenty-Second
      Symposium (International) on Combustion (The Combustion Institute,
      Pittsburgh 1988) p. 395

4.8   D.B Olson, H.F. Calcote: "Ions in Fuel-rich and Sooting Acetylene and
      Benzene Flames", in Eighteenth Symposium (International) on Combus-
      tion (The Combustion Institute, Pittsburgh 1981) p. 453

4.9   P. Michaud, J.L. Delfau, A. Barassin: "The Positive Ion Chemistry in the
      Post-Combustion Zone of Sooting Premixed Acetylene Low-pressure Flat
      Flames", in Eighteenth Symposium (International) on Combustion (The
      Combustion Institute, Pittsburgh 1981) p. 443

4.10 Ch. Knaus: "Untersuchung der Hochtemperaturpyrolyse von 1,3-Butadiin
      im Temperaturbereich von 1150 bis 1710 K mit gaschromatographischer und
      massenspektrometrischer Analyse"; Ph.D. Thesis, Technische Hochschule
      Darmstadt (1991)

4.11 S. Löffler, K.-H. Homann: "Large Ions in Premixed Benzene-Oxygen Fla-
      mes", in Twenty-Third Symposium (International) on Combustion (The
      Combustion Institute, Pittsburgh 1990) p. 355

4.12 J.D. Bittner, J.B. Howard: "Composition Profiles and Reaction Mecha-
      nisms in a Near-sooting Premixed Benzene/Oxygen/Argon Flame", in Eigh-
      teenth Symposium (International) on Combustion (The Combustion Insti-
      tute, Pittsburgh 1981) p. 1105

4.13 Th. Baum, S. Löffler, Ph. Löffler, P. Weilmünster , K.-H. Homann: Ber.
      Bunsenges. Phys. Chem. **96**, 841 (1992)

4.14 M. Hausmann, P. Hebgen, K.-H. Homann: "Radicals in Flames: Analysis
      Via Scavenging Reaction", in Twenty-Fourth Symposium (International)
      on Combustion (The Combustion Institute, Pittsburgh 1992) p. 793

4.15 R. Wegert, K.-H. Homann: "Grössenverteilungen und Wachstum im An-
      fangsstadium der Russbildung in Vormischflammen", in VDI-Berichte Nr.
      922, (VDI-Verlag, Düsseldorf 1991) p. 161

4.16 Q. L. Zhang, S. C. O'Brien, J.R. Heath, Y. Liu, R.F. Curl, H.W. Kroto,
      R.E. Smalley: J. Phys. Chem. **90**, 525 (1986)

Discussion

Warnatz: What is the difference between large PAH and soot you talked
about?

Homann: From the behavior in flames one might distinguish roughly be-
tween low-mass PAH ions ($m \leq 600$ u) and high-mass species ($600 \leq m \leq 2500$ u). The former do not disappear completely when soot is formed and
are re-formed partly in the burned gas, while those of high mass are only

present during the first stage of soot formation and then are consumed. What concerns low-mass PAH the answer is clear. They are individual molecules, mainly planar, occasionally containing 5-membered rings which cause some arching, but they are essentially two-dimensional molecules. They are too small for van der Waals-forces to play a role for coagulation towards soot at flame temperature.

The high-mass species are more difficult to distinguish from the first soot particles. The mass spectra of a sooting acetylene flame indicate that there possibly is a coagulation step between PAH of about $2 \cdot 10^3$ u and the smallest soot particles of approximately $4 \cdot 10^3$ u. Therefore, the soot particles are regarded as three-dimensional species, consisting of two or a few more or less arched high-mass PAH, held together mainly by van der Waals-forces or occasionally by a C-C bond or chain. Because of the relatively weak van der Waals-forces the aromates first are still mobile in the particle, and at the high temperature shift along each other. The latter ideas are hypothetical, since the mass spectrometer gives little information on the particle structure.

Warnatz: That means there is no difference?

Homann: The differences betwen the high-mass PAH and the first soot particles might be those that are caused by coagulation of two or more PAH giving a three-dimensional particle. There is then the possibility for the PAH to react with each other within the particle, for example, to reduce the number of active sites for surface growth or to form C-C bonds which hinders re-evaporation of PAH at flame temperature. Re-evaporation would imply a condensation equilibrium between large PAH and small soot particles which has never been observed. Another difference is that the ionisation might take place more efficiently with the particles.

Warnatz: Could I ask a second short question. You have this little bump or this little discontinuity in the mass spectra which you are interpreting as an effect of coagulation. I suppose that coagulation is a continuous process. Why should it only occur for masses between 1000 and 2000 and not below?

Homann: Mass distributions in the range of some thousand u could only be determined for ions and charged particles so far. Therefore, one might argue that the minimum at about $2 \cdot 10^3$ u is a purely ionic phenomenon. It is reasonable to assume that the formation of large PAH$^+$ in acetylene flames takes place through growth reactions with small unsaturated aliphatics, for example C_2H_2, C_4H_2, C_4H_4 and not by charge acquisation of the large PAH (see answer to Calcote's question). Typical ion-molecule interactions, that is ion-induced dipole forces, however, are comparatively weak in this case: the charge is on the larger molecule, and moreover it is delocalized in most cases, while the polarizability of molecules such as C_2H_2 is comparatively low $(\alpha_{(C_2H_2)} = 3.7 \cdot 10^{-40}$ F m^2 , $\alpha_{(benzene)} = 11.5 \cdot 10^{-40}$ F m$^2)$. Therefore, charge cannot play a role in the coagulation of large PAH$^+$ with neutral PAH such as to cause completely new phenomena.

If coagulation is mainly based on van der Waals interaction, it is understandable that there should be a lower limit with respect to mass. If sticking together of a large PAH ion and a neutral PAH is more like the recombination of two radicals, whereby a C-C bond is formed, the mass should play a minor role. In the case of high mass PAH it might be difficult to distinguish between the two mechanisms. The fraction of radicalic aromates increases with mass, both for neutral and ionic species, in particular if the PAH contain 5-membered rings. We interpret the valley in the mass distribution as a consequence of a stronger influence of "pure" van der Waals-coagulation when the high-mass PAH have reached about $2 \cdot 10^3$ u.

Wagner: Coagulation is only efficient when particles stick together. I think that's the point.

Calcote: I enjoyed your presentation as always. I would like to go through some of your arguments, if I may, and see if they are correct. I believe you said that if you extrapolate the neutral species to larger molecular mass they could not account for soot. Then you looked at the ions and observe that as the ions get larger, regardless of how they get larger, they could account for soot - for charged soot. Then you made a jump in faith and you said that the ions are a mere image what the neutrals. I don't know of any mechanism by which you could produce very large ions from large neutral species, except by thermal ionization and that has been demonstrated not to work. Charge transfer will not work because you would have to transfer a charge from a small ion to very large neutral species. The concentration of small ions is very low and the concentration of large neutral species is very low. So, if you do a simple rate calculation you discover that the rate is too slow to do anything in the time available in the flame.

Homann: I know that we disagree on this point for quite a long time. You argue that the ionic mechanism, that is the growth of hydrocarbon ions by the reaction with smaller neutral hydrocarbons, is mainly responsible for soot formation in general, not only for charged soot. One aspect of our study of the growth of ions is that we can learn something about the route to soot, even if soot formation is not a merely ionic problem. I might mention that at the stage, where we can see the first charged soot particles in low pressure C_2H_2 flames, only about 1% of the total soot is charged and 99% is uncharged. I do not say that ions are a mere image of that what the neutrals are. There are prominent PAH ions, such as $C_{13}H_9^+$ in fuel-rich acetylene flames, which have no pendant among the neutrals. Up to masses of about 500 u the odd-carbon-numbered PAH ions are more abundant than their even-numbered neighbors, whereas among the neutral PAH the even numbered outweigh the odd-numbered at least up to $C_{24}H_{12}$. There is certainly no thermal ionisation equilibrium between neutral and ionized PAH. I estimated once that proton transfer from small ions to neutral PAH could only account for protonated PAH not larger than phenylacetylene.

However, I do not see a reason why growth of charged PAH should not be comparable to that of neutral molecules, even if there are no crosslinks between similar species by proton transfer, charge exchange or whatsoever. Hydrocarbon-ion-molecule addition reactions, as far as they are known, do not follow paths which are completely different from those of neutral reactants. The larger the molecules grow, the smaller will be the kinetic influence of a charge, the more so, if the charge is delocalized as in the case of PAH. There is no evidence that C-H or C-C bonds are any weaker than in the neutrals. In your arguments in favor of an ionic mechanism, if I understand correctly, you do not imply a completely different growth mechanism, only that ions do everything faster.

Calcote: Let me aks you another question. What was the diffference in temperature between the benzene and acetylene flame?

Homann: The maximum temperature of the benzene flame was about 2200 K and the maximum temperature of the acetylene flame was about 2080 K. This is for flames that are comparable with respect to the final soot volume fraction.

Calcote: Then, does it not strike you as strange that in the benzene flame you observe very few ions and in the acetylene flame you observe a lot of ions? In fact, a number of years ago we looked at this problem and the ions decay at just about the same rate as the soot particles form. The only way you can account for the initially charged soot particles is to assume that the equilibrium ratio of charged to neutral soot is approached from the ion side, not from the neutral side. Otherwise, it seems to me that there is no reason why there should be difference between benzene and acetylene.

Homann: There are a number of differences in the ion mass spectra of acetylene and benzene flames, but I think you are referring to the fact, that the concentration of PAH^+ with $m \geq 1200$ u and of small charged soot particles is unmeasureably small in benzene flames whereas in acetylene flames they carry a major part of the positive charge at the end of the oxidation zone (see Fig. 4.4). This may be somewhat different in very rich benzene flames ($C/O > 0.93$) in which some charged soot was detected but in which PAH^+ mass spectra have not yet been taken. Without implying any mechanism, one might say merely from the charge distribution in the late part of the oxidation zone in the benzene flame that the role of the charge carrier is transferred to the fullerenes. It looks that there is even some chemi-ionisation connected with fullerene formation in benzene flames.

The absence of very large PAH^+ and charged soot in benzene flames ($C/O < 0.93$) is indeed very strange. However, it cannot be explained by the direct formation of neutral soot particles from PAH^+ with essentially $m \leq 10^3$ u. The number density of the first neutral soot particles (some 10^{11} cm^{-3}) is orders of magnitude larger than that of the PAH^+ in the range of 500 - 1000 u, and why should the soot particles lose their charge at once. The first soot particles in low-pressure benzene flames are distinctly smaller than in acetylene flames, and they grow much slower and do not become ionized. It suggests itself that the latter fact is connected with their small size and therefore relatively high ionization potential. From the ionization behaviour of soot in acetylene flames and assuming the material not to be very different with respect to ionization, the first soot particles cannot be very much heavier than $2 \cdot 10^3$ u. On the other hand, the fact that they grow much slower, although the temperature is higher and the acetylene concentration is not very much smaller than in the burned gas of comparable C_2H_2 flames, indicates chemical differences.

Wagner: I think this very interesting discussion will accompany us for some time and I think here is an area for a large number of very nice experiments which will tell us many details.

Spectroscopic and Chemical Characterization of Soot Inception Processes in Premixed Laminar Flames at Atmospheric Pressure

Andrea D'Anna [1], Antonio D'Alessio [2],

Patrizia Minutolo [1]

[1]Istituto Ricerche sulla Combustione, CNR,
[2]Dipartimento di Ingegneria Chimica, Università Frederico II,
I-80125 Napoli, Italy

Abstract: Absorption, laser induced fluorescence and laser light scattering measurements have been performed in a slightly-sooting methane/oxygen flame at atmospheric pressure. Sampling and analysis has been also carried out with the specific aim of measuring and characterizing all of the condensable material formed during the combustion process. The combined use of scattering and extinction measurements in the ultra-violet (266 nm) and in the visible (532 nm) allowed to determine simultaneously the volume fraction, the average size and the number density of high molecular weight soot precursors and soot particles. The results of the optical measurements were in good agreement with the concentration measured by the sampling procedure. The formation of "particles" early in the flame with typical size of about 3-4 nm with no absorbtion and fluorescence in the visible could be shown. These "transparent" particles were considered to be precursors of soot particles on the basis of their decreasing concentration during soot inception. The results presented suggest that there is an initial fast polymerization process the building bricks of which are aromatic compounds with few condensed rings (not more than 2-3 rings) connected by aliphatic and oxygen bonds. Particle inception is primarily controlled by the internal arrangement of these polymers leading to structures with more condensed aromatic rings and more compact three dimensional shape thus forming the first soot nuclei.

5.1 Introduction

It is well known that soot formation takes place through a first inception stage where low molecular mass fragments of the fuel grow into high molecular mass compounds or soot nuclei and a subsequent heterogeneous stage where low molecular mass compounds, mainly C_2H_2, are added to soot nuclei. Although the formation of soot through inception accounts only for a limited amount of the total soot, it determines the critical C/O ratio and temperature range for the occurence of soot in a flame.

Much experimental work has been devoted to the understanding of the complex chemistry of soot inception in rich flames. Some years ago *Howard* et al. [5.1, 5.2] have extensively studied the early stages of soot formation in rich acetylene/oxygen flames at low pressure employing a molecular beam sampling system followed by electron microscopy and mass spectrometric analysis of the ionized high molecular mass species. They found that transparent particles with typical size of 1.5 nm (2000 u) are formed very early in the soot inception region of the flame. Their concentration showed a peak just before the appearance of the first soot nuclei. Therefore, they have attributed the role of precursors of soot nuclei to these compounds/particles the formation of which could be explained by binary coagulating collisions of high molecular weight species.

More recently, mass spectra of ionized species in the inception region of acetylene/oxygen flames have been obtained by *Homann* et al. [5.3, 5.4]. They found that soot inception takes place just after the disappearance of structures with molecular masses between 400 and 2000 u. In this way they were able to give an order of magnitude estimate of the molecular mass of the soot precursors.

Sampling and chemical analysis of low pressure flames has been carried out by *Bockhorn* et al. [5.5] employing gas chromatographic analysis (GC). They have been able to identify compounds with molecular masses up to 300 u. These compounds were mainly polycyclic aromatic hydrocarbons (PAH) and their concentrations showed a marked maximum just before soot inception followed by a more gradual growth in the burned gas region of the flame.

More recent work has shown that PAH ranging from naphthalene to coronene is just a limited amount of the material that can be sampled and condensed in the initial region of rich flames, the residual part being a tarry-like yellow-brown material difficult to identify chemically. Recently, efforts have been devoted toward a better characterization of these carbon structures that are heavier than the PAH detectable by GC but lighter than soot particles (molecular masses between 300 and 3000 u) [5.6-9]. It has been found that these structures have concentrations noticeably higher than PAH and comparable with soot particles in near-sooting conditions.

Soot inception has also been studied by optical diagnostics employing absorption measurements, laser induced fluorescence and laser light scattering. It has been noticed in earlier work that the ultra-violet-visible absorption spectra in the initial region of rich flames were noticeably different from those of solid soot particles [5.10-12]. The different profiles of the absorption coefficients in the ultra-violet and near infra-red were assigned to high mass molecular precursors and soot particles, respectively. This procedure has been systematically employed by *Weiner* et al. [5.13], who attributed the ultra-violet absorption to structures with molecular mass around 900 u. The maximum of absorption in the ultra-violet, found in the preinception zone of the flame, was allotted to the transformation of these substances to soot particles.

Laser induced fluorescence has also been applied to the study of soot inception without giving conclusive results [5.14-5.16]. It was impossible to attribute the broad-banded spectra arising from excitation wavelengths in the visible to specific compounds. Furthermore, this kind of fluorescence exhibited a large anti-Stokes component, so that it was difficult to distinguish it from the Rayleigh scattering in the early zones of the flames.

Recently, the limitations of laser induced effects have been overcome by employing pulsed Nd-YAG lasers both in the ultra-violet and visible. In the ultra-violet the induced emission spectra exhibit distinct peaks which can be attributed to a specific class of aromatic structures [5.17]. On the other hand, systematic measurements of the scattering and extinction coefficients in the ultra-violet and in the visible at two different wavelengths allowed the application of scattering/extinction methods that have already been used to follow soot coagulation and growth [5.15, 5.18]. In this way, it has been possible to measure simultaneously the volume fraction of soot precursors and soot particles. The results obtained by optical measurements were in quantitative agreement with the direct sampling measurements of the condensed species [5.9].

The aim of this work is to extend the laser light scattering and extinction measurements carried out in rich ethylene/oxygen flames employing pulsed lasers in the ultra-violet and in the visible to methane/oxygen flames where soot inception takes place more gradually compared with ethylene flames. In this way it is possible to follow the size and number density of the soot precursors in an extended range of residence times and C/O ratios and thus to establish their growth and coagulation kinetics. From a chemical point of view, a more refined characterization of the sampled material was performed in order to distinguish between PAH detectable by gas chromatography and residuals of high molecular weight.

5.2 Experimental

The employed optical methods were applied to flat premixed flames of methane/oxygen stabilized on a porous plate burner. Fuel and oxidant were fed to the center of the burner and the gas velocity was kept constant at 4.74 cm s^{-1} (STP). A general view of the experimental apparatus is given in Fig. 5.1.

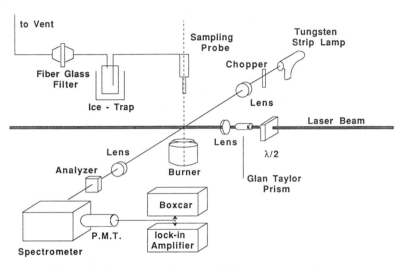

Fig. 5.1. Schematic diagram of the experimental set-up

The light source for scattering measurements was a Nd-YAG laser operated at a frequency of 3 Hz with 15 ns pulse duration and a pulse energy up to 50 μJ on the second harmonic at $\lambda = 532$ nm. The laser was operated at 532 and 266 nm. The energy of the light pulses was attenuated below 50 mJ employing a variable attenuator consisting of a $\lambda/2$ plate and a Glan-Taylor prism. The beam was than focused by a 300 mm cylindrical lens into the flame. The light scattered at 90° to the direction of the incident laser beam was focused on the entrance slit of an f/8.8 spectrometer with a focal length of 500 mm; a polarization analyzer was placed in front of the monochromator. A photomultiplier was mounted on the exit slit of the monochromator the pulsed signal of which was processed by a boxcar averager.

Extinction measurements were made using a tungsten strip lamp. The chopped image of the lamp was focused at the center of the flame and at the entrance slit of the monochromator. The signals were processed and digitized by a lock-in system to avoid interference by the emission from the flame.

Sampling of the particulate (soot and condensed hydrocarbon species) along the flame axis was performed isokinetically by means of a stainless steel water-cooled probe (inner diameter 2 mm) on a slightly-sooting flame with an equivalence ratio $\phi = 2$. Gaseous compounds were analyzed by gas chromatography applying a thermal conductivity detector for the analysis of CO, CO_2 and O_2, and a flame ionization detector for the analysis of light hydrocarbons. Carbon and hydrocarbon species were collected on a fiber glass filter and in a cold trap placed before the filter. The particulate was extracted with dichloromethane in order to separate the condensed hydrocarbon species (CHS) from the carbonaceous material defined as soot. The dried and weighed CHS were dissolved in hexane. At this stage a yellow-brown tarry material appeared that was soluble in dichloromethane and non soluble in hexane. This material will be referred to as the residue fraction. The fraction soluble in hexane was analyzed by gas chromatography with a massspectrometry detector (GC/MS) for the identification of the main components of the CHS fraction and with a flame ionization detector (GC/FID) for the quantitative determination of the main polycyclic aromatic hydrocarbons ranging from two- to seven-ring species. Ultra-violet-visible and infra-red absorption and fluorescence emission measurements were systematically performed on the CHS collected at different conditions and on its soluble and residue fractions.

5.3 Results

A comparative view of the absorption and laser induced emission spectra (elastic scattering and fluorescence) in the soot preinception region at $h = 3$ mm and in the soot growth zone at $h = 6$ mm of a slightly sooting methane/oxygen flame ($\phi = 2$) is given in Fig. 5.2. The absorption coefficients at $h - 3$ mm are much higher in the ultra-violet than in the visible where they attain values of about 10^{-3} cm^{-1}. The absorption spectrum at $h = 6$ mm shows a broad peak in the ultra-violet and the absorption in the visible is about one order of magnitude higher than that measured in the soot preinception zone. The emission spectra induced by the laser radiation at $\lambda_0 = 266$ nm exhibit a Rayleigh scattering peak clearly distinct from the fluorescence spectrum. The fluorescence spectrum does not show an anti-Stokes component in this range of wavelengths in contrast to excitation in the visible. The absence of the anti-Stokes component around the excitation wavelength allows the determination of both the polarized (Q_{VV}) and depolarized (Q_{HV}) scattering coefficients without interferences from the fluorescence background. The fluorescence emission excited at $\lambda_0 = 266$ nm exhibits a prominent maximum around 330 nm while a less intense broader band develops in the visible. This second emission increases with inreasing residence time. These spectra were obtained with laser pulses with very low energy in order to avoid photofragmentation effects leading to C_2-emission [5.17].

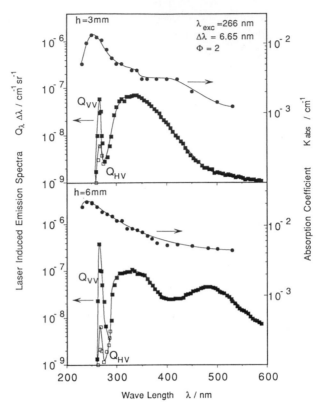

Fig. 5.2. Absorption and laser induced emission spectra (elastic scattering and fluorescence) excited in the ultra-violet at 266 nm in the soot preinception zone ($h = 3$ mm) and in the soot growth zone ($h = 6$ mm) of a slightly-sooting ($\phi = 2$) premixed methane/oxygen flame

The variation of the absorption and scattering coefficients in the ultra-violet ($\lambda_0 = 266$ nm) and in the visible ($\lambda_0 = 532$ nm) is reported in Fig. 5.3 as a function of the height above the burner. The absorption coefficient in the ultra-violet is always higher than that in the visible, the latter increases steeply up to 5 mm above the burner. The vertically polarized scattering coefficients (Q_{VV}) always increase in the flame and even in the initial region they are noticeably higher than those due to scattering from gaseous compounds computed on the basis of their measured composition and temperature (peak flame temperature is about 1800 K). The depolarization ratio ($\gamma_V = Q_{HV}/Q_{VV}$) in the ultra-violet is about 10^{-1} in the initial region of the flame and decreases to values of the order of $2 \cdot 10^{-2}$ in the soot growth region of the flame. The depolarization ratio early in the flame is much higher than either that computed for the major gas components ($\gamma_V = 1.2 \cdot 10^{-2}$) or that of the soot particles ($\gamma_V = 2 \cdot 10^{-2}$).

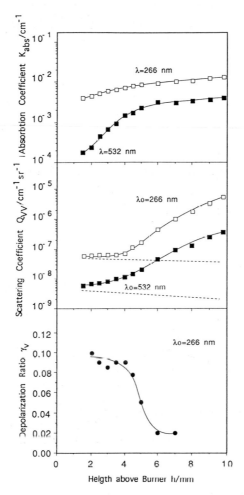

Fig. 5.3. Absorption and scattering profiles in the ultra-violet and visible and depolarization ratio in the ultra-violet as a function of the residence time in a slightly-sooting ($\phi = 2$) premixed methane/oxygen flame; dashed lines: scattering due to gaseous compounds

In order to attribute the measured fluorescence and scattering coefficients to chemical species, sampling of the combustion products was performed at different heights above the burner. The concentration profiles of the total sampled material soluble in dichloromethane, defined as condensed hydrocarbon species (CHS), and that non soluble in dichloromethane, defined as soot, are given in Fig. 5.4. The concentration of CHS is all the time higher than the soot concentration and these species are formed much earlier in the flame. The onset of soot formation occurs at about 5 mm and is preceeded by a decrease of the CHS concentration which accounts almost completely for the initial soot build-up. After the onset of soot formation CHS concentration increases again in the soot growth region of the flame. In the lower part of the figure the concentration of two of the most abundant polycyclic aromatic hydrocarbons, namely acenaphthylene and pyrene, and the total concentration of the material detected by gas chromatography are also reported. This material will be referred to as $\sum$PAH since it is mainly constituted of PAH ranging from naphthalene to coronene. It is worthwhile to remark that the PAH follow the same trend as the CHS material, but their concentration accounts for not more than 30% of the total sampled material. The residual part of CHS is non soluble in hexane and is not separable by common gas chromatographic techniques.

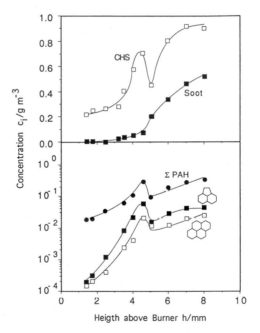

Fig. 5.4. Concentration profiles of soot, condensed hydrocarbon species (CHS), PAH separable by gas chromatography ($\sum$PAH) and the most abundant PAH, i.e. acenaphthylene and pyrene in the slightly-sooting ($\phi = 2$) premixed methane/oxygen flame

To have a better qualification of the material not separable by gas chromatography, the CHS fraction was dissolved in hexane. At this stage a tarry-like yellow-brown material not soluble in hexane but in dichloromethane was separated. This material was classified as residue. The concentration profiles of the soluble and residual fraction of the CHS is reported in Fig. 5.5. It is evident that the formation of the residual part of the CHS fraction occurs in the flame much earlier than the soluble fraction and the residue fraction accounts for most of the CHS up to 3 mm above the burner. At increasing residence time the concentration of the residue fraction increases slowly whereas the concentration of the soluble fraction increases steeply. At about 5 mm in the soot inception zone the concentration of both the soluble and residual fraction of the CHS rapidly declines. Thereafter the concentration of both of them increases again in the soot growth region of the flame.

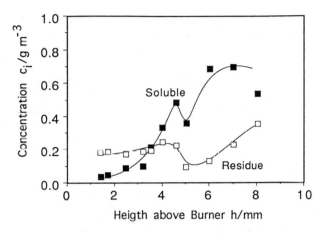

Fig. 5.5. Concentration profiles of the soluble and residual fraction of the CHS in the slightly-sooting ($\phi = 2$) premixed methane/oxygen flame

Condensed hydrocarbon species were analysed by infra-red spectroscopy, elementary analysis, absorption and fluorescence spectroscopy. This material showed infra-red peaks characteristic of both aromatic and aliphatic bonds as well as C=O groups. Its H/C atomic ratio was about 0.7 which is higher than that of very large PAH indicating that aromatic rings and aliphatic functional groups are present in these species. Ultra-violet-visible absorption of the CHS was determined and the values of the imaginary part of the refractive index of the CHS in the ultra-violet (266 nm) and visible (532 nm) are plotted in Fig. 5.6. The absorbtion coefficients of the CHS are very low up to 3-4 mm above the burner. Thereafter they increase attaining a rather constant value at about 5-6 mm both in the ultra-violet and visible.

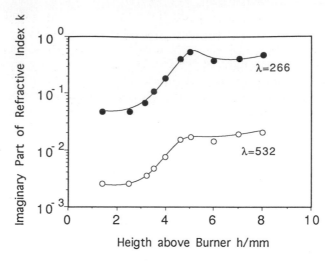

Fig. 5.6. Variation of the imaginary part of the refractive index in the ultra-violet and visible for the condensed hydrocarbon species collected along the flame axis in the slightly-sooting ($\phi = 2$) premixed methane/oxygen flame

5.4 Discussion

The spectroscopic data have to be interpreted in the framework of the photophysics of aromatic structures [5.19-21]. In fact, these types of spectra extending from the ultra-violet to the visible can be attributed only to aromatic species. Absorption and fluorescence in this spectral range are due to electronic transitions of the delocalized π-electrons. The energy difference between the ground and the first excited states decreases as the number of condensed aromatic rings increases. Soot particles significantly contribute to the absorption measured in flames at high C/O ratio and long residence time where soot is formed whereas fluorescence is exclusively due to aromatic molecules. Therefore, the increase of fluorescence in the visible with increasing residence time (Figs. 5.2, 5.3), is a clear indication of the formation of structures with a larger number of aromatic rings. On the other hand, the prevalence of the fluorescence peak near 330 nm and the almost complete absence of a relevant absorption in the visible at low residence time can be ascribed to aromatic structures with a limited number of aromatic rings (2-3 condensed rings) [5.17, 5.22]. PAH were measured along the flame axis and among them, naphthalene, acenaphthylene and pyrene are the most abundant PAH (Fig. 5.4). Therefore, it seems reasonable to attribute the absorption and fluorescence in the ultra-violet in the early region of the flame to these compounds. However, their concentrations do not justify the absorption in the ultra-violet measured early in the flame since the absorption coefficient computed on the basis of the PAH concentrations in the flame and their molar absorbivity [5.23] is more than one order of magnitude lower than that measured in the initial region of the flame. Also the elastic scattering in both the ultra-violet and in the visible is always appreciably higher than that due to gas phase compounds (Fig. 5.3) and also higher than that computed for the amount of PAH collected in flames. Therefore another class of scattering and absorbing species should be responsible for this.

One way for evaluating quantitatively the contribution of these additional scattering and absorbing species is to consider these structures as Rayleigh scatterers as it has been done previously for soot particles [5.18]. Doing this two classes of particles have to be considered: particles that are absorbing in the ultra-violet and are practically transparent in the visible and soot particles that absorb in both spectral regions. The first step is the estimation of the volume fraction of the two kinds of absorbing particles from the absorption measurements in the ultra-violet and visible and the comparison of these results with those obtained from sampling and chemical analysis.

The ultra-violet and visible absorption coefficients for Rayleigh scatterers are given by the equations:

$$K_{abs}^{uv} = \frac{\pi^2}{\lambda_0^{uv}}[f(m_p^{uv})N_p d_p^3 + f(m_s^{uv})N_s d_s^3] \qquad (5.1)$$

$$K_{abs}^{vis} = \frac{\pi^2}{\lambda_0^{vis}}[f(m_p^{vis})N_p d_p^3 + f(m_s^{vis})N_s d_s^3] \qquad (5.2)$$

In (5.1) and (5.2) $f(m) = -Im\{(m^2 - 1)/(m^2 + 2)\}$ and the subscripts p and s refer to the "transparent" particles and soot, respectively. N and d are the number density and the mean diameter and $m = n - ik$ is the complex refractive index for the two classes of particles.

From the solution of the system of equations (5.1) and (5.2) the volume fraction $f_V = N\pi d^3/6$ for "transparent" particles and soot are obtained if the values of the complex refractive index are known both in the ultra-violet and visible.

The complex refractive index for soot particles in the visible has been determined by Lee et al. [5.24]; later Vaglieco et al. [5.25] have estimated the soot refractive index in the ultra-violet. The values for the soot refractive index used in this calculation are 1.9 - i 0.55 in the visible and 1.2 - i 0.3 in the ultra-violet.

The corresponding values for the transparent particles have been evaluated more recently by D'Alessio et al. [5.9] measuring the scattering and absorption coefficients in the early zones of a non-sooting ethylene/oxygen flame ($\phi = 1.9$). At low C/O ratios and residence times the soot concentration is negligible and the scattering and absorption coefficients can be simplified as in the following expressions:

$$Q_{VV} = \frac{\pi^4}{4\lambda^4}\left|\frac{m_p^2 - 1}{m_p^2 + 2}\right|^2 N_p d_p^6 \tag{5.3}$$

$$K_{abs} = -\frac{\pi^2}{\lambda_0} Im\left\{\frac{m_p^2 - 1}{m_p^2 + 2}\right\} N_p d_p^3 \tag{5.4}$$

Therefore, "in-situ" measurements of the scattering and absorption coefficients in the visible at 532 nm and in the ultra-violet at 266 nm along with the absorption measurements performed on the collected material allow the computation of the four unknowns N_p, d_p, n_p^{uv}, n_p^{vis}. The imaginary part of the refractive index was determined by the absorption measurements performed in the ultra-violet and visible on the sampled CHS. From (5.3) and (5.4) then the values of 1.1 and 1.6 in the ultra-violet and visible, respectively, result for the real part of the complex refractive index.

By solving (5.1) and (5.2) the volume fractions for the transparent particles and soot can be computed. The profiles of the optically determined concentration of soot and transparent particles are reported in Fig. 5.7 applying $\rho_s = 1.8$ g cm^{-3} and $\rho_p = 1.0$ g cm^{-3}. In Fig. 5.7 the concentration obtained by sampling and chemical analysis is also given. The agreement between the concentration of the condensed hydrocarbon species and soot obtained by direct sampling and the concentration profiles of the transparent particles and soot obtained by optical measurements is reasonably good. The concentration of soluble species determined by both methods exhibits a slow increase followed by a rapid decline in the soot nucleation zone and a subsequent increase in the soot growth zone. The curve obtained from sampling is shifted about 1 mm with respect to the curve from the optical measurements. This shift can be attributed to a disturbance of the flame by the sampling probe.

The consistency of the results from the two methods indicates that the condensed hydrocarbon species are not an artifact of the sampling procedure. They are present in the flame and their mass concentration is much higher than that of the low molecular mass PAH. Furthermore, from the analysis of the profiles of the concentration of CHS (Fig. 5.4) it can be deduced that they are the precursors of the first soot nuclei.

More details of the properties of the soot precursors and their relationship to soot are obtained when considering the scattering coefficients in the ultra-violet and visible:

$$Q_{VV}^{uv} = \frac{\pi^4}{4(\lambda_0^{uv})^4}[f(m_p^{uv})N_p d_p^6 + f(m_s^{uv})N_s d_s^6] \tag{5.5}$$

$$Q_{VV}^{vis} = \frac{\pi^4}{4(\lambda_0^{vis})^4}[f(m_p^{vis})N_p d_p^6 + f(m_s^{vis})N_s d_s^6] \tag{5.6}$$

where $f(m) = |(m^2 - 1)/(m^2 + 2)|^2$.

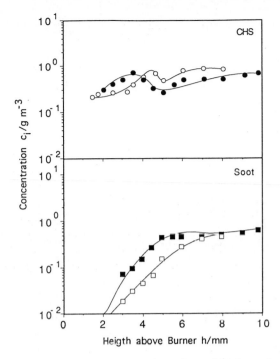

Fig. 5.7. Concentration profiles of soot and condensed hydrocarbon species along the flame axis of the slightly-sooting ($\phi = 2$) premixed methane/oxygen flame; full symbols: optical measurements; empty symbols: sampling and chemical analysis

With (5.1, 5.2) and (5.5, 5.6) it is possible to evaluate the average size and the number density separately. The results of this analysis are given in Fig. 5.8. It appears that in the preinception region of the flame the precursors have a rather constant size around 3-4 nm (5000-10000 u) and their number density passes through a maximum of 10^{12} cm^{3}. Later on soot particles are formed and their number density increases in correspondence to the decrease of the number density of soot precursors. The size of soot particles increases up to about 30 nm; this growth of soot particles is due to the combined effects of surface growth and coagulation, because their volume fraction increases and their number density decreases.

In the light of these considerations, the ultra-violet absorption and fluorescence spectra detected in the flame suggest that these precursors are not giant aggregates of large condensed PAH, but they contain mainly small aromatic subunits connected by aliphatic or, eventually, oxygen bonding. This model is in agreement with the chemical characterization of the sampled material: the measured H/C ratio is noticeably higher than that expected from a very large condensed polyaromatic structure and the infrared spectrum indicates aliphatic bonds and carbonyl groups in addition to the aromatic ones. Carbonyl groups could be simulated by oxygen adsorbed on

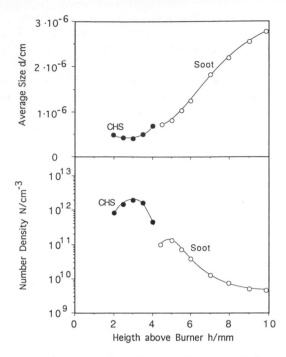

Fig. 5.8. Average size and number density of soot and "transparent" particles evaluated by scattering and absorption measurements in the slightly-sooting ($\phi = 2$) premixed methane/oxygen flame

the sample after it has been extracted from the flame or in the probe itself. However, the speculation that oxygen may be a constituent of this polymer during its formation can not be ruled out.

The increasing absorption and fluorescence in the visible with increasing residence time can be related to an internal arrangement of the precursors structures leading to species with more condensed aromatic rings and more compact three-dimensional shape. This is in agreement with the decrease of the depolarization ratio (Fig. 5.3). The depolarization ratio measured in the flame in the ultra-violet is noticeably higher early in flames ($\gamma_V = 10^{-1}$) and tends progressively to a much lower value in the soot growth region of the flame ($\gamma_V = 2 \cdot 10^{-2}$). The depolarization ratio is a measure of the anisotropy of the particle shape and of their intrinsic polarizability [5.15]. From the decrease of the depolarisation ratio can be inferred that the precursor structure evolves from a relatively loose and planar-like form to a more compact three-dimensional shape. Interestingly enough, the depolarization ratio decreases also before the bulk of soot forms thus suggesting that the reduction of the morphological and optical anisotropy of the polymeric structure occurs not necessarily simultaneous with the formation of soot. The internal rearrangement of the precursor structures leading to species

with more condensed aromatic rings is also in agreement with the spectroscopic characterization of the condensed material which shows a strong
increase of the absorption coefficient just before the first appearance of soot
nuclei (Figs. 5.4, 6).

The combined use of scattering and extinction measurements in the
ultra-violet and in the visible was also applied to flames below the soot
threshold in order to follow the evolution of the transparent particles under
non sooting conditions. A non-sooting flame ($\phi = 1.9$) has been considered
and the variation of the scattering and extinction coefficients in the ultra-
violet ($\lambda_0 = 266$ nm) and in the visible ($\lambda_0 = 532$ nm) as a function of the
height above the burner are shown in Fig. 5.9. The absorption coefficient in
the ultra-violet is always higher than that in the visible, and the polarized
scattering coefficients (Q_{VV}) always increase with residence time and, even
in this non-sooting flame, are higher than those due to gaseous scatterers.

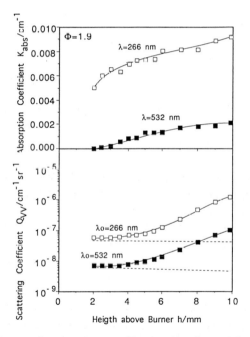

Fig. 5.9. Absorption and scattering profiles in the ultra-violet and visible as a
function of the residence time in a non-sooting ($\phi = 1.9$) premixed methane/oxygen
flame; dashed lines: scattering due to gaseous compounds

The volume fractions, number densities and diameters of both soot and
transparent particles were evaluated using the same approach discussed
above. The results are shown in Fig. 5.10. Below the soot threshold ($\phi
= 1.9$) the volume fraction of the transparent particles is still appreciable
and passes through a maximum. However, most of the intermediates were

destroyed, either oxidized or pyrolyzed, rather than growing into soot. The average size of the transparent particles slightly increases from 3 to 4 nm and correspondingly their number density shows a very slow decrease with the height. Below the soot threshold these particles have a much slower coagulation rate compared to those above the soot threshold.

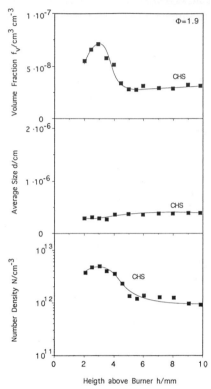

Fig. 5.10. Volume fraction, mean diameter and number density of "transparent" particles from scattering and absorption measurements in a non-sooting ($\phi = 1.9$) premixed methane/oxygen flame

5.5 Final Discussion

By combining scattering and extinction measurements in both the ultraviolet and visible it has been possible to obtain volume fractions and sizes of high molecular weight material present in flames at residence times below the onset of soot formation.

In addition the present results confirm the agreement between concentrations obtained by optical measurements with those by direct sampling [5.9]. The size of the precursors as well as their number density profiles are remarkably similar to those found by *Howard* et al. [5.1, 5.2] in their analysis of low pressure premixed flames employing molecular beam sampling followed by electron microscopy and mass spectrometric analysis of the charged species. It is also worthwhile to remark the agreement with recent results of *Dobbins* et al. [5.26], who carried out thermophoretic sampling in the early regions of diffusion flames. They found by electron microscopy analysis the presence of particles more transparent than soot particles and with typical size around 3 nm.

The chemical characterization of this kind of precursors is of great interest in order to outline a mechanism of soot formation. The analysis of the sampled material has shown that at the onset of soot inception only 30% of the sampled material was composed of PAH which could be identified by GC/MS. Again this result appears consistent with that obtained recently by *Mc Kinnon* et al. [5.8] in their low pressure flame study. However a direct chemical characterization of this material is rather difficult with the present analytical procedures and some indications of its properties could be inferred by spectroscopic methods both "in situ" and on the sampled material.

The conceptual model developed for the interpretation of our experimental results gives new insight into the soot inception process. The presently discussed kinetic models assume fuel pyrolysis, the formation of polycyclic aromatic hydrocarbons, their planar growth and coagulation into spherical particles, and, finally, the surface growth and oxidation of the particles. On the other hand, the results presented in this work suggest that there is an initial fast polymerization process building bricks of which are aromatic compounds with few condensed rings (no more than 2-3 rings). Soot inception is primarily controlled by the internal arrangement of these polymers leading to structures with more condensed aromatic rings and more compact three dimensional shape thus forming the first soot nuclei.

References

5.1   B.L. Wersborg, J.B. Howard, G.C. Williams: "Physical Mechanism in Carbon Formation in Flames", in Fourteenth Symposium (International) on Combustion (The Combustion Institute, Pittsburgh 1972) p. 929

5.2   B.L. Wersborg, A.C. Yeung, J.B. Howard: "Concentration and Mass Distribution of Charged Species in Sooting Flames", in Fifteenth Symposium (International) on Combustion (The Combustion Institute, Pittsburgh 1974) p. 1439

5.3   P. Gerhardt, K.H. Homann, S. Löffler, H. Wolf: "Large ionic species in sooting acetylene and benzene flames", in Proc. PEP Symp. Chania, Crete, Agard Conf. Proc. 422, 1987, pp. 22.1-11

5.4   P. Gerhardt, S. Löffler, K.H. Homann: "The Formation of Polyhedral Carbon Ions in Fuel-Rich Acetylene and Benzene Flames", in Twenty-Second Symposium (International) on Combustion (The Combustion Institute, Pittsburgh 1988) p. 395

5.5   H. Bockhorn, F. Fetting, H.W. Wenz: Ber. Bunsenges. Phys. Chem. **87**, 1067 (1983)

5.6   F.W. Lam: "The formation of polycyclic aromatic hydrocarbons and soot in a jet-stirred/plug-flow reactor"; Ph.D. Thesis, Massachusetts Institute of Technology (1988)

5.7   J.T. McKinnon: "Chemical and physical mechanism of soot formation"; Ph.D. Thesis, Massachusetts Institute of Technology (1989)

5.8   J.T. Mc Kinnon, J.B. Howard: "The Role of PAH and Acetylene in Soot Nucleation and Growth", in Twenty-Fourth Symposium (International) on Combustion (The Combustion Institute, Pittsburgh 1992) p. 965

5.9   A. D'Alessio, A. D'Anna, A. D'Orsi, P. Minutolo, R. Barbella, A. Ciajolo: "Precursor Formation and Soot Inception in Premixed Ethylene Flames", in Twenty-Fourth Symposium (International) on Combustion (The Combustion Institute, Pittsburgh 1992) p. 973

5.10   R.C. Millikan: J. Phys. Chem. **66**, 794 (1962)

5.11   U. Bonne, K.H. Homann, H.Gg. Wagner: "Carbon Formation in Premixed Flames", in Tenth Symposium (International) on Combustion (The Combustion Institute, Pittsburgh 1965) p. 503

5.12   S.C. Graham, J.B. Homer, J.L.J. Rosenfeld: Proc. R. Soc. Lond. **A344**, 259 (1975)

5.13   A.M Weiner, S.J. Harris: Combust. Flame **77**, 261 (1989)

5.14   B.S. Haynes, H. Jander, H.Gg. Wagner: Ber. Bunsenges. Phys. Chem. **84**, 585 (1989)

5.15   A. D'Alessio: "Laser light scattering and fluorescence diagnostics of rich flames produced by gaseous and liquid fuels", in Particulate Carbon: Formation During Combustion, ed. by D.C. Siegla, G.W Smith, (Plenum Press, New York 1981) p. 207

5.16   A. Di Lorenzo, A. D'Alessio, V. Cincotti, S. Masi, P. Menna, C. Venitozzi: "UV Absorption, Laser Excited Fluorescence and Direct Sampling in the Study of the Formation of Polycyclic Aromatic Hydrocarbons in Rich CH_4/O_2 Flames", in Eighteenth Symposium (International) on Combustion (The Combustion Institute, Pittsburgh 1981) p. 485

5.17   F. Beretta, A. D'Alessio, A. D'Orsi, P. Minutolo: Combust. Sci. Technol., in press

5.18   A. D'Alessio, A. DiLorenzo, A.F. Sarofim, F. Beretta, S. Masi, C. Venitozzi: "Soot Formation in Methane-Oxygen Flames", in Fifteenth Symposium (International) on Combustion, (The Combustion Institute, Pittsburgh 1974) p. 1427

5.19   J.B. Birks: "The spectroscopy of the π-electronic states of aromatic hydrocarbons", in Organic Molecular Photophysics, Vol.1, ed. by J.B. Birks, (J. Wiley, New York 1973) p.1

5.20   E. Clar: Polycyclic Hydrocarbons, Vol.1 and 2 (Academic Press, London 1974)

5.21   I.B. Berlman: Fluorescence Spectra of Aromatic Molecules (Academic Press, New York 1971)

5.22   M. Aldèn: "Applications of laser techniques for combustion studies"; Ph. D. Thesis, Lund Institute of Technology (1982)

5.23   L. Petarca, F. Marconi: Combust. Flame **78**, 308 (1989)

5.24   S.C. Lee, C.L Tien: "Optical Constants of Soot in Hydrocarbon Flames", in
       Eighteenth Symposium (International) on Combustion, (The Combustion
       Institute, Pittsburgh 1981) p. 1159
5.25   B.M. Vaglieco, F. Beretta, A. D'Alessio: Combust. Flame **79**, 259 (1990)
5.26   R.A. Dobbins, H. Subramaniasivam: "Soot Precursor Particles in Flames",
       this volume, sect. 17

Discussion

Santoro: I think I have some disagreement with you. You said that you can
trace the depolarization ratio back to a shape factor in the molecules. The
question I am asking is how do you differentiate fluorescence from depolarization? When we made measurements in the lower region of a diffusion
flame, we did fluorescence measurements by looking at non-resonant transitions. We excited at one wavelength and measured the fluorescence at a
non-resonant wavelength. Then we did depolarization measurements in a
region where there were no particles. In this case we measured scattering
at the resonant wavelength, turned the polarizer horizontally for incident
vertically polarized light and got exactly the same profile. The conclusion
was that we were not looking at depolarization due to any species, but
we were just looking at fluorescence which is totally depolarized. Then we
moved to the particle region. There you truly see depolarization that is, if
you measure non-resonant scattering you do see nothing, but if you measure resonant scattering you see of 1 or 2% depolarization. So, when we saw
depolarization of about 20% we said we are looking at fluorescence.

D'Alessio: The point is that we made our fluorescence and scattering measurements in the ultra-violet. This is the difference. When we do these measurements in the visible we have fluorescence from larger structures which
have Stokes and anti-Stokes components. We don't have anti-Stoke components in the ultra-violet. Fluorescence is decreasing in this spectral region
and there we have only depolarizing scattering and we do not need to correct
for fluorescence. In the visible we can not distinguish between fluorescence
and depolarized scattering. My point is that in this case we can do it because we are in the ultra-violet and we don't have anti-Stokes emission. So,
I feel confident that what we measured is just true depolarization.

Bockhorn: How do you explain the decrease in volume fraction of these
"transparent particles" with simultaneously increasing size and decreasing
number densities?

D'Alessio: This profile is the result of the approach discussed above for the
evaluation of our scattering measurements. First we thought of some errors
in the measurements. However, we have to consider that we look at a kind
of tarry material. From sampling measurements and chemical analysis we
find the same profiles for this tarry material and the results of the optical

measurements agree very well with the results from sampling measurements. The decrease in volume fraction means that part of this structures is growing more compact, more aromatic and eventually forms the first soot particles. However, we are speaking about something which is not yet soot.

Sarofim: I was a little confused by your interpretation of the ultra-violet scattering because you showed non-elastic scattering. It is not the wave-lenght λ_0 and obviously you also need optical constants. The second point that I would like to raise is that much of your conclusions are based on the fact that you say large molecules show fluorescence in the visible. As I talked to all my chemist friends at MIT they have no way of predicting what molecules will have fluorescence in the visible. Model compounds don't show systematic variations. Can you support that conclusion?

D'Alessio: The spectra of Fig 5.2 exhibit a peak at $\lambda_0 = 266$ nm, enlarged somehow by the monochromator bandwidth. This peak is attributed to elastic scattering. Then there are two other maxima: the first one in the ultra-violet and the second one in the visible. We attribute the ultra-violet peak around 330 nm to fluorescence emitted by small PAH with no more than two rings. The peak in the visible is attributed to larger PAH on the basis of an extensive literature on this subject. However we do not claim to identify a specific aromatic compound responsible for the fluorescence emitted in the visible. The argument that small PAH absorb and fluorescence in the ultra-violet while larger ones absorb and fluorescence in the visible is firmly based on the quantum theory of the energy states of the aromatic compounds.

Homann: You presented a figure with volume fractions, diameters and number densities and you said this is not soot?

D'Alessio: It is not soot.

Homann: So, what is it? You have very high number densities, about 10^{12} cm^{-3} and apparently low coagulation rates. Could you comment on this?

D'Alessio: We measured the scattering coefficients in the ultra-violet. To derive from this number densities, sizes and volume fractions - as I discussed - we have to estimate an extinction coefficient. This is not a great problem because you can take this material from the flames, take it into a spectrophotometer and get the imaginary part of the refractive index which in this case is about 0.2 and the real part is 1.2 or something like this because we are close to the resonance absorption of this material. I think the results obtained by this approach are reasonably good. The point is why this sort of particles coagulate so slowly in spite of high number densities. I have no answer for it. However, I think that this structures do not behave like soot that we are talking about normally.

Homann: What do you think it is? Droplets of tar or something else?

D'Alessio: I am thinking of structures of tar which later on become more compact and more aromatic. My point of view is that you pass a region in the flame where polymerization is very fast. However, the aromatic part of this polymer is not so large. We think of two-ring structures connected together with some non aromatic, some aliphatic bridges. You may also think about oxygen bridges but I don't have any proof for this at the moment.

Smyth: I find this bimodal distribution of fluorescence surprising. It hasn't shown up in diffusion flame work that I have done. My quick comment on this: the strongest ultra-violet fluorescence in CH_4/air diffusion flames does not have anything to do with precursor formation. But the question I want to ask you is, have you checked for grating effects? If you run a spectrum from 200 to 600 nm, a strong feature at 250 nm will also show up at 500 nm.

D'Alessio: The bimodal distribution of fluorescence is not a spectroscopic artifact. In fact the relative intensity of the ultra-violet peak to the visible one increase regularly as the C/O ratio and residence times decreases. On the other hand, we suspect that the ultra-violet fluorescence in the early region of diffusion flames is due to the same kind of precursors, because we have followed the same bimodal fluorescence spectra in the fuel-side of coflowing diffusion flames just near the soot formation region.

Haynes: Your results remind me on some measurements we did on the depolarization of the scattering from soot in benzene flames. In fact, there is none. And this was a very different result from many of the aliphatic fuel flames. I have no further comment to make than that there exist very big differences between benzene flames and flames with aliphatic fuels and perhaps this relates to some of the differences that we have been hearing about from Professor Homann.

D'Alessio: The difference in your measurements was, if I remember right, that you made them with a cw laser and you were measuring in the visible. There you have this type of fluorescence and you find both Stokes and anti-Stokes fluorescence. Again I have to state that it is difficult for cw measurements in the visible to isolate depolarization from fluorescence. But anyway, I agree with you in the sense that I am very curious to see systematic application of this sort of diagnostics to benzene flames and I expect different results.

General Discussion on PAH Formation in Sooting Flames: Experimental Investigation and Kinetics

chaired by

Heinz Gg. Wagner, Meredith B.Colket

Wagner: I don't think I should try to review the papers which have been presented. I would like to ask my collegue Colket to start the general discussion. He has a kind of structure for the general discussion that we can go through point by point. There will be many problems which will not be solved here, but these are probably the interesting ones.

Colket: There are a lot of potential issues that we could talk about. Rather than starting off and trying to summarize everything in great detail I shall put down a list of some of the areas with potential uncertainties to get the discussion going.

One of the questions here is, in terms of ring formation processes, the importance of C_4 plus C_2 processes versus C_3 plus C_3 reactions with the appropriate hydrogens on them. A second area is PAH formation and growth. I really don't think we have seen much on the mechanisms and descriptions of what these mechanisms are. One might ask the question whether this is important but I don't think we really have good understanding right now about the specific processes.

For example there is the potential of addition off light species. We often focus on addition via acetylene and much evidence of that is given through the growth processes of carbon and from post flame processes. I don't think we can necessarily eliminate other reactive species in the early growth process of PAH.

There are other open questions about the growth. For example the PAH coagulation processes, dimerisation, radical recombination, radical addition processes with loss of H atoms. These steps are obviously very temperature dependent. We still don't know when any one of these effects is important and when it dominates in flames. That might be an important issue, especially when somebody wants to build a model.

Another big issue was thermochemistry. M. Frenklach told us that the uncertainty in model predictions is very dependent on the thermochemistry one assumes. Most of us are using the Stein predictions and calculations using group additivity but there are uncertainties and we don't know how large these uncertainties might be.

And finally, we should talk about the potential uncertainties in experiments, so that we can make them a little bit more precise or provide the information that is needed for some other models in future. What are the critical

experiments remaining to be done and potentially are there rate constants that are critical to be identified? If they are important at all.
To start off any further discussion I want to give M. Frenklach an opportunity to comment.

Frenklach: The issue of C_4 chemistry versus C_3H_3 chemistry developed quickly in the last couple of years. Jim Miller published a paper in the Annual Review of Physical Chemistry where he is saying the C_4 reactions don't even exist and that it is obvious that C_3H_3 is making all the benzene. We proposed a long time ago that there are two channels of C_4, C_4H_3 and C_4H_5. They both occur under some conditions which are a little bit different. We think that the C_4H_3 always dominates in a high temperature zone whereas the C_4H_5 dominates in a lower temperature zone. Jim Miller claims that we have missed an isomerization reaction which is not true. We tested this isomerisation reaction and it doesn't make any difference. He also claimes that C_4 chemistry doesn't predict benzene and this is simply because he uses thermochemistry which is completely different from everyone elses. Now, we performed the following tests. We took into account the C_3H_3 combination forming benzene with a rate coefficient of about $1.0 \cdot 10^{13}$ and calculated two flames. One atmospheric ethylene flame which is S. Harris' flame and another one at lower pressure, that of Westmoreland. We can see that even if we add the C_3H_3 channel to the reaction chemistry it doesn't make any difference for the prediction of benzene in Harris' flame. Benzene is formed predominantly by C_4H_3 plus acetylene. If we go to the other flame we see that it is an important channel under these conditions but it is overpredicting benzene. The message I want to convey is that C_3H_3 is not a solution for all the conditions.

Colket: I have a slightly different viewpoint. I show here some predictions of the model Steve Harris put together using the same data. Again this is Harris's premixed laminar acetylene flame. Only adding this one reaction and using his C_3 chemistry doubles the amount of benzene formation. This is due to the uncertainty of the C_3H_3 chemistry not just the reaction itself. I think that one has to be quite careful when one adds this reaction.

Homann: A further question that could be added to the problem C_3 plus C_3 or C_4 plus C_2 is recombination against addition. For instance the model builders very often calculate with C_4H_5. We used the radical scavenger method to get a profile of this C_4H_5 and we found a comparatively low concentration in acetylene flames. The concentration is close to the detection limit. We detected a lot of C_4H_3 with maximum concentration at high temperature where the formation of benzene is way down. So there are some things that do not fit together and I think it is time now that the model builders not only compare the stable species but also the radical species in their calculations and do not trust the mass spectrometric methods only, which are full of fragmentation problems. If you sample from a low pressure flame at high temperatures the fragmentation pattern might be quite dif-

ferent from what you get if you make a calibration measurement at room
temperature for the same species.

Glassman: May be I am not following this basic argument but it troubles
me to see that many of these discussions are based on premixed flames where
the process is very complex. I'd like to try to put it into a different context of
what the importance is between the two fundamental mechanisms. It really
depends upon what fuel you start with. There is no way that you can explain
the great sooting tendency of propene other than the C_3 mechanism. If you
take butene you have to go through a C_4 plus C_2. I think we should put
it in the context of the fuel we start with. Our models are so complex that
we are forgetting to see what is dominant according to the actual practical
situation.

Pfefferle: We have some experiments that span from both the high oxygen
concentrations to the very low oxygen concentrations in the pyrolysis of allene and ethylacetylene. And I think that this is very illustrative of some
of these problems. When you have very little oxygen present, just traces,
you have the same sort of profiles for the higher hydrocarbons as we have
in the pyrolysis, only shifted towards lower temperature. The C_3H_3 peak is
also higher at low temperatures. At the mid range temperatures we were
seeing C_4H_5. I think again the uncertainties in present model parameters
mean that conclusions about C_3/C_3 versus C_4/C_2 pathways are meaning-
less without radical profiles, along with stable species. And the reason why
we are using vacuum photo ionisation is that for a lot of these species fragmentation is extremely small, less than 0.1%. Fragmentation is greater for
oxygen-bearing and saturated side chains.
However, when you have significant of oxygen present things are very different. The production processes are not strictly governing the higher hydrocarbon masses in the mid range PAH, but also the relative oxidation
efficiency because you are destroying the ones that are more easily oxidized.
Biaryl reaction products were also more prominent with oxygen present.
Again combining broad spectrum radical measurements with neutral measurements across this equivalence ratio and temperature ranges is very important. Because as we increased temperature the C_4H_5 diminishes almost
to nothing and the C_4H_3 increased even for acetylene flames. In a diffusion
flame with little oxygen at low temperatures C_3H_3 pathways could be very
important. As you go into high temperature towards more oxygen it seems
that the C_4 processes are more important. It is clear however, that we do
not have enough data to implicate a particular pathway for aromatic formation from aliphatic fuels.
So, I think it is important that we span both of these regimes and combine
the radical measurements and the neutral measurements in order to get a
better picture. Also we should look at the conditions that are applicable in
practical burners which, in most cases, are diffusion flames.

Santoro: I always appreciate detailed discussions on which radical routes are dominant in a premixed flame because I believe that in our flame systems, whether it is diffusion or premixed, that the hydrocarbon fuels break down to some common radical pool. Dr. Vovelle showed this morning in their experimental results that he didn't change the C_2, C_3, or C_4 radical species pool between his kerosene and decane flames and yet he produced a lot more aromatics which had to come through some other route. He has argued from an aromatic decomposition route but I am sure that this is still open to analysis. In my thinking a new question mark has arisen. When we go to other systems like a kerosene system, which is really of great interest to practical combustion people, the route to the first aromatic structure, which we all think is very imporant, may not be quite as clearly identified as involving the same set of 3 or 5 reaction channels. That's new to me. The other thing that is new to me is the much more extensive results that Professor Homann presented here on the ion mechanism. Although I have never been a fan of an ion mechanism, I have never been convinced that you can say it isn't an effective route. I almost thought that we were going to get an argument at the end of the talk that ions were the major route. It is still not clear what is happening with the neutrals. Although there is a lot of interest in arguing between these various routes I'd like to make sure that we don't lose track of radically new observations in the soot formation process.

Colket: I think we ought to focus our discussions on the PAH aspects and also some questions that we might have in terms of what future work and directions might be going in.

Pfefferle: I just wanted to make a comment about Santoro's comment about other mechanisms starting from higher species. We have also done some studies in our system with toluene and cyclohexene mixture, cyclohexene by itself, and toluene by itself. It is clear that there are other mechanisms involving the ring structures that are not simply broken down and built up again from C_2, C_3 and C_4 compounds.

Colket: I would like people to suggest some experiments or some areas which are still open and where not enough information is available or where there are uncertainties that make modeling very difficult.

Wagner: There has been one remark about the recombination of C_3H_3 to benzene and I would like to clarify this situation. The result of this remark was that the recombination product is not mainly benzene and I would like to ask Professor Homann to give us details.

Homann: We prepared the C_3H_3 by the reaction of sodium vapour with the bromide of C_3H_3 and we studied the reaction products of recombination of C_3H_3. You get about five different products. Four of them were aliphatic and one was of course benzene. You only got the aliphatic products if the first recombination product could be deactivated fast enough to get a linear arrangement. So, certainly the primary step gives you a linear particle. But under the conditions in the flame where you have high temperatures these C_6H_6 aliphatic species isomerize very rapidly to benzene.

Bockhorn: If there is no further question on this C_3H_3 mechanism I would like to come to the coagulation of small size PAH. What puzzles me is the difference in the coagulation behaviour in the pyrolysis experiments of allene that Lisa Pfefferle presented this morning and your experiments on ions. Is there any explanation why smaller PAH-ions do not coagulate and why they start coagulating if they have grown to a much larger size than in the experiments in pyrolysis.

Homann: One difference was the reactor. She used a microjet reactor and we used a low pressure flame. The number density of allene in the microjet reactor was orders of magnitude larger than the densitiy of C_3H_4 in the flame. To your question of the coagulation of medium-sized PAH. If you do some calculations for instance, by trying to produce a C_{40} PAH by coagulation of a C_{20}-ion with a C_{20} neutral, you rarely find reactions between ions, so you do not have to take them into account. The rate constants are not high enough to explain the formation of the C_{40} PAH. Only if you include $C_{19} + C_{21}$, $C_{18} + C_{22}$, etc. you get more reasonable results but they are still not fast enough to explain the growth of the large PAH. So you have to assume the addition of smaller hydrocarbons to the PAH-ions to make a great deal of growth.

Wagner: Another interesting and old question is how we leave the plane when we form soot particles. The smallest soot particles observed with an electromicroscope are definitively not planar particles. Professor Homann mentioned the mass range in which we apparently form particles which have the masses comparable to small soot particles. But I don't think these particles are totaly polycyclic aromatics. We suggested that the formation happens by a kind of chemical bonds (radicals) which fit things out of plane together for some time. How is the public opinion today?

Homann: Professor Wagner calculated once how large a particle has to be to just stick together by van der Waals forces. He came to the result that it needs some 2000 to 3000 mass units to do it. The situation is still the same. The first soot particles that are found as charged species have an average mass of about 4000 mass units. And the largest PAH that we find have about a mass of 2000 to 2500 mass units. So, we are in the same range.

Wagner: From what is known from measurements about the rate of soot formation at the very beginning, these particles must be extremely reactive. Is there any reason why only that range of particles shows that extreme activity?

Colket: I just wanted to add a comment on these dimerisation processes. Wagner's calculations at least for van der Waals forces predict several thousand mass units and H. Miller predicts 1600 mass units for particles to stick together. And Professor Pfefferle's work is indicating even smaller molecules. I think in some of these processes you can have adducts that are formed by radical addition processes or recombinations and it's very difficult to determine a priori whether they are held together with van der Waals forces or some sort of a chemical bond. From the modeling reactions such as phenyl plus benzene building biphenyl and a rapid acetylene addition afterwards leading to something like phenanthrene we know that these reactions can contribute to the dimerisation significantly. So those radical addition processes parallel to the Van der Waals forces can be very important depending on the conditions that you have in the flame.

Santoro: I think the interesting thing that H. Miller has done in his calculations shown at the 23rd Symposium was looking at the collision process that goes on between these large aromatic structures. They can actually stay within a reasonable distance of each other for fairly long times depending on the impact parameter. They seem to orbit one another for some residence time as I understand the argument. When this question came up in Göttingen I remember we were wondering how did the chemical bonds ever get established? I think at least phenomenologically the approach H. Miller has taken provides some indications of that. If the species can be held together long enough, i.e. the chemical bonding time can occur, then the bonds will be established. I think that is a slightly different view of things since we have discussed this in Göttingen. Whether it will extend all the way down to species of 200 mass units, I don't know. The word coagulation was used in Lisa Pfefferle's talk which certainly is going down to lower limits than I've seen before. But this seems to be another approach to look at things besides the Van der Waals force.

Sarofim: When Lisa Pfefferle was talking about dimerisation, I had an image of a large molecule being formed which is different to a dimerisation which involves stacking by Van der Waals forces. I wondered if that is not the operational difference between a soot particle and a large PAH. I would like to ask Professor Homann how he distinguishes between a stacked PAH from a larger PAH molecule?

Homann: Of course, with the mass spectrometer we can not distinguish a stacked two layers PAH from one extended monolayer PAH. But I would like to say a few words regarding dimerisation. This is not a pure dimer as for instance in polymer chemistry people think of dimers. We studied the pyrolysis of C_4H_2 and its trimer is already a benzene ring with several C_2H side chains on it. So, one mustn't think that these polymers or dimers or trimers retain the same chemical properties as the monomers. They are quite different species.

D'Alessio: I don't really understand what we mean when we speak about PAH. The term PAH is going to be a bit confusing to me. I am speaking about something which all aromatics will connect together. So, the thing Professor Homann is saying is if you consider something with two or three rings connected with some other aliphatics, then you can expect that the optical problems, the spectroscopical problems will be the problem of the monomer units inside the polymer. This is going to happen in a soot forming flame.

Frenklach: To support what Professor D'Alessio just said, we certainly see not only regular soot particles when we do pyrolysis. We see material which was identified as carbyne and which has nothing to do with PAH at all. Furthermore, I just heard last week that someone looked inside of soot particles and found graphite inclusions. So, what I would like to say, we heard in your talk today that particles may have different optical properties. Another point is, of course, I certainly expect that the particle optical properties and identities will change as a function of time in the flame. So, young particles may look more like some kind of other matter.

PAH Formation in Sooting Flames
Global Dependence and Modelling

One approach to flame chemistry in fuel rich flames is the separation of elementary steps from the complex system of reactions that occur in combustion of hydrocarbons under fuel rich conditions. A second approach is to monitor measurable properties in flames and to reproduce these properties by numerical simulations with self consistent models. The papers contained in part II provide the experimental basis for the second approach and present models of that kind: Experimental determination of radical concentrations in sooting hydrocarbon diffusion flames is complemented by new methods of on line detection of large aromatic hydrocarbons in sooting flames and investigation of temperature and pressure dependence of PAH and soot concentrations over a wide range of pressures and temperatures. A detailed model for PAH and soot formation concludes this part followed by a summarising discussion.

Radical Concentration Measurements in Hydrocarbon Diffusion Flames

Kermit C. Smyth [1], Thomas S. Norton [1],

J. Houston Miller [2], Mitchell D. Smooke [3], Rahul Puri [4],

Marlow Moser [4], Robert J. Santoro [4]

[1]Building and Fire Research Laboratory, NIST, Gaithersburg,
Maryland 20899, USA

[2]Department of Chemistry, George Washington University,
Washington, District of Columbia 20052, USA

[3]Department of Mechanical Engineering, Yale University,
New Haven, Connecticut 06520, USA

[4]Department of Mechanical Engineering, PA State University,
University Park, Pennsylvania 16802, USA

Abstract: Two investigations of radical concentrations in hydrocarbon diffusion flames are discussed: (1) a comparison of experimental measurements on OH, H atom, O atom, CH, and CH_3 with detailed computations of flame structure in a laminar, CH_4/air flame and (2) new measurements of OH concentrations in the soot oxidation region of both CH_4/air and C_2H_4/air diffusion flames. For the comparison study the concentration profiles have been plotted in terms of the local mixture fraction, after matching the scalar dissipation rate at the stoichiometric surface. The overall agreement in the shape and location of the profiles is best for the major radicals OH, H atom, and O atom, but less satisfactory for the hydrocarbon radicals. For the OH measurements in the soot oxidation region, profile data have been successfully obtained in the presence of significant amounts of soot. These quantitative OH concentrations will help test and guide the development of soot formation models, all of which currently must rely upon calculated or assumed values.

7.1 Introduction

The key to controlling soot formation in combustion systems lies in an understanding of the processes which determine the rates of particle inception, surface growth, and oxidation as a function of local flame conditions. The achievement of this objective requires moving from a general understanding of the relative effects of important operating parameters, such as the temperature and pressure, to more detailed representations which incorporate the basic chemical and physical processes important in soot formation. Resolution of the critical questions presently preventing such a comprehensive understanding of soot formation involves developments in both experimental and theoretical directions. Additionally, emphasis must be given to basic flame studies which help to bridge the gap to practical combustion conditions. In particular, investigations of laminar and turbulent diffusion flames present important opportunities to expand our present understanding of soot formation under conditions where mixing effects are significant, if not dominant.

In approaching such studies, it is important to emphasize that the amount of soot formed in combustion systems is controlled by the competition between particle mass addition (particle inception and surface growth) and particle oxidation. These processes may occur sequentially or concurrently depending on the local flame conditions. Many recent investigations of soot formation have focussed on the particle inception and surface growth processes. Considerable effort has also been devoted to fuel concentration and temperature effects, as well as to the properties and morphology of the soot particles. In contrast, little attention has been paid to the soot oxidation step, which is the key link between the amount of soot produced *within the flame* and the smoke yield *from the flame*. This important gap in our understanding of the controlling factors in soot formation and evolution requires consideration of the relationships between the amount of soot produced within a hydrocarbon flame, the temperature field as it is affected by radiation losses, and the amount of smoke which leaves the flame.

A key development in providing a unifying understanding of soot formation based upon recent progress lies in the formulation of appropriate soot formation models. In the last two years at least three different groups have independently undertaken to construct models of the formation and evolution of soot in hydrocarbon diffusion flames, both laminar and turbulent. These approaches include those of *Kennedy* et al. [7.1-3], *Leung* et al. [7.4], and *Moss* et al. [7.5-7]. Although there have been earlier attempts along similar lines [7.8-11], the new efforts are much more significant. There now exists a sufficient body of experimental data to characterize each of the major stages in soot production, so that a given model can use these results to establish values for the unknown parameters - at least for a certain set of combustion conditions.

Although the various approaches to modelling soot production and destruction in hydrocarbon diffusion flames differ a good deal, there are several common threads. The new investigations utilize such concepts as the local mixture fraction, global kinetic expressions, property maps, and state relationships to simplify the analysis. The models also include a calculation of the temperature field, the flow field, and the mixture fraction field to allow for comparisons to be made with experiment. Several key variables have been identified which are important for describing the local flame conditions (see also [7.12]); these include the temperature, the mixture fraction, and the residence time. In addition, it is anticipated that the fuel structure will be important. All of the new studies cited above have reported at least one successful comparison between experimental measurements and calculated results. After a given model has been developed and fit to a particular experiment, it is clearly of great interest to see how well it performs for other combustion conditions. In particular, the ability to predict soot formation rates for different hydrocarbon fuels is a critical test of any proposed model.

This paper focusses on two areas: (1) How well do current detailed mechanisms and computations of flame structure predict the key radical concentrations in the simplest hydrocarbon diffusion flame (laminar, CH_4/air), and (2) what are the OH concentrations in the soot oxidation regions of both CH_4/air and C_2H_4/air diffusion flames?

7.2 Comparison of Experimental and Computed Species Concentration Profiles

Significant advances have been made in our knowledge of both the chemical reactions which occur under high temperature combustion conditions and the associated fluid mechanical mixing processes. As a consequence, increasing attention is now focussed on improving our understanding of the strong coupling between chemical heat release and turbulent mixing, i.e. chemistry-turbulence interactions. To model turbulent diffusion flames, one approach for including chemical reactions is to utilize libraries of strained laminar flame calculations [7.13-16] which incorporate detailed reaction mechanisms [7.17, 7.18]. An alternative avenue is to carry out a direct simulation in which the number of chemical steps in the basic oxidation mechanism for combustion has been reduced systematically to as few as possible (typically three or four), while still providing an adequate description of the major species concentration and temperature profiles [7.19-21]. In both approaches one can envisage extensions to include chemical growth reactions leading to particle inception as well as soot oxidation processes.

Few studies have addressed the question of how well flame structure calculations reproduce experimental results. Recent work at Sandia National Laboratories, mostly in turbulent jet diffusion flames of diluted methane, has compared measured concentrations of the OH radical and the major stable species to values computed for strained (counterflow) laminar flames [7.22-24]. With respect to developing reduced chemical mechanisms, most of the effort has used a detailed flame structure calculation as the point of reference [7.25]. Thus far only limited comparisons have been made between experimental measurements and detailed flame structure computations for laminar diffusion flames, particularly for the radical species [7.24].

In the last several years comprehensive experimental measurements of species concentrations, velocity, and temperature have been reported for a laminar, CH_4/air diffusion flame [7.26]. In addition to mass spectrometric detection of stable molecules, radical concentration profiles have been obtained for CH_3 [7.27, 7.28], OH [7.29], H atom [7.30], O atom [7.30], and CH [7.31]. The details of the various approaches used in these profile measurements are given in the respective papers. Beyond these results, profile data for radicals in laminar hydrocarbon diffusion flames are scarce. *Melvin* et al. [7.32] have reported an OH profile in a N_2-diluted CH_4/air flame, *Stepowski* et al. [7.33] and *Garo* et al. [7.34] have measured OH concentration profiles in a CH_4/air flame, *Bastin* et al. [7.35] have used molecular beam mass spectrometry to sample numerous radicals in a 4 kPa C_2H_2/O_2/Ar flame, and *Barlow* et al. [7.24] have obtained OH profiles in both counterflow and co-flow air-diluted methane flames.

For the present comparison all of the temperature, velocity, mass spectrometric and optical species profiles have been measured in an overventilated, laminar methane/air diffusion flame burning on a Wolfhard-Parker slot burner at atmospheric pressure. Measurements have been made at heights (h) from 1 to 21 mm above the burner surface; most of our data have been obtained around $h = 9$ mm. This has been a convenient region in which to analyze hydrocarbon growth chemistry [7.36, 7.37] because the concentrations of acetylenic and aromatic hydrocarbons are appreciable, yet significant amounts of soot have not yet formed [7.26].

It has become customary to present model results as functions of a geometry-independent coordinate, such as the local equivalence ratio (ϕ) or the mixture fraction (f). This choice has been motivated primarily by the search for a simplified description of laminar diffusion flame chemistry for use in calculations of turbulent reacting flows [7.14-16, 7.20, 7.21, 7.38-41]. For the purposes of validating a model by comparison with experimental data, the coordinate transformation from a measured laboratory distance to a derived mixture fraction is not ideal, because f is determined from the local species concentrations and therefore is not a truly independent variable. Despite this limitation, the mixture fraction representation remains useful as a means for comparing model results with experimental data from different burner geometries and different flame sizes.

The mixture fraction may be defined in the form

$$f = \frac{(\beta - \beta_2)}{(\beta_1 - \beta_2)} \qquad (7.1)$$

where β represents any linear combination of conserved scalars and the subscripts 1 and 2 indicate limiting values in the fuel and oxidizer streams, respectively. In the present investigation

$$\beta = 2(W_C/M_C) + 0.5(W_H/M_H) - (W_O/M_O) \qquad (7.2)$$

where W_i is the mass fraction of element i and M_i is the molecular weight of species or element i, and the subscripts C, H, and O denote the elements carbon, hydrogen, and oxygen, respectively [7.20].

Substantial research effort has also been involved in deriving universal state relationships for species concentrations and temperature as functions of a single variable, the mixture fraction, in order to create flamelet libraries for use in the modeling of turbulent combustion. However, despite some indications of success [7.26, 7.42-44], it has been found in general that a second variable is needed because the supposedly universal relationships are often not observed [7.26, 7.43] and species concentrations may vary with the local stretch or strain rate in a flame [7.16, 7.38, 7.40]. Therefore, a logical approach involves matching the mixture fraction and a second variable, such as the strain rate, in order to make a meaningful comparison between experimental data and modelling results [7.45].

The experimental profile measurements which are now available in the undiluted CH_4/air diffusion flame enable a much more complete comparison to be made with flame structure calculations than has been previously possible. Our initial focus has been on the radical species H, O, OH, CH, and CH_3 [7.46], since minor species concentrations are in general a more critical test of mechanism and model predictions [7.47]. How well can a detailed mechanism for methane combustion predict radical concentrations for a diffusion flame wherein significant radiation losses due to soot particles do not occur?

Our experimental data on five radical species are compared to flame structure computations of a co-flowing CH_4/air diffusion flame, incorporating a basic methane oxidation mechanism which consists of 26 species and 78 reversible reactions [7.39]. For each radical species two computed profiles are presented, which have been obtained at flame locations where the scalar dissipation rates at the stoichiometric surface are slightly larger and smaller than that of the experimental measurements [7.45]. In this way the sensitivity of the radical's concentration to the local strain rate can be assessed.

This comparison of radical species profiles is similar in spirit to the study of *Westmoreland* et al. [7.47], wherein the predictions from published mechanisms were compared with experimental measurements in a low-pressure premixed $C_2H_2/O_2/Ar$ flame. On balance our findings indicate that the flame structure calculations in a hydrocarbon diffusion flame are comparably successful to predictions in premixed flames, at least for the relatively simple case of CH_4/air combustion. The overall agreement in the shape and the location of the profiles is excellent, with peak concentration positions appearing in the order O atom, OH, H atom, CH, and CH_3 going from lean to rich mixture fractions. For the present studies, the peak concentrations of the radical species occur near the stoichiometric surface. Increasing lateral positions correspond to leaner mixture fractions, while decreasing lateral positions correspond to richer mixture fractions. The experimental locations of the maximum concentrations are somewhat more spread out than the calculated positions. Specific observations regarding individual radical species are discussed in the following paragraphs.

7.2.1 OH, H-atom, and O-atom

Combined laser absorption and laser-induced fluorescence measurements on OH have yielded absolute concentration profile data which are believed to be accurate to better than ±10% over a dynamic range of ten [7.29]. The experimental concentrations, shown in Fig. 7.1., lie above the full equilibrium values and below a partial equilibrium estimate which can be derived from the two chain branching reactions (7.3) and (7.4),

$$H \ + \ O_2 \ \rightleftharpoons \ OH \ + \ O \tag{7.3}$$

$$O \ + \ H_2 \ \rightleftharpoons \ OH \ + \ H \tag{7.4}$$

$$[OH] = (K_1 K_2 [H_2][O_2])^{1/2} \tag{7.5}$$

These results are expected [7.29] on the basis of detailed experimental measurements carried out in low pressure, premixed flames by *Biordi* et al. [7.48] and *Bittner* [7.49]. Examination of Fig. 7.1. reveals that the experimental peak OH concentration exceeds the total equilibrium maximum value by a factor of 2.3; for both lean and rich flame conditions the $[OH]_{experimental}/[OH]_{total\ equilibrium}$ ratio increases rapidly.

Figure 7.2. shows the comparison between the experimental OH profiles and those determined from the flame structure computations. The computed OH profiles are slightly narrower than the experimental measurements; otherwise the agreement is excellent for the peak concentration ($1.8 \pm 0.2 \cdot 10^{16}$ cm^{-3} or mole fraction $5.0 \cdot 10^{-3}$) and the peak location in terms of the local mixture fraction. In addition, our experimental results show little variation in the peak OH concentration for the range of flame heights examined (h from 1 to 21 mm [7.26]), and thus the OH levels are insensitive to the local strain and scalar dissipation rates [7.45], in agreement with the calculated profiles.

For both the H atom and O atom measurements only relative concentration profiles have been obtained, since multiphoton excitation methods

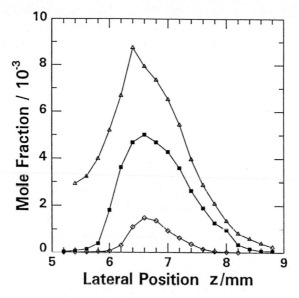

Fig. 7.1. Experimental OH concentration (■) as a function of position at a height $h = 9$mm above the Wolfhard-Parker burner; also shown are the calculated OH profiles assuming total equilibrium (◇) and partial equilibrium (△, Eq. (7.5), see [7.29]); data are presented only for the right-hand side of the flame

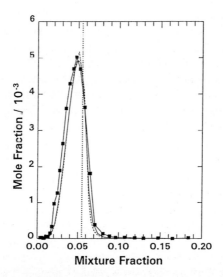

Fig. 7.2. Quantitative OH measurements (solid squares) versus flame structure calculations; the computations are presented for two values of the scalar dissipation rate at the stoichiometric surface: $\epsilon_s = 0.64$ s^{-1} (dotted line) and $\epsilon_s = 1.4$ s^{-1} (solid line); the vertical dotted line denotes the location of the stoichiometric mixture fraction, $f_s = 0.055$

were utilized in their detection [7.30]. Special care was taken in these experiments to avoid the photolytic production of the species of interest. The maximum H-atom concentration has been estimated from our analysis of partial equilibrium in this flame [7.29], which shows that the fast bimolecular reaction (usually designated (7.6))

$$\mathrm{OH} \quad + \quad \mathrm{H_2} \quad \rightleftharpoons \quad \mathrm{H_2O} \quad + \quad \mathrm{H} \tag{7.6}$$

is equilibrated. This reaction controls the peak H-atom concentration in the primary reaction zone and may be used to place the experimental relative H-atom profile on an absolute basis using the species concentration data on $\mathrm{H_2}$, $\mathrm{H_2O}$, and OH combined with the temperature measurements. The resulting peak H-atom mole fraction is $2.5 \cdot 10^{-3}$. Figure 7.3. presents the experimental H-atom profile results scaled to this prediction and corrected for the significant variation in the electron detection sensitivity as a function of flame position [7.50]. Good agreement with the computed maximum H-atom concentration is obtained, as well as with the overall shape of the profile. It should be noted, however, that the calculated results are shifted toward slightly leaner flame positions compared to the experimental data.

The unwanted photolytic production of O-atoms is the primary obstacle for determining accurate relative profiles, as in the H-atom experiments. Our measurements revealed that laser-induced production of O-atoms occurred in rich flame regions for the minimum useable laser intensity of $1.7 \cdot 10^8 \ \mathrm{Wcm}^{-2}$ [7.30]. The tail into the rich flame region of the experimental O-atom profile (see Fig. 7.4.) is thus suspect and also is the only significant point of disagreement with the calculated profiles. The peak O-atom concentration has been scaled arbitrarily to a mole fraction of $2 \cdot 10^{-3}$, which is between two partial equilibrium estimates which can be made from our flame data ($0.80 \cdot 10^{-3}$ and $5.3 \cdot 10^{-3}$ [7.29]). The location and shape of the experimental and calculated O-atom profiles are in excellent agreement for mixture fraction values as large as $f = 0.06$.

The calculated peak concentrations of both H-atom and O-atom are more sensitive to the local scalar dissipation rate than is OH. However, experimental profiles have not been obtained over a wide range of heights above the burner for comparison.

7.2.2 CH, CH$_3$

Our laser-induced fluorescence data on CH give the narrowest profile for any species measured in this $\mathrm{CH_4}$/air diffusion flame [7.31]. Nevertheless, Fig. 7.5. shows that the experimental results are considerably broader than the calculated profile. In addition, the computed peak concentrations are higher than the experimental estimate and occur at leaner flame locations. The maximum CH concentration is found to be quite sensitive to the local

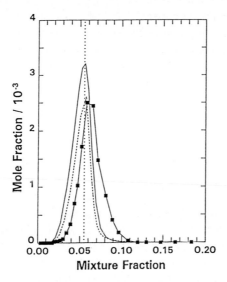

Fig. 7.3. H-atom experimental measurements scaled to the predicted maximum concentration with reaction (7.6) in equilibrium (solid squares, see text) versus flame structure calculations; the computations are presented for two values of the scalar dissipation rate at the stoichiometric surface: $\epsilon_s = 0.64$ s^{-1} (dotted line) and $\epsilon_s = 1.4$ s^{-1} (solid line); the vertical dotted line denotes the location of the stoichiometric mixture fraction, $f_s = 0.055$

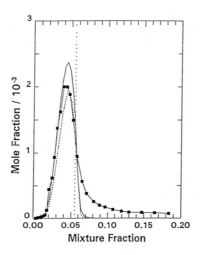

Fig. 7.4. Relative O-atom experimental measurements (solid squares) scaled to a peak mole fraction of $2.0 \cdot 10^{-3}$ (see text) versus flame structure calculations; the computations are presented for two values of the scalar dissipation rate at the stoichiometric surface: $\epsilon_s = 0.64$ s^{-1} (dotted line) and $\epsilon_s = 1.4$ s^{-1} (solid line); the vertical dotted line denotes the location of the stoichiometric mixture fraction, $f_s = 0.055$

scalar dissipation rate, decreasing as the strain rate decreases, in both the experimental data and the computations.

No quantitative measurements have been reported for the CH radical in diffusion flames. Our analysis of the chemical reactions which produce and destroy CH gives a range of values for the ratio $[^3CH_2]_{max}/[CH]_{max}$ between 13 and 300 [7.31]. Using this variation in the $[^3CH_2]_{max}/[CH]_{max}$ ratio and the calculated range of peak $^3CH_2$ mole fractions $(27.5 \cdot 10^{-6} - 29.4 \cdot 10^{-6})$, the maximum CH concentration falls in the range of $0.09 \cdot 10^{-6} - 2.3 \cdot 10^{-6}$ at a height of 9 mm above the burner. In Fig. 7.5. the relative experimental profile has been scaled to the mid-point of this range. The calculated CH peak concentration is expected to be high, since the only CH destruction processes incorporated in the kinetic mechanism are reactions with O_2 and O atom, and at the location of calculated peak [CH], $f = 0.06$, the experimental O_2 concentration is about two times higher than the computed value. Furthermore, our analysis has shown that CH reactions with H_2O, CH_4, C_2H_2, and CO_2 are also important loss channels [7.31].

Figure 7.6. compares our experimental data on the methyl radical from both multiphoton ionization [7.51] and quartz microprobe scavenger sampling experiments [7.28] with the calculated profiles. The optical and mass spectrometric measurements agree well with each other in terms of their location and their profile shapes. These experimental data have been scaled to the same peak value, which is 20 times higher than a lower bound estimate of $1.1 \cdot 10^{-4}$ derived from the CH_3I concentration measured in an I_2 scavenging experiment [7.28]. A second mass spectrometric estimate of the maximum CH_3 mole fraction of $5.5 \cdot 10^{-4}$ at $h = 9$ mm has been made from methyl radical recombination to produce C_2H_6 [7.27]. In contrast, the flame structure calculations give much larger values for the maximum CH_3 mole fraction of $2.3 \cdot 10^{-3}$ to $3.2 \cdot 10^{-3}$. Although the experimental measurements are strictly lower bounds, the wide discrepancy of at least a factor of four is surprising.

The experimental location of the peak methyl radical concentration occurs at a slightly richer mixture fraction than the calculated maximum, while the shapes of the experimental and computed profiles agree closely. For the computed profiles the shape and peak concentrations are quite sensitive to the scalar dissipation rate. However, the multiphoton ionization data and the mass spectrometric measurements do not show decreasing methyl radical concentrations with increasing height in the flame (i.e., decreasing scalar dissipation rate). Clearly, an absolute experimental measurement of the CH_3 concentration is desirable.

The comparison between the experimental and calculated profiles for the CH and CH_3 radicals is less satisfactory than is the case for OH, H atom, and O atom. The computed concentrations of these hydrocarbon radicals are particularly sensitive to the local strain rate. In view of the limited number of reactions included in the chemical mechanism used in the flame

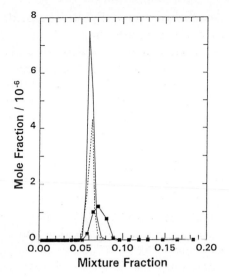

Fig. 7.5. Relative CH experimental measurements (solid squares) scaled to a peak mole fraction of $1.2 \cdot 10^{-6}$ (see text) versus flame structure calculations; the computations are presented for two values of the scalar dissipation rate at the stoichiometric surface: $\epsilon_s = 0.64 \text{ s}^{-1}$ (dotted line) and $\epsilon_s = 1.4 \text{ s}^{-1}$ (solid line); the vertical dotted line denotes the location of the stoichiometric mixture fraction, $f_s = 0.055$

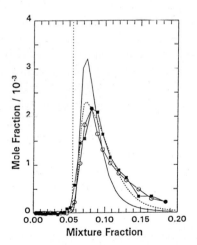

Fig. 7.6. Quantitative mass spectrometric lower limit CH_3 measurements (solid squares) multiplied by a factor of 20 and relative multiphoton ionization data scaled to the same peak concentration (open circles) versus flame structure calculations; the computations are presented for two values of the scalar dissipation rate at the stoichiometric surface: $\epsilon_s = 0.64 \text{ s}^{-1}$ (dotted line) and $\epsilon_s = 1.4 \text{ s}^{-1}$ (solid line); the vertical dotted line denotes the location of the stoichiometric mixture fraction, $f_s = 0.055$

structure calculation, it seems likely that the computations overestimate the peak concentrations of CH and CH_3, and perhaps $^3CH_2$ as well. Reduced mechanisms have been shown to be unsuccessful in establishing proper steady-state expressions for CH [7.31].

7.3 OH Measurements in the Soot Oxidation Region

In the oxidation region of laminar, hydrocarbon diffusion flames there exists a competition for OH between CO and soot particles, since OH is a key oxidizer for each. Furthermore, the oxidation of soot itself leads to enhanced levels of CO. Previous work has shown that the emission of CO from diffusion flames is closely related to the observed soot concentrations within the flame, which also control smoke emission levels [7.52]. Our objective is to unravel this competition for OH. Most of the earlier laser-induced fluorescence studies of OH in hydrocarbon diffusion flames have reported measurements in regions where soot concentrations were negligible [7.29, 7.33, 7.53]. Only *Garo* et al. [7.34] have measured OH concentrations in the presence of soot particles. *Lucht* et al. [7.54] have also made OH concentration measurements in sooting premixed flames.

The flame structure calculations shown in Fig. 7.2. provide an excellent prediction of the OH concentrations over the entire profile in the absence of radiation losses due to soot particles. However, when radiation losses are included, predictions of the temperature field and the degree of radical superequilibrium are difficult to make. The present work is intended to specifically establish how the OH concentration varies as the soot concentration is changed. To do this, quantitative measurements of the soot [7.55], OH, and CO concentrations as well as the surface area of the soot particles (or the primary particle size) and the temperature field in the oxidation region are needed. In addition, data on O_2 concentrations are required to address the question of whether OH or O_2 is more important for the oxidation of soot particles. These data are essential input for an integrated soot formation model; the oxidation step for all of the soot models proposed thus far rests upon calculated or assumed quantities.

7.3.1 Experimental Approach

Laser-induced fluorescence (LIF) has been used to measure relative OH concentration profiles in a series of methane and ethylene laminar diffusion flames burning in air [7.56]. The coannular burner consists of a 1.11 cm diameter fuel tube surrounded by a 10.18 cm diameter air annulus [7.55]. Volumetric flow rates were 7.7 cm^3s^{-1} for methane and 3.85, 4.6, and 4.9 cm^3s^{-1} for ethylene. The air flow rate for the methane flame and the 3.85 cm^3s^{-1} ethylene flame was 1060 cm^3s^{-1}; it was 1300 cm^3s^{-1} for the higher ethylene flow rate flames. No soot is observed to survive the flame tip for the ethylene flame having the lowest fuel flow rate (non-smoking), while soot is clearly observed to issue from the flame for the largest fuel flow case (smoking). For the intermediate 4.6 cm^3s^{-1} flow rate, the flame is quite close to the smoke point condition (incipient smoking; [7.57]).

The LIF experiments were carried out with a Nd-YAG pumped dye laser whose output was frequency doubled to produce light at 278.83 nm and 283.55 nm. These wavelengths excite the $S_{21}(8)$ and the $Q_1(8)$ lines of the $A^2\Sigma^+ \leftarrow X^2\Pi_i(1,0)$ band of OH, respectively, and were selected to minimize the variation of the rotational population with temperature. Fluorescence was measured at 90° using longpass glass filters which minimized elastically scattered light while transmitting the $(0,0)$ and $(1,1)$ band emission of OH at 308 nm and 314 nm, respectively.

In order to calibrate the peak OH concentrations in the coannular burner, point profile measurements were first made in the CH_4/air flame stabilized on the rectangular slot Wolfhard-Parker burner, where previous absorption measurements have established the maximum OH concentration [7.29]. Next, the LIF signal levels in the CH_4/air flame on the rectangular slot burner were compared with those from the CH_4/air flame on the axisymmetric burner; agreement was within 13%. Point profile measurements were also made on the coannular burner for the ethylene flame with the lowest flow rate.

For the fluorescence measurements on the coannular burner, light detection was also accomplished using a gated, cooled CCD detector. Both 1-d line and 2-d planar images of the resulting fluorescence were recorded. The 1-d line images were obtained by exciting the $S_{21}(8)$ OH line, while the 2-d planar images were obtained by exciting the much stronger $Q_1(8)$ OH line. The results presented here have been corrected for the variations in the laser pulse energy and the local quenching rate [7.56]; temperatures used for the latter analysis were obtained from radiation-corrected thermocouple measurements. Uncertainties in the absolute OH concentrations are estimated to be ±50%.

7.3.2 OH Profile Results

An example of a typical fluorescence/scattering profile is shown for the non-smoking ethylene/air flame in Fig. 7.7. Distinct contributions are observed from OH, soot particles, and polycyclic aromatic hydrocarbon (PAH) species [7.26]. When the laser wavelength was tuned off resonance with respect to the OH rotational line, the two peaks furthest from the burner centerline disappeared, while the features attributed to the soot scattering and the PAH fluorescence remained.

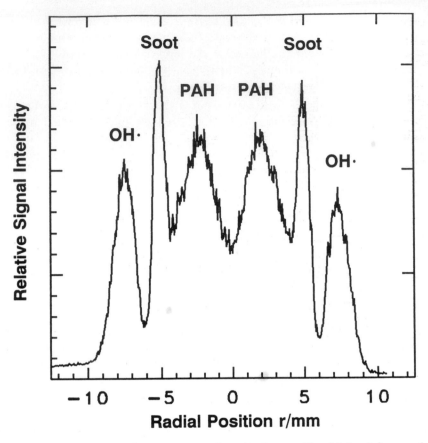

Fig. 7.7. Laser-induced fluorescence and scattering profile obtained from a 1-d line image in the axisymmetric non-smoking ethylene/air flame; the fuel flow rate is 3.85 cm^{3}s^{-1}, and the axial position $h = 7$ mm from the fuel tube exit; the visible flame height $h = 88$ mm

Figure 7.8. shows a series of OH concentration profiles for several axial positions in the axisymmetric CH$_4$/air diffusion flame. The soot and PAH signals have been eliminated by subtracting the off-resonance profile from the on-resonance profile at each height. These data show that the maximum OH concentration decreases while the spatial extent of the profile increases with increasing height in the flame. These results are qualitatively similar to the profiles reported by *Garo* et al. [7.34] for a methane/air flame, and the evolution of the shape of the OH profile with height above the burner also agrees with the flame structure computations of *Smooke* et al. [7.58] on a CH$_4$/air flame. Our results show similar trends for the ethylene flame having a fuel flow rate of 3.85 cm^{3}s^{-1} and nearly an order of magnitude more soot [7.56].

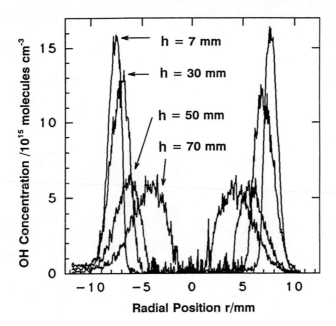

Fig. 7.8. OH concentration profiles in the axisymmetric CH_4/air flame at several axial positions above the fuel tube exit; off-resonance profiles have been subtracted from on-resonance profiles (see text); the visible flame height $h = 79$ mm

When exciting the weak $S_{21}(8)$ transition, the OH LIF signals are difficult to observe in the heavily sooting regions of the ethylene flames with higher fuel flow rates (and consequently greater soot concentrations). This is due to problems associated with the subtraction of two large soot signals relative to a much smaller OH signal. The 2-d planar image data were obtained using the stronger $Q_1(8)$ line and thus showed much larger OH signals relative to the soot scattering signals. Figure 7.9. (top) presents the profiles obtained with the laser tuned both on and off resonance for the ethylene sooting flame at a height of 100 mm. Figure 7.9. (bottom) shows that the net OH signal can be obtained by subtracting the off-resonance profile from the on-resonance profile; OH is present throughout the region where soot breaks out of the flame. Note that the laser beam is propagating from left to right, and that significant absorption and scattering occur. These results demonstrate the capability to measure OH concentration variations in the region of the diffusion flame tip where soot and CO oxidative steps compete for OH.

The OH concentration profiles have been measured near the flame tip for non-smoking, smoke point, and smoking conditions [7.56]. Our results indicate that quantitative OH measurements should be possible in flames which contain a wide range of soot concentrations. Future work will utilize these new data on OH concentrations with previously measured soot, ve-

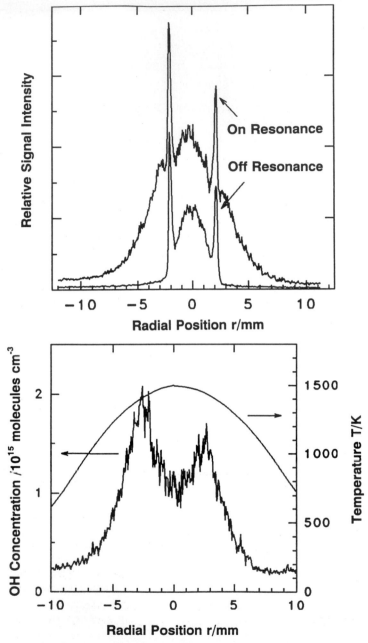

Fig. 7.9. Top: on-resonance and off-resonance profiles in the soot oxidation region of the smoking ethylene/air flame; the fuel flow rate is 4.9 cm^3s^{-1}, and the axial position above the fuel tube exit $h = 100$ mm; bottom: net OH laser-induced fluorescence profile and radiation-corrected thermocouple temperature measurements in the same flame

locity, and temperature fields to establish quantitative oxidation rates for soot particles by OH.

7.4 Summary

The results described above demonstrate that a number of radical species can be measured prior to soot particle inception under diffusion flame conditions. Current models of the chemical structure of diffusion flames are capable of quantitatively predicting the concentration and spatial evolution of some of these species, particularly OH, H atoms, and O atoms. Improvements for hydrocarbon radicals, such as CH and CH_3, must await quantitative experimental measurements and are likely to require extensions to the reaction mechanisms utilized in the modelling studies. For the case of OH, measurement capabilities have been demonstrated in soot particle laden regions of diffusion flames. These results hold the potential to provide the critical concentration information needed to examine soot oxidation processes in diffusion flame environments. Extensions to the measurement of other radical species in sooting regions, such as O atoms and H atoms, remain to be achieved, but would also be of value.

References

7.1   I.M. Kennedy, W. Kollmann, J.-Y. Chen: Combust. Flame **81**, 73 (1990)
7.2   I. Kennedy, W. Kollmann, Y. Chen: AIAA Paper No. 90-0459 (1990)
7.3   M. Metternich, W. Kollmann, I.M. Kennedy: Paper submitted to AIAA Aerospace Sciences Meeting, Reno (1991)
7.4   K.M. Leung, R.P. Lindstedt, W.P. Jones: Combust. Flame **87**, 289 (1992))
7.5   J.B. Moss, C.D. Stewart, K.J. Syed: "Flowfield Modelling of soot formation at elevated pressure", in Twenty-Second Symposium (International) on Combustion (The Combustion Institute, Pittsburgh 1988) p.413
7.6   K.J. Syed, C.D. Stewart, J.B. Moss: "Modelling soot formation and thermal radiation in buoyant turbulent diffusion flames", in Twenty-Third Symposium (International) on Combustion (The Combustion Institute, Pittsburgh 1990) p.1533
7.7   C.D. Stewart, K.J. Syed, J.B. Moss: Combust. Sci. Technol. **75**, 211 (1991)
7.8   P.A. Tesner, T.D. Snegiriova, V.G. Knorre: Combust. Flame **17**, 253 (1971)
7.9   D.E. Jensen: Proc. R. Soc. Lond. **A 338**, 375 (1974)
7.10   B.F. Magnussen, B.H. Hjertager: "On mathematical modelling of turbulent combustion with special emphasis on soot formation and combustion", in Sixteenth Symposium (International) on Combustion (The Combustion Institute, Pittsburgh 1976) p. 719
7.11   B.F. Magnussen, B.H. Hjertager, J.G. Olsen, D. Bhaduri: "Effects of turbulent structure and local concentrations on soot formation and combustion in C_2H_2 diffusion flames", in Seventeenth Symposium (International) on Combustion (The Combustion Institute, Pittsburgh 1978) p. 1393

130   Kermit C. Smyth et al.

7.12 J.H. Kent, D.R. Honnery: Combust. Sci. Technol. **75**, 167 (1991)

7.13 S.K. Liew, K.N.C. Bray, J.B. Moss: Combust. Sci. Technol. **27**, 69 (1981)

7.14 S.K. Liew, K.N.C. Bray, J.B. Moss: Combust. Flame **56**, 199 (1984)

7.15 N. Peters: Prog. Energy Combust. Sci. **10**, 319 (1984)

7.16 N. Peters: "Laminar flamelet concepts in turbulent combustion", in Twenty-First Symposium (International) on Combustion (The Combustion Institute, Pittsburgh 1986) p. 1231

7.17 B. Rogg, F. Behrendt, J. Warnatz: "Turbulent non-premixed combustion in partially premixed diffusion flamelets with detailed chemistry", in Twenty-First Symposium (International) on Combustion (The Combustion Institute, Pittsburgh 1986) p. 1533

7.18 D.C. Haworth, M.C. Drake, R.J. Blint: Combust. Sci. Technol. **60**, 287 (1988)

7.19 G. Paczko, P.M. Lefdal, N. Peters: "Reduced reaction schemes for Methane and Propane flames", in Twenty-First Symposium (International) on Combustion (The Combustion Institute, Pittsburgh 1986) p. 739

7.20 N. Peters, R.J. Kee: Combust. Flame **68**, 17 (1987)

7.21 R.W. Bilger, S.H. Starner, R.J. Kee: Combust. Flame **80**, 135 (1990)

7.22 R.S. Barlow, R.W. Dibble, S.H. Starner, R.W. Bilger: "Piloted diffusion flames of Nitrogen-diluted Methane near extinction: OH measurements", in Twenty-Third Symposium (International) on Combustion (The Combustion Institute, Pittsburgh 1990) p. 583

7.23 R.S. Barlow, R.W. Dibble, S.H. Starner, R.W. Bilger, D.C. Fourguette, M.B. Long: AIAA Paper No. 90-0732 (1990)

7.24 R.S. Barlow, A. Collignon: AIAA Paper No. 91-0179 (1991)

7.25 Reduced Kinetic Mechanisms and Asymptotic Approximations for Methane-Air Flames, Lecture Notes in Physics, Vol. 384, ed. by M.D. Smooke (Springer Verlag, Heidelberg, Berlin, New York 1991)

7.26 K.C. Smyth, J.H. Miller, R.C. Dorfman, W.G. Mallard, R.J. Santoro: Combust. Flame **62**, 157 (1985)

7.27 K.C. Smyth, P.H. Taylor: Chem. Phys. Lett. **122**, 518 (1985)

7.28 J.H. Miller, P.H. Taylor: Combust. Sci. Technol. **52**, 139 (1987)

7.29 K.C. Smyth, P.J.H. Tjossem, A. Hamins, J.H. Miller: Combust. Flame **79**, 366 (1990)

7.30 K.C. Smyth, P.J.H. Tjossem: "Relative H-atom and O-atom measurements in a laminar, Methane/Air diffusion flame", in Twenty-Third Symposium (International) on Combustion (The Combustion Institute, Pittsburgh 1990) p. 1829

7.31 T.S. Norton, K.C. Smyth: Combust. Sci. Technol. **76**, 1 (1991)

7.32 A. Melvin, J.B. Moss: "Structure in Methane-Oxygen diffusion flames", in Fifteenth Symposium (International) Combustion (The Combustion Institute, Pittsburgh 1975) p. 625

7.33 D. Stepowski, A. Garo: Appl. Opt. **24**, 2478 (1985)

7.34 A. Garo, G. Prado, J. Lahaye: Combust. Flame **79**, 226 (1990)

7.35 E. Bastin, J.-L. Delfau, M. Reuillon, C. Vovelle: J. Chim. Phys. **84**, 415 (1987)

7.36   J.H. Miller, W.G. Mallard, K.C. Smyth: "Chemical production rates of intermediate Hydrocarbons in a Methane/Air diffusion flame", in Twenty-First Symposium (International) on Combustion (The Combustion Institute, Pittsburgh 1986) p. 1057

7.37   K.C. Smyth, J.H. Miller: Science 236, 1540 (1987)

7.38   M.C. Drake: "Streched laminar flamelet analysis of turbulent H_2 and $CO/H_2/N_2$ diffusion flames", in Twenty-First Symposium (International) on Combustion (The Combustion Institute, Pittsburgh 1986) p. 1579

7.39   I.K. Puri, K. Seshadri, M.D. Smooke, D.E. Keyes: Combust. Sci. Technol. 56, 1 (1987)

7.40   K. Seshadri, F. Mauss, N. Peters, J. Warnatz: "A flamelet calculation of Benzene formation in coflowing laminar diffusion flames", in Twenty-Third Symposium (International) on Combustion (The Combustion Institute, Pittsburgh 1990) p. 559

7.41   M.D. Smooke, P. Lin, J.K. Lam, M.B. Long: "Computational and experimental study of a laminar axissymmetric Methane-Air diffusion flame", in Twenty-Third Symposium (International) on Combustion (The Combustion Institute, Pittsburgh 1990) p. 575

7.42   R.W. Bilger: Combust. Flame 30, 277 (1977)

7.43   K. Saito, F.A. Williams, A.S. Gordon: J. Heat Transfer 108, 640 (1986)

7.44   Y.R. Sivathanu, G.M. Faeth: Combust. Flame 82, 211 (1990)

7.45   T.S. Norton, K.C. Smyth, J.H. Miller, M.D. Smooke: Combust. Sci. Technol., in press

7.46   J.H. Miller, K.C. Smyth: Paper No. 37, Eastern States Combustion Institute Meeting, Orlando, FL (1990)

7.47   P.R. Westmoreland, J.B. Howard, J.P. Longwell: "Tests of published mechanisms by comparison with measured laminar flame structure in fuel-rich acetylene combustion", in Twenty-First Symposium (International) on Combustion (The Combustion Institute, Pittsburgh 1986) p. 773

7.48   J.C. Biordi, C.P. Lazzara, J.F. Papp: "An examination of the partial equilibration hypothesis and radical recombination in 1/20 atm Methane flames", in Sixteenth Symposium (International) on Combustion (The Combustion Institute, Pittsburgh 1977) p. 1097

7.49   J.D. Bittner: "A Molecular Beam Mass Spectrometer Study of Fuel-Rich and Sooting Benzene-Oxygen Flames"; D.Sc. Dissertation, M. I. T. (1981)

7.50   K.C. Smyth, P.J.H. Tjossem: Appl. Opt. 29, 4891 (1990)

7.51   K.C. Smyth, T.S. Norton: unpublished results (1990)

7.52   R. Puri, R.J. Santoro: "The Role of Soot Particle Formation on the Production of Carbon Monoxide in Fires", in Fire Safety Science-Proceedings of the Third International Symposium (Elsevier Applied Sciences, London 1991) p. 595

7.53   H.G. Wolfhard, W.G. Parker: Proc. Phys. Soc., London A 65, 2 (1952)

7.54   R.P. Lucht, D.W. Sweeney, N.M. Laurendeau: Combust. Sci. Technol. 42, 259 (1985)

7.55   R.J. Santoro, H.G. Semerjian, R.A. Dobbins: Combust. Flame 51, 203 (1983)

132   Kermit C. Smyth et al.

7.56   R. Puri, M. Moser, R.J. Santoro, K.C. Smyth: "Laser-induced Fluorescence
       Meassurements of OH--concentrations in the Oxidation Region of Laminar,
       Hydrocarbon Diffusion Flames", in <u>Twenty-Fourth Symposium (Internatio-
       nal) on Combustion</u> (The Combustion Institute, Pittsburgh 1992), p. 1015
7.57   K.P. Schug, Y. Manheimer-Timnat, P. Yaccarino, I. Glassman: Combust.
       Sci. Technol. **22**, 235 (1980)
7.58   M.D. Smooke, R.E. Mitchell, D.E. Keyes: Combust. Sci. Technol. **67**, 85
       (1989)

Discussion

Kent: In your viewgraphs you had two OH concentration profiles. In one of them you had the OH concentration going down to very low values where the soot formation started but in the other graph you showed a dip in the profile where you said the soot formation started. The first graph seems to me to imply that all the OH was being consumed by the soot at that point but the second graph didn't. Could you just comment on that?

Smyth: I would say something like this: Soot inception occurs in a kind of pyrolysis soup that contains lots of acetylene, lots of benzene, lots of other hydrocarbons. Therefore, OH that starts moving into the fuel rich region has many hydrocarbons that will react with it. If you follow these kinds of OH profiles and if you are looking at the soot scattering with height, the soot scattering signal goes up by a factor of a hundred in just a few millimeters. There isn't a lot of oxidation of soot by OH but rather it is simply being depleted in the fuel rich flame region by reactions with other hydrocarbons. At the top of the flame, of course, there is a high-temperature region where essentially the reaction zone encompasses the flame tip. At the top of the flame – and this was the profile from the latter viewgraph – you are in a situation where essentially there is a bath of OH throughout the oxidation region that you must penetrate with the soot particles.

Moss: Can you tell us where in that flame the stoichiometric contour is at about 100 mm. The reason why I ask is that we have also done some laminar diffusion flame experiments trying to look at the oxidation region at the top. We were quite worried about the computation of this cusp region very near to the top of the flame, because all of the flow is being squeezed through a very narrow stream tube. From the modeling point of view it is not parabolic if that is the code you are using for the calculations. The principle gradients are in right angles to the flow rather than in parallel to the flow direction. To study soot oxidation by arguing with results from computations it seems to be a very unsuitable flame region.

Smyth: In this kind of flame for this oxidation study we have no computational results available at the moment and we don't have presently all the mass spectrometric measurements that we need to define the stoichiometric surface at the height you asked for.

Sarofim: I noticed that you had a partial equilibrium reaction which generates OH. Looking at the characteristic times of that and looking at the sink of soot plus OH could you not get a partial answer as to whether soot is going to change the OH concentration?

Smyth: Well, you know that soot will change the OH concentration. How fast a sink that is and how fast the radical pool adjusts to that in the oxidation region we don't know. If we had to guess today we would say that the super equilibrium OH is higher in the oxidation region than lower in the flame. You actually have a situation where you create a sink in OH plus soot reactions which is not adjusted very quickly by the radical pool. You shoot higher in super equilibrium because your calculated OH concentration goes down as the temperature drops faster than the measurements.

Haynes: I would like to press Kermit a little on the location of the stoichiometric stream line. Can we presume that it has reached the center line at the height of 100 mm ?

Smyth: I don't think so.

Haynes: So, the temperature profile would presumably also be showing this cusp behaviour. We don't have a uniform temperature or a smooth temperature profile through here. Put it in another way, is the peak temperature at the center line at this location?

Santoro: Yes.

Smyth: That's right. But if I remember what Houston Miller presented recently which is a computation done by D. Honnery on an ethylene flame, the stoichiometric contour does not reach the center line.

Kent: I expect the stoichiometric contour to be lower than that.

Detection of Large Aromatics in Flames by REMPI-MS

Reinhold Kovacs [1], Silke Löffler [2], Klaus H. Homann [1]

[1]Institut für Physikalische Chemie, T. H. Darmstadt,
64287 Darmstadt, Fed. Rep. of Germany
[2]present address: BKA, 65193 Wiesbaden, Fed. Rep. of Germany

Abstract: A new experimental technique for detecting large polycyclic aromatic hydrocarbons (PAH) is described. This technique complements presently employed methods that derive information about large PAH in the mass range between 500 and 1500 u from studying PAH ions. First experiments have been carried out in flat premixed laminar flames at low pressure. The experimental technique comprises sampling of the flames by means of a differentially pumped molecular beam sampling system. The PAH present in the molecular beam are ionized in a resonant multiphoton ionization process (REMPI) at 259 nm. The PAH ions are analyzed by a time-of-flight mass spectrometer. REMPI time-of-flight mass spectrometry offers several advantages: it is very specific, i.e. only distinct groups of molecules are ionized at a fixed wavelength. Fragmentation can be almost completely avoided. In the first experiments which are presented in the paper PAH up to $C_{19}H_{12}$ could be observed. The experimental technique offers the possibility of studying large aromatic hydrocarbons that are likely soot precursors and offers thereby some more insight into the mechanism of soot formation.

8.1 Introduction

Polycyclic aromatic hydrocarbons (PAH) are a particularly interesting group of compounds because they are the most likely soot precursors [8.1-2]. So far, several different approaches have been carried out to obtain information about PAH in flames, e.g. condensation of flame gases and subsequent GC-MS analysis [8.2-3] and also the investigation of flame ions [8.1,8.4]. The investigation of flame ions brought about a great deal of information about PAH-, oxo-PAH- (PAH ions which contain oxygen atoms) and fullerene ions [8.5] but elucidated clearly that additional information about large neutral species in flames is necessary. For example, we could follow the growth of positive PAH ions up to the first small soot ions in acetylene flames. However, this is not possible in benzene flames, because no small soot ions are formed in benzene flames. In the soot formation zone of benzene flames almost the entire electric charge in the flame is concentrated on fullerene ions and not on PAH ions. Another interesting question, which could not be fully answered by the investigation of ions, is: what is the relationship between PAH and fullerenes? Fullerenes are formed both in non-sooting flames and in the oxidation and the soot forming zone of sooting flames. The only suitable building blocks for fullerenes existing in non-sooting flames are PAH. We feel that more information about neutral PAH in the mass range from 500 to 1500 u, which has not been covered by the above mentioned GC-MS experiments, and fullerenes would be useful here. In order to achieve this aim we have developed an experimental set-up that enables us to investigate large aromatics in flames by REMPI time-of-flight mass spectrometry. REMPI-TOF-MS offers several advantages:

- It is an on-line method with no biasing of the results by interfering chemical reactions which can occur in sampling and chemical analysis.

- It is very specific. At a fixed wavelength only one group of particles will be ionized, e.g. PAH of a certain mass range but no aliphatic molecules.

- There will be only negligible (if any) fragmentation if the laser energy is adjusted properly.

- There is no restriction concerning the mass range as in the case of GC analysis.

- If instead of the linear TOF-MS, which we employed in the first experiments with this technique, a reflectron is used, the mass resolution will be very high.

A disadvantage of this method is the necessity to calibrate it extensively for a very large number of compounds in order to obtain quantitative results. In addition, the wavelength and energy of the laser has to be adjusted not only for the respective mass range but also for the types of molecules, e.g. aromatic molecules, aliphatic molecules, aromatic molecules with side chains and so on. Thus, the specific ionization by REMPI also has its drawbacks, since information about the gas components is not comprised in one mass spectrum but has to be put together from many individual mass spectra, each of which describes only part of the complicated mixture of hydrocarbons. However, REMPI mass spectra should reflect the nature of the hydrocarbons much better than the rather unspecific electron bombardment ionization.

8.2 Experimental

The experimental set-up consists of a differentially pumped molecular beam sampling system (Fig. 8.1) which is coupled with a linear TOF-MS.

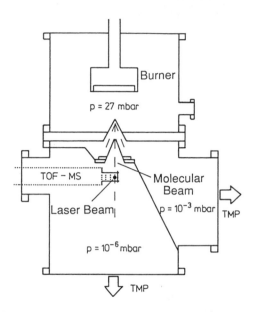

Fig. 8.1. Molecular beam sampling system, side view (TMP = turbo molecular pump)

Flat premixed laminar acetylene-oxygen flames burnt at low pressure (2.7 kPa). The molecular beam was formed by a conical quartz probe followed by a skimmer. The molecular beam was intersected by the laser beam in the ion source. The laser beam was unfocused to avoid fragmentation. Its width was about 1 mm. This resulted in a considerable uncertainty in the local origin of the ions and impaired the mass resolution of the TOF-MS. This drawback will be eliminated by the use of a reflectron. With the reflectron installed, the set-up can be operated either as a linear TOF-MS or as a reflectron TOF-MS, as indicated in Fig. 8.2.

The laser beam was generated by an excimer laser pumping a dye laser. The applied laser wavelength was 259 nm which is very close to the absorption maximum of benzene at 258.9 nm. Up to now optimization of the wavelength with regard to larger PAH has not been carried out. At 259 nm we were able to detect PAH up to $C_{19}H_{12}$ (240 u).

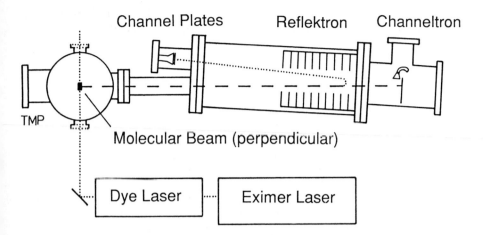

Fig. 8.2. Photoionization and detection equipment, top view

8.3 Results

In the following, first results from some experiments with the new experimental set-up are presented. Figure 8.3 shows a mass spectrum, which has been taken from the oxidation zone of a sooting acetylene-oxygen flame. The spectrum is averaged over 2000 laser pulses at a pulse frequency of 20 Hz. The final laser energy was 0.15 mJ/pulse. Prominent peaks in the spectrum are those of benzene (C_6H_6) and naphthalene ($C_{10}H_8$). A group of C_{16}- to C_{19}-PAH is clearly measurable.

The height of the signals decreases with increasing mass. Many of the mass peaks can be attributed to aromatic species known to be present in sooting flames (Fig. 8.4). The partial superposition of the signals in the upper range of the mass spectrum is due to the much poorer resolution of the linear TOF-MS, since for these first measurements the reflectron has not yet been installed. Therefore, results for PAH with the same number of C-atoms but a different hydrogen content cannot yet be reported.

However, there are also some peculiarities:

The peak ratio of C_6H_6 and $C_{10}H_8$ (≈ 3.5) is much smaller than the concentration ratio of the two aromates in flames which is about 20 [8.3, 8.6]. Other signal ratios, for example that of phenylacetylene (C_8H_6) and pyrene or fluoranthene ($C_{16}H_{10}$), are as well smaller than the respective concentration ratios in a similar flame [8.3]. This shows that extensive calibration measurements for each compound have to be carried out in order to obtain quantitative results. On the other hand, signals of some PAH

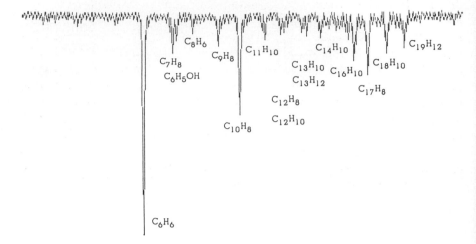

Fig. 8.3. Mass spectrum from the oxidation zone of a sooting flame

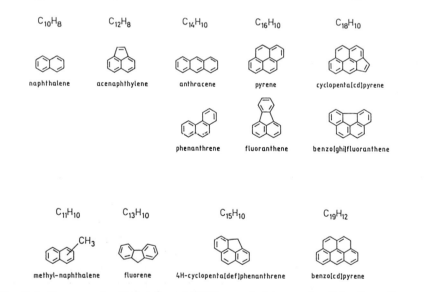

Fig. 8.4. Even- and odd-numbered PAH reported in condensibles from flames

which are present in the flame are unexpectedly weak or even missing. Among these are acenaphthylene ($C_{12}H_8$), methylanthracene ($C_{15}H_{12}$) and 4H-cyclopenta[def]-phenanthrene ($C_{15}H_{10}$).

Because a conical quartz probe is used for molecular beam sampling, the flame is disturbed considerably at sampling positions close to the burner. Values measured there are always too high, since stagnation flame gas from the side is sucked into the nozzle. Therefore, the profiles of the mass spectrometric signals of several compounds shown in Figs. 8.5 to 8.8 start at some distance from the burner (3 mm). Since calibration and flame temperature measurements are still missing, the ordinate is given in arbitrary intensity units which are the same for all species. For reference, the peak mole fraction of benzene in this flame is $2 \cdot 10^{-4}$, that of naphthalene $1.1 \cdot 10^{-5}$ [8.6]. All PAH profiles have a maximum within the oxidation zone, then decrease towards the end of the oxidation zone where the temperature is maximum, and mostly show a re-increase in the burned gas. This re-increase is in part due to the increase in absolute gas density because of the decrease in temperature. Species having methyl side chains, for instance toluene, or oxygen-containing substances (oxo-PAH) like phenol or benzaldehyde are only formed when O_2 is present, but not in the burned gas. The decrease to a minimum is comparatively stronger for the lower-mass PAH.

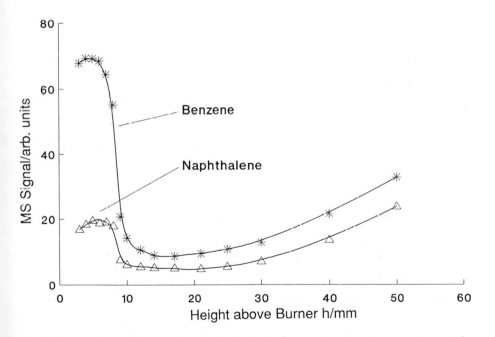

Fig. 8.5. Profiles of benzene and naphthalene in a sooting flame, $C/O = 1.0$, $p = 2.7$ kPa, $v_u = 42$ cm/s

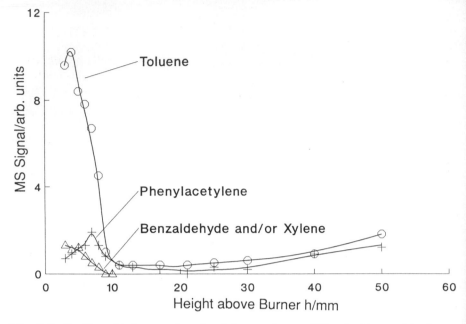

Fig. 8.6. Profiles of toluene, phenylacetylene and mass 106 u (xylene and/or benzaldehyde) in a sooting flame, $C/O = 1.0$, $p = 2.7$ kPa, $v_u = 42$ cm/s

Figure 8.7 reveals the influence of the molecular mass and the hydrogen content on the location of the maximum. Usually, the maxima move away from the burner with increasing mass. However, there are some exceptions of this rule, for example $C_{18}H_{12}$ (m = 228 u). Despite of the higher mass, this signal peaks earlier than expected which is certainly effected by its rather high H/C-ratio.

The profiles of acenaphthylene ($C_{12}H_8$) in two fuel-rich flames with a $C/O = 0.8$ and a $C/O = 1.0$, respectively, of which the latter is sooting are plotted in Fig. 8.8. Evidently in the leaner flame, the maxima appear closer to the burner, are less intense, and the re-increase in the burned gas is insignificant or often not detectable.

These characteristics for PAH profiles have also been reported from other experiments where conventional electron bombardment ionization has been applied or from gas chromatographic analyses of condensibles from similar flat flames [8.3, 8.7].

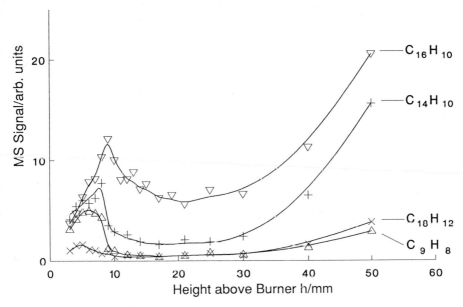

Fig. 8.7. Profiles of C_9H_8 (indene), $C_{14}H_{10}$ (anthracene, phenanthrene), $C_{16}H_{10}$ (pyrene, fluoranthene), and $C_{18}H_{12}$ (m = 228 u) in a sooting flame, $C/O = 1.0$, $p = 2.7$ kPa, $v_u = 42$ cm/s

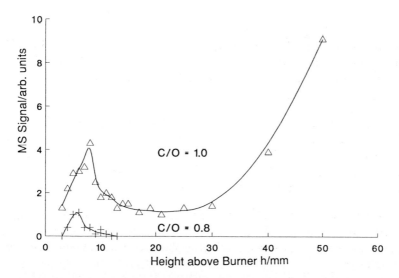

Fig. 8.8. Comparison between the profiles of acenaphthylene ($C_{12}H_8$, m = 152 u) in a fuel-rich non-sooting flame ($C/O = 0.8$) and a sooting flame ($C/O = 1.0$), $p = 2.7$ kPa, $v_u = 42$ cm/s

8.4 Conclusions

In this paper we have presented two-photon ionization mass spectrometry which is a new experimental technique to study large PAH in flames. Although the first experiments still suffer from many initial imperfections which will certainly be eliminated in the near future, we are encouraged by our first results and feel confident that this method will help to learn more about the chemistry of PAH in sooting flames.

8.5 Acknowledgements

This work is part of the Arbeitsgemeinschaft TECFLAM and is supported by the Bundesministerium für Forschung und Technologie. The support by the Bundesland Hessen is gratefully acknowledged.

References

8.1   P. Gerhardt, K.-H. Homann: J. Phys. Chem. **94**, 5381 (1990)

8.2   J.Bittner, J. Howard: "Composition Profiles and Reaction Mechanisms in a Near-Sooting Premixed Benzene/Oxygen/Argon Flame", in Eighteenth Symposium (International) on Combustion (The Combustion Institute, Pittsburgh 1981) p. 1105

8.3   H. Bockhorn, F. Fetting, H.W. Wenz: Ber. Bunsenges. Phys. Chem. **87**, 1067 (1983)

8.4   D.B. Olson, H.F. Calcote: "Ions in Fuel-Rich and Sooting Acetylene and Benzene Flames", in Eighteenth Symposium (International) on Combustion (The Combustion Institute, Pittsburgh 1981) p. 453

8.5   P. Gerhardt, S. Löffler, K.-H. Homann: "The Formation of Polyhedral Carbon Ions in fuel-rich Acetylene and Benzene Flames", in Twenty-Second Symposium (International) on Combustion (The Combustion Institute, Pittsburgh 1988) p. 395

8.6   M. Hausmann, P. Hebgen, K.-H. Homann: "Radicals in Flames: Analysis via Scavenging Reactions", in Twenty-Fourth Symposium (International) on Combustion (The Combustion Institute, Pittsburgh 1992) p. 793

8.7   B.D. Crittenden, R. Long: Combust. Flame **20**, 359 (1973)

Discussion

Jander: If I remember right you are able to detect PAH in the mass range up to 600 u by REMPI-MS.

Löffler: You are correct. However, the resolution is still very poor.

Bockhorn: The concentration ratio of phenylacetylene and naphthalene is puzzling me. Generally, the concentration ratio of phenylacetylene and naphthalene in sooting flames is the inverse of the signal ratio for the respective masses you showed. In sooting flames you have a lot of phenylacetylene and less naphthalene. Did you make calibrations for phenylacetylene and could this be an effect of the absorption coefficient of phenylacetylene and naphthalene.

Löffler: Yes, I am sure it is. Another example is benzene and naphthalene. Even if we assume an equal absorption coefficient for benzene and naphthalene the concentration ratio is not correct and naphthalene seems to be present in much higher concentration than it actually is. So, this is a problem of calibration. I also mentioned that we do not see aromatic compounds with conjugated side chains at this wavelength and phenylacetylene is the first representative of these.

Frenklach: My question concerns the signal ratio of acenaphthalene and naphthalene. If I recall Bockhorn's analysis, in all flames the concentration of acenaphthalene is very high and comparable to toluene. Does the same reasoning that you just gave apply to this problem?

Löffler: Yes. I also mentioned that we had this problem with the species with an odd number of carbon atoms. We are the first applying this technique and from the experiments presently performed there are not enough data to explain all the phenomena. We have to do much more experiments on optimizing the wavelength and the laser energy etc.

Calcote: I look forward to seeing results at large masses. It would be interesting to compare your results with those of Delfau and Vovelle who measured the same profiles in pretty much the same flame. That might be a means of calibrating your system because they used a different mass spectrometer and ionization source. My question, however, is: you describe some of these peaks as being multiple peaks. I would have a hard time believing that. Do you really believe that?

Löffler: What do you mean by multiple peaks? Do you mean the concentration profiles consisting of different compounds or do you mean the mass spectra?

Calcote: You had concentration profiles from your mass spectra which had several peaks for the same species. I find that hard to belive. I would suspect it's in the noise or possibly in the sampling system. When the cone moves into the flame it could disturb it.

Löffler: There are several compounds that have a perfectly normal profile even if they are present in very low concentrations. For example toluene or phenol. For others in much higher concentrations we detected these oscillating profiles. I am very sure that this has nothing to do with possible variations in the experimental conditions, for example variation in the laser energy. We measured the laser energy at every point and it was constant over the whole range. In spite of this we have for example for C_{16} three maxima but for C_{18} we have only one maximum. So this kind of profile does not depend on the size of the molecule or on the concentration. It is still puzzling.

Calcote: We will see.

Note added in proof: The question of Dr. Calcote about 'multiple peaks' in the profiles refers to some figures shown on slides which have turned out to be artifacts caused by the poor resolution of neighboured mass peaks. These figures have been deleted from the paper.

Haynes: I think we all agree that the method promises some information that we all would like to have, but I just wonder how long it is going to take to calibrate all of those peaks for all of those temperatures and compositions.

Pressure Dependence of Formation
of Soot and PAH
in Premixed Flames

Heidi Böhm, Mathias Bönig, Christian Feldermann,

Helga Jander, Gabriele Rudolph, Heinz Georg Wagner

Institut für Physikalische Chemie, Universität Göttingen,
37077 Göttingen, Fed. Rep. of Germany

Abstract: The influence of pressure on the stable products of a C_2H_4-air flame with special regard to PAH has been investigated by optical methods and chemical analysis for pressures up to $70 \cdot 10^5$ Pa. PAH could be analyzed up to the mass range of dicoronyl, $C_{48}H_{22}$, 598 u. The identified PAH in the mass range above coronene consist mainly of pure 6-ring-systems, systems containing 6- and 5-rings, and aza-aromates. They are relatively stable, that means that their concentrations are kinetically controlled. The carbon density of the PAH, $\rho_{c\Sigma(PAH)}$, shows similarly to soot a p^2-dependence in the pressure regime up to $10 \cdot 10^5$ Pa. Above $10 \cdot 10^5$ Pa this dependence changes into an approximately linear one. The dependence of the PAH concentrations on mixture composition and temperature is similar to the dependence of the final soot volume fraction $f_{V\infty}$ on those parameters: one finds bell shaped curves for the temperature dependence of the PAH concentrations and $\rho_{c\Sigma(PAH)} = [(\frac{C}{O})_{actual} - (\frac{C}{O})_{crit}]^n$ with $2.5 \leq n \leq 3$.

9.1 Introduction

The formation of polycyclic aromatic hydrocarbons (PAH) in technical combustion systems and diesel engines has attracted increasing attention due to their part in soot formation and growth and due to their role as pollutants [9.1-13].

In premixed sooting C_2H_4-air flames burning at pressures between $1 \cdot 10^5$ Pa and $10 \cdot 10^5$ Pa a substantial fraction of the carbon of the original fuel is converted into soot (about 1.5 to 7.5 %) and PAH (≤ 0.07 %). That means that in the postflame gases of sooting hydrocarbon flames soot and PAH are simultaneously present.

From measurements in low pressure hydrocarbon/oxygen flames information about the formation of PAH can be obtained [9.1-2, 9.4-5, 9.7, 9.10]: Under low pressure conditions two regions of PAH formation are observed. A rapid production within the oxidation zone is followed by a relatively slow formation of PAH in the burned gas region [9.2, 9.7, 9.10]. In atmospheric and high pressures flames the measured PAH concentration profiles hint to only one formation process or the two formation processes seem to coincide [9.6, 9.8-9]. It is, therefore, of interest to study the formation of PAH in flames over a wide range of reaction conditions.

In the present study measurements of PAH concentration profiles in atmospheric C_2H_4-air flames are reported in order to obtain information about their appearance in the early phase of the combustion process. In addition, PAH and soot were studied in the postflame gases for different mixture compositions, temperatures and pressures. The latter data should give information about the growth of soot particles and PAH in the later phase of the combustion process.

9.2 Experimental

The measurements were performed on a porous plate flat flame burner at pressures up to $70 \cdot 10^5$ Pa. This burner is described in detail in [9.14].

The sampling probe used at elevated pressures consisted of a small quartz tube surrounded by a thermostated metal tube. The diameter of the probe orifice was less than 0.3 mm. In order to prevent blocking of the orifice by soot particles and to maintain a steady flow, it was necessary to continuously clean the orifice of the probe. For that purpose a fine wire could be inserted into the orifice keeping the orifice unblocked. The probe was mounted horizontally. The flame gases were expanded overcritically within the probe.

At the exit of the sampling probe a soot filter was mounted. The sampling line behind the soot filter could be linked either directly to the sample loop of a gas chromatograph or to two cold traps cooled with liquid nitrogen or with a mixture of ice and water. The exit of the sampling probe and all tubings were kept at a temperature of 280°C to prevent condensation of water and high molecular PAH.

After preparation of the sample - weighing of the sample, adding the first internal standard, extracting the sample several times with CH_2Cl_2, concentrating the organic phase to a definite volume, adding the second internal standard - the organic solution was analyzed. The PAH with molecular masses up to coronene were determined in a GC/MS device, the PAH with molecular masses in the mass range above coronene in an HPLC apparatus.

The flame temperatures of the sooting C_2H_4-air flames were measured applying the *Kurlbaum* method [9.15].

In the burnt gases of the C_2H_4-air flame the following stable products were analyzed:

1. The water gas components: CO, CO_2, H_2
2. The main aliphatic hydrocarbons: $C_2H_2, CH_4, C_2H_4, C_nH_2(n = 4, 6, 8)$
3. The polycyclic aromatic compounds (PAH) :
3.1. quantitative recording of PAH up to coronene, $C_{24}H_{12}$
3.2. qualitative or half-quantitative analysis of PAH in the mass range higher than coronene
4. Soot
4.1. Soot volume fraction f_V (cm^3 soot per cm^3 flame volume)
4.2. Particle number density N (cm^{-3})
4.3. Particle radius R (nm)

Soot volume fraction, particle number density and particle radius were calculated from the results of laser light scattering and absorption measurements [9.16] using the complex refractive index given by *Lee* et al. [9.17].

9.3 Experimental Results

9.3.1 PAH in the Postflame Region of a C_2H_4-air Flame at $1 \cdot 10^5$ Pa

9.3.1.1 Product Spectrum of PAH

The spectrum of PAH from an atmospheric sooting C_2H_4-air flame shows the typical groups of PAH which are also obtained from other hydrocarbon fuels and flames at lower and higher burning pressures:

(i)   pure condensed 6-ring compounds;
(ii) compounds which contain 6-rings and also 5-rings in the condensed ring systems
(iii) ring systems with side chains. The side chains mostly consist of a methyl-, vinyl- or ethinyl-group.

The concentrations of the PAH depend on mixture composition, temperature and pressure. PAH with side chains show always lower concentrations in the burned gases than those without side chains. This behavior can be found for PAH in the mass range up to coronene, $C_{24}H_{12}$ (atomic mass units ≤ 300).

In general, this qualitative picture for the PAH concentrations does not change for the PAH analyzed in the mass range above coronene. Figure 9.1 shows the HPLC spectrum of a sample from a C_2H_4-air flame at $1 \cdot 10^5$ Pa with a C/O ratio of 0.79. The sample was taken at 10 mm height above the burner. The part of the spectrum for the PAH in the mass range below coronene has a much smaller scale than that part for masses in the range above coronene.

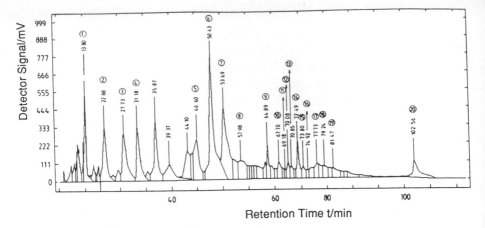

Fig. 9.1. HPLC of a sample taken at 10 mm height above the burner in a C_2H_4-air flame at $1 \cdot 10^5$ Pa, C/O ratio $= 0.79$; the first part of the spectrum up to coronene has a much smaller scale and the peaks are not identified; for identification of the single peaks in the mass range above coronene compare Table 9.1

Table 9.1. PAH in the mass range above coronene from a sample taken from a C_2H_4-air flame at 10 mm height above the burner; $p = 1 \cdot 10^5$ Pa, C/O ratio $= 0.69$.

Peak No.	Compound	Formula	u
1	dibenzo[a,b]anthracene	$C_{22}H_{14}$	278
2	coronene*	$C_{24}H_{12}$	300
3	bibenzo[a]pyrene	$C_{40}H_{22}$	502
4	dibenzo[a,e]pyrene	$C_{24}H_{14}$	302
5	benzo[a]coronene	$C_{28}H_{14}$	350
6	naphthocoronene	$C_{30}H_{14}$	374
7	acridine	$C_{13}H_9N$	179
8	ovalene	$C_{32}H_{14}$	398
9	periflanthene	$C_{32}H_{16}$	400
10	phenanthridine	$C_{13}H_9N$	179
11	1-azafluoranthene	$C_{15}H_9N$	203
12	benz[a]acridine	$C_{17}H_{11}N$	229
13	isochinoline	C_8H_6N	116
14	benzo[c]chinoline	$C_{13}H_9N$	179
15	4-azapyrene	$C_{15}H_9N$	203
16	7-azafluoranthene	$C_{15}H_9N$	203
17	dibenzo[a,j]acridine	$C_{21}H_{13}N$	279
18	dicoronyl	$C_{48}H_{22}$	598
19	anthraceno[2,3,a]coronene	$C_{36}H_{16}$	448
20	quarterrylene	$C_{40}H_{18}$	489

* molefraction of coronene (2): $3 \cdot 10^{-6}$

Coronene was determined quantitatively and its mole fraction is indicated at the respective peak. From this value the order of magnitude of the concentration of the other compounds can be obtained. Most of the PAH could be identified by comparing their retention time with different injected standards or by the correlation log $M_{(PAH)}$ versus capacity factor k' which is a straight line for the different PAH standards [9.18].

In Fig. 9.1 only some of the identified PAH are labelled with their structures and formulas in order to characterize the different types of compounds. Similarly to the mass range below 300 u one finds the pure condensed 6-ring systems and the systems containing 6- and 5-rings in the high mass range. The compound with the highest mass which we could identify at the moment is dicoronyl, $C_{48}H_{22}$. Another type of compounds, which have been identified, are the "aza aromates", nitrogen containing heterocyclic compounds, for example 1-azafluranthene [9.18].

9.3.1.2 PAH Mole Fraction Profiles

Concentration profiles of PAH in atmospheric and higher pressure flames show approximately the same shape. This is demonstrated in Fig. 9.2 where only two PAH, naphthalene and benzo[g,h,i]fluoranthene, are given as representatives of all the determined PAH.

Figure 9.2 contains as well the total emission of light from this sooting C_2H_4-air flame. The blue emission of the main oxidation zone and the yellow luminosity of the soot particles are indicated. High pressure flames exhibit very similar profiles. The PAH which are present in the postflame gases appear at the end of the oxidation zone. In the particle formation zone the concentration of PAH increases up to a maximum value. In the post flame region some PAH decrease in concentration, others increase slightly. The position of the maxima of the PAH concentration profiles in flames at different pressure appear at about the same relative flame height.

9.3.2 Dependence of Soot Volume Fraction and PAH Concentrations on Temperature in the Pressure Range up to $10 \cdot 10^5$ Pa

It is well known that temperature has a strong influence on final soot volume fraction $(f_{V\infty})$, compare [9.14, 9.19-21]. This can be seen from Fig. 9.3 where the "final" carbon density ρ_{ci} [g cm^{-3}] of soot, of some of the determined PAH, the sum of PAH and of the main aliphatic compounds CH_4, C_2H_2, C_2H_4 is given as a function of the temperature at 10 mm height above the burner T_M for flames at $10 \cdot 10^5$ Pa and a C/O ratio of 0.72. The concentrations are measured at 30 mm height above the burner. The carbon density of soot is calculated from $f_{V\infty}$ with the density of soot assumed to be 2 g cm^{-3}.

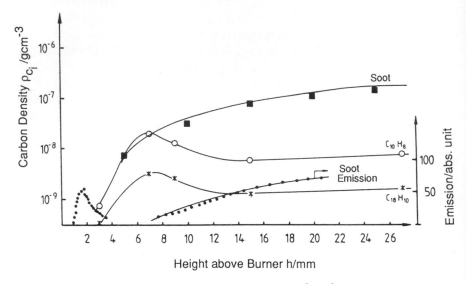

Fig. 9.2. Concentration profiles of anthracene, benzo[g,h,i]fluoranthene, and soot in an ethylene air flame with a C/O ratio of 0.72 at $1 \cdot 10^5$ Pa, cold gas velocity: 3.74 cm s^{-1}; the total emission of light (dotted line) from this flame is also indicated

Two types of dependence are obvious: the linear dependence for CH_4, C_2H_4 (the carbon density of C_2H_2 shows a slightly nonlinear behavior) and the bell-shaped curves for PAH and soot. The temperature dependence of the PAH and, therefore, also the sum of PAH is similar to that of the soot volume fraction: Their carbon density increases from low temperatures, passes through a maximum between 1650 to 1700 K. At higher temperature ($T > 1750$ K) the concentration of the PAH decreases rapidly to negligibly small values.

The lower threshold of soot formation at these reaction conditions is at about 1370 K [9.23]. Below that threshold for soot formation the PAH are already present in measurable concentrations.

The carbon density of soot depends on temperature in a similar way as the PAH concentrations. Its maximum, however, is shifted towards higher temperatures and its maximum concentration at $10 \cdot 10^5$ Pa is higher by a factor of 60 than the sum of all analyzed PAH (see Fig. 9.3).

The shapes of the ρ_{ci} versus temperature plots of soot and PAH at $10 \cdot 10^5$ Pa are similar to those at $1 \cdot 10^5$ Pa. However, the absolute concentrations and the concentration ratios are quite different. The fraction of carbon in soot increases approximately with the square of pressure, see Fig. 9.5 and ref. [9.23]. For the same C/O ratio the carbon density of soot at the soot maximum at $10 \cdot 10^5$ Pa is two orders of magnitude larger than at $1 \cdot 10^5$ Pa. The carbon density of C_2H_2 at $10 \cdot 10^5$ Pa is noticeably lower than that of

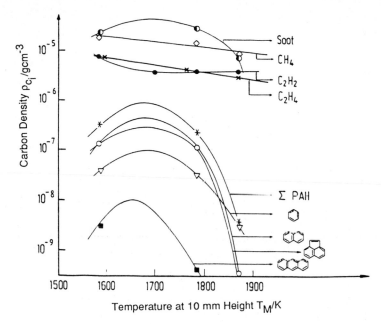

Fig. 9.3. Carbon density ρ_{ci} of several PAH and soot as a function of flame temperatures at a burning pressure of $10 \cdot 10^5$ Pa, C/O ratio $= 0.72$

CH_4 in contrast to the $1 \cdot 10^5$ Pa results where C_2H_2 is the major hydrocarbon product in the postflame gases, see Fig. 9.5.

9.3.3 Dependence of Soot Volume Fraction and PAH Concentrations on C/O ratio in Pressure Range up to $10 \cdot 10^5$ Pa

The C/O ratio has a strong influence on the carbon density of soot. This is demonstrated in Fig. 9.4 where ρ_{ci} is plotted as a function of the C/O ratio for pressures $1 \cdot 10^5$ Pa and $10 \cdot 10^5$ Pa. The carbon density of the sum of the PAH and of C_2H_2 at these two pressures are also shown. The thick bar on the abszissa shows the shift of the threshold of soot formation with pressure.

Above the threshold of soot formation, the carbon density of soot increases strongly. At higher C/O ratios the increase in the carbon density with increasing C/O ratio flattens. This holds for both pressures. However, the dependence at $10 \cdot 10^5$ Pa is much stronger than that at $1 \cdot 10^5$ Pa. That means that increasing amounts of the fuel carbon are converted into soot with increasing pressure: soot formation is favoured by higher pressure. The carbon densities of PAH are about two orders of magnitude lower than those of soot. The dependence of the carbon densities of soot and the sum of PAH

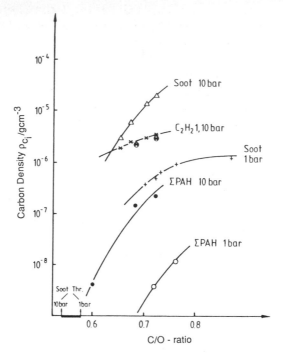

Fig. 9.4. Carbon density ρ_{ci} of C_2H_2, of the sum of PAH and soot at different C/O ratios at $1 \cdot 10^5$ Pa and $10 \cdot 10^5$ Pa

on the C/O ratio at different pressures again are very similar. From the results obtained in the pressure range from $1 \cdot 10^5$ Pa to $10 \cdot 10^5$ Pa the following expression can be derived:

$$\rho_{c\,soot,\Sigma PAH} \approx [(C/O) - (C/O)_{crit}]^n \quad \text{for} \quad T > 1700 \text{ K}, \ 2.5 \leq n \leq 3.0 \tag{9.1}$$

In equation 9.1 $(C/O)_{crit}$ means the C/O ratio at the threshold of soot formation.

Completely different from the behaviour of soot and the sum of PAH is that of C_2H_2. C_2H_2 is present in noticeable concentrations far below the threshold of soot formation of the C_2H_4-air flames. In the stronger sooting flames, C/O ratio > 0.65, the final C_2H_2 concentration varies only weakly. This has been observed as well for low pressure flames, compare ref. [9.7]. The most evident effect is the decrease of the C_2H_2 concentration between $1 \cdot 10^5$ Pa and $70 \cdot 10^5$ Pa, see Fig. 9.5.

9.3.4 Dependence of Soot Volume Fraction, PAH Concentrations, and Concentration of Aliphatic Species on Pressure in the Pressure Range above $10 \cdot 10^5$ Pa

Figure 9.4 reveals that the influence of pressure on the formation of PAH and soot is quite strong. In Fig. 9.5 the carbon density of several stable species from the postflame zone of C_2H_4-air flames with a cold gas velocity of 6 cm s^{-1} are plotted as a function of pressure. The carbon density of the respective species i, ρ_{ci}, is normalized with the carbon density in the fuel, $\rho_{c\ fuel}$. The figure shows that most of the fuel carbon is converted into the carbon containing water gas compounds CO and CO_2. In the measured pressure range this fraction amounts to about 72 to 87 %. Taking into consideration the differences in the flame temperatures, the carbon fraction in CO and CO_2 shows a slightly increasing tendency with increasing pressure. That means that at $70 \cdot 10^5$ Pa more fuel carbon is converted into CO and CO_2 than at $1 \cdot 10^5$ Pa at otherwise identical conditions.

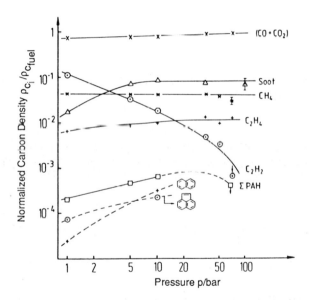

Fig. 9.5. Carbon densities ρ_{ci} of different species normalized with the carbon density of the fuel for C_2H_4-air flames at different pressures; experimental conditions: $v_{cold\ gas} = 6$ cm s^{-1}; 1600 K $\leq T_{10mm} \leq$ 1750 K; C/O = 0.69

In the pressures range above $10 \cdot 10^5$ Pa the main fraction of the non oxidized carbon in the postflame gases of the C_2H_4-air flames appears in soot. The pressure dependence of $\rho_{ci}/\rho_{c\ fuel}$ changes at high pressures and $\rho_{ci}/\rho_{c\ fuel}$ becomes nearly constant between around $10 \cdot 10^5$ Pa and $70 \cdot 10^5$ Pa. In atmospheric C_2H_4-air flames C_2H_2 is the major hydrocarbon species in the postflame gases. For pressures above $5 \cdot 10^5$ Pa the major hydrocarbon compound is CH_4. Interestingly, $\rho_{c\ CH_4}/\rho_{c\ fuel}$ is not dependent on pressure in the covered pressure range. The most striking result is the strong decrease of the C_2H_2 concentration above $10 \cdot 10^5$ Pa. At $70 \cdot 10^5$ Pa the C_2H_2 concentration is below the detection limit.

The PAH concentrations follow those of soot. They are shifted slightly towards higher pressures. In general, the carbon density of the sum of the PAH is more than one order of magnitude lower than that of soot. The PAH show also a p^2-dependence in the pressure range up to $10 \cdot 10^5$ Pa.

9.4 Discussion

From low pressure premixed sooting hydrocarbon flames [9.2, 9.7, 9.10, 9.24] we know that two phases of formation of PAH exist during combustion. The first one is within the main oxidation zone and includes PAH roughly up to coronene, $C_{24}H_{12}$ [9.25]. The second one starts at the beginning of the postflame zone [9.25].

In atmospheric hydrocarbon flames the PAH appear at the end of the main oxidation zone, see Fig. 9.2. One may ask where the PAH in atmospheric and high pressure flames reported here originate. Do they form in the first phase or in the second phase, or is there a basic change in the formation mechanism of PAH with increasing pressure?

Measurements of Wenz [9.24] in a C_3H_8/O_2 low pressure flame indicate that there is a continuous transition from two totally resolved peaks in the PAH concentration profiles to a concentration profile similar to that in Fig. 9.2 when changing the pressure from 5 kPa to 30 kPa. From this we can conclude that the two formation phases of PAH coincide at higher pressures.

Obviously two effects contribute to these results:
 i) The pressure effect reduces the extension of and the distance between the single formation zones of PAH within the flame. The extension of the main oxidation and of the particle formation zone decreases approximately proportional to the increase in pressure.
 ii) The kinetics of PAH formation according to the results of Wenz [9.24] should be pressure dependent. In particular, the reactions for the formation of PAH in the second phase become faster so that the second increase of the PAH concentrations appears at shorter reaction times.

These two effects overlap so that the two PAH forming processes can not be resolved at burning pressures above 30 kPa. Therefore, the PAH profiles in atmospheric flames are possibly a superposition of both formation processes or, more probably, they result only from the second one.

Far in the postflame gases of the flames the concentration of PAH attain quasi constant values. It seems worth-while to look at the PAH concentrations obtained from the experimental invstigation from a thermodynamic point of view. *Stein* et al. [9.26] and recently *Alberty* et al. [9.27] calculated PAH concentrations in equilibrium with C_2H_2 and H_2. These authors argued that soot formation might occur through those PAH which represent the most favourable thermodynamic path. The most stable species on this path are peri-condensed PAH with occasional 5 membered rings at the periphery.

We have calculated the concentrations of some typical PAH for conditions in the burnt gas region of the investigated flames at elevated pressures. Figure 9.6 shows the carbon density of C_6H_6, $C_{10}H_8$ and $C_{14}H_{10}$ in dependence on temperature at $10 \cdot 10^5$ Pa in equilibrium with C_2H_2 and H_2. The employed concentrations of C_2H_2 and H_2 are the experimentally determined values.

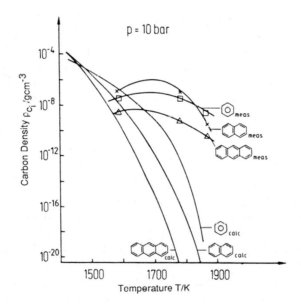

Fig. 9.6. Calculated equilibrium concentrations of several PAH in dependence on temperature at $10 \cdot 10^5$ Pa, C/O ratio $= 0.72$ compared with measured values; the concentrations are given in terms of carbon density ρ_{ci}

The computed concentrations of peri-condensed PAH decrease strongly at the prevailing temperatures. At low temperatures around 1400 to 1500 K the computed carbon densities amount to 10^{-4} to 10^{-6} g cm^{-3}. However, with increasing flame temperature these "equilibrium concentrations" decrease very rapidly to negligible small values. The calculated PAH concentrations pass through a maximum at very low temperatures, not indicated in Fig. 9.6. In contrary to the calculation, the measurements obviously demonstrate that the PAH seem to be pretty stable at higher temperatures. At $10 \cdot 10^5$ Pa PAH could be detected at temperatures above 1900 K in a temperature range where the calculated PAH concentrations are very small. Therefore, in the postflame gases of hydrocarbon flames the formation of PAH is kinetically controlled like the formation of acetylenes.

The experimentally determined carbon densities $\rho_{c \, soot \, \infty}$ and $\rho_{c\Sigma(PAH)}$ exhibit a similar dependence on temperature, pressure, and C/O ratio and it seems worth-while to ask for reasons for this. A short summary of the data at different conditions is given in Table 9.2.

Table 9.2. Dependence of $f_{V\infty}$ and $\rho_{c\Sigma(PAH)}$ on pressure, temperature, and mixture composition

Variable	$p/10^5 Pa$	T/K	
			Dependence on Pressure
$f_{V\infty}$	1–10	>1700	$f_{V\infty} \propto p^2$
$f_{V\infty}$	10–100	>1700	$f_{V\infty} \propto p$
$\rho_{c\Sigma PAH}$	1–10	>1700	$\rho_{c\Sigma PAH} \propto p^2$
$\rho_{c\Sigma PAH}$	10–100	>1700	$\rho_{c\Sigma PAH} \propto p$ or less
			Dependence on C/O ratio
$f_{V\infty}$	1–10	>1700	$f_{V\infty} \propto [C/O-(C/O)_{crit}]^n$
$\rho_{c\Sigma PAH}$	1–10	>1700	$\rho_{c\Sigma PAH} \propto [C/O-(C/O)_{crit}]^n$
			$2.5 \leq n \leq 3.0$
			Dependence on Temperature
$f_{V\infty}$	1–10	1400–1900	bell shaped
$\rho_{c\Sigma PAH}$	1–10	1400–1900	bell shaped

From the PAH concentration profiles given in Fig. 9.2 we can conclude that the PAH appearing in the postflame gases are formed at the end of the oxidation zone. The results also clearly demonstrate that the origin of PAH and soot particles is pretty close together: The PAH are formed shortly before soot appears. Part of the PAH are consumed during the coagulation of the PAH [9.10] and also later on. Although the PAH and soot appear close together, it is still an open question how far the inception of the PAH and soot determines the values of $f_{V\infty}$ and $\rho_{c\Sigma(PAH)}$. On their way through the flame the soot particles experience very reactive surroundings and PAH may be oxidized - in particular at the end of the oxidation zone - or may be formed via gas phase reactions or at the surface of particles as long as those particles are sufficiently young and reactive.

Concentration measurements in the burnt gases of flames just below the sooting limit at different pressures [9.28] show that with increasing pressure the C_2H_2 concentration in the burnt gases decreases relatively to that at normal pressure [9.22] whereas the C_2H_2 maximum concentration seems to increase proportional to p. This means that in the flames at the soot limit more C_2H_2 must be consumed by oxidation e.g. via $C_2H_2 + OH \longrightarrow CH_2CO + H$. It seems very probable that this tendency will continue from non-sooting conditions into sooting conditions so that even the strong consumption of C_2H_2 in high pressure flames, compare Fig. 9.5, is not completely due to the increase in soot volume fraction, but to some extent to oxidation of C_2H_2. This is strongly supported by numerical simulations of high pressure flames performed by [9.29].

9.5 Conclusions

The stable species in the burnt gas region of C_2H_4-air flames burning at pressures up to $70 \cdot 10^5$ Pa and temperatures between 1400 and 1900 K and mixture compositions of $0.52 \leq C/O \leq 0.72$ have been determined. The water gas compounds, the aliphatic compounds C_2H_2, CH_4, C_2H_4 and soot and, in particular, a number of PAH were analyzed.

PAH could be detected in the mass range above coronene up to dicoronyl ($C_{48}H_{22}$). In general, the concentrations of PAH between benzene and dicoronyl decrease with increasing mass. Some exceptions of this tendency are dibenzo[a,b]anthracene $C_{22}H_{14}$, naphthocoronene $C_{30}H_{18}$ and quarterrylene $C_{40}H_{18}$ which are present in higher concentrations.

The PAH appear in atmospheric pressure flames at the end of the oxidation zone and their concentration profiles do not exhibit two formation processes which are obvious from low pressure flames. They are present in the postflame gases together with soot, the aliphatic compounds and the water gas compounds. They are more stable than calculated thermodynamically. Their concentration, therefore, seems to be kinetically controlled. They show strikingly similar dependence on the burning conditions as the final soot volume fraction. This indicates a similar dependence of their formation and consumption on temperature, mixture composition and pressure.

9.6 Acknowledgements

The financial support by the Commission of the European Communities within the frame of the JOULE Programme, by the National Swedish Board for Industrial and Technical Development (NUTEK), and by the Joint Research Committee of European automobile manufacturers (Fiat, Peugeot SA, Renault, Volkswagen and Volvo) within the IDEA Programme is gratefully acknowledged.

References

9.1   E.E. Tompkins, R. Long: "The Flux of Polycyclic Aromatic Hydrocarbons and of Insoluble Material in Pre-Mixed Acetylene-Oxygen flames", in Twelfth Symposium (International) on Combustion (The Combustion Institute, Pittsburg 1969) p. 625

9.2   B.D. Crittenden, R. Long: Combust. Flame **20**, 359 (1973)

9.3   U. Bonne, K.-H. Homann, H.Gg. Wagner: "Carbon Formation in Premixed Flames", in Tenth Symposium (International) on Combustion (The Combustion Institute, Pittsburgh 1965) p. 503

9.4   K.-H. Homann, H.Gg. Wagner: "Some New Aspects of the Mechanism of Carbon Formation in Premixed Flames", in Eleventh Symposium (International) on Combustion (The Combustion Institute, Pittsburgh 1967) p. 371

9.5   J.B. Bittner, J.B. Howard: "Pre-particle chemistry in soot formation", in Particulate Carbon: Formation during Combustion, ed. by D.C. Siegla, W.G. Smith, (Plenum Press, New York 1981) p. 109

9.6   A. D'Alessio, A. Di Lorenzo, A.F. Sarofim, F. Beretta, S. Masi, C. Venitozzi: "Soot Formation in Methane-Oxygen Flames", in Fifteenth Symposium(International) on Combustion (The Combustion Institute, Pittsburgh 1975) p. 1427

9.7   H. Bockhorn, F. Fetting, H.W. Wenz: Ber. Bunsenges. Phys. Chem. **87**, 1067 (1983)

9.8   F.W. Lam, J.B. Howard, J.P. Longwell: "The Behaviour of Polycyclic Aromatic Hydrocarbons during the Early Stages of Soot Formation", in Twenty-Second Symposium (International) on Combustion (The Combustion Institute, Pittsburgh 1988) p. 323; F.W. Lam, J.P. Longwell, J.B. Howard: "The Effect of Ethylene and Benzene Addition on the Formation of Polycyclic Aromatic Hydrocarbons and Soot in a Jet-Stired Plug-Flow Combustor", in Twenty-Third Symposium (International) on Combustion (The Combustion Institute, Pittsburgh 1990) p. 1477

9.9   J.B. Howard: "Carbon Addition and Oxidation Reactions in Heterogenous Combustion and Soot Formation", in Twenty-Third Symposium (International) on Combustion (The Combustion Institute, Pittsburgh 1990) p. 1107

9.10   S. Löffler, K.H. Homann: "Large Ions in Premixed Benzene-Oxygen Flames", in Twenty-Third Symposium (International) on Combustion (The Combustion Institute, Pittsburgh 1990) p. 355

9.11   Polycyclic Aromatic Hydrocarbons and Astrophysics, ed. by A. Léger et al., (D. Reidel Publishing Co., Dordrecht, Boston, Lancester, New York 1987)

9.12   M.S. Akhter, A.R. Chughtai, D.M. Smith: Appl. Spectros. **39**, 154 (1985)

9.13   J.C. Fetzer, W.R. Biggs: Chromatographia **21**, 439 (1986)

9.14   H. Böhm, D. Hesse, H. Jander, B. Lüers, J. Pietscher, H.Gg. Wagner, M. Weiss: "The Influence of Pressure and Temperature on Soot Formation in Premixed Flames", in Twenty-Second Symposium (International) on Combustion (The Combustion Institute, Pittsburgh 1988) p. 403

9.15   A.G. Gaydon, H.G. Wolfhard: Flames: Their Structure, Radiation and Temperature, (Chapman and Hall, New York 1978)

9.16   B.S. Haynes, H. Jander, H.Gg. Wagner: "The Effect of Metal Additives on the Formation of Soot in Premixed Flames", in Seventeenth Symposium (International) on Combustion (The Combustion Institute, Pittsburgh 1979) p. 1365

9.17   S.C. Lee, C.L. Tien: "Optical Constants of Soot in Hydrocarbon Flames", in Eighteenth Symposium (International) on Combustion (The Combustion Institute, Pittsburgh 1981) p. 1159

9.18   G.P. Blümer, M. Zander: Fresenius Z. Anal. Chem. **288**, 277 (1977)

9.19   R.C. Millikan: J. Phys. Chem. **66**, 794 (1962); R.C. Millikan, W.I. Foss: Combust. Flame **6**, 210 (1962)

9.20   J. Flossdorf, H.Gg. Wagner: Z. Phys. Chem. NF **54**, 113 (1967)

9.21   B.S. Haynes, H.Gg. Wagner: Prog. Energy Combust. Sci. **7**, 229 (1981)

9.22   M. Bönig, Chr. Feldermann, H. Jander, B. Lüers, G. Rudolph, H.Gg. Wagner: "Soot Formation in Premixed C_2H_4 Flat Flames at Elevated Pressure", in Twenty-Third Symposium (International) on Combustion (The Combustion Institute, Pittsburgh 1990) p. 1581

9.23   B. Lüers, H. Jander, G. Rudolph: in VDI-Bericht Nr. 922 (VDI-Verlag, Düsseldorf 1991) p. 191

9.24   H.W. Wenz: "Untersuchungen zur Bildung von höhermolekularen Kohlenwasserstoffen in brennerstabilisierten Flammen unterschiedlicher Brennstoffe und Gemischzusammensetzungen"; Ph.D. Thesis, Technische Hochschule Darmstadt (1983)

9.25   H. Jander, H.Gg. Wagner: "Soot Formation in Combustion" - an International Round Table Discussion, (Vandenhoeck and Ruprecht, Göttingen 1990)

9.26   S.E. Stein, D.M. Golden, S.W. Benson: J. Phys. Chem. **81**, 314 (1977)

9.27   R.A. Alberty, A.K. Reif: J. Phys. Chem. Ref. Data **18**, 241 (1988); R.A. Alberty, M.B. Chung, A.K. Reif: J. Phys. Chem. Ref. Data **18**, 77 (1989)

9.28   Th. Heidermann: "Messungen der Abgaszusammensetzung in fetten Kohlenwasserstoff-Luft-Flammen bei hohen Drücken"; Diplom-Thesis, Universität Göttingen (1991)

9.29   H. Böhm, Chr. Feldermann, Th. Heidermann, H. Jander, B. Lüers, H.Gg. Wagner: "Soot Formation in Premixed C_2H_4-Air Flames for Pressures up to 100 bar", in Twenty-Fourth Symposium (International) on Combustion (The Combustion Institute, Pittsburgh 1992) p. 991

9.30   Th. Richter: "Untersuchungen an der unteren Rußgrenze von vorgemischten Ethen-Luft-Flammen"; Ph.D. Thesis, Universität Göttingen (1992)

Discussion

Warnatz: Again I have to pose the question what the difference between PAH and soot is? If you make a difference between PAH and soot you must have a reason for this, or don't you have?

Jander: The PAH in the mass range up to 598 u that we identified at the present are planar PAH. Soot is a three dimensional structure with a H/C ratio $0 \leq H/C \leq 1$. A species with a mass of 600 u is not soot.

Santoro: I have two questions. As you are going up in pressure and looking at the effect of temperature I noticed that the soot curve on that graph on the log-scale looked flatter at $10\cdot10^5$ Pa than it did at $1\cdot10^5$ Pa. There seemes to be less temperature sensitivity to the soot formation process. I would be interested in a comment on that point. A second question concerns your observation that when you went up in pressure, the acetylene concentration remains fairly constant while soot goes up dramatically. We normally think of acetylene as a key surface growth species and, in fact, it is always present in excess. Do you have any evidence as you go up in pressure that the surface growth species are changing or do you think the constant acetylene concentration is just because of the production and destruction terms are in balance?

Jander: Indeed, the experimental results of $f_{V\infty}$ vs. temperature at different pressures show that with increasing pressure the $f_{V\infty}$-curves show a wider maximum which means that they are less sensitive to temperature. Therefore at $10\cdot10^5$ Pa or higher pressures a much wider temperature regime exists for considerable soot formation than at $1\cdot10^5$ Pa.

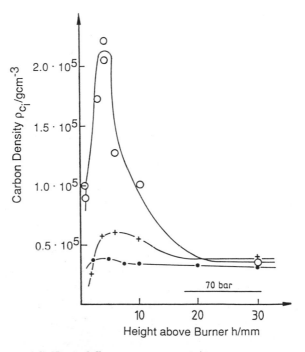

Fig. 9.7. Profiles of C_2H_2 at different pressures; C/O ratio 0.68; $v_{cold\ gas}$: 6 (cm/s). ● $1\cdot10^5$Pa, $+$ $3\cdot10^5$Pa, ○ $10\cdot10^5$Pa

What are the reasons for this favoured soot formation?
The analysis of C_2H_2-concentration profiles did show that the carbon den-

sity $\rho_{c\ C_2H_2}$ contained in C_2H_2 in the burned gases is nearly constant, only in the pressure range up to $10 \cdot 10^5$ Pa (see fig. 9.7). Above $10 \cdot 10^5$ Pa it decreases and at $70 \cdot 10^5$ Pa it lies below our detection limit (< 5 ppm).
This is different from the maximum C_2H_2 carbon density near the main reaction zone of the flame (see fig. 9.7).
For a given C/O ratio the total carbon density rises proportional with increasing pressure. Model computations of C_2H_4-air flames up to higher pressures indicate that the maximum C_2H_2 carbon density rises even stronger with pressure. A fraction of C_2H_2 is consumed by oxidation while the remaining C_2H_2 can be used for soot formation. At high pressure the C_2H_2 is consumed nearly completely, this limits the main soot growth process. The experiments indicate that in these C_2H_4-air flames C_2H_2 stays the main surface growth species at high pressures.
Parallel to the rise of the C_2H_2 maximum with pressure is the formation of soot precursors, soot inception increases rapidly with increasing pressures, with obvious consequences.

D'Alessio: Did you look for this tar like material that we have found in ethylene flames in addition to the PAH. This tar like material it is not soot and it is not PAH.

Jander: Yes, we have observed this tar like material which looks brown and is not soot. At the moment we have no quantitative information about the composition of this material or how large the concentration of individual species contained in this material is besides the data of Flossdorf [9.20] and Long [9.1]. We have analysed PAH up to masses of about 600 u and we are going to extend the mass range towords higher masses because now we know how to do it. The mole fractions of coronene or naphthocoronene are about 10^{-6}. From this we conclude that species with higher masses have mole fractions much smaller than this.

Lepperhoff: I was very impressed by your results which cover the pressure range from $10 \cdot 10^5$ Pa to $70 \cdot 10^5$ Pa and which in my opinion are very important to engineering applications. There is a shift in the temperature for the maximum PAH concentration relative to the temperature for the maximum soot volume fraction with increasing pressure. At $1 \cdot 10^5$ Pa this temperature is about 1600 K for the maximum PAH conncentration and about 1700 K for the maximum soot volume fraction. At $10 \cdot 10^5$ Pa the temperature for the maximum PAH concentration is about 1700 K whereas the temperature for the maximum soot volume fraction is not effected by the pressure change. Is there any reason for this?

Jander: At present I have no explanation for this.

Homann: To my knowledge the detection of nitrogen containing PAH from a fuel that does not contain nitrogen is a new result. Do you have an idea how these nitrogen containing PAH are formed? Is NO involved in the formation of these species?

Jander: Yes I think so. We have hydrocarbon-air flames and, therefore, hydrocarbon radicals which can break the N-N bond, so we have the nitrogen in there and we have NO formation via the prompt NO-channel that involves also NCO. NCO is a very reactive species and via NCO nitrogen is easily incorporated into aromatic compounds.

Haynes: Just at this moment before you move on I want to remark on a couple of things. I think the nitrogen species in the gas phase is most likely to be hydrogen cyanide and this opens a very interesting question about the growth species. If we could look at a flame which contained a lot of hydrogen cyanide and see what happens to these aza-polycyclics this could give us some insight into what is happening because it is much more difficult for the hydrogen cyanide with its triple bond to grow continuously in the same way as the acetylene species can. This would actually be a distinction which we might be able to use to our advantage. So, I too was very interested to see that new result on the aza-compound.

Colket: I'll start up by saying that I agree very much with those last comments. I think it is a very interesting area especially with respect to the NO_x control problems, NO_x soot interactions, and with respect to the problem to simultaneously control NO_x and soot emission which is important for all combustion systems. I have a question on this potential change in mechanism for soot formation from second to first order with pressure as you increase pressure. It looks like a second order mechanism below $10 \cdot 10^5$ Pa and above it was more like first order. Could it be just the fact that it is still an acetylene addition process rather than methane addition. It looks very tempting to consider that and what is happening above $10 \cdot 10^5$ Pa where you have a limitation of how much acetylene is formed.

Wagner: You mentioned that acetylene addition is essential for soot formation. At lower pressures soot formation ceases not because of a lack of a growth species that could be added like C_2H_2 but due to the fact that the particles lose their reactivity.

Colket: At what pressures?

Wagner: At pressures from $1 \cdot 10^5$ Pa to $10 \cdot 10^5$ Pa. When going to higher pressure the situation changes. As Dr. Jander showed in one of the slides that at $10 \cdot 10^5$ Pa a reasonable amount of acetylene and similar species are being consumed in a very short time. That corresponds approximately to a one to one relation to the increase of the soot mass. However, we are still in the situation where the soot growth ceases with the reactivity of the soot particles. At higher pressure soot is growing so rapidly that all the acetylene that is formed as an intermediate is consumed before the reactivity of the particles decreases. Then soot formation levels off because of a lack of a growth species which can be added in the time available. The addition of methane or its direct reaction products to the soot particles is a much slower process. The time scale for this process is some orders of magnitude higher. This is the reason why acetylene and the polycyclic aromatics disappear when going to high pressures. You know, in high pressure premixed flames the soot cleans up the burnt gas and you end up with the water gas components which do not change very much with pressure and with soot and with methane and the rest is negligible. This is completely different from pyrolysis or diffusion flames. In diffusion flames as in the diesel engine the situation is much more complicated because you have a large range of C/O ratios. In premixed flames we have one single C/O ratio.

This phenomenology is also consistent with shock tube pyrolysis experiments where the reaction times are shorter than in our experiments described here. The particles, after a short reaction time, do not grow and they do not coagulate any longer. What you observe at high pressures is similar to what Professor Bockhorn observes when he looks at particles with metal ions within them. If particles collide they either stick or they do not. If the particles are tempered without growing by surface growth reactions as is the case for particles with electric charge they do not like to stick together. Coagulating these big molecules or particles requires some glue to keep them together as I remarked to Professor Homann this morning. That may be a mechanistic problem or it may have some other chemical origin. At high pressure the soot particles stay fairly small and all growth processes are determined by the particle growth itself and that makes little difference whether you start from aromatic or other hydrocarbons.

Frenklach: I think I might have a possible explanation for the difference in temperatures for the maximum in the PAH concentrations and for soot which you noticed. That was interesting to me also. One possibilty is that it is diffusion. In our laminar premixed flame simulations we noticed diffusion of PAH, even if it is of small significance. While PAH condense to three dimensional soot particles - I will reflect in my talk on the difference between soot and PAH - they are still diffusing, they are reacting and diffusing; until they grow into soot. At high pressure the diffusion length is much smaller and so the temperature maxima appear closer to each other.

Glassman: It seems to me that whenever we go to high pressures and experimental configurations like in your flat flame premixed burner we have to be concerned about - and this is a little reflection on Professor Frenklach's statement - how structure of the flame front changes itself and how close the flame gets to the burner and all those other extraneous effects. The pressure effect on a diffusion flame and the pressure effect on a premixed flame is quite different and the phenomena could be very different. I want you to comment on that. Second, you showed one slide where you showed the variation of PAH as a function of temperature and the lowest measuring point you had was about 1600 K. Did you try to make any measurements lower than 1600 K or do you find nothing at temperatures less than 1600 K?

Jander: Indeed the primary effect on premixed and diffusive flames is different. The transport coefficients k and D change proportional to p^{-1}. This changes the flame structure in diffusive flames. For premixed laminar flames the reaction zone thickness Δ and the mean reaction time τ decrease approximately proportional to p^{-1}. The distance to the burner and the length of the soot induction zone changes therefore also proportional to p^{-1}. It is in fact the main problem of those experiments to have burners which allow to burning of good flat flames.

Concerning your second question: we measured at normal pressure down to 1400 K at C/O ratios between 0.72–1.2. As you know from the measurements about threshold of soot formation vs temperature at $1 \cdot 10^5$ Pa, in C_2H_2-air flames soot formation ceases at temperatures < 1400 K. So we tried to measure the main stable postflame products at these low flame temperatures.

We started in the non sooting flame at $T = 1390$ K, C/O $= 0.8$, $p = 1 \cdot 10^5$ Pa and went on by increasing the flame temperature up to sooting conditions and at 1600 K to a heavily sooting flame. For details see ref. [9.30].

Homann: Did you mean that if the soot particles stop growing by surface growth because the flame runs out of acetylene they also lose their activitiy for coagulation?

Jander: Yes.

Homann : I think I have another example for that.

Wagner: The first thing that you observe is a number density which is proportional to the soot volume fraction and the particles are approximately of the same size. Inspecting these particles they do really have the appearance of old and tempered soot particles. At these temperatures and under these conditions they have no reason to stick together and there is no more glue which can be put in between the particles which fills the holes and tries to keep particles together.

Homann : I think that is a very important observation for the people who are concerned about surface growth of soot particles.

Detailed Mechanism and Modeling of Soot Particle Formation

Michael Frenklach, Hai Wang

Department of Material Science and Engineering,
Pennsylvania State University,
University Park, PA 16802, USA

Abstract: A detailed chemical reaction model for the growth of polycyclic aromatic hydrocarbons and soot particle nucleation and growth is presented. The model begins with fuel pyrolysis, followed by the formation of polycyclic aromatic hydrocarbons, their "planar" growth and coagulation into spherical particles, and finally, surface growth and oxidation of the particles. The surface processes are described in terms of elementary chemical reactions of surface active sites. The method of moments is used to express the mathematical formalism of the undergoing chemical and physical processes. A new submodel is presented which is capable of calculating the optical properties of an arbitrary ensemble of soot particles. Computer simulations with this model are in quantitative agreement with experimental results from several laminar premixed hydrocarbon flames. The model predicts the classical picture of soot particle inception and the classical description of soot particle structure.

10.1 Introduction: General Mechanistic Concepts

For a long time, *Homann* et al. [10.1] have advocated the radical nature of chemical reactions responsible for the formation of soot. Calcote argued against it, suggesting that reactions of neutral species cannot possibly explain the fast growth rates, whereas those of ions can [10.2]. Recent advances in detailed kinetic modeling, ours [10.3] and others [10.4], demonstrate that radical reactions are rapid enough to account for experimentally observed rates of soot formation in flames.

Many suggestions have been made as to the nature of possible soot precursors, e.g., acetylene, polyacetylenes, allene, butadiene, polycyclic aromatic hydrocarbons (PAH), etc.; these proposals were reviewed by *Calcote* [10.2]. There is growing evidence – experimental [10.5], thermodynamic [10.6] and kinetic [10.7] – that the formation of soot proceeds via

PAH. *Stehling* et al. [10.8] suggested that the molecular growth involves reactions between aromatics and acetylenic species. *Bittner* et al. [10.9] further supported this proposal, suggesting several possible chemical reactions. Analysis of the bell-like dependence of soot yields on reaction temperature observed in shock-tube studies [10.10-11] indicated that the overall kinetics is consistent with the critical role of aromatic-acetylenic interactions. Analysis of the shock-tube data obtained with many different hydrocarbons led us to propose that the reaction sequence for the build-up of PAH should be the same in all cases; what makes the difference starting with different fuels is the initiation of this sequence [10.10]. The results of subsequent modeling studies have been found to support this conceptual view.

Recently, we developed a detailed chemical kinetic model of soot particle nucleation and growth in laminar premixed flames [10.3]. The model predicts the classical picture of soot particle inception and is found to be in accord with the classical description of soot particle structure. The objective here is to review the principal elements of the model and to summarize its mathematical and numerical details. A new numerical extension to the model for calculation of the optical properties of an arbitrary ensemble of soot particles is also presented. It is shown that in this way, the model predictions can be compared directly to the experimental data on soot particle laser scattering and absorption collected in flame studies, thus excluding the current ambiguity associated with interpretation of the optical data.

10.2 Overall Structure of the Model

The overall model of soot formation can be thought of as consisting of four major processes: *initial PAH formation*, which includes the formation of the first aromatic ring in an aliphatic system; "planar" PAH growth, comprised of replicating-type growth; *particle nucleation*, consisting of coalescence of PAH into three-dimensional clusters; and *particle growth* by coagulation and surface reactions of the forming clusters and particles. Our primary attention in this discussion is on the last three processes, although some comments are pertinent concerning the formation of the first aromatic ring.

10.3 Formation of the First Aromatic Ring

The formation of the first aromatic ring in flames of non-aromatic fuels begins usually with vinyl addition to acetylene. At high temperatures, it forms vinylacetylene followed by acetylene addition to n-C_4H_3 radical formed by the H-abstraction from the vinylacetylene (Fig. 10.1). At low temperatures, the addition of acetylene to vinyl results in n-C_4H_5, which upon addition of acetylene produces benzene. Benzene and phenyl are converted to one another by the H-abstraction reaction and its reverse. (In low-pressure high-temperature flames other reaction channels may become important; this will be addressed in a later publication.)

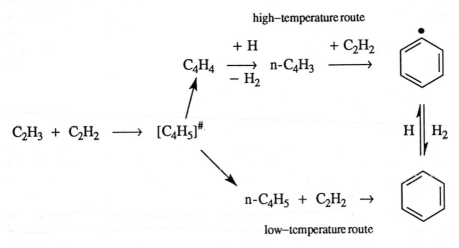

Fig. 10.1. The two reaction pathways for the formation of the first aromatic ring

In a recent review article on chemical kinetics and combustion modeling, *Miller* et al. [10.12] suggested that the above cyclization reactions cannot be responsible for the formation of the first aromatic ring because the concentrations of n-C_4H_3 and n-C_4H_5 radicals should be low since reactions

$$H + n\text{-}C_4H_3 \rightleftharpoons H + i\text{-}C_4H_3 \qquad (10.1)$$
$$H + n\text{-}C_4H_5 \rightleftharpoons H + i\text{-}C_4H_5 \qquad (10.2)$$

deplete the concentrations of the n-isomers required for the cyclizations. Reactions (10.1) and (10.2) were not included in our model initially, because the computational results indicated that the n- and i-isomers are already equilibrated by several other reactions in the system. Nonetheless, to test the Miller's suggestion, we performed additional simulations of the three laminar premixed flames we analyzed previously [10.13]. Reactions (10.1) and (10.2) were now included in the simulations assuming rate coefficients 1×10^{14} mol^{-1} cm^3 s^{-1} for the exothermic directions. The results of these simulations for all the three flames tested in *Frenklach* et al. [10.13] indicated that the inclusion of reactions (10.1) and (10.2) – even with upper-limit rate coefficient values – does not make a difference on the computed profile of benzene.

As an alternative, *Miller* et al. [10.12] suggested that benzene is formed by combination of propargyl radicals producing benzene or phenyl. A similar proposal was made by *Alkemade* et al. [10.14], *Kern* et al. [10.15] and *Stein* et al. [10.16]. However, the results of flame simulations with such reaction included using the rate coefficient of $5 \times 10^{12} \, \text{mol}^{-1} \, \text{cm}^3 \, \text{s}^{-1}$ suggested by *Stein* et al. [10.16] indicated by *Frenklach* [10.13] that the inclusion of this additional cyclization channel does not always increase the production rate of benzene. Although the C_3H_3 recombination has long been implicated in the formation of higher molecular weight hydrocarbons [10.17] including benzene [10.11], the precise mechanism and the rate coefficient of this reaction remain unclear. For instance, recent QRRK calculations of *Westmoreland* et al. [10.18] show that the main products of the self-reaction of C_3H_3 at flame conditions should be linear C_6H_6 species and not benzene.

10.4 Growth of Aromatics Beyond the First Ring

Once formed, aromatic rings grow by a sequential two-step process: H-abstraction which activates the aromatic molecules, and acetylene addition which propagates molecular growth and cyclization of PAH (Fig. 10.2). We will refer to this H-abstraction–C_2H_2-addition reaction sequence as HACA. Starting with an aromatic fuel, a "direct combination" of the intact aromatic rings becomes important. For example, in the case of high-temperature pyrolysis of benzene the reactions shown in Fig. 10.3 were found to dominate the initial stages of PAH growth [10.19]. However, as the reaction progresses, the initial benzene molecules decompose forming acetylene. As the concentration of acetylene approaches that of benzene, which occurs shortly after the initial reaction period, the PAH growth switches to the HACA mechanism discussed above. In other words, the reaction system *relaxes* to the acetylene-addition pathway. The relaxation is faster in oxidation [10.20] as compared to pyrolysis and in mixtures of hydrocarbons [10.21] as compared to individual fuels.

Generalizing these results, we conclude the following. Although in pyrolysis of hydrocarbon fuels, reactions of aromatic rings with species other than acetylene (like vinyl in the pyrolysis of 1,3-butadiene [10.19]) may be important initially, as the pyrolysis progresses, the aromatic growth becomes dominated by the HACA reaction sequence. On these grounds, the proposed growth of aromatic rings by the addition of C_3H_3 [10.16], for instance, is unlikely to take place in flames, as the concentration of acetylene is much larger than that of C_3H_x species within and after the main reaction zone. In support of this conclusion, C_3-substituted aromatics have not been detected experimentally [10.22].

Some of the acetylene addition reactions in the HACA sequence form particularly stable aromatic molecules, like pyrene, coronene, etc. The change of Gibb's free energy in these reactions is so large that they become practically irreversible. This, in turn, has an effect of "pulling" the reaction sequence forward, towards formation of larger PAH molecules. Other acetylene addition steps are highly reversible, in which case the rate of the

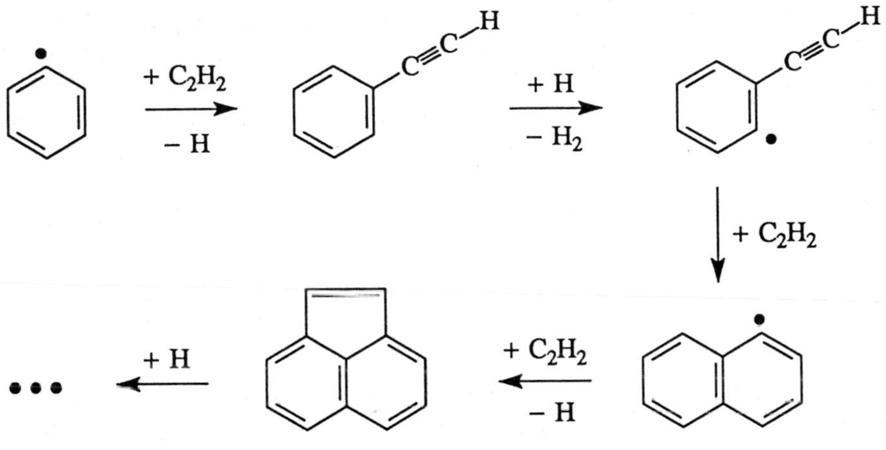

Fig. 10.2. H-abstraction–C_2H_2-addition reaction pathway of PAH growth

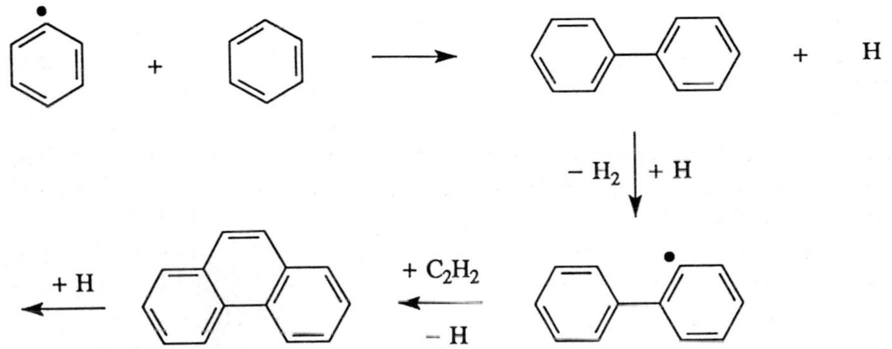

Fig. 10.3. PAH growth initiated by aromatic "combination"

forward direction is nearly balanced by the rate of its reverse. These steps with tightly balanced reaction fluxes create a thermodynamic barrier to PAH growth. It is this thermodynamic "resistance" which is responsible for the appearance of most stable, condensed aromatic structures, as opposite to open shell carbon clusters leading to fullerenes [10.23]. For instance, due to the relatively small differences in reaction enthalpies, the reaction flux from phenanthrene to benzo[ghi]perylene shown on the left of Fig. 10.4 was computed to be faster than the one on the right hand side of Fig. 10.4 by an order of magnitude [10.7].

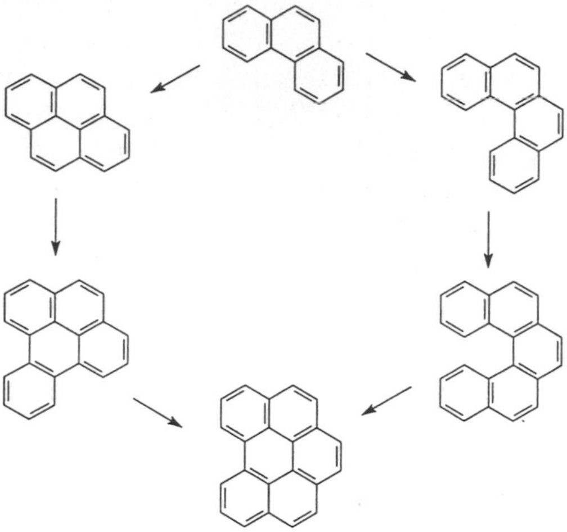

Fig. 10.4. Comparison of two pathways of PAH growth

The main kinetic features of PAH growth after a certain PAH size, i_0, are revealed by considering an analytical solution with the smallest set of reactions that represent the principal elements of the HACA sequence [10.24]. This minimal reaction set is given as

$$A_i + H \rightleftharpoons A_i \bullet + H_2 \tag{10.3}$$

$$A_i \bullet + C_2H_2 \rightleftharpoons A_iC_2H_2\bullet \tag{10.4}$$

$$A_iC_2H_2\bullet + C_2H_2 \to A_{i+1} + H \tag{10.5}$$

where A_i denotes an aromatic molecule containing i fused aromatic rings ($i = i_0, i_0 + 1, \ldots, \infty$), $A_i\bullet$ is an aromatic radical formed by the abstraction of an H-atom from A_i, and $A_iC_2H_2\bullet$ is a radical formed by the addition of C_2H_2 to $A_i\bullet$. It is assumed that reactions (10.3) and (10.4) are reversible and reaction step (10.5) is irreversible. The rate of PAH mass accumulation is proportional to

$$\frac{K_3 \frac{[H]}{[H_2]}}{\frac{1}{K_4 k_4 [C_2H_2]^2} + \frac{1}{k_5[C_2H_2]} + \frac{1}{k_{-3}[H_2]}} \int r_0 \, dt \tag{10.6}$$

where t is the reaction time, r_0 the rate of irreversible formation of A_{i_0} by initiation reactions, k_j the rate coefficient of the jth reaction, and $K_j = \frac{k_j}{k_{-j}}$ the equilibrium constant of the jth reaction.

In (10.6), term $\int r_0 dt$ represents the contribution of the initiation reactions, i.e., those leading to the formation of the first few PAH. Term $K_3 \frac{[H]}{[H_2]}$ accounts for the "equilibrium position" or, in more rigorous terms, reaction affinity of (10.3). It represents the superequilibrium of H-atoms – the first kinetic factor responsible for PAH growth. Term $k_5[C_2H_2]$ is the effective rate constant of the irreversible addition of acetylene, reaction (10.5), forming particular stable PAH molecules – the second kinetic factor responsible for PAH growth. Terms $k_4[C_2H_2]$ and $K_4[C_2H_2]$ specify kinetic and thermodynamic factors, respectively, of the reversible addition of acetylene, reaction (10.4); the latter expresses the thermodynamic resistance to PAH growth. And finally, term $k_{-3}[H_2]$ accounts for the effective rate constant of the H-abstraction, reaction (10.3); to illustrate it, consider the limit of $k_{-3}[H_2] \to 0$ under which condition the ratio in (10.6) reduces to $k_3[H]$.

At high pressures, reaction

$$A_i \bullet + H \to A_i \tag{10.7}$$

should contribute to the overall balance of the $A_i \bullet$ radical, and at a high concentration of hydrogen atoms, such that $k_7[H] \gg k_{-3}[H_2]$, we obtain an interesting limit of the PAH growth rate being independent of the H concentration.

An additional kinetic factor, $k_8[O_2]$, where k_8 is the rate coefficient of oxidation reaction

$$A_i \bullet + O_2 \to \text{products}, \tag{10.8}$$

is introduced in oxidative environments. Although many oxidative reaction channels are possible, it appeared, as a result of our kinetic analysis of shock-tube oxidation [10.20] and flame [10.3,10.25] environments, that PAH removal by oxidation occurs predominantly via molecular oxygen attack on aromatic radicals. Also, due to reactions of O_2 with C_2H_3 and C_4H_3 radicals (and of OH with C_2H_2 etc.), the concentrations of these critical intermediates are decreased in oxidative environments, which in turn reduces the formation rate of the first aromatic ring. At the same time, the presence of O_2 in the mixture has a promoting effect on aromatics formation because of the accelerated chain branching leading to enhanced fuel pyrolysis and thus increased production of critical intermediates and hydrogen atoms. The balance of all of these factors determines the net effect of oxygen addition.

10.5 Linear Lumping of PAH Growth

The growth of PAH beyond a prescribed size, acepyrene in this study, was described by the technique of linear (or chemical) lumping [10.26-29]. This is a mathematically rigorous method, a method of moments, in which the kinetics of an infinite sequence of polymerization-type reactions of PAH growth is described by a small number of differential equations developed for the moments of the PAH distribution function.

The specific reactions and associated rate coefficients for the replicating reaction sequence are given in Table 10.1. The last reaction in this table

Table 10.1. Reaction mechanism of PAH growth — linear lumping

$$k = AT^n \, e^{-E/RT}$$

no.	reaction	A (cm^3/mol s)	n	E (kJ/mol)	references
L1	$A_{i-1}\bullet + C_2H_2 \rightarrow A_i + H$	4.0(13)		42.3	10.30
L1a	$A_i + H \rightleftharpoons A_i\bullet + H_2$	2.5(14)		66.9	10.31
L1b	$A_i + OH \rightleftharpoons A_i\bullet + H_2O$	2.1(13)		19.1	10.32
L1c	$A_i\bullet + H \rightleftharpoons A_i$	4.3(15)	−0.78	5.5	90 torr, 10.33
		4.2(11)	0.48	−0.3	760 torr, 10.33
L1d	$A_i\bullet + O_2 \rightarrow$ products	2.0(12)		31.3	10.34
L2	$A_i\bullet + C_2H_2 \rightleftharpoons A_i C_2H + H$	4.0(13)		42.3	10.30
L2a	$A_i C_2H + H \rightleftharpoons A_iC_2H\bullet + H_2$	2.5(14)		66.9	10.31
L2b	$A_i C_2H + OH \rightleftharpoons A_iC_2H\bullet + H_2O$	2.1(13)		19.1	10.32
L2c	$A_iC_2H\bullet + H \rightleftharpoons A_iC_2H$	4.3(15)	−0.78	5.5	90 torr, 10.33
		4.2(11)	0.48	−0.3	760 torr, 10.33
L2d	$A_iC_2H\bullet + O_2 \rightarrow$ products	2.0(12)		31.3	10.34
L3	$A_iC_2H\bullet + C_2H_2 \rightleftharpoons A_{i+1}\bullet$	4.0(13)		42.3	10.30
L3a	$A_{i+1} + H \rightleftharpoons A_{i+1}\bullet + H_2$	2.5(14)		66.9	10.31
L3b	$A_{i+1} + OH \rightleftharpoons A_{i+1}\bullet + H_2O$	2.1(13)		19.1	10.32
L3c	$A_{i+1}\bullet + H \rightleftharpoons A_{i+1}$	4.3(15)	−0.78	5.5	90 torr, 10.33
		4.2(11)	0.48	−0.3	760 torr, 10.33
L3d	$A_{i+1}\bullet + O_2 \rightarrow A_i +$ products	2.0(12)		31.3	10.34
L4	$A_{i+1}\bullet + C_2H_2 \rightleftharpoons A_{i+2} + H$	4.0(13)		42.3	10.30
L4a	$A_{i+2} + H \rightleftharpoons A_{i+2}\bullet + H_2$	2.5(14)		66.9	10.31
L4b	$A_{i+2} + OH \rightleftharpoons A_{i+2}\bullet + H_2O$	2.1(13)		19.1	10.32
L4c	$A_{i+2}\bullet + H \rightleftharpoons A_{i+2}$	4.3(15)	−0.78	5.5	90 torr, 10.33
		4.2(11)	0.48	−0.3	760 torr, 10.33
L4d	$A_{i+2}\bullet + O_2 \rightarrow A_{i+1}+$ products	2.0(12)		31.3	10.34
L5	$A_{i+2}\bullet + C_2H_2 \rightarrow A_i + H$	4.0(13)		42.3	10.30

represents the lumping reaction step. The numbering of the reactions reflects the structure of the replicating sequence – the reactions marked with whole (integer) numbers are those which increase the molecular size (i.e., those that add carbon atoms). The rate coefficients for these reactions are taken from the reference sources listed in Table 10.1 without change. The fall-off for the H-atom combinations with aromatic radicals (reactions LIc, I = 1,2,3,4) are calculated using the code of *Gilbert* [10.35] with the parameters of *Rao* et al. [10.33].

In the method of linear lumping, the differential equations describing the evolution of individual PAH concentrations are regrouped into differential equations that describe the evolution of the moments of the PAH distribution. We define these moments in two ways. First are the *concentration* moments,

$$M_r^{\text{PAH}} = \sum_{i=i_0}^{\infty} m_i^r N_i^{\text{PAH}}, \tag{10.9}$$

where M_r^{PAH} is the rth concentration moment of the distribution of the PAH species concentrations, i_0 the initial PAH size from which the lumping begins and the summation is understood as taken over all PAH species above this size, and m_i and N_i^{PAH} the mass and number concentration, respectively, of the PAH species of size class i. For the problem of PAH growth addressed here, we chose to represent the PAH mass by the number of carbon atoms, ignoring the hydrogen atoms. This assumption simplifies the lumping equations without essentially affecting the results.

The second type of moments are the *size* rth moments,

$$\mu_r^{\text{PAH}} = \frac{M_r^{\text{PAH}}}{M_0^{\text{PAH}}}, \tag{10.10}$$

which are in essence the normalized concentration moments, since $M_0^{\text{PAH}} = \sum_{i=i_0}^{\infty} N_i^{\text{PAH}}$ and hence $\mu_0^{\text{PAH}} = 1$. The meaning of the two types of moments is revealed by comparing the first moments: the first concentration moment gives the total PAH *mass*, i.e. the total number of carbon atoms accumulated in the PAH species (per unit volume), while the first size moment gives the average PAH size, i.e. the average number of carbon atoms per PAH.

The equations for the concentration moments are summarized below

$$\frac{dM_0^{\text{PAH}}}{dt} - r_0 \tag{10.11 - 0}$$

$$\frac{dM_1^{\text{PAH}}}{dt} = m_0 r_0 + \sum_{\ell=1}^{\ell_c} \Delta m_\ell R_\ell^{[0]} \tag{10.11 - 1}$$

$$\frac{dM_2^{\text{PAH}}}{dt} = m_0^2 r_0 + \sum_{\ell-1}^{\ell_c} \Delta^{[2]} m_\ell R_\ell^{[0]} + 2\Delta_c m \sum_{\ell=1}^{\ell_c} \Delta m_\ell R_\ell^{[1]} \tag{10.11 - 2}$$

$$\cdots$$

$$\frac{dM_r^{\text{PAH}}}{dt} = m_0^r r_0 + \sum_{j=0}^{r-1} \binom{r}{j} (\Delta_c m)^j \sum_{\ell=1}^{\ell_c} \Delta^{[r-j]} m_\ell R_\ell^{[j]} \tag{10.11 - r}$$

where

$$\Delta^{[p]} m_\ell = (m_\ell - \Delta_c m)^p - (m_{\ell-1} - \Delta_c m)^p, \qquad \ell = 1, 2, \dots, \ell_c,$$

$$\Delta_c m = \sum_{\ell=1}^{\ell_c} \Delta m_\ell = \sum_{\ell=1}^{\ell_c} (m_\ell - m_{\ell-1}) = m_{i+\ell_c} - m_i.$$

In these equations, r_0 is the rate of formation of the first lumped species (the rate of reaction L0 in Table 10.1), m_0 is the number of carbon atoms contained in the first species of the lumping reaction sequence (A_i in Table 10.1), ℓ counts the mass-addition reactions of the replicating sequence with the total number ℓ_c of such reactions, m_ℓ is the number of carbon atoms of the reactant PAH in the ℓ's mass-addition reaction, $\Delta m_\ell = \Delta^{[0]} m_\ell$ is the number of carbon atoms added in the ℓ's mass-addition reaction , $\Delta_c m$ is the number of carbon atoms added during the entire replicating cycle, and $R_\ell^{[j]}$ is the net flux of j-lumped reaction ℓ.

The latter are defined in the following manner. Solution of the differential equations describing the reaction set in Table 10.1 gives directly rate terms $R_\ell^{[0]}$. For instance, in the case of the lumped reaction (L1) in Table 10.1,

$$R_1^{[0]} = k_{L1}[C_2H_2] \sum_{i=0}^{\infty}[A_{i_0+3i}\bullet] - k_{-L1}[H] \sum_{i=0}^{\infty}[A_{i_0+3i}C_2H\bullet]. \qquad (10.12)$$

The terms $\sum_{i=0}^{\infty}[A_{i_0+3i}\bullet]$ and $\sum_{i=0}^{\infty}[A_{i_0+3i}C_2H\bullet]$ are the lumped species concentrations, or the zeroth concentration moments of the distribution functions of species $A_i\bullet$ and $A_iC_2H\bullet$, respectively. Thus, $R_\ell^{[0]}$ can be defined as *zeroth-moment rates*. The summation defining an individual lumped species is taken over all chemically similar replicating species above size i_0; for example, $\sum_{i=0}^{\infty}[A_{i_0+3i}\bullet] = [A_{i_0}\bullet] + [A_{i_0+3}\bullet] + [A_{i_0+6}\bullet] + \dots$.

The j-moment rates, $R_\ell^{[j]}$ are obtained by solving equations similar to those that determine the zeroth-moment rates, but in which the zeroth concentrations moments are replaced by the moments

$$S_{\{PAH_{i'}\}}^j = \sum_{i=0}^{\infty}(i+1)^j \, [PAH]_{i'+3i}, \qquad (10.13)$$

where i' identifies a lumped PAH within the (first) replicating cycle. In addition, the term $k_{L4}[C_2H_2]S_{\{A_{i+2}\bullet\}}^j$ in the rate equation for the very first lumped species in the replicating reaction sequence, A_i in Table 10.1, is replaced by

$$k_{L4}[C_2H_2] \sum_{k=0}^{j} \binom{j}{k} S_{\{A_{i+2}\bullet\}}^k.$$

An example of (10.13) is

$$S_{\{A_i\bullet\}}^j = \sum_{i=0}^{\infty}(i+1)^j \, [A_{i_0+3i}\bullet]$$

$$= [A_{i_0}\bullet] + 2^j[A_{i_0+3}\bullet] + 3^j[A_{i_0+6}\bullet] + \dots$$

for the case of species $A_i\bullet$.

The present model uses six concentration moments (defining 5 PAH-size moments) with $m_0 = 18$, $\ell_c = 4$, $\Delta m_\ell = 2$ for $\ell = 1, 2, 3, 4$, and $\Delta_c m = 2 \times 4 = 8$. The changes in the concentrations of the "monomer" species – H, H_2, OH, H_2O and C_2H_2 – occurring due to the lumping reactions were properly accounted for. When combined with the equations describing soot particle formation and growth, as done in the present work, equations (10.11) for the PAH moments are subtracted additional terms to account for the disappearance of PAH due to PAH coagulation and PAH condensation on the particle surface. This will be described in the following section.

10.6 Nucleation, Growth and Oxidation of Soot Particles

The formation and evolution of soot particles was mathematically described using the method of moments [10.36]. The soot particle moments were defined, similarly to the PAH moments [cf. (10.9) and (10.10)], as

$$M_r^{\text{soot}} = \sum_{i=1}^{\infty} m_i^r N_i^{\text{soot}} \tag{10.14}$$

and

$$\mu_r^{\text{soot}} = \frac{M_r^{\text{soot}}}{M_0^{\text{soot}}}, \tag{10.15}$$

where M_r^{soot} and μ_r^{soot} are the rth concentration and size moments, respectively, of the soot particle distribution, and m_i and N_i^{soot} the mass and number density, respectively, of the soot particles of size class i. As in the case of the PAH moments, and to be consistent with them, the soot particle mass was represented by the number of carbon atoms.

The specific equations for the concentration moments, which account for nucleation, coagulation, surface growth and oxidation of soot particles, are summarized below

$$\frac{\mathrm{d}M_0^{\text{soot}}}{\mathrm{d}t} = R_0 - G_0 \tag{10.16 - 0}$$

$$\frac{\mathrm{d}M_1^{\text{soot}}}{\mathrm{d}t} = R_1 + W_1 \tag{10.16 - 1}$$

$$\frac{\mathrm{d}M_2^{\text{soot}}}{\mathrm{d}t} = R_2 + G_2 + W_2 \tag{10.16 - 2}$$

$$\cdots$$

$$\frac{\mathrm{d}M_r^{\text{soot}}}{\mathrm{d}t} = R_r + G_r + W_r \tag{10.16 - r}$$

where R, G and W are the nucleation, coagulation and surface rate terms, respectively.

The formation of soot particles was assumed to be initiated by PAH coagulation, i.e., by a coalescence of two PAH species into a dimer. The formation rate of all possible dimers is given by

$$R_r = \frac{1}{2} \sum_{i=i_0}^{\infty} \sum_{j=i_0}^{\infty} \beta_{i,j}^{\mathrm{PAH}} (m_i + m_j)^r N_i^{\mathrm{PAH}} N_j^{\mathrm{PAH}}, \qquad r = 0, 1, 2, \ldots, \quad (10.17)$$

where

$$\beta_{i,j}^{\mathrm{PAH}} = 2.2 \sqrt{\frac{\pi k_{\mathrm{B}} T}{2 \mu_{i,j}}} (d_i + d_j)^2 \qquad (10.18)$$

is the frequency of PAH collisions, k_{B} the Boltzman constant, T the reaction temperature, $\mu_{i,j}$ the reduced mass of the ith and jth PAH species, d_i the collisional diameter of the ith PAH, and the multiplier 2.2 is the van der Waals enhancement factor [10.37-38]. We assumed that the PAH collision diameter is related to its carbon atom content as

$$d_i = d_{\mathrm{A}} \sqrt{\frac{2 m_i}{3}}, \qquad (10.19)$$

which expresses a geometric relationship for the most condensed PAH series: benzene, coronene, circumcoronene, etc. d_{A} in (10.19) is the size of a single aromatic ring, equal $1.395\sqrt{3}$ Å, where 1.395 Å is the length of an aromatic C–C bond [10.39].

Our previous calculations [10.3,10.27] indicated that the computed PAH distribution is rather narrow, i.e., $\sigma_{\mathrm{PAH}} \ll \mu_{\mathrm{PAH}}$, where $\sigma_{\mathrm{PAH}} = \mu_2^{\mathrm{PAH}} - (\mu_{\mathrm{PAH}})^2$ is the PAH standard deviation and $\mu_{\mathrm{PAH}} = \mu_1^{\mathrm{PAH}}$ the PAH mean size. Based on this, the rates in (10.17) can be rewritten in terms of PAH moments following Method I of *Frenklach* et al. [10.36], in which the collision frequency (10.18) is expanded into a Taylor series about point $(i = \mu_{\mathrm{PAH}}, j = \mu_{\mathrm{PAH}})$. Then, performing the summations, we obtain for the second-order approximation of the PAH ensemble-average collision coefficient

$$\langle \beta^{\mathrm{PAH}} \rangle = 2.2 \frac{2 d_{\mathrm{A}}^2}{3} \sqrt{\frac{\pi k_{\mathrm{B}} T \mu_{\mathrm{PAH}}}{2 m_{\mathrm{C}}}} \left(4\sqrt{2} + \frac{D_{\mathrm{PAH}} - 1}{2\sqrt{2}} \right)$$

where m_{C} is the mass of a carbon atom and $D_{\mathrm{PAH}} = \dfrac{\mu_2^{\mathrm{PAH}}}{(\mu_1^{\mathrm{PAH}})^2}$ the dispersion of the PAH distribution. Equations (10.17) then take the form

$$R_r = \frac{1}{2} \langle \beta^{\mathrm{PAH}} \rangle \left(M_0^{\mathrm{PAH}} \right)^2 \sum_{k=0}^{r} \binom{r}{k} \mu_k^{\mathrm{PAH}} \mu_{r-k}^{\mathrm{PAH}}, \qquad r = 0, 1, 2, \ldots. \quad (10.20)$$

The coagulation terms G in (10.16) are defined as previously [10.36], using the interpolation techniques (Method II in [10.36]) to evaluate the time derivatives of the soot particle moments:

$$G_0 = \frac{1}{2} \langle \varphi_{0,0} \rangle \left(M_0^{\text{soot}} \right)^2$$

$$G_2 = \langle \varphi_{1,1} \rangle \left(M_0^{\text{soot}} \right)^2$$

$$\cdots$$

$$G_r = \frac{1}{2} \left(M_0^{\text{soot}} \right)^2 \sum_{k=1}^{r-1} \binom{r}{k} \langle \varphi_{r,r-k} \rangle.$$

Here $\langle \varphi_{x,y} \rangle = L_{1/2} \left({}^0\varphi_{x,y}, {}^1\varphi_{x,y}, {}^2\varphi_{x,y}, {}^3\varphi_{x,y} \right)$ for $(x,y) = (0,0), (1,1),$
$(1,2), (2,2)$, $\langle \varphi_{x,y} \rangle = L_{1/2} \left({}^0\varphi_{x,y}, {}^1\varphi_{x,y}, {}^2\varphi_{x,y} \right)$ for $(x,y) = (1,3), (2,3),$
and $\langle \varphi_{1,4} \rangle = L_{1/2} \left({}^0\varphi_{1,4}, {}^1\varphi_{1,4} \right)$, where the operator L_p represents Lagrange
logarithmic interpolation with respect to p between the points indicated in
the argument list. The expressions $\langle {}^\ell \varphi_{x,y} \rangle$ are defined as

$$\langle {}^\ell \varphi_{x,y} \rangle = 2.2 \sqrt{\frac{6 k_B T}{\rho}} \left(\frac{3 m_C}{4 \pi \rho} \right)^{1/6} \langle {}^\ell f_{x,y} \rangle,$$

where ρ is the density of soot particle material, assumed here equal
1.8 g/cm^3, and

$$\langle {}^\ell f_{x,y} \rangle =$$

$$\left(\frac{1}{M_0^{\text{soot}}} \right)^2 \sum_{i=1}^{\infty} \sum_{j=1}^{\infty} (m_i + m_j)^\ell \left(m_i^{1/3} + m_j^{1/3} \right)^2 m_i^{x-1/2} m_j^{y-1/2} N_i^{\text{soot}} N_j^{\text{soot}}$$

for $\ell = 0, 1, 2, 3$. The latter summations are resolved in terms of the particle
size moments, whole and fractional, positive and negative. The whole-order
positive moments are determined by the solution of (10.16) and fractional-
order positive moments by interpolation between the whole positive moments, i.e.,

$$\mu_p^{\text{soot}} = L_p \left(\mu_0^{\text{soot}} \equiv 1, \mu_1^{\text{soot}}, \mu_2^{\text{soot}}, \mu_3^{\text{soot}}, \mu_4^{\text{soot}}, \mu_5^{\text{soot}} \right), \qquad p > 0. \quad (10.21)$$

The fractional-order negative moments can be determined by interpolation

$$\mu_p' = L_{2p} \left(\mu_{-\infty}', \mu_0' \equiv 1, \mu_1' \right), \qquad p < 0,$$

where

$$\mu_r' = \frac{\mu_r^{\text{soot}}}{\left(m_1^{\text{soot}} \right)^r}$$

and

$$\mu_{-\infty}' = \lim_{r \to -\infty} \frac{\mu_r^{\text{soot}}}{\left(m_1^{\text{soot}} \right)^r} = \lim_{r \to -\infty} \sum_{i=1}^{\infty} \frac{m_i^r}{(m_1)^r} \frac{N_i^{\text{soot}}}{M_0^{\text{soot}}} = \frac{N_1^{\text{soot}}}{M_0^{\text{soot}}}.$$

Here m_1^{soot} and N_1^{soot} are the number of carbon atoms and number density, respectively, of the first, smallest PAH dimer. The value of N_1 is determined by solution of an additional differential equation,

$$\frac{dN_1^{soot}}{dt} = \beta_{1,1}^{PAH} \left(N_1^{PAH}\right)^2 - \langle\varphi_i\rangle N_1^{soot} M_0^{soot} - W_{N_1},$$ (10.22)

where $\beta_{1,1}^{PAH}$ is determined by (10.18) for $i = j = 1$, and

$$\langle\varphi_1\rangle = 2.2\sqrt{\frac{6k_BT}{\rho}} \left(\frac{3m_C}{4\pi\rho}\right)^{1/6} \langle f_1\rangle,$$

where $\langle f_1\rangle$ is determined by interpolation between functions

$$\langle^\ell f_1\rangle = \frac{1}{\sqrt{m_1^{soot}}M_0^{soot}} \sum_{j=1}^{\infty} \left(m_1^{soot} + m_j\right)^\ell \left[\left(m_1^{soot}\right)^{1/3} + m_j^{1/3}\right]^2 m_j^{-1/2} N_j^{soot}$$

at $\ell = 0, 1, 2, \ldots$. W_{N_1} in (10.22) represents the rate of disappearance of the first dimer due to the surface reactions. The latter are described below.

In the present work, the fractional negative moments were determined by extrapolation of the first three moments, namely

$$\mu_p' = L_p\left(\mu_0^{soot}, \mu_1^{soot}, \mu_2^{soot}\right), \qquad \text{for} \qquad p < 0.$$

The surface growth was assumed to occur due to reactions of C_2H_2 with surface radicals and condensation of PAH on the particle surface, and the surface oxidation due to reactions of O_2 with surface radicals and reactions of OH with the particle surface. The surface reactions of H, C_2H_2 and O_2 were assumed to be analogous, on the per-site basis, to the corresponding gaseous reactions of PAH species [10.3]

$$C_{soot}\text{-}H + H \rightleftharpoons C_{soot}\bullet + H_2$$ (10.23)
$$C_{soot}\bullet + H \rightarrow C_{soot} - H$$ (10.24)
$$C_{soot}\bullet + C_2H_2 \rightarrow C_{soot} - H + H$$ (10.25)
$$C_{soot}\bullet + O_2 \rightarrow products$$ (10.26)

where C_{soot}-H represents an arm-chair site on the soot particle surface and $C_{soot}\bullet$ the corresponding radical. Since little is known on the oxidation of PAH by OH, the oxidation of soot particles by OH,

$$C_{soot}\text{-}H + OH \rightarrow products$$ (10.27)

was described following *Neoh* et al. [10.40], using the probability of reaction, $\gamma_{OH} = 0.13$, upon a collision of OH with the particle surface.

Applying the standard description for the surface growth kinetics [10.36], we obtain

$$W_r^{C_2H_2} = k_{25}[C_2H_2]\alpha\chi_{C_{soot}\bullet}\pi C_s^2 \sum_{\ell=0}^{r-1} \binom{r}{\ell} M_{\ell+2/3}^{soot} 2^{r-\ell} \qquad (10.28)$$

$$W_r^{O_2} = k_{26}[O_2]\alpha\chi_{C_{soot}\bullet}\pi C_s^2 \sum_{\ell=0}^{r-1} \binom{r}{\ell} M_{\ell+2/3}^{soot} (-2)^{r-\ell} \qquad (10.29)$$

$$W_r^{OH} = \gamma_{OH}[OH]\sqrt{\frac{\pi k_B T}{2m_{OH}}}\pi C_s^2 \sum_{\ell=0}^{r-1} \binom{r}{\ell} M_{\ell+2/3}^{soot} (-2)^{r-\ell} \qquad (10.30)$$

$$W_r^{PAH} = 2.2\sqrt{\frac{\pi k_B T}{2m_C}} \sum_{\ell=0}^{r-1} \binom{r}{\ell} \left(C_h^2 M_{r-\ell+1/2}^{PAH} M_\ell^{soot} \right.$$
$$\left. + 2C_h C_s M_{r-\ell}^{PAH} M_{\ell+1/3}^{soot} + C_s^2 M_{r-\ell-1/2}^{PAH} M_{\ell+2/3}^{soot} \right) \qquad (10.31)$$

for $r = 1, 2, \ldots$, where $W_r^{C_2H_2}$, W_r^{PAH}, $W_r^{O_2}$ and W_r^{OH} are the contributions to the surface growth terms W_r in (10.16) due to the addition of mass by C_2H_2 and PAH and oxidation by O_2 and OH, respectively. In the derivation of these equations, it was assumed that the reduced mass of the colliding pair is equal to the mass of the much lighter gaseous species, because the mass of the soot particles increases rapidly.

Constants C_h and C_s are defined as

$$C_h = d_A\sqrt{2/3}$$
$$C_s = \left(\frac{6m_C}{\pi\rho}\right)^{1/3}.$$

Parameter α appearing in (10.28) and (10.29) accounts for the fraction of surface sites available for the corresponding reactions [10.3]; its value will be discussed in Section 10.9 below. $\chi_{C_{soot}\bullet}$ is the number density of surface radicals, which is determined assuming a steady state for $C_{soot}\bullet$ in reactions (10.23)–(10.26),

$$\chi_{C_{soot}\bullet} = \frac{k_{23}[H]}{k_{-23}[H_2] + k_{24}[H] + k_{25}[C_2H_2] + k_{26}[O_2]}\chi_{C_{soot}-H},$$

where $\chi_{C_{soot}-H}$ is the number density of $C_{soot}-H$ sites on the particle surface; the value of 2.3×10^{15} cm^{-2}, estimated previously [10.3], was used in the present simulations.

To properly account for the mass added in the surface reactions, equations (10.11) describing the PAH concentration moments are subtracted the following terms:

$$2R_0 + 2.2\sqrt{\frac{\pi k_B T}{2m_C}} \left(C_h^2 M_{1/2}^{PAH} M_0^{soot} + 2C_h C_s M_0^{PAH} M_{1/3}^{soot} + C_s^2 M_{-1/2}^{PAH} M_{2/3}^{soot} \right)$$
$$(10.32)$$

from (10.11-0) for the total PAH concentration, and

$$
\begin{aligned}
R_r + 2.2 \sqrt{\frac{\pi k_\mathrm{B} T}{2 m_\mathrm{C}}} \Bigg(& C_\mathrm{h}^2 M_{r+1/2}^{\mathrm{PAH}} M_0^{\mathrm{soot}} + 2 C_\mathrm{h} C_\mathrm{s} M_r^{\mathrm{PAH}} M_{1/3}^{\mathrm{soot}} \\
& + C_\mathrm{s}^2 M_{r-1/2}^{\mathrm{PAH}} M_{2/3}^{\mathrm{soot}} \Bigg)
\end{aligned}
\tag{10.33}
$$

from the rest of the moments, $r = 1, 2, \ldots$, respectively. The first terms in (10.32) and (10.33) account for the PAH–PAH coagulation and the rest for the PAH condensation on the soot particle surface. The terms R_r are those given by (10.20). The multiplier 2 in the first term of (10.32) is a stoichiometric coefficient: formation of *one* particle due to PAH coalescence removes *two* PAH species. Similar corrections are made to the rates of the concentration moments of the individual lumped PAH species, $S^j_{\{\mathrm{PAH}_{i'}\}}$, i.e., to equations like (10.12).

10.7 Optical Properties of Particle Ensemble

The optical properties of a particle ensemble – light scattering, extinction, and radiation – were modeled by the Mie theory [10.41] reformulated in terms of the moments of soot particle distribution function. Let us consider, as before, a system of N particles per unit volume with an arbitrary particle size distribution. Let N_i be the number density of particles having size i, i.e., $N = \sum_{i=1}^{\infty} N_i$.

The ensemble monochromatic scattering coefficient (vertical-vertical in this example) is given by the sum of the contributions from the individual size classes,

$$
Q_{\mathrm{vv}} = \sum_{i=1}^{\infty} C_{\mathrm{vv},i} N_i.
\tag{10.34}
$$

Using the *Penndorf* [10.42] expansion of the Mie-theory expression for particle monochromatic scattering coefficient, (10.34) takes the form

$$
Q_{\mathrm{vv}} = \sum_{i=1}^{\infty} \frac{9 N_i}{4 k^2} \left(K_6 a_i^6 + K_8 a_i^8 + K_9 a_i^9 + K_{10} a_i^{10} + \ldots \right)
\tag{10.35}
$$

where $k = \frac{2\pi}{\lambda}$ is the wave number, $a = \frac{\pi d}{\lambda}$ the particle size parameter, λ the light wavelength, d the particle diameter, and K's are the coefficients determined by the optical properties (i.e., by the complex refractive index m) of the particle material. These coefficients are specified in the Appendix. Upon substitution of k and λ into (10.35) and truncating the expansion, we obtain

$$Q_{vv}$$

$$\approx \frac{9}{16} \sum_{i=1}^{\infty} N_i \left[K_6 \left(\frac{\pi}{\lambda} \right)^4 d_i^6 + K_8 \left(\frac{\pi}{\lambda} \right)^6 d_i^8 + K_9 \left(\frac{\pi}{\lambda} \right)^7 d_i^9 + K_{10} \left(\frac{\pi}{\lambda} \right)^8 d_i^{10} \right] \tag{10.36}$$

Performing the summation in (10.36) term by term and taking into account relationship

$$\rho \frac{\pi d^3}{6} = m_c m_i,$$

we obtain

$$Q_{vv} = N \left(Y_1 \mu_2 + Y_2 \mu_{8/3} + Y_3 \mu_3 + Y_4 \mu_{10/3} \right), \tag{10.37}$$

where

$$Y_1 = \frac{9}{16} K_6 \left(\frac{\pi}{\lambda} \right)^4 \left(\frac{6m_c}{\pi \rho} \right)^2$$

$$Y_2 = \frac{9}{16} K_8 \left(\frac{\pi}{\lambda} \right)^6 \left(\frac{6m_c}{\pi \rho} \right)^{\frac{8}{3}}$$

$$Y_3 = \frac{9}{16} K_9 \left(\frac{\pi}{\lambda} \right)^7 \left(\frac{6m_c}{\pi \rho} \right)^3$$

$$Y_4 = \frac{9}{16} K_{10} \left(\frac{\pi}{\lambda} \right)^8 \left(\frac{6m_c}{\pi \rho} \right)^{\frac{10}{3}}.$$

Here μ's are the soot particle moments defined by (10.14)–(10.15). As before, m_c the atomic mass of carbon atom, m_i is the number of carbon atoms contained in the particle of size i, and ρ the mass density of soot particle material.

A similar analysis leads to the result for the extinction coefficient,

$$k_{ext} = N \left(X_1 \mu_1 + X_2 \mu_{5/3} + X_3 \mu_2 + X_4 \mu_{7/3} \right), \tag{10.38}$$

where

$$X_1 = K_1 \frac{\lambda}{4\pi} \left(\frac{\pi}{\lambda} \right)^3 \left(\frac{6m_c}{\pi \rho} \right)$$

$$X_2 = K_3 \frac{\lambda}{4\pi} \left(\frac{\pi}{\lambda} \right)^5 \left(\frac{6m_c}{\pi \rho} \right)^{\frac{5}{3}}$$

$$X_3 = K_4 \frac{\lambda}{4\pi} \left(\frac{\pi}{\lambda} \right)^6 \left(\frac{6m_c}{\pi \rho} \right)^2$$

$$X_4 = K_5 \frac{\lambda}{4\pi} \left(\frac{\pi}{\lambda} \right)^7 \left(\frac{6m_c}{\pi \rho} \right)^{\frac{7}{3}}$$

and the K's are the optical coefficients given in the Appendix.

The radiation of particles at wavelength λ can be characterized by the monochromatic emission coefficient [10.43]

$$J_\lambda = K_{abs} B_\lambda(T),$$

where K_{abs} is the absorption coefficient and $B_\lambda(T)$ is the Planck function,

$$B_\lambda(T) = \frac{2hc^2/\lambda^5}{e^{\frac{hc}{\lambda k_B T}} - 1}.$$

The absorption coefficient is defined as

$$K_{abs} = \sum_{i=1}^{\infty} \frac{\pi d_i^2}{4} Q_{abs,i} N_i,$$

where $Q_{abs,i}$ is the absorption efficiency. Expressing the latter as the difference between the extinction and scattering efficiencies,

$$Q_{abs,i} = Q_{ext,i} - Q_{sca,i},$$

and performing developments similar to those done above for Q_{vv} and k_{ext}, we obtain

$$K_{abs} = N \left(L_1 \mu_1 + L_2 \mu_{5/3} + L_3 \mu_2 + L_4 \mu_{7/3} + L_5 \mu_{8/3} \right), \qquad (10.39)$$

where

$$L_1 = K_1 \frac{\pi}{4} \left(\frac{\pi}{\lambda} \right) \left(\frac{6m_c}{\pi\rho} \right)$$

$$L_2 = K_3 \frac{\pi}{4} \left(\frac{\pi}{\lambda} \right)^3 \left(\frac{6m_c}{\pi\rho} \right)^{\frac{5}{3}}$$

$$L_3 = (K_3 - 6K_6) \frac{\pi}{4} \left(\frac{\pi}{\lambda} \right)^4 \left(\frac{6m_c}{\pi\rho} \right)^2$$

$$L_4 = K_5 \frac{\pi}{4} \left(\frac{\pi}{\lambda} \right)^5 \left(\frac{6m_c}{\pi\rho} \right)^{\frac{7}{3}}$$

$$L_5 = -K_8 \frac{3\pi}{2} \left(\frac{\pi}{\lambda} \right)^6 \left(\frac{6m_c}{\pi\rho} \right)^{\frac{8}{3}}$$

and the K's are the optical coefficients given in the Appendix.

Integration of (10.16) provides the moments M^{soot} and through (10.15) and (10.21) all the required here moments μ^{soot}. Substitution of the latter into (10.37)–(10.39) determined the optical properties of the evolving soot particle cloud.

10.8 Computational Details

The chemical reaction mechanism, comprised of a total of 337 reactions and 70 species, was essentially the same as used in our previous study [10.3]. The reactions and associated rate and thermochemical data for pyrolysis and oxidation of small hydrocarbon molecules, responsible for the main flame structures, were taken from [10.44]. The formation and growth of aromatics followed the reaction scheme described in Sections 10.3 and 10.4, with the rate coefficients listed in Table 10.1. This reaction mechanism is capable of predicting near-quantitatively the measured species profiles, including those of aromatics, for several laminar premixed flames [10.13,10.45].

The Sandia burner code [10.46] was used for the flame modeling, simulating PAH formation and growth up to coronene. The computed profiles of H, H_2, C_2H_2, O_2, OH, H_2O and the rate of formation of the first lumped species (the formation rate of acepyrene, reaction L0 in Table 10.1) were then used as an input for the particle nucleation and growth simulations. The latter were accomplished with an in-house kinetic code. The value of the complex refractive index was taken from [10.47], $m = 1.57 - 0.56i$. The computations were performed on an IBM 3090/600S main-frame computer.

10.9 Numerical Results and Discussion

Computer simulations were performed for several laminar premixed flames [10.3]. Here we will address the flame denoted as Flame B in [10.3], i.e., flame 1 of *Wieschnowsky* et al. [10.48]: a 25.4% C_2H_2-19.6% O_2–Argon mixture at a pressure of 12 kPa. The model predictions are compared to the experimental data for the transmitted and scattered light properties in Fig. 10.5.

We note that using the new optical algorithm, the computational results are compared in Fig. 10.5 not to the *derived* properties (like particle number density and average size, see Fig. 1 of [10.3]) but directly to the actual physical entities measured. Comparing Fig. 10.5 with Fig. 1 of [10.3], we note that the match between the model and experiment, in the sense of the profile shapes, is much improved for the direct comparison. Since the physical model used in the both cases is essentially the same, this observation indicates that the size distribution of soot particles is not self-preserving, as usually assumed in interpretation of the optical data collected in soot forming regions of a flame. This is further demonstrated in Fig. 10.6, by plotting the dispersion of the computed soot particle distribution, $D_{\text{soot}} = \mu_2^{\text{soot}}/(\mu_1^{\text{soot}})^2 = \langle d^6 \rangle / (\langle d^3 \rangle)^2$. As can be seen in this figure, the value of D_{soot} is substantially larger than that characteristic of a self-preserving distribution (about 2 [10.49]) within the particle inception

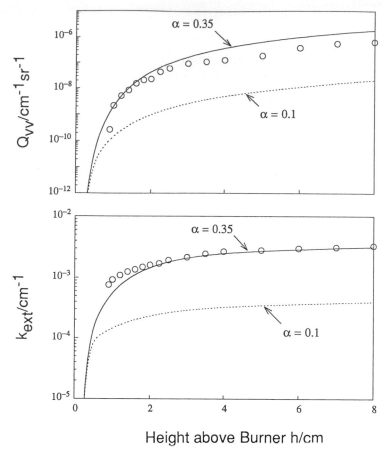

Fig. 10.5. Comparison of model predictions, lines, with experimental data, circles, for the scattering coefficient, Q_{vv}, and extinction coefficient, k_{ext}, at $\lambda = 488$ nm; the experimental points are derived from the data reported by *Wieschnowsky* et al. [10.48] for their flame 1: a 25.4% C_2H_2-19.6% O_2-Argon mixture at a pressure of 12 kPa; the lines are computed by (10.37) and (10.38), respectively, for the two values of the steric parameter α

zone of the flame. Our computational results showed that the extent of this overshoot depends upon the strength and duration of the nucleation source.

Further inspection of Fig. 10.5 indicates that a quantitative agreement between the model and experiment is now computed with the steric parameter α equal 0.35, significantly larger than the value of 0.1 assumed in our previous simulations for the same conditions [10.3]. This implies that the steric phenomena represented by α, i.e. the temperature dependence of the fraction of C_{soot}-H sites that can participate in reactions (10.23)–(10.26), is less pronounced than was estimated earlier. This would be in accord with the conclusion of *Colket*, cf. section 28. However, in light of the present uncertainty in the kinetic and thermodynamic parameters, this issue needs further investigation.

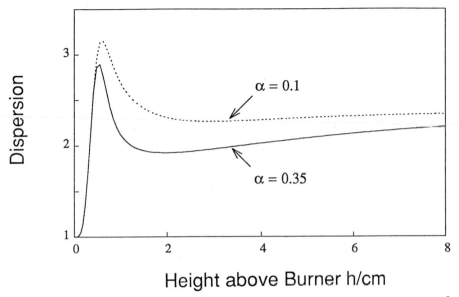

Fig. 10.6. Dispersion of the soot particle size distribution, $D_{\text{soot}} = \mu_2^{\text{soot}} / \left(\mu_1^{\text{soot}} \right)^2$ $= \langle d^6 \rangle / \left(\langle d^3 \rangle \right)^2$, computed as a function of the flame height for the conditions of Fig. 10.5

The rest of the computational results obtained in the present study were essentially the same as presented and discussed in [10.3]. To complete our discussion here, we list some of the major conclusions of our present and previous [10.3] modeling studies:

(i) The computed rate of nucleation is balanced by the rate of coagulation throughout the particle inception zone, however, the nucleation rate decays more slowly with flame height than is usually deduced from experiment.

(ii) Particle inception is primarily determined by PAH coagulation, initiated and controlled by PAH coalescence into dimers.

(iii) While the average soot particle is computed to contain $10^3 - 10^5$ carbon atoms, the corresponding average PAH size, μ_1^{PAH}, is only 20 to 50 carbon atoms. This indicates that the "crystallites" comprising incipient soot particles should be on the order of 7 to 12 Å, in agreement with experiment [10.23].

(iv) The oxidation of soot particles by OH and O_2 is quite insignificant in the particle inception zone and in the initial part of the post-flame zone.

(v) The surface growth of soot mass is primarily determined by two processes: acetylene addition via the H-abstraction/C_2H_2-addition reaction sequence, and PAH condensation on the particle surface. The relative contribution of each of these processes appears to change with experimental conditions. The main contribution of the PAH condensation occurs at the early stages of PAH coagulation.

(vi) The model is in accord with the classical structure of soot particles: less dense particle core, composed of randomly oriented PAH oligomers, and a more dense concentrically-arranged particle shell.

(vii) Surface processes can be understood in terms of elementary chemical reactions of surface active sites. The number density of these sites is determined by the chemical environment.

(viii) Determination of soot particle properties by light absorption and scattering measurements may be (and probably is for the particle inception zone) in error, caused by the deviation of the particles size from the usually assumed self-preserving distribution.

10.10 Acknowledgements

The computations were performed using the facilities of the Penn State Center for Academic Computing. The research was sponsored by the Air Force Office of Scientific Research, Air Force Systems Command, USAF, under grant No 91-0129. The US Government is authorized to reproduce and distribute reprints for Governmental purposes notwithstanding any copyright notation thereon.

10.11 Appendix

The coefficients K used in Section 10.7 are defined as follows:

$$K_1 = 6A_{1,3}^R$$
$$K_3 = 6A_{1,5}^R + 6B_{1,5}^R + 10A_{2,5}^R$$
$$K_4 = 6A_{1,6}^R$$
$$K_5 = 6A_{1,7}^R + 6B_{1,7}^R + 10A_{2,7}^R + 10B_{2,7}^R + 14A_{3,7}^R$$
$$K_6 = \left(A_{1,3}^R\right)^2 + \left(A_{1,3}^I\right)^2$$
$$K_8 = 2\left(A_{1,3}^R A_{1,5}^R + A_{1,3}^I A_{1,5}^I\right)$$
$$K_9 = 2\left(A_{1,3}^R A_{1,6}^R + A_{1,3}^I A_{1,6}^I\right)$$
$$K_{10} = 2A_{1,3}^R\left(A_{1,7}^R - \frac{5}{3}B_{2,7}^R - \frac{7}{6}A_{3,7}^R\right) + 2A_{1,3}^I\left(A_{1,7}^I - \frac{5}{3}B_{2,7}^I - \frac{7}{6}A_{3,7}^I\right)$$

Table 10.2. Elements of Mie coefficient expansion from [10.42]

			j	
	3	5	6	7
$A_{1,j}^{R}$	$\frac{2}{3}P_2$	$\frac{2}{5}(P_1Q_2+P_2Q_1)$	$\frac{4}{9}(P_1{}^2-P_2{}^2)$	$\frac{1}{175}(P_1R_2+P_2R_1)$
$B_{1,j}^{R}$		$\frac{1}{45}V_2$		$\frac{1}{945}W_2$
$A_{2,j}^{R}$		$\frac{1}{15}S_2$		$-\frac{1}{21}T_2$
$B_{2,j}^{R}$				$\frac{1}{1575}V_2$
$A_{3,j}^{R}$				$\frac{1}{1575}U_2$
$-A_{1,j}^{I}$	$\frac{2}{3}P_1$	$\frac{2}{5}(P_1Q_1-P_2Q_2)$	$-\frac{8}{9}P_1P_2$	$\frac{1}{175}(P_1R_1-P_2R_2)$
$-B_{1,j}^{I}$		$\frac{1}{45}V_1$		$\frac{1}{945}W_1$
$-A_{2,j}^{I}$		$\frac{1}{15}S_1$		$-\frac{1}{21}T_1$
$-B_{2,j}^{I}$				$\frac{1}{1575}V_1$
$-A_{3,j}^{I}$				$\frac{1}{1575}U_1$

The terms A and B are defined in Table 10.2, and the entities $P, Q, R, S, T,$ U, V and W appearing in this table are specified in Table II of *Penndorf* [10.42].

References

10.1   K.H. Homann, H.Gg. Wagner: "Some New Aspects of the Mechanism of Carbon Formation in Premixed Flames", in Eleventh Symposium (International) on Combustion (The Combustion Institute, Pittsburgh 1967) p. 371

10.2   H.F. Calcote: Combust. Flame **42**, 215 (1981)

10.3   M. Frenklach, H. Wang: "Detailed Modeling of Soot Particle Nucleation and Growth", in   Twenty-Third Symposium (International) on Combustion (The Combustion Institute, Pittsburgh 1991) p. 1559

10.4   M.B. Colket: "Successes and Uncertainties in Modeling Soot Formation in Laminar, Premixed Flames", this volume, sect. 28; F. Mauss, B. Trilken, H. Breitbach, N. Peters: "Soot Formation in Partially Premixed Diffusion Flames at Atmospheric Pressure", this volume, sect. 21

10.5   B.S. Haynes, H. Gg. Wagner: Prog. Energy Combust. Sci. **7**, 229 (1981)

10.6   S.E. Stein: J. Phys. Chem. **82**, 566 (1978)

10.7   M. Frenklach, D.W. Clary, W.C. Gardiner, Jr., S.E. Stein: "Detailed Kinetic Modeling of Soot Formation in Shock-Tube Pyrolysis of Acetylene", in Twentieth Symposium (International) on Combustion (The Combustion Institute, Pittsburgh 1985) p. 887

10.8   F.C. Stehling, J.D. Frazee, R.C. Anderson: "Mechanisms of Nucleation in Carbon Formation", in Eighth Symposium (International) on Combustion (Williams and Wilkins, Baltimore 1962) p. 774

10.9   J.D. Bittner, J.B. Howard: "Composition Profiles and Reaction Mechanisms in a Near-Sooting Premixed Benzene/Oxygen/Argon Flame", in Eighteenth Symposium (International) on Combustion (The Combustion Institute, Pittsburgh 1981) p. 1105

10.10 M. Frenklach, S. Taki, R.A. Matula: Combust. Flame 49, 275 (1983)

10.11 M. Frenklach, S. Taki, M.B. Durgaprasad, R.A. Matula: Combust. Flame 54, 81 (1983)

10.12 J.A. Miller, R.J. Kee, C.K. Westbrook: Annu. Rev. Phys. Chem. 41, 345 (1990)

10.13 M. Frenklach, H. Wang: "Aromatics Growth Beyond the First Ring and the Nucleation of Soot Particles", in Preprints of the 202nd ACS National Meeting (American Chemical Society, Washington, D.C. 1991) Vol. 36, No. 4, p. 1509

10.14. U. Alkemade, K.H. Homann: Z. Phys. Chem. N. F. 161, 19 (1989)

10.15 R.D. Kern, K. Xie: Prog. Energy Combust. Sci. 17, 191 (1991)

10.16 S.E. Stein, J.A. Walker, M.M. Suryan, A. Fahr: "A New Path to Benzene in Flames", in Twenty-Third Symposium (International) on Combustion (The Combustion Institute Pittsburgh, 1991) p. 85

10.17 J.H. Bradley, K.O. West: J. Chem. Soc., Faraday Trans. 71, 967 (1975); C.E. Canosa-Mas, M. Ellis, H.M. Frey, R. Walsh: Int. J. Chem. Kin. 16, 1103 (1984)

10.18 S.D. Thomas, F. Communal, P.R. Westmoreland: "C_3H_3 Reaction Kinetics in Fuel-Rich Combustion", in Preprints of the 202nd ACS National Meeting (American Chemical Society, Washington, D.C. 1991) Vol. 36, No. 4, p. 1448

10.19 M. Frenklach, D.W. Clary, W.C. Gardiner, S.E. Stein: "Effect of Fuel Strukture on Pathways to Soot", in Twenty-First Symposium (International) on Combustion (The Combustion Institute, Pittsburgh 1988) p. 1067

10.20 M. Frenklach, D.W. Clary, T. Yuan, W.C. Gardiner, Jr., S.E. Stein: Combust. Sci. Technol. 50, 79 (1986)

10.21 M. Frenklach, T. Yuan, T.M.K. Ramachandra: Energy Fuels 2, 462 (1988)

10.22 K.H. Homann: "Formation of Large Molecules, Particulates and Ions in Premixed Hydrocarbon Flames; Progress and Unresolved Questions", in Twentieth Symposium (International) on Combustion (The Combustion Institute, Pittsburgh 1985) p. 857

10.23 M. Frenklach, L.B. Ebert: J. Phys. Chem. 92, 561 (1988)

10.24 M. Frenklach: "On the Driving Force of PAH Production", in Twenty-Second Symposium (International) on Combustion (The Combustion Institute, Pittsburgh 1989) p. 1075

10.25 M. Frenklach, J. Warnatz: Combust. Sci. Technol. 51, 265 (1987)

10.26 M. Frenklach, W.C. Gardiner Jr.: J. Phys. Chem. 88, 6263 (1984)

10.27 M. Frenklach: Chem. Eng. Sci. 40, 1843 (1985)

10.28 M. Frenklach: "Modeling of Large Reaction Systems", in Complex Chemical
       Reaction Systems: Mathematical Modelling and Simulation, ed. by J. War-
       natz, W. Jäger (Springer-Verlag, Berlin 1987) p. 2

10.29 M. Frenklach: "Reduction of Chemical Reaction Models", in Numerical
       Approaches to Combustion Modeling, ed. by E.S. Oran and J.P. Boris
       (American Institute of Aeronautics and Astronautics, Washington, D.C.
       1991) p. 129

10.30 A. Fahr, S.E. Stein: "Reactions of Vinyl and Phenyl Radicals with Ethyne,
       Ethene and Benzene", in Twenty-Second Symposium (International) on
       Combustion (The Combustion Institute, Pittsburgh 1989) p. 885

10.31 J.H. Kiefer, L.J. Mizerka, M.R. Patel, H.C. Wei: J. Phys. Chem. **89**, 2013
       (1985)

10.32 S. Madronich, W. Felder: J. Phys. Chem. **89**, 3556 (1985)

10.33 V.S. Rao, G.B. Skinner: J. Phys. Chem. **92**, 2442 (1988)

10.34. C.Y. Lin, M.C. Lin: in Twentieth Fall Technical Meeting of the Eastern
       States of the Combustion Institute (Gaithersburg, MD 1987), p. 7-1

10.35 R.G. Gilbert, S.C. Smith: Theory of Unimolecular and Recombination Re-
       actions (Blackwell, London 1990)

10.36 M. Frenklach, S.J. Harris: J. Colloid Interface Sci. **118**, 252 (1987)

10.37 S.J. Harris, I.M. Kennedy: Combust. Sci. Technol. **59**, 443 (1988)

10.38 J.H. Miller: "The Kinetics of Polynuclear Aromatic Hydrocarbon Agglomer-
       ation in Flames", in Twenty-Third Symposium (International) on Combus-
       tion (The Combustion Institute, Pittsburgh 1991) p. 91

10.39 CRC Handbook of Chemistry and Physics, ed. by R.C. Weast, M.J. Astle
       (CRC Press, Boca Raton, Florida, 60th Edition) p. F-216

10.40 K.G. Neoh, J.B. Howard, A.F. Sarofim: "Soot Oxidation in Flames",
       in Particulate Carbon: Formation During Combustion, ed. by D.C. Siegla,
       G.W. Smith (Plenum, New York 1981) p. 261

10.41 H.C. van de Hulst: Light Scattering by Small Particles (Dover, New York
       1981) Chap. 9; A. D'Alessio: "Laser Light Scattering and Fluorescence
       Diagnostics of Rich Flames Produced by Gaseous and Liquid Fuels",
       in Particulate Carbon: Formation During Combustion, ed. by D.C. Siegla,
       G.W. Smith (Plenum, New York 1981) p. 207

10.42 R.B. Penndorf: J. Optical Soc. Am. **52**, 896 (1962)

10.43 G.B. Rybicki, A.P. Lightman: Radiative Processes in Astrophysics (J. Wi-
       ley, New York 1979) Chap. 1

10.44 M. Frenklach, H. Wang, M.J. Rabinowitz: Prog. Energy Combust. Sci. **18**,
       47 (1992)

10.45 H. Wang, M. Frenklach: "Modeling of PAH Profiles in Premixed Flames",
       in Fall Technical Meeting of the Eastern States Section of the Combustion
       Institute (Albany, New York, October 1989), p 12-1

10.46 R.J. Kee, J.F. Grcar, M.D. Smooke, J.A. Miller: Sandia Report No. SAND
       85-8240, December 1985

10.47 W.H. Dalzell, A.F. Sarofim: J. Heat Transfer **91**, 100 (1969)

10.48 U. Wieschnowsky, H. Bockhorn, F. Fetting: "Some New Observations
       Concerning the Mass Growth of Soot in Premixed Hydrocarbon-Oxygen
       Flames", in Twenty-Second Symposium (International) on Combustion
       (The Combustion Institute, Pittsburgh 1989) p. 343

10.49 A. Robinson, S.C. Graham: J. Aerosol Sci. **7**, 261 (1976)

Discussion

Warnatz: You said that you performed linear and nonlinear lumping simultaneously. From your transparencies I recognized that you first make the linear lumping and then you start somewhere with A_5 or A_7 with the nonlinear lumping. So, it is sequential and not simultaneous.

Frenklach: No, in our modeling studies, linear and nonlinear lumping schemes are implemented simultaneously, beginning with acepyrene, A_4R5. We can start it from benzene if we want. For physical reasons we chose not to do so.

Bockhorn: What I want to stress is that there is some discrepancy between the calculated and measured evolution of the particle size with time. The model predicts initially slowly growing particles and then the growth accelerates where it should stop. Another point is that in our results which we presented at the 20th Symposium we attributed surface growth mainly to acetylene because we calculated the possible contributions of all hydrocarbon species to surface growth and those of PAH are negligible compared with those of acetylene.

Frenklach: The problem we face when comparing the model to experiment is that, we think, the assumption of log-normal distribution is not adequate. From your measurements of Q_{vv} and k_{ext}, you derived properties like particle number density, surface area, etc. Instead of comparing to these derived properties, we tried to predict strictly what you see experimentally, i.e. Q_{vv} and k_{ext}. Then we do not need to deal with various assumptions that enter into evaluation of measured quantities. And if we do that, make a direct comparison between predicted and experimental Q_{vv} and k_{ext}, the agreement is much better than when we compare the derived properties.

Bockhorn: So, you dropped the assumption of log-normal distribution?

Frenklach: We don't have it anywhere, yes.

Bockhorn: However, we measured log-normal distributions at larger particle sizes.

Frenklach: First, our prediction for the deviation from a log-normal distribution is for the particle inception zone. Second, in your experiments, the sample was withdrawn from a flame by a probe. If using a probe you see a log-normal distribution within the particle inception zone, for which we predict the dispersion to be significantly larger than 2, the conclusion must be that the probe does not sample what optical methods see.

To your second point: I referred to your work presented at the 22nd Symposium where you came to the conclusion that the surface growth is not proportional to the surface area and you suggested that it may possibly be due to aromatics. For the soot particle inception zone of your flames, our model predicts that the surface growth occurs by PAH primarily and not acetylene.

D'Alessio: I have a very short question. You showed somewhere in your viewgraphs the agreement between your predictions and Nagle-Strickland-Constable kinetics. The Nagle-Strickland-Constable kinetics is not a totally empirical one, and there is a change of regimes because of a change in the reactivity of the solid phase. That is the idea behind the semiempirical Nagle-Strickland-Constable kinetics. Are you making the same type of assumption or do you just consider the gas phase kinetics?

Frenklach: This is a good point. Our assumption is that a surface radical is formed by H abstraction and this radical is oxidized with a rate comparable to the oxidation of a gaseous aromatic radical. I am aware of the assumptions of Nagle and Strickland-Constable: two types of reactive sites. We did not fit their rate expression. We calculated from their rate expression what the oxidation rate must be and compared it to our first-principles predictions, and this is the agreement I stressed.

Peters: Somewhere you mentioned the importance of PAH oxidation in the early stages of soot formation. Am I right that your linear lumping method does not take into account the oxidation of lower PAH?

Frenklach: In our linear lumping we consider five classes of oxidation. Among them the only one which really matters is oxidation of aromatic radicals by O_2. And these oxidation reactions are included for all PAH sizes.

Santoro: I want to make some comments rather than to put questions at this stage. The first one is that there has been a pretty intense discussion following the last couple of papers and I have a sort of rule of thumb that the intensity of the discussion is inversely proportional to our understanding of a problem and I think some of that is reflected in our discussion: but I would really like to stress the process that the last papers are bringing to the modeling area. I was surprised Michael didn't put up his slide that he sometimes shows containing a comment from Howard Palmer's review, that we will have come a long way in understanding soot formation when somebody can write down a model from fuel to particles and get reasonable predictions. I think for the last three years we have been in a situation where people are willing to tackle that question and put numbers on a page and then put data next to it. We should not underestimate this remarkable of progress. So, I think these kind of models are very useful.

For example in Michael's presentation his α is an adjustable parameter. In my interpretation of it, this has a lot of similarities to other approaches. Brian Haynes' paper in Combustion and Flame a few months ago on active site distribution as a function of temperature had some similarity to Michael's work. So, I think the important thing is what questions are the modelers asking the experimentalists that we can now go back and challenge. Michael confronted me a couple of times on the log-normal distribution assumption and we don't agree on this point. But the test is for me to do a better measurement and then he can do a better calculation. So, I think we should look at these models for identifying critical problem areas.

The other thing that we also have to keep in mind is the ultimate goal here. The people who are most interested in my work are people in diesel engine. And we have to remember that companies really want to use these models and we have to be able to provide models that they can incorporate into very much more complicated situations. So, I urge the model development people to use every trick they can to keep it simple and accurate, because the people who are going to use it really have lot of other things to model in that problem as well.

So, I think the presentation has identified some critical measurements that are needed: measurements in the inception zone and we need to understand the surface chemistry of the soot particles better than we do right now.

Frenklach: I appreciate your point of view, but I disagree with the basic presumption that modeling at this stage is just some kind of activity of academics and has nothing to do with reality. What I expect from this model is not that all the numbers are correct, but that every process that model is describing in rigorous terms is predicted right. I believe that if we are predicting certain facts they are there, and if you want to dismiss this as an experimentalist, you will work from a wrong direction. I do not agree with the complete discount of model predictions at this stage.

Sarofim: My question is: I find it very hard to believe that under very fuel rich conditions the O_2 would be the dominant oxidant. When you gave good agreement with the Nagle-Strickland-Constable kinetics for O_2 how did you find under those conditions that O_2 became more important than OH?

Frenklach: Our conclusion that O_2 is the dominant oxidant refers to the oxidation of PAH, which is dominated by the reaction of O_2 with aromatic radicals. For soot particles, our numerical results show that their oxidation by OH is faster than by O_2.

General Discussion on PAH formation in Sooting Flames: Global Dependence and Modelling

chaired by

Irvin Glassman, Brian S. Haynes

Wagner: M. Frenklach pointed out the importance of hydrogen atoms in connection with the soot particles. I am not so sure whether the H-atoms play the role under all the conditions which have been applied. Firstly, we could not find a correlation between the soot growth and H-atoms in the burned gases when you have slow soot formation. Secondly, when you go to pyrolysis of any kind of hydrocarbons there is hardly any influence of pressure. If the H-atoms would be important their concentration should strongly depend on pressure in a certain phase of the process and that can not be observed even if you go up to several hundred bar. It may be different when you are at low pressure. In general in low pressure flames you can get a strong influence of these atoms because here you really get high H-atom concentrations. I think this is an important point also with respect to particle growth. It would be nice if we could say no H-atoms, no soot. You can however also get soot growth when you have no hydrogen present at all and it is not much different from what you observe here.

Haynes: There will of course be much more discussion tomorrow on surface growth process.

D'Alessio: Are you sure that in order to understand and model the inception of soot particles we have to put together all these aromatic rings? The experimental evidence for looking only at aromatic structures in soot is because people normally had just measured PAH by gaschromatography so far. There is a lot more structures in a flame that has not been analysed yet because we didn't have the technical capability of doing this but now we have. So, modelers should also explore other aspects in their models.

Haynes: I guess in defense of the modelers I have to say here that they are already accused of manufacturing truth when they try to fit data when they have no experiments.

Frenklach: I don't want to give an impression that we see the H-atoms as the only cause of that effect. We are basically saying that for the conditions of S. Harris's flame we found that it can be explained by H-atom decay but we also think the parameter alpha that we have introduced should be a function of temperature and a function of size also. Two factors should be taken into account: the H-atoms and the surface morphology.

Wagner: You say you add a few PAH and then they agglomerate or coagulate and form soot. I think today we have methods to get very detailed information about the internal structure of even very small particles. I do know it's a very laborious work and I remember the time when Homann tried to go through that business. But the methods today are so much better than they have been in the past and that's why I think somebody should take the time to collect young soot particles and inspect the internal structure. Then we should give them a little heat treatment and look at them again. I have the feeling you will find a very large variety of possible structures of these small particles. I don't think it's so simple that one can say we just add particles of a certain size because you have branches where really a good deal of soot mass grows by coagulation. There are other situations where you can forget coagulation after 5% and you grow 95% by what is called surface growth. You can even identify the "growth" species. They don't need to be polycyclic aromatics.

Peters: If all these particles have a different size and different molecular weight they will have different diffusion coefficients. Can you take that into account in your lumping and moment methods?

Frenklach: We can take diffusion into account for PAH. Unfortunately, we didn't develop our own computer codes for flames and we have to make some simplifications. So diffusion for particles is not taken into account.

Warnatz: In the version I presented here generated by the group in Berlin it's not possible to do that. But in principle it's possible to make weight coefficients which are dependent on the polymerisation degree and to have diffusion coefficients which are dependent on it. That was just what I wanted to demonstrate that we don't have all these limitation which are in the earlier models. The second problem is to get the data. We try to make models from the molecules or atoms up to the particles. I know that this is not realistic at this point in time since we don't have the data for comparison. But that is the only chance we have if we wish to interpret measurements like those from Göttingen.

Colket: I was wondering whether M. Frenklach could give me some ideas about the effect that diffusion has on a higher molecular species. Did you compare the profiles of your full detailed chemistry model with higher order aromatics when you put in low diffusion coefficients.

Frenklach: As a test, we set diffusion to zero. We got a big effect even on a large size PAH which we didn't expect. So, diffusion is still playing a role even on larger sizes.

Wagner: Could you tell us the computation cost for your results?

Warnatz: If you take a detailed mechanism for small hydrocarbons then you have to take into account about 60 species and for the model you have to add 10 lumped species.

Santoro: I just wanted to follow up on a point that Professor Wagner was making about the need to look at particles in general. This problem has been recently brought to my attention through the diesel community which is now looking at the effects that soot have on lubricants in the engine itself, that is the soot particles get from the cylinder into the lubricant and degrade its performance rather rapidly. In fact they are finding as they move to higher compression ratios that this problem is becoming much more severe. The explanation from the companies that make additives is that the surface chemistry of the soot is different. We all know the carbon black industry makes 200 different kinds of carbon black for sale for various applications. I fully agree with Professor Wagner's question that maybe we are missing that there are some important differentiations we should be looking at. Fortunately, people in the oil companies said if I give them the soot they'll start looking at it to see if there are any differences as you change fuels and pressures. They are quite anxious to do that.

Lepperhoff: There are a lot of different processes going on in the diesel fuel area. Especially the interaction between lubricant and soot formation at the wall. I agree we have to simulate these processes under the boundary condition as they are in diesel engines. That means high pressure, high temperature and very short time scales. So we have to get more information about the sensitivity of different parameters. We hope to get this information from the modelers because they can vary parameters much easier than we can do in the diesel engine.

Glassman: I'll make a standard remark which will upset people I am sure. When I hear models I am reminded of the statement that I would never fly on an airplane if it's jet engines were designed strictly by models. And I don't think most of us would either. For the first time I think I agree with my colleague Santoro. I really don't feel that the people designing diesel engines are going to use models such as we have heard of. The process in the diesel engines is just too complex. In the first place you are injecting a liquid spray and we are not even sure how this spray burns yet. We don't know whether it's single droplet burning or whether the droplets evaporate first and you have gaseous jet. The engine is working in an unsteady condition and I don't think any of these effects are included in the models yet. I think we need much simpler concepts that we derive from this. Maybe I'll give away one of my points that I wanted to give in my talk but I was going to show in my talk 3 or 4 examples where practical situations show you that you can eliminate a soot formation particularly in gas furnances. We simply vary the oxygen concentration so that the temperature drops below a certain point and you don't get any soot. I doubt these mechanisms calculate that temperature. It doesn't matter how you measure a sooting tendency you find that butane soots less than allene which soots less than benzene. If you look at the hydrogen equivalence ratio you see that generally you are going to get more H-atoms from the butane pyrolysis than from the other two. I feel that what's controling the process is basically the amount of the first two rings that you form and everything after that becomes the same. You want to look at the total mass coming out and I think that's governed by the total amount of the soot particles being formed and that's what controls the overall soot coming out of a system. The extensive growth in mass patterns can be calculated. We can already predict what is the total mass of soot forming according to the particular fuel constituent you start with.

Soot in Flames
Mass Growth and Related Phenomena

The largest part of soot in flames is formed from surface growth reactions. Therefore, the understanding of the mass growth of soot by surface reactions is a key to the understanding of important phenomena during soot formation. The first paper of part III focuses on the question whether mass growth of soot or oxidation of soot particles in diffusion flames is a function of the local mixture composition or a function of the history of soot particles along their trajectories in the flames. This is followed by an investigation of temperature and concentrations effects on mass growth of soot in diffusion flames where temperature and concentrations have been varied independently from other flame properties. Active site approach for surface growth reactions versus recovery of surface sites by reactions with species from the gas phase is discussed extensively for different flame systems and finally - before concluding discussion of that part - experimental evidence for micro droplets as soot particle precursors is presented.

Soot Mass Growth in Laminar Diffusion Flames - Parametric Modelling

John. H. Kent, Damon R. Honnery

Department of Mechanical Engineering,
University of Sydney,
Sydney, N.S.W. 2006 Australia

Abstract: Spatially resolved soot mass growth rates are measured in a range of laminar ethylene and ethane diffusion flames. The surface specific mass growth rates are correlated with local temperature and mixture fraction to obtain a simple functional relationship for each fuel. The flames are modelled using the soot growth function and an oxidation model. Good predictions of soot concentrations are obtained for most conditions. The smoke-point effect is predictable, but an improved oxidation model is needed.

12.1 Introduction

A method of modelling soot concentrations in practical diffusion flames is to use soot growth rate functions determined from laminar diffusion flames [12.1-2]. The aim of the present work is to obtain a soot mass growth model for laminar diffusion flames in terms of relatively simple parameters. The model is evaluated by its degree of consistency in terms of the chosen parameters over a range of flame conditions and it is tested in a computational flame model by predicting soot concentrations for all our experimental conditions.

There have been several approaches to formulating soot models in diffusion flames. Particle inception and surface growth models have been proposed by *Tesner* et al. [12.3], *Magnussen* et al. [12.4], *Moss* et al. [12.5] and *Kennedy* et al. [12.6]. Soot concentrations in terms of a scalar variable have been used by *Kent* et al. [12.7] and *Gore* et al. [12.8] for laminar and turbulent flames. Other models are found in *Jander* et al. [12.9]. Soot mass growth rates in terms of local stoichiometry and local temperature in the flame field has been investigated by *Honnery* et al. [12.10] and *Kent* et al. [12.11].

The present investigation aims to model soot surface specific mass growth rate in terms of local temperature and mixture fraction. Mixture fraction is the mass fraction of fuel element at any point in the flame [12.12]. Soot specific surface growth rates are obtained from measurements in the flames and are mapped as functions of local temperature and stoichiometry. The maps for a particular fuel at each flame condition are combined to obtain an overall function for that fuel. The validity of the function can be tested by the amount of data scatter on the individual and on the composite maps for all flame conditions.

Two fuels, ethylene and ethane, are used for the experimental and modelling programs. The flames are laminar round vertical flames issuing into still air. Flame sizes range from 4 to 95 cm in order to achieve characteristic time scales of 20 to 200 ms.

The measurements consist of optical extinction and scattering for soot volume fraction and number density, together with thermocouple temperatures. These are combined with the numerical computations of the flame velocity field and the mixture fraction field. Soot particle trajectories obtained in this way allow the local growth rates to be calculated and to be related to the local environment or to the particle history.

The resulting soot growth function for each fuel is tested in a general flame computational model to predict the soot concentrations over the range of laminar flame conditions. A soot oxidation model is included to predict burnout and a radiative transfer model is added to couple flame temperatures with soot concentrations.

12.2 Experiment

The experimental apparatus and conditions are the same as described previously [12.13-14]. Pure ethylene and ethane laminar diffusion flames burn in still air on a 10.7 mm diameter vertical nozzle. The flames are surrounded by two concentric wire screens which assist in stabilising the long flames. Table 12.1 shows the experimental conditions. The time scale is for particle trajectories originating at the base of the flame to the height of maximum soot concentration.

Soot measurements are made by traversing the flame horizontally across a laser beam at 488 nm or 633 nm and measuring beam extinction and 90° scattered light intensity [12.15]. The traverse is automated and measurements are made continuously. The extinction measurements are inverted to obtain local soot volume fractions and are combined with the scattering measurements to get particle number densities. Rayleigh regime absorption and extinction for monodispersed particles is assumed and the soot index of refraction is taken to be 1.94 - 0.54i [12.16]. Traverses are made at flame height intervals of 0.25 to 4.0 cm depending on flame length. The radial

Table 12.1. Flame Conditions; lengths and residence times are calculated

Flame No.	Fuel	Flow/mls^{-1}	Stoich. Length/cm	Residence Time t/ms
F0	C_2H_4	2.3	4.1	25
F1	C_2H_4	4.8	8.8	45
F2	C_2H_4	11.6	24.0	92
F4	C_2H_4	43.7	88.1	204
F5	C_2H_6	12.2	27.6	95
F6	C_2H_6	19.1	44.9	141
F7	C_2H_6	40.2	95.2	181

profiles are smoothed for subsequent differentiation to obtain local growth rates.

Temperature measurements are made by a rapid insertion technique [12.17] with bare thermocouples corrected for radiation losses.

12.3 Flame Model

The flame model is required for two purposes. It predicts the velocity field and the mixture fraction field to deduce particle trajectories and so to extract local soot growth rates from the measured soot concentrations. It is also used later to predict the soot concentrations in the flames when the soot growth rate function is inserted.

The model, described previously [12.11,12.13], solves transport equations for axial momentum and mixture fraction in axisymmetric stream function coordinates with boundary layer assumptions.

$$\partial\psi = \rho u r \partial r \tag{12.1}$$

$$\frac{\partial u}{\partial x} = \frac{\partial}{\partial\psi}\left(\rho u r^2\mu\frac{\partial u}{\partial\psi}\right) + \frac{1}{\rho u}a_g(\rho_\infty - \rho) \tag{12.2}$$

$$\frac{\partial\Phi}{\partial x} = \frac{\partial}{\partial\psi}\left(\rho u r^2\Gamma_\Phi\frac{\partial f}{\partial\psi}\right) + \frac{1}{\rho u}S_\Phi \tag{12.3}$$

Here Φ represent enthalpy or mixture fraction. Mixture fraction (f) is the mass fraction of fuel element at any point in the flame field disregarding the fuel mass in the soot. At the stoichiometric contour $f = 0.064$ for ethylene, in the fuel stream $f = 1$ and in the air stream $f = 0$. An assumption that the diffusivities of all major species is the same allows the species concentrations and gas molecular weight to be related to the mixture fraction via a partial equilibrium model.

The transport coefficients, viscosity, thermal conductivity and diffusivity are calculated as functions of composition and temperature. Mixture fraction diffusivity is taken from the mixture thermal conductivity assuming unity Lewis number.

The equations are parabolic and a finite difference marching computational scheme is used to obtain a solution. Typically 1000 axial steps and 70 radial intervals are used. The velocity predictions have been validated against the measurements of Santoro et al. [12.18] as reported previously.

Measured temperatures are used when solving the flame field for purposes of soot data reduction. This avoids sensitivity to radiation losses from the soot. When predicting the complete flame field with soot concentrations later, radiative transfer in the horizontal plane is modelled [12.19].

$$\zeta \frac{\partial I}{\partial r} - \frac{\eta}{r} \frac{\partial I}{\partial \Theta} = -K_{abs} I + K_{abs} \frac{\sigma_{SB} T^4}{\pi} \qquad (12.4)$$

Soot is regarded as the predominant radiating species for the short path length here. The total absorption coefficient is obtained by calibrating the radiation model against measured temperatures in flames where the soot concentration field is known. It is taken to be [12.11]

$$K_{abs} = 1600 f_V T \qquad (12.5)$$

Soot particles are considered to have negligible diffusivity. They are transported by convection and by thermophoretic forces. The transport equation for soot mass is

$$\frac{\partial W_s}{\partial x} = -\rho v_{th} R \frac{\partial W_s}{\partial \psi} + \frac{1}{\rho u} S_s \qquad (12.6)$$

Thermophoresis can displace particles horizontally by 0.3 mm during their trajectories. The thermophoretic velocity is given by

$$v_{th} = \frac{-3R}{4(1 + \pi \alpha/8)} \frac{\partial \ln T}{\partial r} \qquad (12.7)$$

Numerical differentiation of Eq. (12.6) between measurement planes provides the local soot mass growth rates S_s. The field is discretised into intervals of 0.1 mm in the radial direction. Soot particle surface area is obtained from the experimental data.

When predicting soot concentrations S_s is the sum of the soot mass growth rate and the soot oxidation rate. Generally the two processes do not overlap much with major soot growth occurring on the fuel side of stoichiometry while burnout occurs near the stoichiometric contour and on the air side.

The soot oxidation model uses the mechanism and rates of Bradley et al. [12.20] for graphite oxidation in methane-air flames. The mechanism includes the Nagle and Strickland- Constable [12.21] expression for oxidation by molecular oxygen plus reactions involving H_2O, CO_2, H, O and OH. The concentrations of these species are assumed to be the equilibrium values at the local temperature and mixture fraction. The soot surface area is for spherical particles from experimental data correlations with soot volume fraction.

12.4 Experimental Results

The measured soot volume fractions and temperatures for a large ethylene flame are shown in Fig. 12.1. There is no soot burnout shown within the

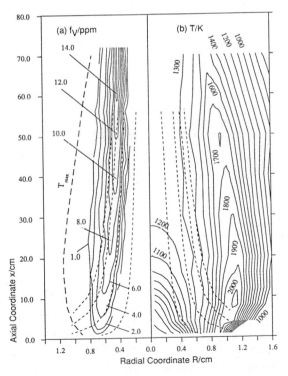

Fig. 12.1. Measured soot volume fraction and temperature and computed particle trajectories for ethylene flame F4; ———— temperature or soot volume fraction; - - - - - particle trajectories

measurement region. The locus of maximum temperature is shown relative to the soot bearing zone and temperatures decrease with increasing height due to radiation losses. Some computed particle trajectories are shown. Particles are swept into the fuel rich region as they are convected upwards.

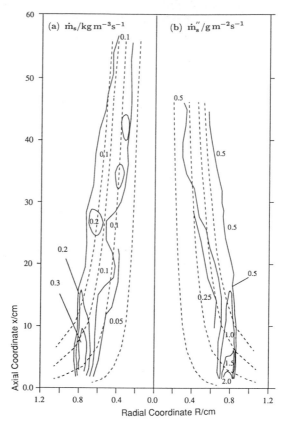

Fig. 12.2. Soot mass growth rate (kg m^{-3}s^{-1}) and surface specific growth rate (g m^{-2}s^{-1}) for ethylene flame F4

The derived mass growth rates (kg m^{-3}s^{-1}) and surface specific growth rates (g/s soot/m^2 particle area) are shown in Fig. 12.2. This flame exhibits fairly uniform growth rates unlike the smaller flames where there is a greater variation between high values at the base and the low or negative values at the flame tip as shown for ethane in Fig. 12.3.

Particle surface areas for the specific growth rates are deduced from soot volume fractions and number densities assuming monodispersed spherical particles. The surface areas are found to correlate well with soot volume

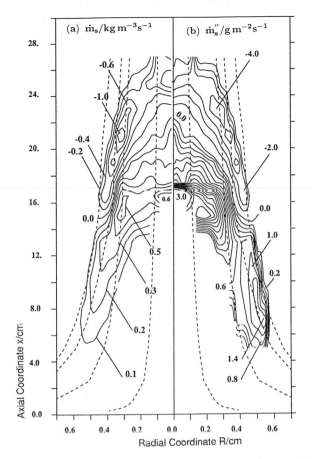

Fig. 12.3. Soot mass growth rate (kgm−3s−1) and surface specific growth rate $(g\,m^{-2}s^{-1})$ for ethane flame F5

fraction except at low values (< 2 ppm) for each fuel as shown in Fig. 12.4. A straight line fit to the data is indicated. This correlation is convenient because it avoids the necessity of predicting number densities when modelling soot growth.

Figure 12.5 shows soot surface specific growth rates against local temperature and mixture fraction for each ethane flame condition. The maps are generated by grouping data into bins with intervals of 0.02 in mixture fraction and 50 K in temperature. Longer flames have lower temperatures for the same mixture fractions, because of radiation losses and so there is a progressive shift to lower temperatures in the mapped regions for these flames. The regions of the maps which do overlap between flames show reasonable similarity in growth rates.

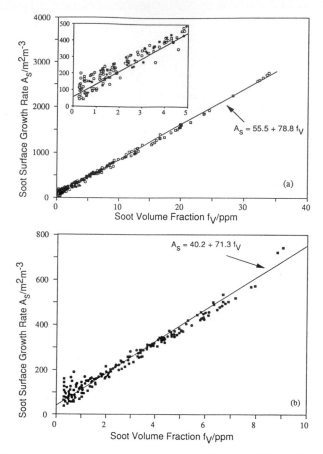

Fig. 12.4. Soot surface area against soot volume fraction; (a) ethylene □ F1, ○ F2, ∗ F4; (b) ethane ● F5, □ F6

Figure 12.6 shows the root mean square scatter of the data for the individual flame conditions and for all the data combined on one map. Generally the scatter has values of 0.1-0.3 about means of 0.6-1.0. The scatter of the data when combined has not significantly changed, indicating that the combined map is a valid representation for the range of flame sizes. The number of entries per bin for the combined map is typically 15 in regions of individual map overlap.

Figure 12.7 shows the ethane specific growth rates against temperature at three selected mixture fractions. Good correlations are shown except at $f = 0.09$ where oxidation is probably a significant factor.

No correlation of the data in a mixture fraction - temperature bin with particle age was found. This is evident in Fig. 12.8 where specific growth rates at three temperature - f combinations against particle age from the

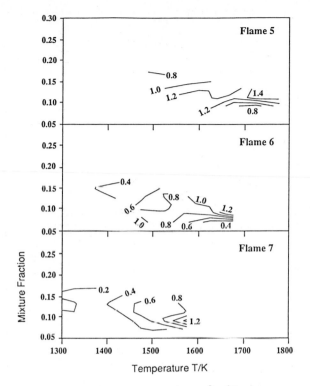

Fig. 12.5. Soot surface specific growth rate (g m^{-2}s^{-1}) against temperature and mixture fraction for individual ethane flames

trajectory origin at the reaction zone are shown. The above characteristics are similar for the ethylene flames reported elsewhere [12.22].

The overall combined maps for mean specific surface growth rates against temperature and mixture fraction for both fuels are shown in Fig. 12.9 and are tabulated in Table 12.2. The bounds of the data collection regions are indicated by the dashed lines. Soot growth may be possible outside these bounds, but the trajectories of the present flame configurations did not provide data beyond these regions.

Growth rates for the ethylene flames have peak values of about 3 g m^{-2}s^{-1} and for the ethane flames about 1.5 g m^{-2}s^{-1}. Also, the mixture fraction range of data for the ethane flames is narrower than for ethylene. Peak growth rates occur at about 1600 to 1700 K and f from 0.12 to 0.16.

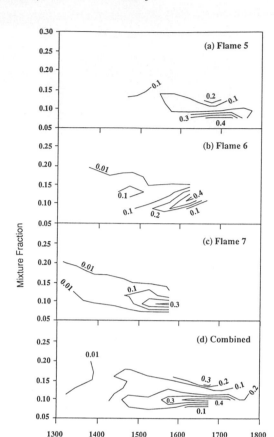

Fig. 12.6. Root mean square scatter of surface specific growth rates $(g\ m^{-2}s^{-1})$ in temperature, mixture fraction bins for individual and combine ethane flame conditions

12.5 Soot Modelling

The maps in Fig. 12.9 now become the soot mass source term in the flame model described above to predict soot concentrations for all the flames. A test of the generality of the growth rate function is how well the single function predicts soot for all conditions. The oxidation rate, as described, is also added to the source term.

The correlation of Fig. 12.4 provides the surface area. An initial "inception" surface area of 70 m^2/m^3 (per gas volume) is selected. There is some uncertainty in the initial surface area and a resultant sensitivity in

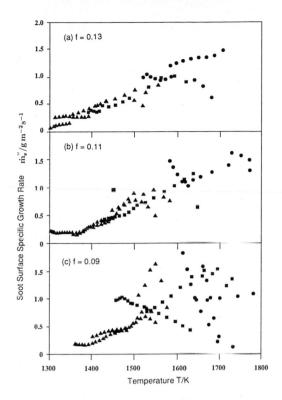

Fig. 12.7. Soot surface specific growth rates against temperature at three selected mixture fractions for ethane flames; ● F5, ▢ F6, △ F7

Table 12.2a. Tabulated values of Fig. 9 for ethylene

Mixture Fraction	1325	1375	1425	1475	1525	1575	1625	1675	1725	1775
0.29				0.387	0.751					
0.27					1.054					
0.25		0.340		0.500		1.313				
0.23		0.644	0.400	0.631			1.690			
0.21	0.043	0.663	0.640		1.178			1.926		
0.19	0.125	0.396	0.747	1.224	1.366	2.395				
0.17	0.184	0.364	0.563	1.423	1.835	2.798	2.553	2.499		1.764
0.15	0.138	0.258	0.596	1.204	1.558	3.033	2.741	2.415	2.423	
0.13		0.230	0.626	0.985	1.163	1.837	2.542	2.443	2.525	1.977
0.11			0.666	0.707	0.648	1.049	1.435	1.638	2.465	2.013
0.09			1.047	0.560	0.622	0.758	0.96	1.471		
0.07				0.693						
0.05										

Temperature T/K

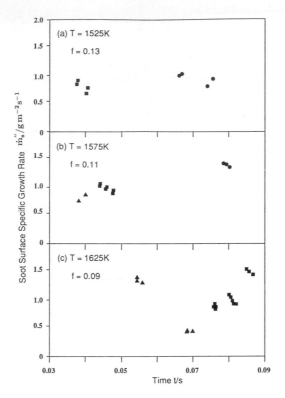

Fig. 12.8. Soot surface specific growth rates at selected temperature-mixture fraction combinations against particle age; ● F5, □ F6, △ F7

Table 12.2b. Tabulated values of Fig. 9 for ethane

Mixture Fraction	1325	1375	1425	1475	1525	1575	1625	1675	1725	1775
0.21										
0.19	0.081									
0.17	0.183	0.201	0.323	0.814	0.764					
0.15	0.257	0.374	0.451	0.579	0.844	0.941	0.965			
0.13	0.186	0.322	0.460	0.589	0.865	1.032	1.171	1.051	1.491	
0.11	0.201	0.22	0.404	0.681	0.743	1.092	1.063	1.244	1.506	1.467
0.09		0.174	0.322	0.640	0.826	0.971	1.134	1.026	0.951	1.088
0.07				0.827	0.826	0.628	0.302	0.218		
0.05					0.441					

Temperature T/K

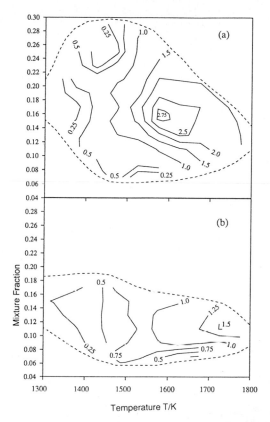

Fig. 12.9. Mean soot surface specific growth rates (g m^{-2}s^{-1}) against temperature and mixture fraction; (a) ethylene; (b) ethane; - - - - - bounds of measurements

the predictions to be discussed below. The only other input conditions to the model are the fuel type, flow rate and nozzle diameter. Temperature is calculated from the radiation model and the calculated soot concentrations.

Figure 12.10 shows the predicted and measured integrated soot mass flows (kg/s) with height for all flames. In the soot formation regions the predictions are generally quite good except for the very small flame F0. This agreement over a wide range of flame sizes is unusual for a single function.

Flames F0 and F1 for ethylene and F5 for ethane are below their smoke points as shown by the complete burnout in measured soot concentrations. The predictions show complete burnout for F0, partial burnout for F1, and little burnout for F5. Thus, the smoke-point effect is predicted, but not accurately. Oxidation rates are quite temperature sensitive and they depend on the coupling between soot concentrations and radiation losses.

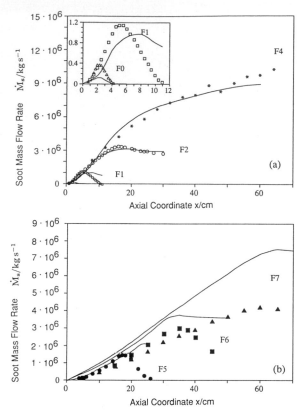

Fig. 12.10. Predicted and measured integrated soot mass flow rates; (a) ethylene △ F0, ▢ F1, ○ F2, * F4; (b) ethanc • F5, ▢ F6, △ F7; predictions

Generally the predicted oxidation rates are less than observed. The reasons for the discrepancy, apart from the kinetics assumptions, can be attributed to incorrect temperatures, incorrect gas phase species concentrations and incorrect particle surface areas. Predicted temperatures are quite close to measurements. The most likely reason is that super-equilibrium concentrations of O and OH are found in these flames [12.23,12.24] with resultant effects on burnout rates.

Figure 12.11 shows a comparison of predicted and measured soot volume fraction contours for an ethylene flame. The values and their locations are well predicted. A selection of radial profiles of predicted and measured soot volume fractions and temperatures are shown in Figs. 12.12 and 12.13. Temperatures are predicted well. Soot volume fractions are predicted low for the small flame, but are otherwise satisfactory for both fuels.

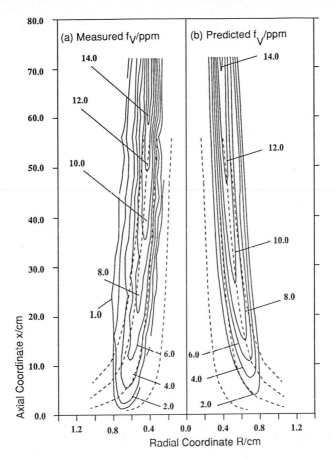

Fig. 12.11. Measured and predicted soot volume fractions (ppm) for ethylene flame F4

Figure 12.14 shows prediction sensitivity to initial surface area using different starting values. Clearly there is a multiplier effect when the initial area is varied. Increased surface area produces more soot which in turn produces more surface area. A better understanding of the dynamics controlling available surface area is required in the initial regions.

12.6 Conclusions

Soot surface specific growth rate as a function of local temperature and mixture fraction shows a good degree of consistency over a wide range of flame sizes for two fuels. The single function, for each fuel, is more successful than any previous model in predicting soot concentrations for all these

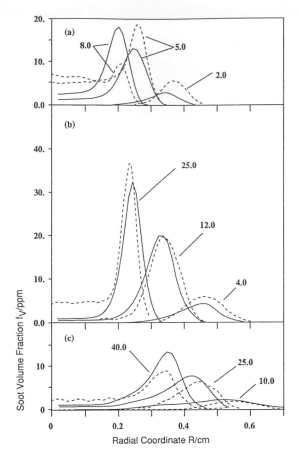

Fig. 12.12. Measured and predicted radial profiles of soot volume fraction at heights (cm) for flames (a) F1, (b) F2 and (c) F6; —— predictions, - - - - - - measurements

conditions. The model depends on the correct specification of surface area. Sensitivity to initial conditions means that the factors which determine surface area and its relationship to soot mass need more investigation.

The smoke-point effect can be predicted by this model, but not accurately because the predicted soot burnout rates are generally lower than measured. Super-equilibrium concentrations, not accounted for here, of gas-phase oxidant species are a likely cause.

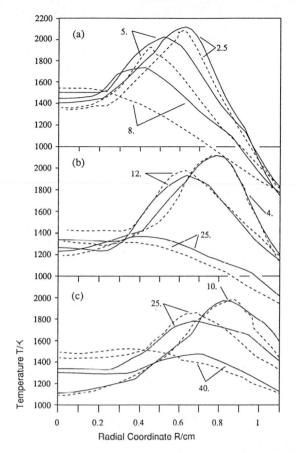

Fig. 12.13. Measured and predicted radial profiles of temperature at heights (cm) for flames (a) F1, (b) F2 and (c) F6; ——— predictions, - - - - - - measurements

12.7 Acknowledgements

The authors gratefully acknowledge the support of the sponsors of this meeting, the Australian Research Council and Aeronautical Research Laboratories.

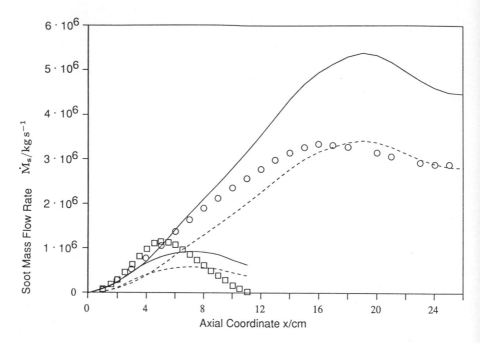

Fig. 12.14. Effect of initial soot surface area on predicted soot mass flow; □ F1, ○ F2; - - - - - $A_s = 56$ m^{2}m^{-3}; —— $A_s = 83$ m^{2}m^{-3}

12.8 Appendix

Nomenclature

A_s	soot surface area per gas volume
a_g	gravitational acceleration
f_V	soot volume fraction
I	radiation intensity
K_{abs}	radiation absorption coefficient
$\dot{m}_s$	soot mass growth rate
R	radial coordinate
S_ϕ	source/ sink for ϕ
T	temperature
u	axial velocity component
v_{th}	thermophoretic velocity
x	axial coordinate
W_s	soot mass fraction
α	accommodation coefficient
Γ_ϕ	exchange coefficient for ϕ

ζ, η direction cosines
θ azimutal angle
μ dynamic viscosity
f mixture fraction
ρ gas density
σ_{SB} Stefan-Boltzmann coefficient
ν kinematic viscosity
ϕ scalar variable
ψ stream function

References

12.1 J. B. Moss: "Modelling Soot Formation for Turbulent Flame Prediction", this volume, sect. 33

12.2 W. Kollmann, I.M. Kennedy, M. Metternich, J.-Y. Chen: "Application of a Soot Model to a Turbulent Ethylene Diffusion Flame", this volume, sect. 31

12.3 T.A. Tesner, T.D. Snegiriova, V.G. Knorre: Combust. Flame **17**, 253 (1971)

12.4 B.F. Magnussen, B.H. Hjertager: "On mathematical modelling of turbulent combustion with special emphasis on soot formation and combustion", in Sixteenth Symposium (International) on Combustion (The Combustion Institute, Pittsburgh 1976) p. 719

12.5 J.B. Moss, C.D. Stewart, K.J. Syed: "Flow field modelling of soot at elevated pressure", in Twenty-Second Symposium (International) on Combustion (The Combustion Institute, Pittsburgh 1988) p. 413

12.6 I.M. Kennedy, W. Kollmann, J.-Y. Chen: Combust. Flame **81**, 73 (1990)

12.7 J.H. Kent, D.R. Honnery: Combust. Sci. Technol. **54**, 383 (1987)

12.8 J.P. Gore, G.M. Faeth: "Structure and spectral radiation properties of turbulent ethylene/air diffusion flames", in Twenty-first Symposium (International) on Combustion (The Combustion Institute, Pittsburgh 1987) p. 1521

12.9 H. Jander, H. Gg. Wagner: "Soot Formation in Combustion"-an International Round Table Discussion (Vandenhoeck and Ruprecht, Göttingen 1990)

12.10 D.R. Honnery, S. Djingga, J.H. Kent: "Soot formation rates in two laminar ethylene diffusion flames" in 10th Australasian Fluid Mechanics Conference (University of Melbourne 1989) p. 14A1-14A4

12.11 J.H. Kent, D.R. Honnery: Combust. Flame **79**, 287 (1990)

12.12 R. W. Bilger: Annu. Rev. Fluid Mech. **21**, 101 (1989)

12.13 D.R. Honnery, J.H. Kent: Combust. Flame **82**, 426 (1990)

12.14 J.H. Kent, D.R. Honnery: Combust. Sci. Technol. **75**, 167 (1991)

12.15 A. D'Alessio, A. Di Lorenzo, F. Beretta, C. Venitozzi: "Optical and chemical investigation on fuel-rich methane-oxygen premixed flames at atmospheric pressure", in: Fourteenth Symposium (International) on Combustion (The Combustion Institute, Pittsburgh 1973) p. 941

12.16 S.C. Lee, C.L. Tien: "Optical constants of soot in hydrocarbon flames", in: Eighteenth Symposium (International) on Combustion (The Combustion Institute, Pittsburgh 1981) p. 1159

12.17 J.H. Kent, H. Gg. Wagner: Combust. Sci. Technol. **41**, 245 (1984)

12.18 R.J. Santoro, T.T. Yeh, J.J. Horvath, H.G. Semerjian: Combust. Sci. Technol. **53**, 89 (1987)

12.19 J.S. Truelove: Differential equations of radiative transfer, Harwell Report AERE-R-8364, U.K. Atomic Energy Authority 1976

12.20 F. Bradley, G. Dixon-Lewis, S. El-Din Habik, E.M. Mushi: " The oxidation of graphite powder in flame reaction zones" in: Twentieth Symposium (International) on Combustion (The Combustion Institute, Pittsburgh 1984) p. 931

12.21 J. Nagle, R.F. Strickland-Constable: "Oxidation of carbon between 1000-2000°C" in: Proceedings of the Fifth Conference on Carbon (Pergamon Press, London 1962) p. 154

12.22 D.R. Honnery, M. Tappe, J.H. Kent: Combust. Sci. Technol. **83**, 305 (1992)

12.23 A. Garo, G. Prado, J. Lahaye: Combust. Flame **79**, 226 (1990)

12.24 K.C. Smyth, P.J.H. Tjossem, A. Hamins, J.H. Miller: Combust. Flame **79**, 366 (1990)

Discussion

Santoro: I am interested in the conclusions about the initial conditions. Ian Kennedy in his model is finding that it is not very sensitive to the inception conditions and I believe even in ethylene flames this is true. Some of us have argued in the past that inception may be the controlling part of the process and that if we know how many particles we make initially then we can predict the final amount of soot in the flame. So it would be interesting to hear a little more about this conclusion on the importance of this initiation region. In your view, is inception controlling what happens further down or is this more of a model boundary condition problem.

Kent: It depends very much on the relationship between surface area and soot volume fraction. If you have a constant diameter then the surface area varies proportional to the soot volume fraction. If you have a constant number density then the surface area varies as the 2/3 power of soot volume fraction. Using the 2/3 power relation and integrating one ends up with an initial soot volume fraction separated from the growth term. In this case we come to the conclusion that the initial condition is not very important. Using the relation for surface area proportional to the soot volume fraction one ends up on integration with an exponential relationship and there is no way to get away from the initial condition. This is the difference between Ian Kennedy's work and what we have been using here. We measured number densities and traced them along the trajectories and did not find that 2/3 power relation and the predictions are sensitive to the initial conditions.

Kennedy: I agree with you that it does depend on that functional relationship. If you take the surface area proportional to the 2/3 power of the soot volume fraction the predictions are not that sensitive to the initial conditions. But I think there is a slight difference in what we call our initial conditions. I start off by looking at some particle inception rate during which coagulation occurs whereas John is talking about initial conditions of a certain surface area. The two are not exactly comparable. When I say the predictions are not sensitive to initial conditions I mean that they are not sensitive with respect to varying the inception rate a great deal. But at the same time the surface area is affected by coagulation. So it is not exactly the same thing when we talk about initial conditions.

Moss: Have you tried to make your expressions explicit rather than having the numerically defined maps. If you are using them for turbulent flame prediction it is obviously much more convenient to have them in some explicit form.

Kent: It is a very simple surface. I just have not done this fitting for the purpose of this paper. I just wanted to show the data. But there is no problem to do this, we are doing this and we will plug it back into the model. The extend of the map is important too. It is not an infinite map, we have just defined it for the temperatures where we made our measurements.

Wagner: I would like to know what temperatures you have at the centerline for the beginning of soot formation in your modelling.

Kent: We have temperatures of about 1400 K at the centerline where the soot volume fraction is about 5ppm.

Wagner: In all cases ?

Kent: No, for the large flame we end up with no soot on the centerline. But there is fluid mechanics to be taken into account for this case too, and the question is what happens to these trajectories and where do they end.

Glassman: I think you have to assume that some of the particles get on the centerline due to thermophoresis.

Kent: Yes, they do drift. It is of the order of about half to one millimeter in the lateral direction. We considered that in the trajectories which we have computed.

Kennedy: John, you mentioned the counter flow diffusion flame. Does that mean you suspect that there may be a third parameter, the scalar dissipation rate or something in the turbulent flame and is that going to complicate things beyond hope?

Kent: Well, in our sort of flames the stretch rates are not very high and I would expect stretch rates to increase as the size of the flame decreases and the gradients get steeper. However, the stretch rates you have in counter flow diffusion flames would be much higher and it would be interesting to see whether this model would predict anything useful in that case or whether extra parameters are needed.

Warnatz: Could you comment on the pressure effect, for instance under diesel engines conditions? I suppose that in this case you have a dependence with the fourth or fifth power.

Kent: Of course, you are quite right. I did not take this into account yet. Bill Flower's experiments found a pressure square dependence. I can not tell you more than what you know already.

Concentration and Temperature Effects on Soot Formation in Diffusion Flames

Robert J. Santoro, Thomas F. Richardson

Department of Mechanical Engineering,
The Pennsylvania State University,
University Park, PA 16802, USA

Abstract: The effects of fuel dilution on soot formation in laminar diffusion flames is investigated. The lower soot formation rate observed is due to a combination of concentration, temperature and residence time effects. Based on local measurements of the temperature, temperature seems to be the most important effect. Modeling results show that concentration effects are mitigated by diffusional processes associated with the diffusion flame structure. Furthermore, dilution increases the time for soot particle inception to occur, resulting in a decreased residence time for soot growth. The observed effects are most important in the soot inception region and appear to decrease in importance as residence time and sooting propensity increase.

13.1 Background

The formation of soot particles in combustion environments involves a complex series of chemical and physical processes which control the conversion of fuel carbon to carbonaceous particles. Since soot formation is well known to be a chemically kinetic limited process, the effects of temperature and concentration are expected to be important. In the case of concentration effects, it would seem straightforward to expect a direct proportionality between fuel concentration and soot formation rates. However, in many situations, mixing and temperature effects can mask the importance of the fuel concentration variations on the amount of soot formed in the combustion environment.

One means to investigate the effect of fuel concentration is through the introduction of an inert diluent into the fuel stream. Diluting the fuel stream has been observed to reduce soot concentration in laminar diffusion

flames [13.1]. This reduction is in part a result of lower flame temperatures achieved in diluted flames as compared to the undiluted case. These
lower temperatures reduce the pyrolysis reaction rates leading to decreased
soot formation rates and consequently lower soot concentrations. Reducing
the fuel concentrations may have an effect on soot concentration through a
proportional reduction in fuel species concentration in the soot formation
region, i.e. reaction rates are decreased by lower species concentration. A
study of the effect of fuel dilution has been undertaken with the objective
of further refining our understanding of the effects of concentration and
temperature. As part of this study the effects of other parameters such as
residence time have also been considered in the analysis. Although reduction
in soot through inert addition has been observed by numerous researchers,
only a few studies exist which investigate the relative roles of fuel concentration and flame temperature. Previous studies of laminar diffusion flames
have employed dilution as a means to study temperature effects on soot formation [13.2-3]. In these cases the effects of dilution were argued to be small
relative to the temperature reduction contribution. *Kent* et al. [13.4] observed that soot concentrations in diluted flames are reduced by an amount
that can not be explained by the reduction in flame temperature alone.
Axelbaum et al. [13.5-6] addressed this issue in a series of experiments in
counterflow flames. The important conclusion from these studies is that
soot formation rates depend linearly on the initial fuel concentration. *Axelbaum* et al. [13.7] recently extended their dilution studies to include coflow
diffusion flames. They report a dilution effect for coflow flames that is consistent with the counterflow flame results; that is, the maximum soot volume
fraction increases linearly with respect to be the fuel concentration at the
burner exit. These results imply that the role of temperature in soot formation, commonly believed to the governing parameter of soot formation
rates, is in some situations less important than fuel concentration.

13.2 Experimental Setup and Operating Conditions

Investigating the role of temperature and concentration on soot formation requires a careful selection of the experimental approach. Following
a method similar to the approach of *Axelbaum* et al. [13.7], dilution is
studied through a series of flames at identical calculated adiabatic flame
temperatures, but which have different initial fuel concentrations. The difference in heat capacities of argon and nitrogen permit comparison of flames
with identical temperatures, but different dilutions. Therefore, the effect of
concentration is systematically isolated from temperature. Similarly, flame
conditions can be selected with identical dilutions, e.g. 50% N_2 and 50% Ar,
but different temperatures, thus, isolating the effect of temperature on soot
formation. Desired flame conditions are calculated with the NASA Chemi-

cal Equilibrium Code [13.8] based on the assumption that the temperatures in the formation region of laminar diffusion flames scale with calculated adiabatic temperatures (T_{ad}).

An atmospheric coannular burner was used to study laminar diffusion flames burning ethene or propane. A laser scattering/extinction system was used to obtain data on the soot particle field in these flames. The burner and light scattering apparatus has been previously described in the literature [13.9] and will be only briefly described. The coannular burner consists of an inner brass fuel tube (1.1 cm inner diameter) surrounded by an outer tube (10.0 cm inner diameter) for the air flow. The fuel or fuel inert mixture is burned in a highly over-ventilated air flow, minimizing the effect of airflow rate [13.10]. Table 13.1. summarizes the flow conditions, which span a range of flow rates, dilutions, and temperatures.

Table 13.1. Flow conditions for the diluted and undiluted laminar flames specifying fuel, fuel flow rate ($\dot{V}$), diluent, initial fuel mole fraction (X_F) and calculated adiabatic flame temperature (T_{ad}); the total integrated soot volume fraction, F_V, and the ratio expressed in percentage of the measured F_V to the F_V for a pure undiluted fuel case is also given

Exp.#	Fuel ($\dot{V}/cm^3s^{-1}$)	Diluent	x_F	T_{ad}/K	F_V/cm^2	% of Pure $(F_V)/(F_V)_{pure}$
1	$C_2H_4(2.75)$	N_2	0.50	2310	$1.66 \cdot 10^{-6}$	46.7
2	$C_2H_4(2.75)$	Ar	0.37	2310	$1.05 \cdot 10^{-6}$	29.6
3	$C_2H_4(4.90)$	N_2	0.50	2310	$2.91 \cdot 10^{-6}$	51.8
4	$C_2H_4(4.90)$	Ar	0.37	2310	$2.12 \cdot 10^{-6}$	37.7
5	$C_2H_4(4.90)$	N_2	0.74	2346	$4.90 \cdot 10^{-6}$	87.2
6	$C_2H_4(4.90)$	Ar	0.64	2346	$4.56 \cdot 10^{-6}$	81.2
7	$C_2H_4(4.90)$	N_2	0.64	2333	$4.08 \cdot 10^{-6}$	72.7
8	$C_2H_4(4.90)$	Ar	0.52	2333	$3.77 \cdot 10^{-6}$	67.2
9	$C_2H_4(6.58)$	N_2	0.64	2333	$5.30 \cdot 10^{-6}$	74.5
10	$C_2H_4(6.58)$	Ar	0.52	2333	$4.48 \cdot 10^{-6}$	63.0
11	$C_2H_4(2.75)$	--	1.00	2369	$3.54 \cdot 10^{-6}$	100.0
12	$C_2H_4(4.90)$	--	1.00	2369	$5.62 \cdot 10^{-6}$	100.0
13	$C_2H_4(6.58)$	--	1.00	2369	$7.10 \cdot 10^{-6}$	100.0
14	$C_3H_8(2.56)$	N_2	0.61	2240	$2.66 \cdot 10^{-6}$	79.8
15	$C_3H_8(2.56)$	Ar	0.50	2240	$2.46 \cdot 10^{-6}$	73.7
16	$C_3H_8(2.56)$	--	1.00	2266	$3.33 \cdot 10^{-6}$	100.0
17	$C_2H_4(3.85)$ $+C_7H_8(0.3)$	N_2	0.33	2248	$8.21 \cdot 10^{-6}$	63.3
18	$C_2H_4(3.85)$ $+C_7H_8(0.3)$	N_2	0.66	2333	$1.14 \cdot 10^{-5}$	88.1
19	$C_2H_4(3.85)$ $+C_7H_8(0.3)$	--	1.00	2362	$1.30 \cdot 10^{-5}$	100.0

The laser light extinction/scattering apparatus used a 4W argon ion laser operating at the 514.5 nm laser line. The laser source was modulated using a mechanical chopper operating at 1 kHz to allow for synchronous detection of the transmitted light signals. The transmitted light was detected using a silicon photodiode with the output of the detector input to a two phase lock-in amplifier interfaced to an IBM-XT computer. Laser light scattering measurements at a scattering angle of 90° with respect to the incident beam could also be obtained with this system. However, in the present experiments such measurements were only used for comparisons with previous studies. Since these measurements are not used in the present results, this part of the system will not be described in detail.

The entire atmospheric burner was mounted on a three-dimensional translating stage system. Computer controlled stepper motors were used to adjust the vertical and one of the horizontal coordinates. This allowed radial profiles of the laser extinction to be achieved at various axial positions in the flame.

Rapid insertion thermocouple measurements provide measurements of the temperature distributions in the flames studied. The rapid insertion thermocouple technique avoids continuous coating with soot by burning off soot formed on the thermocouple surface in the oxidation zone of the flame. Pt/Pt-10% Rh fine wire thermocouples were made from 125 μm diameter wire. The resulting bead diameter as typically 160 μm. The uncoated thermocouples are mounted on a stepper motor stage for quick positioning. This technique provides accurate measurements of relative changes in the temperature as diluent species and concentrations are varied.

13.3 Analysis

In the present study, the quantity of interest is the change in soot volume fraction, f_V, resulting from dilution of the fuel. In previous studies of laminar diffusion flames, tomographic reconstruction approaches have proved useful in providing spatially resolved measurements of f_V from line-of-sight extinction measurements. In the present case, however, the errors associated with the tomographic reconstruction approach detract from its usefulness since small variations in the f_V need to be measured.

An alternate approach is to relate the extinction measurement to an appropriate spatial integral of the soot volume fraction. For soot particles in the Rayleigh size limit $(d/\lambda << 1)$ the extinction (I/I_o) is related to f_V as:

$$\int_{-\infty}^{+\infty} f_V \, dx = -c(\lambda, m) \ln\left(\frac{I}{I_o}\right) \tag{13.1}$$

where x is the direction along which the laser beam propagates, $c(\lambda, m)$ is a constant determined from Rayleigh theory and x is the path length through the flame. A measure of the total integrated soot volume fraction, F_V, can be obtained by integrating Equation (13.1) along the direction perpendicular to x. This integrated volume fraction is a measure of the total amount of soot at a particular axial location and is given by

$$F_V = -\int_{-\infty}^{+\infty} c(\lambda, m) \ln\left(\frac{I}{I_o}\right) dy \qquad (13.2)$$

where y is the direction perpendicular to x and $c(\lambda, m)$ is a constant. The integrated volume fraction, F_V, has been selected because it incorporates any changes in flame diameter as the flow rate is varied.

13.4 Results

These studies address three aspects of the influence of inert additives on the reduction in soot formation: 1) the relation between the degree of dilution and soot reduction; 2) fuel flow rate effects; and 3) potential fuel structure effects. Using the information from these experiments, the relative influence of fuel concentration and temperature are considered for coflowing laminar diffusion flames. The approach used to determine the relation between the degree of dilution (i.e. fuel concentration) on soot formation involved comparing argon and nitrogen diluted flames having the same calculated temperature. Since argon and nitrogen have different molar heat capacities, the initial fuel concentration of these flames must be different if the temperature is to be maintained constant. To ensure that the differences in soot formation rates in these flames of equal temperature are a result of differences in fuel concentration and not temperature, the temperature field was determined from rapid insertion thermocouple measurements. In addition, these measurements allowed a determination of whether calculated adiabatic temperatures may be used to represent the changes in temperature expected with dilution. This may not be self evident since calculated adiabatic temperatures are based on the fuel/diluent mixture being supplied to the burner and not the local concentration of the flames.

Thermocouple measurements were made in 4.90 cm^3/s ethene flames (see Table 13.1) for fuel/diluent flow conditions that correspond to calculated temperatures of 2369 K (flame 12), 2346 K (flames 5 and 6) and 2333 K (flames 7 and 8). The two lower temperature conditions were established individually for nitrogen (flames 5 and 7) and argon (flames 6 and 8) diluents. Uncorrected measured temperatures are shown in Fig. 13.1. for an undiluted and diluted flame (flames 12 and 7) and indicate that the adiabatic flame temperature is not achieved. This observation is a result of two factors. First, the thermocouple measurements were not corrected for radiative losses. Second, temperatures in sooting flames are lower than calculated adiabatic values because of radiation losses from the soot particles as well as effects of conduction and finite chemistry [13.3-4, 13.11]. The radial temperature profiles of flames 12 and 7, shown in Fig. 13.1. as a function of axial position, reveal a trend expected in the temperature. At axial heights less than 50 mm, temperatures are reduced in the diluted flame, while above 50 mm, the diluted flame has a higher temperature. The explanation for this observation lies in the fact that more soot is being formed in the initially hotter undiluted flame, increasing radiative losses, thus cooling the flame to a greater extent [13.3-4].

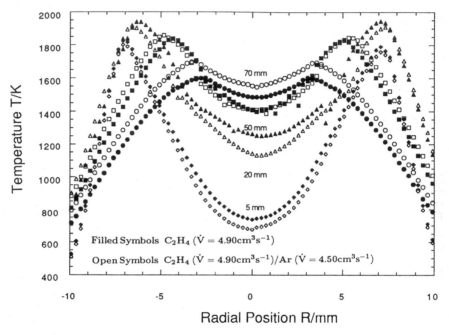

Fig. 13.1. Uncorrected thermocouple measurements of the temperature profiles as a function of radial position for several axial heights in an undiluted (flame 8) and diluted (flame 12) ethene diffusion flame; the ethene flow rate is 4.90 cm^{3}s^{-1} while the argon diluent flow rate is 4.50 cm^{3}s^{-1}

Careful inspection of the temperature profiles reveals that low in the flame, dilution reduces temperatures at the centerline to a greater extent than in the annular region [13.11]. This difference in the effect of the diluent as a function of position is related to the diffusional nature of these flames. For both the N_2 and Ar diluent cases, N_2 from the air stream is diffusing towards the centerline while the *fuel* (diluted or undiluted) diffuses towards the flame front. This interdiffusional aspect of the flame tends to mitigate the initial fuel concentration differences resulting from the addition of the diluent species. This point will be discussed in detail later.

In the annular region, low in the flame (5 mm axial location), the measured temperature differences are similar to the differences in calculated adiabatic flame temperatures (see Fig. 13.2.). Flames in which diluent conditions are chosen to match the temperature show temperatures which differ by less than 10 K in the region near the temperature maximum. As the centerline is approached from the maximum temperature location, differences between the N_2 and Ar diluted flames for identical calculated adiabatic flame conditions increase with the N_2 diluted flames having slightly higher temperatures. In the region where soot is formed this difference can be as large as 40 to 60 K. Positioning uncertainty with respect to the thermocouple measurement location is not believed to be the source for these observed differences, since both the centerline minimum and the symmetric temperature maximums in the annulus show good positional agreement for each of the flames (see Fig. 13.2.).

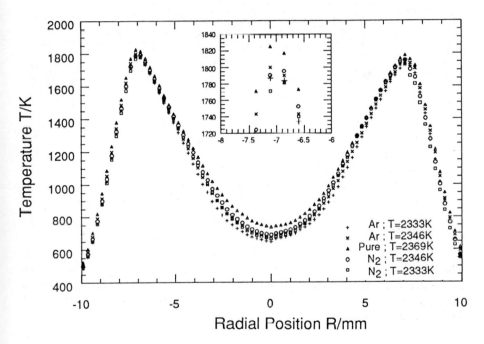

Fig. 13.2. Comparison of uncorrected thermocouple measurements of the temperature profile at an axial height of 5 mm as a function of radial position; results are presented for an undiluted (flame 12) and two pairs of diluted flames (flame 5 and 6 and flames 7 and 8) which have identical calculated adiabatic temperatures (2346 and 2333 K, respectively)

These observations underscore the need to consider the local temperature and concentration, as well as the overall expected behavior of the flame in assessing the effects of fuel dilution. One of the goals of this investigation is to determine the variation in soot formation rates with the initial fuel concentration. This is most easily achieved by comparing the soot yield in flames of similar temperatures but different dilutions. A study of the soot formation in a set of three pairs of flames, each pair at a different temperature, is intended to address this issue. The variation in the integrated soot volume fraction, F_V, at various axial locations is shown in Fig. 13.3. for each of the ethene flames with a flow rate of 4.90 cm^{3}s^{-1}. It is interesting to note that, although diluent addition increases the burner exit velocities, the height at which the peak volume fraction is reached is nearly constant in these flames with possibly the exception of flame 4 which has the highest dilution. The height at which the soot maximum is reached is determined by the location at which oxidizer replaces the fuel at the centerline of the flame [13.12]. Thus, the rate of diffusion of oxidizer into the flame front is only weakly dependent on the amount of fuel dilution. The peak integrated soot volume fraction is used as a readily determined measure of the sooting propensity of these flames. The reduction in soot formed, shown in Table 13.1., is calculated from the peak F_V values. In this comparison, the reduction in soot may be due to both temperature and/or concentration variations. Peak soot concentration ratios show that reducing fuel concentration and temperature reduces the amount of soot formed. The flames which show the largest relative reduction in soot (flames 3 and 4), the most diluted flames, show a dependence on the soot formation rate which is directly proportional to the fuel concentration. This implies that soot formation rates vary with concentration in less than a 1 to 1 linear manner, since the temperature reduction effect on soot formation has not been taken into account.

In order to quantify the effect of fuel concentration on the soot formation rates, the integrated soot volume fraction, F_V, was taken to be related to the initial fuel concentration by:

$$F_V \propto [\text{Fuel}]^b \qquad (13.3)$$

From the measured differences in F_V for the argon and nitrogen cases, a value for b can be obtained and is shown in Table 13.2. along with the pertinent parameters for each case. It should be noted that the value of b is determined for flames having the same calculated adiabatic temperature and further assumes that F_V is directly proportional to initial fuel concentration. A review of the values for b listed in Table 13.2. shows that b varies from approximately 0.4 to 1.5. No systematic variation in b is observed with respect to fuel flow rate. These results can be viewed to be similar to the earlier results of *Axelbaum* et al. [13.5-7] who have argued that concentration effects can be of more significance than temperature effects for fuel dilution studies of laminar diffusion flame systems. However, the deviation of b from unity indicates that the dependence on initial fuel concentration could also be complicated by diffusion and residence time effects.

To further examine the effects of fuel dilution, an additional study was conducted. A series of flames in which a mixture of ethene and toluene were burned in air for different dilution levels was studied. For these studies,

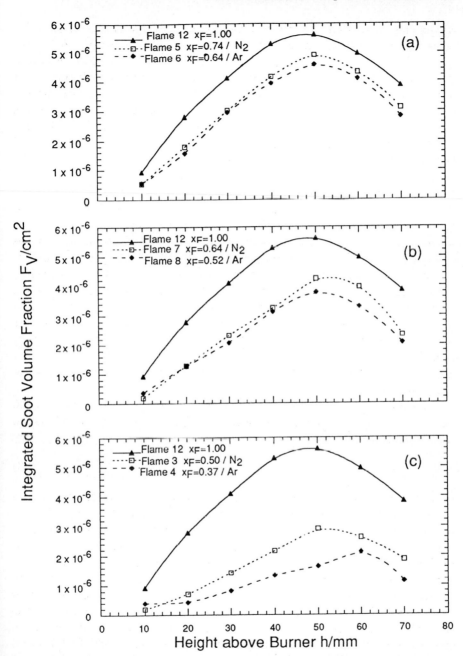

Fig. 13.3. Total integrated soot volume fraction, F_V, as a function of height in the flame; a) flames 12, 5, and 6; b) flames 12, 7, and 8; c) flames 12, 3, and 4; the fuel mole fraction is listed in parentheses next to the symbol identification

Table 13.2. Comparison of fuel concentration effects for flames at similar calculated adiabatic flame conditions

Exp.#	Fuel $(\dot{V}/cm^3s^{-1})$	T_{ad}/K	Δx_F^* / %	ΔF_V^{**} / %	b^{***}
1 & 2	C_2H_4(2.75)	2310	26.0	36	1.51
3 & 4	C_2H_4(4.90)	2310	26.0	27	1.06
5 & 6	C_2H_4(4.90)	2346	13.5	6.9	0.49
7 & 8	C_2H_4(4.90)	2333	18.8	7.6	0.38
9 & 10	C_2H_4(6.58)	2333	18.8	15	0.81
14 & 15	C_3H_8(2.56)	2240	18.0	7.7	0.40

the fuel flow rate was held constant at 3.85 and 0.3 cm^3s^{-1} for ethene and toluene, respectively, while the nitrogen diluent flow rates of 0, 2.1 and 8.4 cm^3s^{-1} were used. The results of these studies are shown in Table 13.1. Again the change in fuel concentration is observed to be significantly greater than the observed variation in soot as measured by ΔF_V. For these toluene studies, the temperature was not held constant and, thus, the observed reduction in soot is due to both temperature and concentration effects.

13.5 Discussion

From the above results, it is clear that fuel dilution studies of soot formation can not be interpreted as simple mechanisms involving well controlled overall variations in temperature and concentration. Rather, as with all practical combustion environments, effects of mixing and residence time may be important to consider. In particular, it is important to ascertain how the local concentration and temperature are affected. The first consideration to be addressed relates to the effects of mixing and asks how diffusion mitigates the effects of the dilution of the fuel. Clearly, when a diluent is introduced, a well defined concentration exists at the fuel tube exit. However, diffusion of species, such as nitrogen from the air flow and combustion products from the reaction zone, begin to alter the concentration. Furthermore, it seems reasonable that the effects of concentration would be most important in the region where soot is formed. To address this question, a laminar diffusion flame code by *Mitchell* et al. [13.13] has been used to investigate the variation of fuel concentration as a function of position in the flame. Turning to the modeling results, Fig. 13.4. shows calculations corresponding to four of the ethene/air flames studied (a) flames 11 and 1, b) flames 13 and 9). In this figure, the fuel mole fraction as a function of height along the flame center line is shown. In each case, undiluted or diluted, the fuel mole fractions quickly obtain similar values at axial locations between 10 to 30 mm above the burner exit depending on the fuel flow rate. This effect is a result of the rapid diffusion of nitrogen from the air to the fuel region mitigating the initial dilution. Consequently, all laminar diffusion flames burning in air undergo *dilution* by nitrogen, rapidly reducing the local fuel concentration. These results are further examined in Fig. 13.5. where radial profiles are presented for flames 11 and 1 at three axial locations (0.3 mm, 2.7 mm and 14.5 mm). Table 13.3. shows results tabulated for other fuel flow rates at the same axial locations for selected radial positions (4.63 mm and the center line). It is important to note that soot formation is initiated well away from the center line of the flame. Previous results show soot is first observed at a radial location about 5 mm from the center line at an axial location between 3 and 5 mm above the burner for flow conditions typical of the present experiments [13.9]. Comparisons shown in Table 13.3. indicate that, in the region where soot is first formed, the local concentration ratio between the 50% diluted and undiluted flame is typically between 0.7 and 0.85. Additionally, the effect of dilution is slightly more pronounced for the lower volumetric fuel flow rate case. At $h = 2.7$ mm and $R = 4.63$ mm, the concentration ratio is 0.76 for the 2.75 cm^3s^{-1} case as compared to 0.85 for the 6.58 cm^3s^{-1} flame.

From the above results, the effects of dilution of the fuel flow would be expected to be more acute for low fuel flow rate cases and to vary in a less than first order manner. Both of these effects are observed in the present experiments (see Table 13.1.). From these results, one can conclude that the local concentration of fuel and combustion products will quickly achieve similar values in both diluted and undiluted flames. Furthermore, concentration effects, based on initial values at the fuel tube exit, will not reflect the true concentration effects. In addition to these observations, the temperature measurements indicate that dilution does not effect the temperature through the flame in a universal manner. Local temperature variations can be observed in regions where the peak temperatures show good agreement in terms of achieving equivalent *adiabatic* temperature conditions. These local temperature differences, as will be seen shortly, could be responsible for significant variations in the amount of soot formed.

The modeling results described above have shown that the ratio of the local concentration values in diluted and undiluted flames can be expected to differ significantly from the initial values at the fuel tube exit. In fact,

Table 13.3. Mitchell flame comparisons for fuel mole fraction X_F in 50% diluted and undiluted ethene diffusion flame at selected radial (R) and axial (x) coordinates

Fuel Flow Rate $\dot{V}/\text{cm}^3\text{s}^{-1}$	R/mm	h/mm	x_F(Pure)	x_F(50%)	$\dfrac{x_F(50\%)}{x_F(\text{Pure})}$
6.58	4.63	0.3	0.482	0.351	0.727
6.58	4.63	2.7	0.256	0.218	0.849
6.58	0.00	14.5	0.504	0.401	0.795
4.90	4.63	0.3	0.410	0.304	0.740
4.90	4.63	2.7	0.207	0.170	0.823
4.90	0.00	14.5	0.383	0.330	0.860
2.75	4.63	0.3	0.298	0.220	0.738
2.75	4.63	2.7	0.132	0.101	0.764
2.75	0.00	14.5	0.215	0.189	0.879

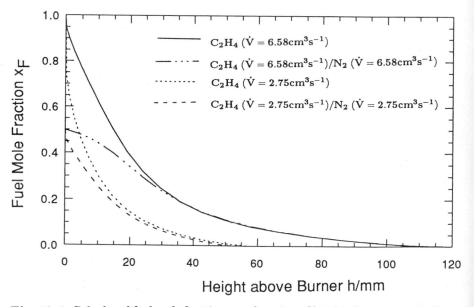

Fig. 13.4. Calculated fuel mole fraction as a function of height along the centerline of a diffusion flame for two ethene fuel flow rates (2.75 cm^3s^{-1} and 6.58 cm^3s^{-1}); calculations compare an undiluted and diluted (50% N_2) flame case; flow rates in cm^3s^{-1} for C_2H_4 and N_2 are listed in parenthesis next to the identification of the line symbols

at sufficiently large distances from the burner exit, the model shows that both the diluted and undiluted flames have similar mole fractions for major species. This raises a question regarding the mechanism responsible for the variation in soot formation observed for these dilution studies. Two additional effects of particular interest are the temperature and residence time.

Previous work by *Kent* et al. [13.4] and *Böhm* et al. [13.14] indicate that temperature changes of 50 K in diffusion or premixed flames can result in approximately a factor of two change in the maximum soot volume fractions observed. In the present flames, comparative differences of approximately 40 K are observed in the sooting forming region of the N_2 and Ar diluted flames. The lower temperatures are observed for the Ar diluted case which exhibit the lower soot formation rates. Consequently, temperature effects could be more pronounced than originally anticipated for flames which have equal adiabatic flame temperatures. These results, again, point to the need for local measurements to properly examine effects based on overall variations of the initial conditions of the diffusion flame.

Residence time effects have also been shown to be important in diffusion flame studies of soot formation [13.15-16]. In fact, a squared or cubed power dependence on the residence time has been observed [13.16]. Thus, changes in the velocity which result from addition of the dilutent species could affect the residence time and the soot formation rate as a result. However, the major observed effect of dilution with respect to residence time is a shift of the location where soot is first observed to higher axial positions. Since the location of the maximum soot volume fraction is largely unaffected by dilution, the total soot residence time is reduced. Note the shift in the location of the initial soot formation region could be a result of temperature, velocity and/or concentration effects. Our preliminary evaluation of this effect estimates that between 5 to 20% of the reduction in soot could be a result of the variation in residence time during which soot growth can occur. These studies will be refined in the near future to better characterize the onset location of soot formation as a function of dilution, since this has proved to be a point of some interest.

13.6 Conclusions

The present study of the effects of fuel dilution on soot formation in laminar diffusion flames indicates a number of important effects are present. The lower soot formation rate observed is due to a combination of concentration, temperature and residence time effects. Based on local measurements of the temperature variations in the soot forming regions of the flame, along with previous measurements of the sensitivity of the soot formation process to temperature variation, we argue that temperature is still the most important effect. The modeling results show that concentration

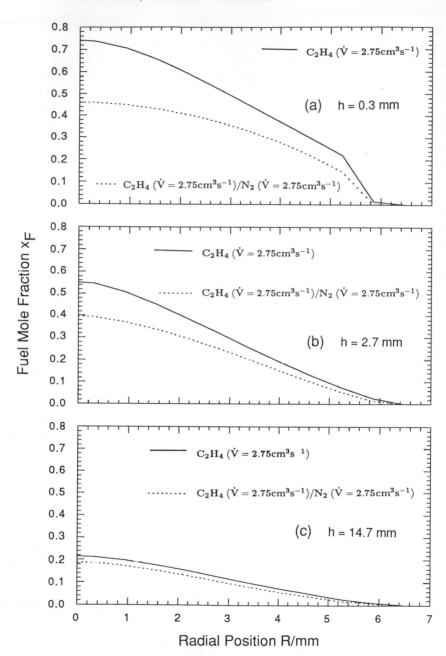

Fig. 13.5. Comparison of the calculated fuel mole fraction as a function of radial position for a diluted (50% N_2) and undiluted flame at heights of a) 0.3 mm, b) 2.7 mm, and c) 14.7 mm above the fuel tube; flow rates in cm^3s^{-1} for C_2H_4 and N_2 are listed in parantheses next to the identification of the line symbols

effects are mitigated by diffusional processes associated with the diffusion flame structure. In particular, diffusion of nitrogen from the surrounding air quickly results in similar concentration fields for both diluted and undiluted flames. Furthermore, dilution appear to increase the time for soot particle inception to occur, resulting in a decreased residence time for soot growth. Our studies indicate that concentration effects can be responsible for some fraction of the observed variation in soot formation as diluent is added. These effects are most important in the soot inception region and appear to decrease in importance as residence time and sooting propensity increase. Additionally, the present studies point to the usefulness of detailed studies to investigate fundamental mechanisms involving soot formation as opposed to global studies. Future studies need to more quantitatively evaluate the contributions of temperature and concentration effects on the soot formation process in these flames. Measurements of species concentrations should be obtained to compare with the model prediction with respect to the effects of dilution. Concentration variations could be further studied by varying the operating pressure of the flames.

13.7 Acknowledgements

The authors would like to acknowledge the support of this research by the Air Force Office of Scientific Research, Air Force Systems Command, USAF, under grant number AFOSR-87-0145 with Dr. Julian Tishkoff serving as contract monitor. The U.S. Government is authorized to reproduce and distribute reprints for Governmental purposes notwithstanding any copyright notation thereon.

References

13.1   P. Deardon, R. Long: J. Appl. Chem. **18**, 243 (1968)
13.2   I. Glassman, P. Yaccarino: "The Temperature Effect in Sooting Diffusion Flames" in Eighteenth Symposium (International) on Combustion (The Combustion Institute, Pittsburgh 1981) p. 1175
13.3   R.J. Santoro, H.G. Semerjian: "Soot formation in diffusion Flames: Flow Rate, Fuel Species, and Temperature Effects" in Twentieth Symposium (International) on Combustion (The Combustion Institute, Pittsburgh 1984) p. 997
13.4   J.H. Kent, H.Gg. Wagner: Combust. Sci. Technol. **41**, 245 (1984)
13.5   R.L. Axelbaum, W.L. Flower, C.K. Law: Combust. Sci. Technol. **61**, 51 (1988)
13.6   R.L. Axelbaum, C.K. Law, W.L. Flower: "Preferential Diffusion and Concentration Modification in Sooting Counterflow Diffusion Flames", in Twenty-Second Symposium (International) on Combustion (The Combustion Institute, Pittsburgh 1988) p. 379

13.7 R.L. Axelbaum, C.K. Law: "Soot Formation and Inert Addition in Diffusion Flames", in Twenty-Third Symposium (International) on Combustion (The Combustion Institute, Pittsburgh 1990) p. 1517

13.8 S. Gordon, B.J. McBride: "Computer Program for Calculation of Complex Chemical Equilibrium Compositions, Rocket Performance, Incident and Reflected Shocks, and Chapman-Jouget Detonations"; NASA SP-273, Interim Revision N78-17724, March 1976

13.9 R.J. Santoro, H.G. Semerjian, R.A. Dobbins: Combust. Flame **51**, 203 (1983)

13.10 R.G. Roper: Combust. Flame **29**, 219 (1977)

13.11 L.R. Boedeker, G.M. Dobbs: Combust. Sci. Technol. **46**, 301 (1986)

13.12 R.G. Roper, C. Smith, A.C. Cunningham: Combust. Flame **29**, 227 (1977)

13.13 R.E. Mitchell, A.F. Sarofim, L.A. Clomburg: Combust. Flame **37**, 227 (1980)

13.14 H. Böhm, D. Hesse, H. Jander, B. Luers, J. Pietscher, H.Gg. Wagner, M. Weiss: "The Influences of Pressure and Temperature on Soot Formation in Premixed Flames", in Twenty-Second Symposium (International) on Combustion (The Combustion Institute, Pittsburgh 1988) p. 403

13.15 R.J. Santoro, T.T. Yeh, J.J. Horvath, H.G. Semerjian: Combust. Sci. Technol. **53**, 89 (1987)

13.16 D.R. Honnery, J.H. Kent: Combust. Flame **82**, 426 (1990)

Discussion

Sarofim: I have a simplistic view of diffusion and premixed flames which tells me that the chemistry is very different. Therefore, your final statement confused me a little bit. I thought that pyrolysis on the fuel rich side would give you very different pathways with a different temperature dependence and concentration dependence than in a premixed flame. I just wish to have a clarification of your statement.

Santoro: You are right that in the diffusion flame the temperature time history that the particle goes through and the concentration profiles are different from the temperature time history and concentration profiles that you have in a premixed flame. But in certain parts of the flame they are going to have very similar concentration conditions, I think. Under those conditions I would expect the temperature effects for example to be very similar. For the surface growth process where we have mainly acetylene addition to the particle there are not radically different behaviors in the two flames. In the inception zones I think you are absolutely right. You can have conditions in premixed flames that are so far from the conditions in diffusion flames that what you are saying is right. But I do think there are some regions of the flame where the local conditions are similar enough that we should expect the same general trends when we look at changes in temperature or changes in pressure. What we need to account for is the change in those local conditions. I think Kent's experiments are very much headed in this direction.

I think the underlying physics is the same and if you pose the proper analysis there should be a unifying element in it. We are now looking at pressure effects on soot formation in diffusion flames because there have been differences observed between premixed and diffusion flame studies and I personally do not understand why. It seems to me that the two flames should have similar behavior.

Kent: Perhaps I can just enlarge on that last point. May be I am answering your question. If we compare the over-all growth rates in a Harris and Weiner flame with one of our diffusion flames which has the closest condition – a C/O-ratio of about 0.76 turns out to be an equivalent mixture fraction ratio of 1.3 – we have for the over-all growth rates about 15 times as much in the diffusion flame as in the premixed flame. On a surface specific basis it is about 4 or 5 times as much growth rate. Is that helpful?

Glassman: I think that Professor Sarofim and I have the same opinion about your last statement. I think that in a premixed flame indeed all the fuels break down to a common species which then build up to soot in the post flame zone. If you form more soot from one flame to another it is the different critical C/O-ratio that has essentially to do with the quantities of soot that are formed. Whereas in a diffusion flame, even though we have a common mechanism, different fuels contribute at a different part of the over all common mechanism. And therefore there is a fundamental difference between the chemistry in the diffusion flame which is a pure pyrolysis starting, let us say, with butadiene and forming the various aromatic rings. By starting with butadiene in a premixed flame I am absolutely convinced we break the fuel down to acetylene during the flame process and then build up large molecules again.

But I would like to pose another problem to you about the inception point. Prof. Wagner was saying earlier that there is the concept that even in diffusion flames particle formation starts at roughly the same temperature. I am going to discuss this in my talk this afternoon. Then the question arises why for different fuels do I obtain different smoke heights. The only argument in my opinion is that different fuels give you different inception particle concentrations even though they build up from a common mechanism as Michael Frenklach and others have argued. I think the number of particles that form is important and, of course, I agree with you that you must look at the temperature profiles in a diffusion flame if you are measuring a smoke height. Because as you change the temperature, the distance at which we think the soot starts to form becomes different and therefore the growth rate is affected. Indeed I think that is why you get a so called smoke height test.

Finally, I have to comment that I am glad that the recommendation I made about dilution about 15 years ago has caused so much controversy.

Colket: I just want to mention that Mitch Smooke has done some similar calculations for constant fuel flow and just diluting with nitrogen. He basically comes up with the conclusion that temperature appears to be the dominant effect in the control of the soot process and that is the primary effect of the dilution.

I also want to ask a question with regards to what we call a sort of inception zone down near the lip of the burner where perhaps the dilution effects and residence time effects may be more significant than they would be further up in the flame. Could you comment on this?

Santoro: It should be pointed out that the residence time shift that I showed could be due either to a concentration change or a temperature change. There is nothing in my measurements that showed that it shifted solely due to temperature. So you're right, low in the flame concentration effects are more significant. You saw in the calculations that I showed that the concentration differences are more severe low in the flame. This is some support, I think, for Kennedy's arguments about inception not being the controlling influence in some flames. In toluene flames we find that dilution has much less effect. And that is because these toluene flames are producing so much soot. It appears that a larger fraction of soot is attributed to the surface growth process. We can dilute toluene flames by large factors and only affect the amount of soot by 10 or 20 %. So that seems to be another trend I have listed in the conclusions.

I want to say a few words about the controversy Irv mentioned. I want to compliment Law and Axelbaum for re-examining this problem. Everybody said, well, we dilute and change temperature. Law and Axelbaum went back and looked at it in a different way. It has been very illuminating for me to look at this problem. I do not happen to agree fully with their conclusions but this is an honest scientific disagreement. I am glad that they and Gülder are asking these questions again and I am learning a great deal from re-examining this entire question.

Mass Growth of Charged Soot Particles in Premixed Flames

Paul Roth, Andreas Hospital

Lehrstuhl für Verbrennung und Gasdynamik,
Universität Duisburg,
47057 Duisburg, Fed. Rep. of Germany

Abstract: The mass growth of charged soot particles was studied in low pressure premixed flat flames of acetylene and oxygen burning with C/O ratios ranging between 0.95 and 1.2. A particle mass spectrometer (PMS) was used to determine the size distribution of charged particles. It extracts a molecular beam from the flame gases and allows to determine the particle kinetic energy as well as the particle velocity thus providing a direct measure of the particle mass $m_P/z\ c$. From different flames the growth behaviour of singly charged soot particles in the mass range $0.5 \cdot 10^{-20}$ g $\leq m_P \leq 50 \cdot 10^{-20}$ g was measured without interference by particle coagulation. The obtained data could be interpreted in terms of a specific mass growth rate.

14.1 Introduction

The formation and emission of soot from hydrocarbon flames is the subject of numerous investigations. It is of technical interest to understand the formation and growth mechanisms of soot and to describe its chemical and physical properties. The main strategy in practical combustion processes is to prevent soot formation or to minimize its emission. Some excellent reviews on soot and particulate carbon have been published, see for example [14.1-3].

In this paper a newly developed particle mass spectrometer (PMS) was applied to determine the mass of charged particles. It consists of a molecular beam sampling system, an electrical deflection and filter system, and electrometric amplifiers of very high sensitivity. The measurements were performed in low pressure premixed flat C_2H_2/O_2 flames with C/O ratios ranging between 0.95 and 1.2. It was possible to measure the mass growth and the agglomeration of soot particles. The results obtained allow to determine a specific surface growth rate $\dot{m}_S''$.

14.2 Particle Classification and Measurement Principles of the PMS

The basic particle properties used to realize particle classification in the PMS are the electrical charge and the inertial behaviour of aerosol particles. For the following considerations a well focused beam of charged particles is assumed to move in x-direction into a vacuum containment. The particles are characterized by the mass m_P which is a distributed property, the uniform velocity $\dot{x}_P = v_P$, and the electrical charge $z\,e$, which means the number z of elementary charge $e = 1.6 \cdot 10^{-19}$ coulomb per particle. The principle situation is illustrated in Fig. 14.1. The particle beam enters an electrical field represented by a capacitor of length l_K and width b_K supplied with a voltage U_K. The lines of force of the electrical field are assumed to be only existing inside the capacitor perpendicular to the surfaces. The field strength is $E = U_K/b_K$. An individual charged particle entering the electrical field is deflected by a force $F_y = z\,e\,U_K/b_K$ which has the direction of the field lines. It causes an acceleration of the particle in the y-direction according to its mass: $\ddot{y}_P = z\,e\,U_K/b_K\,m_P$. It can be assumed that the gravity field does not significantly influence the particle movement, e.a. $\ddot{x} = 0$. The individual particle trajectory and the velocity components inside the capacitor can be described in cartesian cordinates using the time t as parameter:

$$y_P = \frac{1}{2}\frac{z\,e\,U_K}{m_P b_K}t^2 \qquad x_P = v_P t \qquad (14.1a)$$

$$\dot{y}_P = \frac{z\,e\,U_K}{m_P b_K}t \qquad \dot{x}_P = v_P \qquad (14.1b)$$

for $0 \leq x_P \leq l_K$ or $0 \leq t \leq l_K/v_P$. Outside the electrical field the particles move on straight lines as indicated in Fig. 14.1. The direction of the movement with respect to the x-axis results from the condition of constant velocity components $\dot{y}_P$ and $\dot{x}_P$ outside the capacitor. The particle trajectory is given by

$$y_P = \frac{z\,e\,U_K}{m_P v_P^2}\frac{l_K}{b_K}\left(x_P - \frac{l_K}{2}\right) \qquad (14.2)$$

for $l_K \leq x_P$. Every individual particle of mass m_P and charge $z\,e$ has according to Eq. (14.2) its individual trajectory $y_P = f(x_P)$; in other words: the originally uniform and well focused particle beam is now splitted up into a fan-shaped form.

The described situation of a beam dispersed according to the mass and charge of individual particles is very similar to the optical spectroscopy,

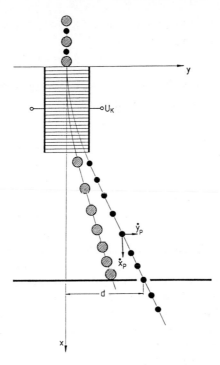

Fig. 14.1. Deflection of charged particles by an electrical field and resulting particle trajectories in a vacuum containment

where light is dispersed by optical elements. Like in spectroscopy a grounded slit is now introduced, which is located at a distance $x = l$ from the origin of the coordinate system having a slit width Δd at a position $y = d$ from the x-axis. It seperates a certain class of particles which can pass the slit whereas the rest is blocked and collected by the grounded plate. From re-arranging Eq. (14.2) introducing $y_P = d$ and $x_P = l$ the kinetic energy selecting operation of the whole arrangement will become obvious.

$$\frac{m_P}{z\,e}\frac{v_P^2}{2} = U_K \frac{l_K}{b_K}\left(\frac{2l}{d} - \frac{l_K}{d}\right) \qquad (14.3)$$

For fixed geometrical conditions, especially for a fixed slit position or beam deflection d, capacitor dimensions l_K and b_K, and for a known distance l between the entrance of the capacitor and the monochromator slit, the voltage U_K selects particles of the kinetic energy $m_P v_P^2/2$ carrying z elementary charges. By varying or tuning the deflection voltage U_K, particles of various kinetic energies can pass the monochromator slit. In other words: U_K is a direct measure of the kinetic energy of such particles which can pass the slit arrangement.

$$U_K = const\,\frac{m_P}{z\,e}\frac{v_P^2}{2} \qquad (14.4)$$

Up to this point the particle beam is classified by the deflection voltage U_K together with the slit into classes of kinetic energy. The aspired proportionality between particle mass $m_P/z\ e$ and the deflection voltage U_K according to Eq. (14.4) can be obtained, if the particle velocity v_P is known. This property can be measured by an electrical chopping system, which is illustrated in Fig. 14.2. The selected beam of charged particles passes two grids supplied with a synchronously pulsed repelling potential. The upper grid forms packages of charged particles in the following way: during the time when the potential is zero the particles can pass the grid and they are repelled during the time where the voltages is suddenly increased to 1 kV and so on. The length of the package l_P depends on the frequency of the grid voltages f and on the particle velocity v_P. Whether the full package of particles or only part of it can pass the second grid and reach the monochromator slit can be estimated by a simple geometrical phase condition. Only very few or no particles pass the second grid if the package length l_P is an odd-numbered part of the grid distance l_G. This results in a series of minima particle current $I_P = I_{P\ \min\ i}$, which occure in the plots at frequencies $f_{\min\ i}$, for the following conditions:

$$v_P = 2\,f_{\min\ i}\,l_{P\ i} \qquad (14.5)$$

$$l_{P\ i} = l_G,\ l_G/3,\ l_G/5,\ \text{etc.}$$

Equation (14.5) represents the measurement instruction for the particle velocity v_P. The frequencies $f_{\min\ i}$ of the repelling potential must be determined which causes a minimum of the particle current I_P. This minimum frequency respectively the particle velocity v_P can be measured for every group of particles selected by the deflection voltage U_K. So Eq. (14.4) can be further simplified to its final form:

$$U_K = const\ \frac{m_P}{z\ e} \qquad (14.6)$$

This is a typical relation known from mass spectroscopy. The deflection voltage U_K of the capacitors is a known direct measure of the particle mass m_P divided by the particle charge $z\ e$. The remaining problem of the PMS is to determine the particle current $I_P(m_P)$ classified by both the deflection voltage U_K and the monochromotor slit. This value is also called the probability density function of the particle mass m_P. The simplest way is to collect the charged particles by a Faraday cup and to convert it into an equivalent electrical current I by a very sensitive electrometric amplifier. This was done in the present case, see Sect. 14.4. The measured electrical current I is, e.g. for singly charged particles ($z = 1$), directly proportional to the particle current I_P.

$$I = f(U_K)$$

$$I \sim I_P = \Delta N_P/\Delta m_P = I_P(m_P) \qquad (14.7)$$

The measured electrical current I represents the probality density function of the particle mass m_P. For particles carrying more than one elementary charge $(z > 1)$ the situation is principally similar.

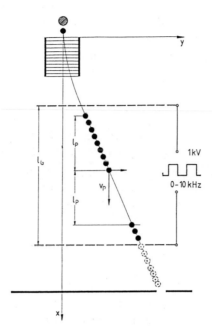

Fig. 14.2. Formation of particle packages and operation of the electrical two-grid-filter supplied with synchronously pulsed high voltage potential

Beside the described method of recording the classified particle current, other physical principles are also applicable. A laser beam crossing the particle beam behind the slit can be used together with an optical system to count the particles flowing through the scattering volume. It is also possible to generate ionic fragments or free electrons from the beam of charged particles behind the slit and to count the electrical pulses by a photomultiplier. In both cases the number of pulses per time is proportional to the probability density function $I_P(m_P)$.

14.3 Experimental

The apparatus used to study the formation of soot particles in low pressure premixed flat C_2H_2/O_2 flames, described earlier [14.4] is shown in Fig. 14.3. It consists of a low pressure burner and the PMS system for the particle analysis.

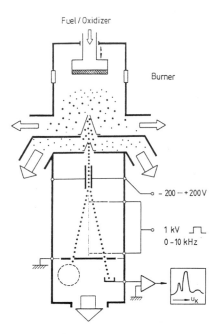

Fig. 14.3. Experimental set-up: low pressure burner and PMS (particle mass spectrometer)

The investigated low pressure flames were stabilized on a 75 mm diameter water-cooled sintered bronze plate. Calibrated flow meters were used to control the flow rates of fuel and oxidizer. In the experiments the C/O ratio was varied between 0.95 and 1.2, and the velocity of the cold unburned gas ranged between 45 cm s^{-1} $\leq v_u \leq$ 60 cm s^{-1}. The pressure in the combustion chamber was held by a 0.025 m^3 s^{-1} mechanical pump together with a controller system at pressure levels between 2.0 kPa and 4.0 kPa. The whole burner can be moved in the vertical direction to vary the distance between burner plate and the fixed nozzle.

The PMS consists of a molecular beam sampling system, by which a probe of the flame exhaust gases was supersonically expanded through a 0.75 mm diameter platin plated and electrically grounded 45° quartz nozzle. This is also indicated in Fig. 14.3. The center of the free jet was ex-

tracted by a 0.75 mm diameter grounded skimmer. The molecular beam
contains charged and uncharged particles. The kinetic energy spectra of the
charged soot particles were determined by passing them through an electric
field. The deflection from the beam axis depends on the number of charges
per particle, on the deflection voltage, and on the kinetic energy of the
particles. For a fixed geometry, the recorded current is a measure of the
number of particles belonging to the selected energy class represented by
the deflection voltage U_K. The particle beam also passes two grids supplied
with synchronously pulsed repelling potentials serving as electrical filters.
By tuning the pulse frequency, this method delivers the time of flight of
the particles between the two grids, which is a measure of the particle velocity. As indicated in the previous section the results of both combined
measurements of particle kinetic energy and particle velocity are necessary
to determine the probability density function of the soot particle mass.

14.4 Results

A typical result obtained with the described PMS is shown in Fig. 14.4. The
burning conditions of the low pressure flame are $C/O = 1.1$, $p = 2.5$ kPa,
$v_u = 52.5$ cm s^{-1} and the sample was taken at 60 mm height above the
burner. The measured positive and negative electrical currents are shown
as a function of the deflection voltage U_K. Both curves have three maxima at
voltages $U_K = 30$ V, 60 V, and 120 V. According to Eq. (14.7) the electrical
current of Fig. 14.4. is proportional to the particle current per mass class
Δm_P and represents the probability density function of m_P. The deflection
voltage U_K is directly proportional to $m_P/z\,e$, see Eq. (14.6). So, in addition
to the U_K-scale in Fig. 14.2., the particle mass and particle size scale for
$z = 1$ are also given. A typical sequence of $U_K \sim m_P / z\,e$ spectra of
young charged soot particles is shown in Fig. 14.5. With increasing height
above the burner, that means increasing time, a shift of the current peaks
to higher U_K values and a broadening of the peaks is obvious. This effect
is accompanied by the appearance of a multi peak structure. The spectra
obtained at $h \geq 40$ mm show a three peak strucutre. The measured
particle velocities decrease with increasing height above burner from about
$v_P = 1050$ m s^{-1} at $h = 20$ mm to $v_P = 970$ m s^{-1} at $h = 60$ mm.

The present measurements were performed under different burning conditions represented by different C/O ratios and different velocities of the
unburned gas v_u. The general shape of the curves remained always the same
although the measured current signals varied about two orders of magnitude
in the peak height. With increasing height above the burner all peaks were
shifted on the U_K axis to higher values.

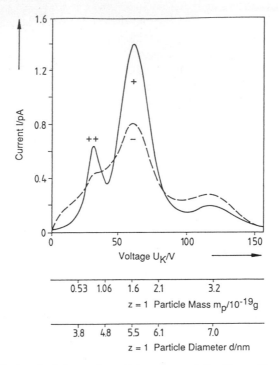

Fig. 14.4. PMS signal of charged particles versus deflection voltage U_K at 60 mm height above the burner; burning conditions: $C/O = 1.1$, $p = 2.5$ kPa, $v_u = 52.5$ cm s^{-1}

14.5 Discussion

The broadening and the shift of the current peaks to higher U_K values with increasing height above the burner (Fig. 14.5.) is a result of the particle growth process. With increasing height above the burner a multi peak structure of the current signals is obvious which is totally developed at a height above the burner of 60 mm. The signals can be understood as a combination of three seperate peaks. According to the spectrum at 60 mm height above the burner of Fig. 14.4., the peak at $U_K = 60$ V must be related to singly charged particles, that at $U_K = 30$ V to doubly charged particles and the last one at $U_K = 120$ V to agglomerates. This interpretation is supported by the increase of the third peak with increasing time respectively height above the burner.

The evaluation of the height above the burner dependent peak sequences of the singly charged particles in the range 0 mm $\leq h \leq$ 60 mm can be done by combining the spectra of Fig. 14.5. with particle velocity measurements. The deflection voltage U_K can than be converted to the mass m_P of individual particles as described earlier. By a detailed evaluation of the spectra

Height above Burner h/mm

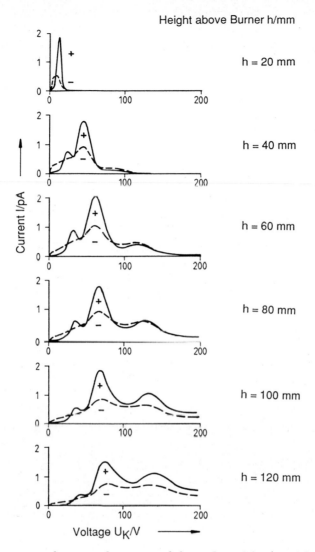

Fig. 14.5. Sequence of measured currents of charged particles (particle mass spectra); burning conditions: C/O $= 1.1$, $p =2.5$ kPa, $v_u = 52.5$ cm s^{-1}

like that of Fig. 14.4., the size distribution of the soot particles, not discussed here in detail, could easily be determined. A log-normal distribution was obtained having a mean geometric standard deviation of $\sigma_g \approx 1.15$. The further evaluation of the spectra was concentrated on the shift in the current peaks to higher mass values. It must be pointed out that the interpretation of the shift in terms of particle mass growth is independent on the particle coagulation. Every successful particle collision results in a new particle of about the double mass, which jumps in the spectrum to the double U_K value.

It only affects the current height in the spectrum but does not falsify the particle mass evaluation. In Fig. 14.6. examples of measured mass growth curves of singly charged particles obtained from the PMS-spectra of different flames are shown as a function of particle time. Two facts are interesting to notice:

a) at early reaction time, the particle mass seems to level off with a tendency to reach a constant value,

b) in nearly all flames, the particle mass increases again significantly at longer reaction time without indicating to approach a final value.

The evaluation of the PMS spectra was stopped, if the individuality of the peaks disappears. This is the reason why not all mass curves are available up to 50 ms.

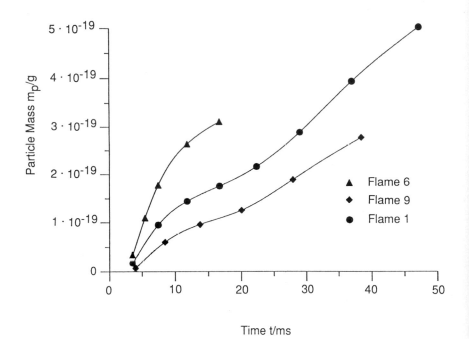

Time t/ms

Fig. 14.6. Measured growth of singly charged primary soot particles; burning conditions: flame 1: $C/O = 1.1$, $p = 2.5$ kPa, $v_u = 52.5$ cm s^{-1}; flame 6: $C/O = 1.2$, $p = 2.5$ kPa, $v_u = 52.5$ cm s^{-1}; flame 9: $C/O = 1.1$, $p = 2.0$ kPa, $v_u = 52.5$ cm s^{-1}

To find the connection to previous work, we have tried to interpret the first part of the particle mass growth by a first order apparent growth law similar to the one proposed by *Haynes* et al. [14.5] for the soot volume fraction.

$$\frac{d\,m_P}{dt} = k_{sg}(m_{P\infty} - m_P) \tag{14.8}$$

In that equation $m_{P\infty}$ and k_{sg} are the final particle mass and the surface growth rate coefficient, respectively. It seems clear that this equation cannot describe the complete observed particle growth behaviour. By variation of the parameters k_{sg} and $m_{P\infty}$, the integrated Eq. (14.8) can be fitted to the first part of the measured particle mass growth curves. Results for the rate coefficient k_{sg} obtained are similar to values reported by *Haynes* et al. [14.5] and by *Bockhorn* et al. [14.6-7]. It is clear and was pointed out by *Haynes* et al. [14.5] and others that an Arrhenius plot of k_{sg} is not strictly allowable in non-isothermal flames. It only indicates that the apparent soot reactivity is a strong function of temperature.

During our PMS measurements we observed a significant further growth of the soot particles after the tendency to level off. This is a remarkable fact which cannot be described by Eq. (14.8) and which must be a result of the temperature and concentration profiles in the flame gases. The mass growth of an individual particle in the free molecular regime is determined either by mass addition or mass abstraction reactions at the particle surface. The problematic properties are the nature of the active reacting species and the formulation of their global reaction probabilities. It is useful to condense the whole complexity of the heterogeneous particle surface kinetics into the simple property:

$$\dot{m}_S'' = \frac{1}{a_P} \frac{d\,m_P}{dt} \qquad (14.9)$$

which is the specific mass growth rate. It has the quality of a heterogeneous reaction rate and depends on temperature, surface structure and reactivity, as described above. It is a big advantage of our PMS measurements to have precise mass growth data which are independent on optical properties and which are not affected by coagulation. Like *Harris* et al. [14.8] did before, we have determined dm_P/dt by graphically differentiating the mass curves of Fig. 14.6. The particle surface a_P was calculated from m_P by assuming spherical particles having a density of about 1.8 g cm^{-3}. Results of specific mass growth rates $\dot{m}_S''$ of singly charged primary particles are shown in Fig. 14.7. as a function of particle time. The different symbols stand for the $\dot{m}_S''$ values obtained from the different flames. The general tendency of the measurements seems to indicate a S-shaped curve for the specific growth rate of the soot particles. It is obvious that $\dot{m}_S''$ decreases dramatically during the first 10 ms and seems to reach a more or less constant level followed again by a tendency to rapidly decrease again. The scattering of the data is relatively low at early reaction time and increases with increasing time. The observed behaviour of $\dot{m}_S''$ is a result of the temperature decrease in the flames, which affects the concentration of the growth compounds as well as their reaction probabilities. It seems clear from the results of Fig. 14.7. that the particle growth process is not finished during the reaction time of 40 ms, it only proceeds with a very low rate.

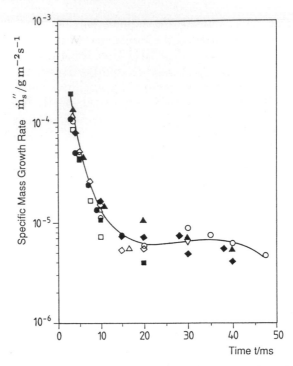

Fig. 14.7. Measured specific mass growth rates of individual singly charged soot particles obtained from different flames

14.6 Acknowledgement

This work originated in the Sonderforschungsbereich 209 of the University of Duisburg. The financial support of the Deutsche Forschungsgemeinschaft is gratefully acknowledged.

References

14.1   H.Gg. Wagner: "Soot Formation in Combustion", in Seventienth Symposium (International) on Combustion (The Combustion Institute, Pittsburgh 1978) p. 3

14.2   Particulate Carbon: Formation During Combustion, ed. by D.C. Siegla, G.W. Smith (Plenum Press, New York 1981)

14.3   F.J. Weinberg: Advanced Combustion Methods (Academic Press, London 1986)

14.4   A. Hospital, P. Roth: " In-situ Mass Growth Measurements of Charged Soot Particles From Low Pressure Flames", in Twenty-Third Symposium (International) on Combustion (The Combustion Institute, Pittsburgh 1990) p. 1573

14.5   B.S. Haynes, H.Gg. Wagner: Z. Phys. Chem. **133**, 201 (1982)

14.6   H. Bockhorn, F. Fetting, A. Heddrich, G. Wannemacher: "Investigation of the Surface Growth of Soot in Flat Low Pressure Hydrocarbon Oxygen Flames", in Twentieth Symposium (International) on Combustion (The Combustion Institute, Pittsburgh 1984) p. 979

14.7   H. Bockhorn, F. Fetting, A. Heddrich: "Investigation of Particle Inception in Sooting Premixed Hydrocarbon Oxygen Low Pressure Flames", in Twenty-First Symposium (International) on Combustion (The Combustion Institute, Pittsburgh 1986) p. 1001

14.8   S.J. Harris, A.M. Weiner: "Some Constraints on Soot Particle Inception in Premixed Ethylene Flames", in Twentieth Symposium (International) on Combustion (The Combustion Institute, Pittsburgh 1984) p. 969

Discussion

Bockhorn: In deriving the rate expression for the change of particle mass with time you assumed that particle mass increases only due to surface growth. The evolution of the volume fraction of soot with time is due to surface growth. However, the main effect that changes the single particle volume may come from coagulation and not from mass addition from the gaseous phase. Can you exclude in your system coagulation of charged particles with neutrals?

Roth: Soot particles grow by surface growth and by coagulation. We observe both in our particle mass spectra. The coagulation of a neutral and a charged particle forms a charged agglomerate, which appears as a hump in the mass spectrum at the double deflection voltage (third peak in the spectrum). The surface growth of the soot particles is a continuous process causing a continuous shift of the deflection voltage in the spectra.

Lahaye: From your current versus voltage diagrams you brought about the conclusion that surface growth and coagulation are consecutive steps. In other words the surface growth is achieved when the coagulation of particle is occuring. If you look at soot particles by phase contrast electron microscopy you notice that the surface growth continues after the coagulation of the particles. Do you have a comment about this aspect?

Roth: You are right, particle coagulation and particle surface growth take place simultaneousely. With our PMS technique we are able to separate both processes. The coagulation process forms new particle classes, which are indicated by additional peaks in the spectra. Simultaneousely the surface growth process shifts the primary particles as well as the agglomerates to higher mass values. Both were observed in our experiments, but the growth of the primary particles appeared more clearly at lower heights above the burner.

Calcote: You showed a very nice peace of work and you have moved into a range of particle masses where measurements are urgently needed. Have you considered measuring neutral particles with your method? If you pass the neutral particles through an electron swarm for example you could produce charged particles and then you have both charged and the neutral particles.

Roth: Thank you for your comment. We are thinking about that and we are working on a charging mechanism so that we also have information about neutral particles. But it is not so easy to charge particles and not to change the conditions in the flame or in the beam.

Lepperhoff: Coagulation means that two particles collide and stick together and this doubles their mass. This is the third peak in your spectra. Do you have coagulation only between two particles not between a third one. In other words, why don't you find further peaks or are they included in the second peak.

Roth: In principle you are right. Coagulation is not limited to coagulation of single primary particles. In our model we also include coagulation of coagulated particles. But we could show that under the prevailing number densities of the soot particles coagulation of the coagulated particles is not a dominant process. This is confirmed by the experiments.

Smyth: If your signal to noise is as high as it appears, then you could narrow the slit and get much higher resolution.

Roth: We can of course narrow the slit and increase the resolution and we are on the way to do this and to determine the size distribution of the particles. Up to now we found mainly log normal distributions. We are also working on the detector side of the instrument. We want to replace the electrometric amplifiers by a counting system to increase the sensitivity of the instrument.

Homann: Could you comment on the differences between the energy spectra of the negatively and positively charged particles, particularly for the very light and the very heavy particles?

Roth: What you can see from the spectra is that the behaviour of the positively charged particles is significantly different from the behaviour of the negatively charged particles. The main difference occurs in the first part of the spectra at deflection voltages between 0 and 20 V, see Fig. 14.4. Here the positively charged particles increase little and the current transported by the negatively charged particle is much higher. This can be interpreted as big particles carrying multiple charges or very light negatively charged particles. We believe that there are large particles carrying multiple charges and to demonstrate this we did the following experiment. We increased the pressure in our high vacuum chamber where we observed the particle beam. By this the dispersion of the particle beam is increased dramatically. However, mainly the small particles are affected by this so that the small particles should disappear and the bigger particles should survive. What we observed is that the first part of the spectra is only weakly affected by an increase in pressure. This is an indication that there are probably big particles carrying multiple negative charges.

Growth of Soot Particles in Premixed Flames by Surface Reactions

Henning Bockhorn [1], Thomas Schäfer [2]

[1]Interdisziplinäres Zentrum für wissenschaftliches Rechnen,
Universität Heidelberg,
69120 Heidelberg, Fed. Rep. of Germany

[2]Institut für Chemische Technologie, T. H. Darmstadt,
64287 Darmstadt, Fed. Rep. of Germany

Abstract: Mean radii, number densities and specific surface areas of the soot particles generated in laminar premixed flames are measured by means of laser light extinction and scattering. Temperatures are obtained from thermal radiation of soot particles and by applying the sodium-D-line reversal method. Gas samples of the flames are analysed by means of mass-spectrometry. The experimental part of this work covers two series of experiments in sooting ethyne-oxygen-argon flames. These experiments were performed to elucidate some hypotheses for soot formation and surface growth:

- The dependence of soot appearance rates on the surface area of the soot particles is investigated. For this flames are seeded with alkali metal solutions to prevent the soot particles from coagulation. These flames are compared with the equivalent unseeded flames. The experimental results give the appearance rates of soot as beeing independent of the outer surface of the soot particle aggregates.
- The influence of chemical composition - in particular the dependence of the soot appearance rate on the ethyne partial pressure - on soot formation has been experimentally investigated. For this, flames with different pressure, flow rates, and composition of the feed have been investigated. The experiments show that the appearance rate of soot is not simply proportional to the ethyne partial pressure as stated in frequently used surface growth models but rather influenced by a number of flame properties.

An explanation of the experimental findings is given on the basis of the frequently discussed HACA-mechanism for surface growth of soot particles with the C-H sites accessible for growth being spread about the inner surface of the soot particle aggregates.

15.1 Introduction

Hydrocarbons tend to form soot when burning under fuel rich conditions. It is obvious that the formation of soot, i.e. the conversion of a hydrocarbon fuel molecule containing few carbon atoms into a carbonaceous agglomerate containing some millions of carbon atoms, is an extremely complicated process in that it is a kind of gaseous - solid phase transition where the solid phase exhibits no unique chemical and physical structure. Therefore, soot formation encompasses chemically and physically different processes, e.g. the formation and growth of large aromatic hydrocarbons and their transition to particles, the coagulation of primary particles to larger aggregates, and the growth of solid particles by picking up growth components from the gas phase.

While the above mentioned processes may be attributed to the formation of the bulk of soot numerous other phenomena connected with soot formation and determining the "fine structure" of soot have to be considered, e.g. the formation of electrically charged soot particles, the formation - charged and neutral - of fullerenes, or the formation of high molecular tarry modifications with optical properties quite different from carbon black.

Much progress has been achieved in understanding of all these processes. However, numerous problems remain unsolved. This is documented in periodically appearing comprehensive discussion and critical review of the state of the art, compare e.g. refs. [15.1-5] for general aspects, refs. [15.3,15.6] for the role of ions, or refs. [15.7,15.8] for formation of fullerenes during soot formation, as well as several workshops devoted to particular problems and phenomena connected with soot formation, compare e.g. [15.9-11].

The ideal system for identifying the most important parameters influencing the over-all processes are one-dimensional ones, such as premixed laminar flat flames, or zero-dimensional ones, such as shock tubes. In these systems chemical reactions leading to soot do not interfere with convective/diffusive transport or molecular dissipation that change the local elementary composition of the reacting mixture.

In premixed flames the appearance rates of soot given as the variation of the soot volume fraction with time can be described by means of an apparent first order rate expression [15.12-14]

$$\frac{\mathrm{d}\, f_V}{\mathrm{d}\, t} \;=\; k_{\mathrm{sg}}\,(f_{V\infty} \;-\; f_V) \tag{15.1}$$

where k_{sg} is a coefficient rather independent of fuel type but depending on temperature with an apparent activation energy of 120 to 160 kJ mole^{-1} [15.13,15.14]. k_{sg} ranges from 30 to 500 s^{-1} depending on the burning conditions and the temperature dependence of k_{sg} has to be interpreted as the variation of k_{sg} in different flames having different maximum temperatures.

Equation (15.1) is demonstrated in Fig. 15.1 for a variety of premixed flames burning at different conditions.

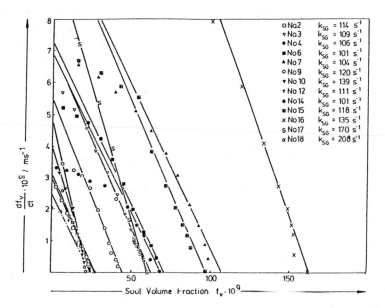

Fig. 15.1. Appearance rates of soot for premixed hydrocarbon flames [15.14]

The influence of the chemical nature of the fuel on the appearance rate of soot is reflected by $f_{V\infty}$ which is the "final" soot volume fraction attained after the rapid formation phase of soot shortly after the main reaction zone of the flame. $f_{V\infty}$ depends on pressure, mixture composition, temperature and nature of fuel. This dependence is investigated comprehensively in refs. [15.15-17]. The influence of the fuel enters via $f_{V\infty} \propto [C/O - (C/O)_{crit}]^n$, where $(C/O)_{crit}$ is the C/O-ratio at the sooting limit, e.g. $(C/O)_{crit} \approx 0.6$ for ethylene, ≈ 0.65 for benzene, or ≈ 0.47 for propane, and the exponent n is e.g. $n \approx 2.5 - 3$ for ethylene, ≈ 4 for propane, or ≈ 2.8 for benzene [15.14-17].

From an over-all view Eq. (15.1) provides a means of predicting appearance rates of soot or soot volume fractions from empirical or experimental input. Integrating Eq. (15.1) with the assumption of constant temperature in the soot forming region one obtains

$$f_V = f_{V\infty} \left(1 - e^{-k_{sg}t}\right) \tag{15.2}$$

which may be substituted into Eq. (15.1) to give

$$\frac{d f_V}{d t} = f_{V\infty} k_{sg} e^{-k_{sg}t} . \tag{15.3}$$

Equation (15.3) gives an exponential decay of the appearance rate of soot with time. If the largest part of soot is formed by surface growth reactions a coarse model for soot formation is a model for surface growth and Eq. (15.1) or Eq. (15.3) are equally well empirical surface growth rate expressions the coefficients of which have to be interpreted chemically.

One plausible approach for surface growth is the decomposition of acetylene, which is the most abundant chemical species in the soot forming region of premixed flames, at the surface of soot particles [15.18-20]. This results in

$$\frac{d\,f_V}{d\,t} = k_{C_2H_2} p_{C_2H_2} A_s \tag{15.4}$$

where A_s is the surface area of the soot particles and $p_{C_2H_2}$ is the partial pressure of acetylene. The rate coefficient $k_{C_2H_2}$ is supposed to be time dependent, viz. $k_{C_2H_2} = k_0\,e^{-\alpha t}$, reflecting the exponential decay of the surface growth rate according to Eq. (15.3) and being interpreted as the decay of the reactivity of the soot particles. Integrating Eq. (15.4) again assuming constant properties in the soot forming zone of the flame - and comparing the results with Eq. (15.2) gives

$$f_{V\infty} = f_{V0} + p_{C_2H_2} A_s \frac{k_0}{\alpha} \tag{15.5}$$

and

$$k_{sg} = \alpha. \tag{15.6}$$

This approach gives the rate coefficient k_{sg} as the rate coefficient for the process of deactivation of the soot surface and $f_{V\infty}$ is determined through the initial soot volume fraction f_{V0}, the partial pressure of acetylene, and the total soot surface area.

A more detailed discussion of the decay of the soot particle reactivity is provided in refs. [15.21-22]. There surface growth is proposed to occur via the reaction of gaseous growth components at active sites on the soot particle surface. Then the surface growth rate is given by

$$\frac{d\,f_V}{d\,t} = \gamma \frac{\bar{c}}{4}(N_s\,a_s)(N_g\,V_g) \tag{15.7}$$

where N_s is the number density of the active sites with the area a_s, N_g the number density of the gaseous growth components which occupy the volume V_g after being fixed at the surface and $\bar{c}$ is the molecular mean velocity of the growth components which are deposited with the reaction probability γ. The number of active sites remains uninfluenced by coagulation but decreases by thermal stabilisation with a rate given by

$$N_s = N_{s0} e^{-k_{da} t}. \tag{15.8}$$

On integration Eq. (15.7) and comparing the result with Eq. (15.2) one obtains

$$f_{V\infty} = \frac{\gamma \frac{\bar{c}}{4}(N_{s0}\, a_s)(N_g\, V_g)}{k_{da}}$$ (15.9)

and

$$k_{sg} = k_{da}\,.$$ (15.10)

Following this approach the rate coefficient k_{sg} is the rate coefficient of the thermal stabilisation reaction of the active sites and $f_{V\infty}$ is related to the initial number density of active sites and the concentration of the gaseous growth component which is preferably acetylene. A more detailed discussion is given in ref. [15.22].

The active site model as well as the acetylene decomposition model provide a chemical interpretation of the exponential decay of the surface growth (appearance) rate of soot given by Eq. (15.3). In employing these models for prediction of soot volume fractions in flames, in addition to the deactivation or thermal stabilisation rates, the partial pressure of acetylene - or the appropriate growth components - has to be provided by combustion modelling. Moreover, the initial soot volume fraction or the initial number density of active sites have to be introduced by reasonable arguments.

Chemical reactions of surface growth are treated in the active site model or the acetylene decomposition model from an over-all point of view. A mechanistic interpretation of surface growth reactions has been introduced by Frenklach [15.23-25] treating the surface growth reactions analogously to the planar growth of polycyclic aromatic hydrocarbons. The basic concept of this approach is the transfer of the H-abstraction C_2H_2-addition mechanism (HACA-mechanism) to the heterogeneous surface growth of soot particles that consequently occurs at edges or ridges offering a PAH-like structure to H-atoms and acetylene, compare Fig. 15.2. This approach has been meanwhile modified and adopted by others, e.g. refs. [15.26-28]. The big impact on modelling soot formation from a mechanistic point of view is that comparable to large reaction systems in polymerisation the growth reactions of PAH encompass reactions between similar classes of particles so that the complex mixture may be described by lumped species classes rather than by single PAH-species. In using this approach to predict soot volume fractions the "surface growth" part has to be embedded into the detailed description of gas phase chemistry, that provides H-atom and acetylene concentration, formation and polymerisation of PAH, and formation and growth of soot particles by particle inception, which basicly is treated as coagulation of PAH, surface growth and possibly other condensation processes.

The aim of this work is to provide some more experimental information to the discussion of hypotheses for surface growth of soot particles, in particular the dependence of the appearance rates of soot on the amount of soot particle surface area and on the chemical environment of the soot particles has been investigated. For this premixed ethyne-oxygen flames are seeded with alkali metal solutions to prevent the soot particles from coagulation. These flames are compared with the equivalent unseeded flames.

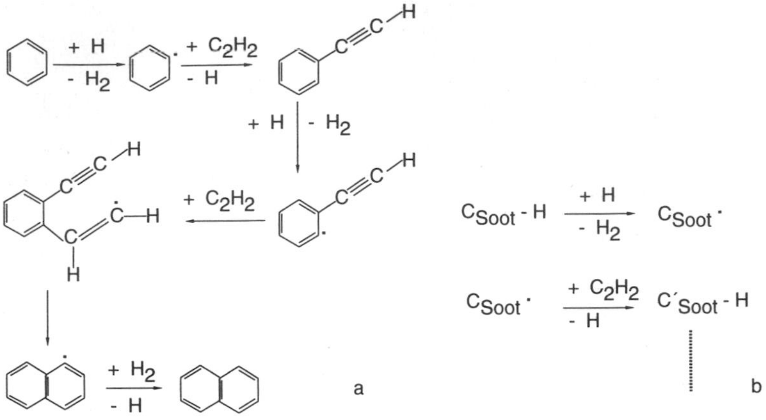

Fig. 15.2. H-abstraction C$_2$H$_2$-addition mechanism for planar PAH growth (a) and extension to the surface growth of soot (b) [15.24-25]

The second aim - the dependence of the soot appearance rates on the ethyne partial pressure - has been experimentally investigated by burning flames with different pressure, flow rates, and composition of the feed.

15.2 Experimental

Premixed laminar flat ethyne-oxygen-argon-flames have been investigated at reduced pressure. The burner system has been equipped with all facilities for optical measurements and for sampling the flames by means of a micro-probe. Soot volume fractions, mean soot particle radii, and number densities were measured by means of laser light scattering and extinction. Scattered and attenuated light intensities were detected with photomultipliers with respect to polarization and wavelength of the incident laser beam. Parti-cle temperatures and gas temperatures of the flames were obtained from thermal radiation. Gas samples of the flames obtained from microprobe sampling were analysed by online mass spectrometry for concentrations of the major chemical species. All experimental procedures are described in detail in refs. [15.14,15.29].

Particle number densities and particle sizes were obtained from Lorenz-Mie-theory for spherical, homogeneous, isotropic particles. In the Rayleigh regime of scattering this yields

$$Q_{VV} = \frac{\lambda}{4\pi^2} \left| \frac{m_s^2 - 1}{m_s^2 + 2} \right| \mu_6 \, N_p \qquad (15.11)$$

and

$$k_{ext} = -\frac{\lambda^2}{\pi} \, \mathrm{Im} \left(\frac{m_s^2 - 1}{m_s^2 + 2} \right) \mu_3 \, N_p \,. \qquad (15.12)$$

In Eqs (15.11) and (15.12) Q_{VV} is the measured monochromatic scattering coefficient for vertical polarization of both scattered and incident light of wavelength λ and k_{ext} is the measured extinction coefficient; m_s means the complex refractive index of the soot particles which is taken from [15.30], $m_s = 1.57 - 0.56i$. The μ_i are the i-th moments of the distribution $P(\alpha)$ of the size parameter $\alpha = \frac{2\pi r}{\lambda}$. Equations (15.11) and (15.12) form a system of equations for particle number density N_p and one moment of the size distribution. As in the Rayleigh regime of scattering Q_{VV} is independent on scattering angle, no further information is obtainable from Q_{VV} and k_{ext}. Hence, the other moment of the size distribution has to be guessed. The size distributions were supposed to be log-normal with a standard deviation $\ln \sigma_g = 0.34$. This value is close to the value that follows from coagulation theory for a free molecular aerosol. The validity of this assumption and of the applied scatterer model for the prevailing experimental conditions is discussed in [15.29]. The above scatterer model with a lognormal size distribution holds equally well for the seeded flames with negatively charged soot particles. However, the absence of coagulation in the region of strong surface growth of the particles results in somewhat narrower size distributions. Thus, a value of $\ln \sigma_g = 0.34$ leads to a slight overestimation of the effect of the size distribution on mean particle size and number densities. A further source of uncertainty is the complex refractive index. The value adopted from [15.30] is not derived from in situ measurements in flames where the refractive index of soot particles is found to be appreciably dependent on particle age [15.31,15.32]. In spite of these uncertainties in the data entering the evaluation of the measured Q_{VV} and k_{ext} using Eqs (15.11) and (15.12), the method is valid for comparing flames with similar properties.

A systematic error of the method discussed above is the absorption of gaseous molecules. These interferences are particularly effective at incipient soot formation where the contributions of the soot particles to extinction are relatively small. This effect is evident e.g. in flat benzene-air flames from an abnormal disperison of the extinction coefficient [15.33]. To account for this possible bias the extinction mesurements were performed at two wavelengths (488 nm and 633 nm) and the soot volume fractions obtained from Eq. (5.12) were compared. For the investigated flames soot volume fractions measured at the two wavelengths do not differ within the statistical limits of error.

To account for other systematic errors in evaluating the scattering measurements, e.g. due to inelastic interactions of gaseous molecules with the incident light, the procedure outlined in [15.14,15.29] has been applied.

Particle temperatures were evaluated by comparing the radiation intensity of the sooting flames at 488 nm and 633 nm with a calibrated tungsten strip lamp the necessary absorbtivity of the flames being obtained from extinction measurements.

Particle temperatures determined in this way may depart from gas temperatures. The particle temperature in the soot formation zone may exceed the gas temperature and decreases more rapidly in the burnt gas. Under consideration of this systematic error the method nevertheless is useful to indicate temperature differences between the single flames.

Measuring of species concentrations of major components was performed by sampling the flames by means of microprobes which have been prepared to minimize disturbances of the flames. The diameters of the probe orifices were about 110 μm. To minimize catalytic activity the microprobes were treated with concentrated nitric acid and 5%-ic hydrofluoric acid. The gas samples were analyzed by means of a quadrupole mass spectrometer. The mole fractions X_i of the single components are derived from the mass spectra of the unresolved gas mixture assuming linear superposition of the components spectra by solving the system of linear equations

$$I_m = \sum_{i=1}^{N} a_{m,i} X_i, \quad m = 1, \ldots, M. \tag{15.13}$$

I_m are the recorded signals at the mth mass unit and the $a_{m,i}$ are the contributions of the components i to the signal at the mth mass unit which have to be determined by calibration measurements for the stipulated working conditions of the mass spectrometer. As usually $M \geq N$, the system of equations (15.13) has to be solved by linear regression which yields the mole fractions X_i of the components and their standard deviation.

The statistical errors for the mole fractions determined in this way were small compared with systematic errors caused by sampling the flames with microprobes. These systematic errors mainly result from disturbances of the flames by the probe which acts as sink for heat and mass. A number of investigations, e.g. refs [15.34-35], give experimental evidence that the disturbance of the flames by the microprobe has the effect of shifting the measured profiles a few millimeters in downstream direction in the region of steep gradients of temperature and mole fractions.

15.3 Results and Discussion

15.3.1 First Series of Experiments: Varying the Surface Area by Seeding the Flames

Laminar flat low pressure ethyne-oxygen-argon flames are seeded with 0.1 m aqueous solutions of caesium chloride and potassium bromide by passing the main feed through a capillary atomizer. The seeded flames are compared with the corresponding unseeded flames diluted by the equivalent amount of water. The data of the investigated flames are given in table 15.1.

Table 15.1. First series of experiments, data of the investigated flames; flow rates of the feed are given in l_N/h

flame	Ar	O_2	C_2H_2	C/O	pressure	additive
1	235.1	86.4	112.3	1.3	12 kPa	0.1 m CsCl
2	235.1	87.2	111.7	1.28	12 kPa	water
3	235.1	87.2	111.7	1.28	12 kPa	0.1 m KBr
4	235.1	82.3	115.3	1.4	12 kPa	0.1 m CsCl
5	235.1	84.0	114.0	1.36	12 kPa	water

The addition of about 10^{19} / m³ caesium atoms to flames 1 and 4 slightly decreases the soot volume fraction. This effect is compensated in flames 2 and 5 by a slightly lower supply of fuel. To account for the variation of the gas volume with temperature, soot volume fraction, particle number densities and specific surface areas are normalized to a gas volume according to 1900 K.

Seeded and unseeded flames show no significant variation in soot volume fraction, temperature and chemical environment, compare Figs 15.3 to 15.5. As evident from figure 15.3, seeded and unseeded flames give approximately identical soot formation rates. The formal rate coefficient k_{sg} from Eq. (15.1) amounts to 140 s^{-1} for flames 1 to 3 and 130 s^{-1} for flames 4 and 5.

Soot particles are assumed to be charged instantaneously by charge transfer from the alkali metals. Thus soot particles are prevented from coagulation and particle number densities remain at a high level, whereas particle radii show only a slight increase compared with the unseeded flames. The soot particles are completely charged when adding caesium chloride. The charge transfer effect when adding potassium bromide is slightly weaker because of the higher ionisation potential of potassium (4.34 eV for K, 3.9 eV for Cs).

Specific surface areas of the soot particles are plotted versus time in figure 15.6. For the unseeded flames 2 and 5 the specific surface area slightly decreases with time. Surface growth and soot particle coagulation nearly compensate with respect to the specific surface area. For the seeded flames 1, 3 and 4 the specific surface area attains more than the twofold value in

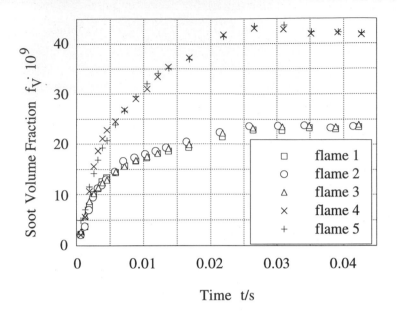

Fig. 15.3. Soot volume fractions of flames 1–5 versus reaction time

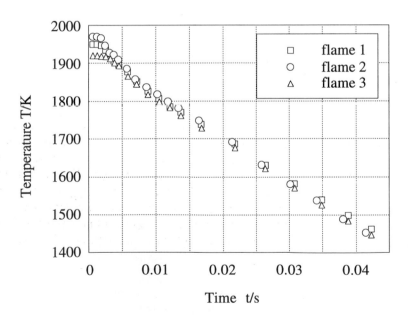

Fig. 15.4. Particle temperatures of flames 1–3 versus reaction time

comparison with the unseeded flames due to the suppression of coagulation. The comparison of the flames seeded with caesium and potassium shows that the specific surface areas of the latter are about 20 % lower, indicating that less soot particles are charged.

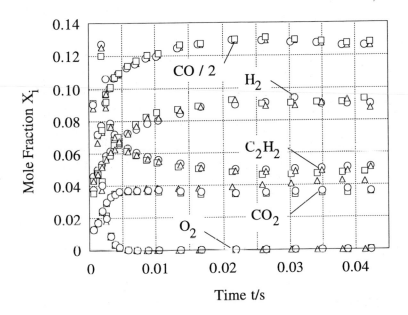

Fig. 15.5. Mole fractions of CO, H_2, C_2H_2, CO_2 and O_2 for flame 1 □, flame 2 ○, flame 3 △ versus reaction time

These experiments lead to the conclusion that surface growth rates of soot particles in premixed low pressure ethyne-oxygen-argon flames do not vary proportional to the variation of the specific surface area of the soot particles in the soot forming region of the flames. For the further discussion it should be noted that surface area of the soot particles has to be interpreted in terms of the surface of an equivalent sphere as seen by scattering experiments. This surface area is the outer surface area of an aggregate and all of the arguments above apply for the outer surface area of an aggregate.

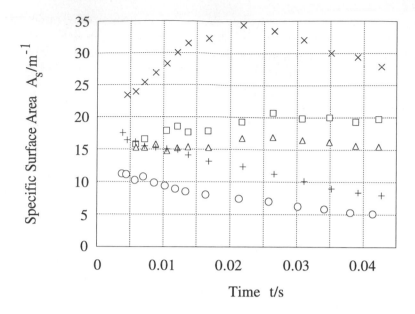

Fig. 15.6. Specific surface area for flame 1 □, flame 2 ○, flame 3 △, flame 4 × and flame 5 + versus reaction time

15.3.2 Second Series of Experiments: Varying Ethyne Partial Pressure

In this series of experiments a number of flames has been investigated at different pressure. To achieve the same final soot volume fraction within each group of flames argon dilution and C/O ratio of the feed are varied to compensate for the change in soot volume fraction with pressure. The pressure increase from 9 to 18 kPa leads to higher soot volume fractions. Therefore, the higher pressure flames burnt at increased flow rates of the inert gas argon and at reduced C/O ratio. The data of the flames are summarized in table 15.2.

Table 15.2. Second series of experiments, data of the investigated flames; flow rates of the feed are given in l_N/h

flame	Ar	O_2	C_2H_2	C/O	pressure
6	267.9	91.8	106.2	1.16	12 kPa
7	150.6	91.8	106.2	1.16	9 kPa
8	297.7	133.2	150.0	1.13	15 kPa
9	150.6	89.3	109.2	1.22	9 kPa
10	297.7	128.4	153.5	1.2	15 kPa
11	357.2	127.3	146.1	1.15	18 kPa

Again soot volume fractions, particle number densities and specific surface areas are normalized to a gas volume according to 1900 K to account for the variation of the gas volume with temperature. The dependence of soot volume fraction on time is obvious from Fig. 15.7. Soot formation rates, which are shown in Fig 15.8, are approximately identical for flames with the same soot volume fraction, although pressure and composition of the feed are remarkably different. The apparent first order rate coefficient k_{sg} can be estimated to about 84 s^{-1} for all these flames.

As evident from figure 15.9 increasing the flow rate and reducing C/O the ratio lead to slightly increased temperatures, although the pressure is increased.

The results of the scattering measurements are summarized in the representation of the profiles of the specific surface areas in Fig. 15.10. As for flames 6 to 8 with a final normalized soot volume fraction of about $26 \cdot 10^{-9}$ particle number density, radius and specific surface areas are approximately identical. For the flames 9 to 11 with a final nomalized soot volume fraction of about $43 \cdot 10^{-9}$ particle radii slightly increase with increasing pressure, particle number densities and specific surface areas slightly decrease. The change of pressure and temperature apparently affect the coagulation kinetics and give rise to slightly different particle sizes.

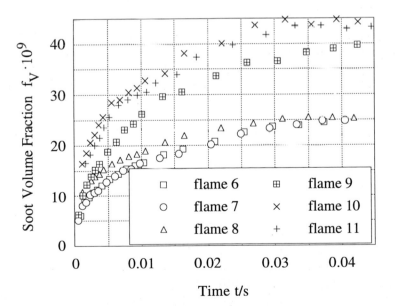

Fig. 15.7. Soot volume fractions of flames 6–11 versus reaction time

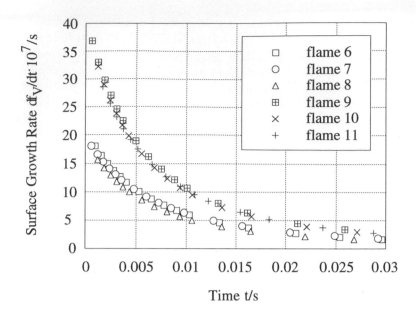

Fig. 15.8. Soot appearance rates for flames 6–11 versus reaction time

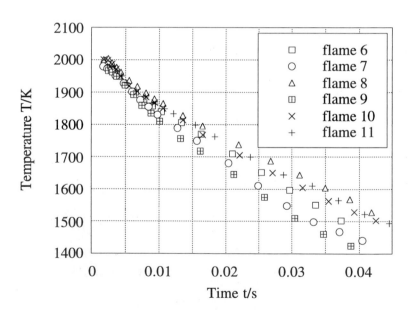

Fig. 15.9. Particle temperatures for flames 6–11 versus reaction time

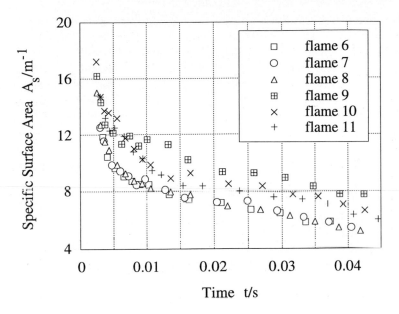

Fig. 15.10. Specific surface areas for flames 6–11 versus reaction time

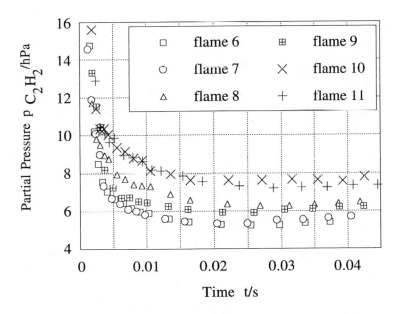

Fig. 15.11. Ethyne partial pressures for flames 6–11 versus reaction time

The chemical atmosphere around the soot particles also varyies with the applied different burning conditions. The partial pressure of ethyne is plotted versus reaction time in Fig. 15.11. Flames 6 and 7 and flames 10 and 11, respectively, have similar ethyne partial pressure. For flames 6 and 7 and for flame 10 and 11 the decrease of pressure and composition of the feed compensate with respect to the partial pressure of ethyne. Differences up to 30% are observed comparing these flames with flame 8 and 9, respectively. As evident from figures 15.9 and 15.11 increasing ethyne partial pressures are accompanied by slight increases of temperature in the flames under consideration. According to the ethyne decomposition model increasing temperature and ethyne partial pressure should give rise to higher surface growth rates. This is not observed in our experimental results, which lead to the conclusion that surface growth is not simply a first order decomposition of ethyne on soot particles surface.

15.3.3 Discussion in Terms of the HACA-Mechanism

Three major items from the above presented results will be discused in more detail in terms of the HACA-mechanism, viz.
  – The dependendence of the surface growth rate on the ethyne partial pressure.

Following the mechanistic interpretation as given by *Frenklach* [15.24-25], Fig. 15.2, surface growth of soot particles, on the per-site basis, can be written according to the following reactions. In addition oxidation of soot particles by OH and O_2 has been taken into account.

$$
\begin{aligned}
1. \quad & C_{soot,i}H + H && \overset{k_{1,s}}{\longleftrightarrow} && C^*_{soot,i} && + H_2 \\
2. \quad & C^*_{soot,i} + H && \overset{k_{2,s}}{\longrightarrow} && C_{soot,i}H \\
3. \quad & C^*_{soot,i} + C_2H_2 && \overset{k_{3,s}}{\longrightarrow} && C_{soot,i+1}H && + H \\
4. \quad & C^*_{soot,i} + O_2 && \overset{k_{4,s}}{\longrightarrow} && \text{products} \\
5. \quad & C_{soot,i}H + OH && \overset{k_{5,s}}{\longrightarrow} && \text{products}
\end{aligned}
$$

$C_{soot,i}H$ represents an arm-chair site on the soot particle surface and $C^*_{soot,i}$ the corresponding radical. If $\chi(C_{soot,i})$ is introduced as the number of $C_{soot,i}H$-sites and $\chi(C^*_{soot,i})$ is replaced by the assumption of quasi stationarity, the rate of surface growth is given by

$$
\frac{df_V}{dt} = \left(\frac{k_3[C_2H_2] \cdot k_{1,f}[H]}{k_{1,b}[H_2] + k_2[H] + k_3[C_2H_2] + k_4[O_2]} - k_5[OH] \right) \cdot \chi(C_{soot,i}) \cdot A_s .
$$
(15.14)

In Eq. (15.14) A_s represents the total surface area accessible for surface growth reactions. In contrast to Eq.(15.4) this equation is of first order in ethyne partial pressure only in the limiting case where the rate of carbon addition is small compared with the rate of radicalic site consuming and oxidation reactions in the nominator of the first term on the right hand side. In the other limiting case at high rates of carbon addition, the rate of hydrogen abstraction is rate limiting and Eq.(15.14) is independent of the ethyne partial pressure. Obviously, in flames 6 through 11 the prevailing conditions favour carbon addition so that hydrogen abstraction and possibly the oxidation reactions are rate limiting. Then the effect of increasing partial pressure of ethyne and increasing temperature on surface growth rate is concealed by the hydrogen abstraction and oxidation reactions. Most flames that have been referenced for the development and validation of rate expression for surface growth according to Eq. (15.4), e.g. ref. [15.36], are ethylene flames where carbon addition reactions are rate limiting.

– The independence of the appearance rates of soot of the "outer surface" of the soot particles.

According to the mechanistic interpretation of surface growth given above carbon addition to soot particles occurs by creating $C^*_{soot,i}$ sites from $C_{soot,i}H$ sites on the soot particle surface. This is the basic mechanism for surface growth accepted in numerous modelling studies, cf. [15.23-28]. During coagulation of soot particles $C^*_{soot,i}$ sites or $C_{soot,i}H$ may not be affected by coagulation because soot particles do not truly coalesce but retain their identity during coagulation into aggreagtes. This point of view is also discussed by [15.22, 15.36]. If sites accessible for surface growth are not depleted by coagulation, then the "internal surface" of a soot particle agglomerate or cluster which is measured from optical methods rather than the "outer surface" is determining the surface growth rate. During the experiments presented in section 15.3.1 the final soot volume fractions of the seeded and unseeded flames were kept identical. Soot particles in the seeded flames are prevented from coagulation thus retaining there initial surface area which is approximately equal to the "inner surface" area of soot particle agglomerates accessible for surface growth in the unseeded flames. Therefore, surface growth rates are identical in the unseeded and seeded flames that differ only with respect to the "outer surface" area.

– The exponential decay of the appearance rate of soot according to Eq. (15.3).

The exponential decay of the appearance rate of soot equivalent to an exponential decay of the surface growth rate is explained in the work of Haynes and Harris by the deactivation of active sites due to a thermal deactivation process, see Eq. (15.8). In the mechanistic interpretation of surface growth given above carbon addition to soot particles occurs by creating $C^*_{soot,i}$ sites from $C_{soot,i}H$ sites on the soot particle surface. No additional process except oxidation consuming or creating active sites is introduced. Therefore, the decay of surface growth rates in terms of the HACA concept may be explained by the decay of the rate of the abstraction reaction or an increase of the rate of its reverse, first reaction of the table of reactions given above for the cases where hydrogen abstraction is rate limiting. Obviously, these conditions are the prevailing ones for the flames investigated in this work. Figure 15.12 gives calculated profiles of ethyne, molecular hydrogen and hydrogen atoms for flames 8 and 9 of the second set of experiments. The calculations have been performed using a one-dimensional flame code employing a mechanism for fuel rich combustion of ethyne according to [15.23]. First of all, the calculated ethyne partial pressure profiles agree well with the measured ones from Fig. 15.11. Further, the profiles of hydrogen atoms peak in the soot forming region of the flames and decrease rapidly in the flame zone where the decay of the surface growth rate is observed. Simultaneously, the partial pressure of molecular hydrogen inreases pushing the initiation reaction for the surface growth reaction scheme in the backward direction.

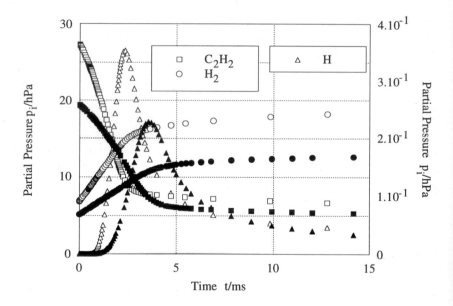

Fig. 15.12. Calculated molecular hydrogen, hydrogen atom, and ethyne partial pressures for flames 8 (open symbols) and 9 (full symbols) versus reaction time

From this picture the explanation of the decay of the appearance rate of soot by shifting the initiation step of the surface growth reactions by the depletion of hydrogen atoms and the increasing concentration of molecular hydrogen is obvious. However, this explanation implies that soot particles continue to grow by surface growth reactions when being exposed to a hydrogen atom rich atmosphere. There are some experimental indications for this, e.g. [5.37], and further experiments validating these hypotheses derived from the extension of the HACA mechanism to surface growth will be conducted in the near future.

15.4 Conclusions

The dependence of soot appearance rates on the surface area of the soot particles has been investigated. For this flames are seeded with alkali metal solutions to prevent the soot particles from coagulation. These flames are compared with the equivalent unseeded flames. The experimental results give the appearence rates of soot as beeing independent of the outer surface of the soot particle aggregates.

The influence of chemical composition - in particular the dependence of the soot appearance rate on the ethyne partial pressure - on soot formation has been experimentally investigated. For this, flames with different pressure, flow rates, and composition of the feed have been investigated. The experiments show that the appearance rate of soot is not simply proportional to the ethyne partial pressure as stated in frequently used surface growth models but rather influenced by a number of flame properties.

An explanation of the experimental findings is given on the basis of the frequently discussed HACA-mechanism for surface growth of soot particles with the C-H sites accessible for growth being spread about the inner surface of the soot particle aggregates. The complex dependence of the surface growth rate of soot on the ethyne partial pressure can be explained as well as the exponential decay of the surface growth rates.

15.5 Acknowledgement

This research was supported by the Commission of the European Communities, the Swedish National Board for Technical Development (STU) and the Joint Research Committee of european car manufacturer (Fiat, Peugeot SA, Renault, Volkswagen and Volvo) within the IDEA program.

References

15.1   B.S. Haynes, H.Gg. Wagner: Prog. Energy Combust. Sci. **7**, 229 (1981)

15.2   I.O. Smith: Prog. Energy Combust. Sci. **7**, 275 (1981)

15.3   K.H. Homann: "Formation of Large Molecules, Particulates and Ions in Premixed Hydrocarbon Flames; Progress and Unresolved Questions", in Twentieth Symposium (International) on Combustion (The Combustion Institute, Pittsburgh 1985) p. 857

272   Henning Bockhorn, Thomas Schäfer

15.4 I. Glassman: "Soot Formation in Combustion Processes", in Twenty-Second Symposium (International) on Combustion (The Combustion Institute, Pittsburgh 1988) p. 295

15.5 J.B. Howard: "Carbon addition and oxidation reactions in heterogenous combustion and soot formation", in Twenty-Third Symposium (International) on Combustion (The Combustion Institute, Pittsburgh 1990) p. 1107

15.6 H.F. Calcote: Combust. Flame **42**, 215 (1981)

15.7 J.B. Howard: "Fullerenes Formation in Flames", in Twenty-Fourth Symposium (International) on Combustion (The Combustion Institute, Pittsburgh 1992) p. 933

15.8 J.B. Howard, A.L. Lafleur, Y. Makarovsky, S. Mitra, C.J. Pope, T.K. Yadav: Carbon **30**, 1189 (1992)

15.9 D.C. Siegla, G.W. Smith (eds.): Particulate Carbon - Formation During Combustion (Plenum Press, New York 1981)

15.10 J. Lahaye, G. Prado (eds.): Soot in Combustion Systems and its Toxic Properties (Plenum Press, New York 1983)

15.11 H. Jander, H.Gg. Wagner: "Soot Formation in Combustion" - an International Round Table Discussion (Vandenhoeck and Ruprecht, Göttingen 1990)

15.12 B.S. Haynes, H.Gg. Wagner: Z. Phys. Chem. N.F. **133**, 201 (1982)

15.13 L. Baumgärtner, H. Jander, H.Gg. Wagner: Ber. Bunsenges. Phys. Chem. **87**, 1077 (1983)

15.14 H. Bockhorn, F. Fetting, A. Heddrich, G. Wannemacher: "Investigation of the Surface Growth of Soot in Flat Low Pressure Hydrocarbon Oxygen Flames", in Twentieth Symposium (International) on Combustion (The Combustion Institute, Pittsburgh 1985) p. 879

15.15 H. Böhm, D. Hesse, H. Jander, B. Lüers, J. Pietscher, H.Gg. Wagner, M. Weiss: "The Influence of Pressure and Temperature on Soot Formation in Premixed Flames", in Twenty-Second Symposium (International) on Combustion (The Combustion Institute, Pittsburgh 1988) p. 403

15.16 M. Bönig, Chr. Feldermann, H. Jander, B. Lüers, G. Rudolph, H.Gg. Wagner: "Soot Formation in Premixed C_2H_4 Flat Flames at Elevated Pressure", in Twenty-Third Symposium (International) on Combustion (The Combustion Institute, Pittsburgh 1990) p. 1581

15.17 H. Böhm, Ch. Feldermann, Th. Heidermann, H. Jander, B. Lüers, H. Gg. Wagner: "Soot Formation in Premixed C_2H_2-Air Flames for Pressures up to 100 bar", in Twenty-Fourth Symposium (International) on Combustion (The Combustion Institute, Pittsburgh 1992) p. 991

15.18 S.J. Harris, A. Weiner: Combust. Sci. Tech. **72**, 67 (1990)

15.19 S.J. Harris: Combust. Flame **66**, 211 (1986)

15.20 J.C. Dasch: Combust. Flame **61**, 219 (1985)

15.21 I.T. Woods, B.S. Haynes: Combust. Flame **85**, 523 (1991)

15.22 I.T. Woods, B.S. Haynes: this volume, sect. 16.

15.23 M. Frenklach, J. Warnatz: Combust. Sci. Tech. **51**, 265 (1987)

15.24 M. Frenklach, H. Wang: "Detailed Modeling of Soot Particle Nucleation and Growth", in Twenty-Third Symposium (International) on Combustion (The Combustion Institute, Pittsburgh 1991) p. 1559

15.25 M. Frenklach, H. Wang: this volume, sect. 10

15.26 M.B. Colket, R.J. Hall: this volume, sect. 27

15.27 F. Mauss, B. Trilken, H. Breitbach, N. Peters: this volume, sect. 21

15.28 F. Mauss, N. Peters, H. Bockhorn: "A Detailed Chemical Model for Soot Formation in Premixed Acetylene- an Propane-Oxygen Flames", in <u>Proceed ings of the Anglo-German Combustion Symposium</u> (The British Section of The Combustion Institute, Cambridge 1993) p. 470

15.29 U. Wieschnowsky, H. Bockhorn, F. Fetting: "Some New Observations Concerning the Mass Growth of Soot in Premixed Hydrocarbon-Oxygen Flames", in <u>Twenty-Second Symposium (International) on Combustion</u> (The Combustion Institute, Pittsburgh 1989) p. 343

15.30 W.H. Dalzell, A.F. Sarofim: J. Heat Transfer **91**, 100 (1969)

15.31 H. Bockhorn, F. Fetting, A. Heddrich, G. Wannemacher: Ber. Bunsenges. Phys. Chem. **91**, 819 (1987)

15.32 A. D'Alessio: "Laser Light Scattering and Fluorescence Diagnostics of Rich Flames Produced by Gaseous and Liquid Fuels", in <u>Particulate Carbon: Formation During Combustion</u>, ed. by D.C. Siegla, G.W. Smith (Plenum Press, New York 1981) p. 207

15.33 B.S. Haynes, H. Jander, H. Gg. Wagner: Ber. Bunsenges. Phys. Chem. **84**, 585 (1980)

15.34 C.P. Lazzara, J.C. Biordi, J.F. Papp: Combust. Flame **21**, 371 (1973)

15.35 J.C. Biordi, C.P. Lazzara, J.F. Papp: Combust. Flame **23**, 73 (1974)

15.36 S.J. Harris: Combust. Sci. Tech. **72**, 67 (1990)

15.37 M. Huth, W. Leuckel: this volume, sect. 23

Discussion

Frenklach: You say that rate coefficients are independent of the index i, i.e. independent of the size. If you sum the equation with the right hand side $N_i A_{si}$ over all i then you will have an expression for the total surface area. But the total surface rate should be proportional to the total surface area. So, where does it disappear in your analysis?

Bockhorn: Here, $N_i A_{si}$ is the external surface of the soot particles. You can express this external surface of the soot particles in coagulating systems by the total number density times the size distributions and integrate it from zero to infinity. Altough the external surface area is changed by coagulation there is no change in reactivity. Surface reactions may occur in the inner part of the particle and PAH may move into and out of the particle by a kind of diffusion process because such particles are very porous. So, this is similar to Bryan Haynes's approach who suggests that the active sites are not destroyed by coagulation. The internal surface area of a structure is responsible for the surface growth of a particle and is not the total external surface area of the particles.

Sarofim: The accessibility of the internal surface area will decrease with time due to diffusion, surface reaction, and pore mouth plugging. We have indeed found on oxidizing soot that you will get a great opening up of its surface area as you oxidize a few percent suggesting you get pore mouth plugging. But when we tried to put constants into pore diffusion model they didn't seem to make sense. I wonder if you have looked at the effect of this factor when you were using your rates.

Bockhorn: I think oxidizing soot particles differs from the formation of soot particles. But what led me to these arguments are some experiments about desorption of PAH from soot particles. If we plot the activation energy of desorption of PAH from soot particles which are sampled at different hights above the burner we find that for very young the soot particles we have a high activation energy for the desorption of PAH. Then it decreases until about 60 - 70 % of the soot is formed and then it increases again. This increase of the activation energy of desorption with growing age of the soot particles means that the diffusion of the PAH out of the soot particle become more difficult with growing age. This is one reason which shows that surface growth may also be dependent on the internal surface area.

Haynes: What sort of growth is internal growth? Internal growth is not going to show up unless we take changes in optical properties into account.

Bockhorn: In my conclusions the internal growth essentially is the same as the external growth. The species which cause the internal growth are the same as the species which cause the external growth. They diffuse into the particle and may react there at active sites or they may cause growth at layered PAH which are at the internal surface of the particles.

Active Sites in Soot Growth

Ian T. Woods, Brian S. Haynes

Department of Chemical Engineering,
University of Sydney,
Sydney, N.S.W. 2006 Australia

Abstract: Soot surface growth rates in premixed flames can be described in terms of acetylene addition to active sites on the soot surface. These sites are conserved in the growth process but are lost with increasing age of the particles. Factors influencing the number of active sites available for the addition of acetylene have been investigated for a range of ethylene/air flames at atmospheric pressure. No correlation is found between local conditions, especially the concentration of hydrogen atoms and the local temperature, and the number of active sites for surface growth. Rather, the number of sites appears to be determined by the number present on the soot particles as they are first formed and by the loss of these sites by purely thermal "tempering" processes. The initial number of active sites is such that a constant fraction of the soot aerosol area is occupied by active sites, over a wide range of conditions. Comparison with other studies reveals this to be true for pressures from $1 \cdot 10^4$ Pa to $1 \cdot 10^5$ Pa; temperatures from 1600 to 2000 K; and fuels such as methane, ethylene, propane, and acetylene. Despite this common factor, it is not possible to predict the initial number of active sites. The failure to establish a correlation between active site density and the local reaction conditions, especially with regard to the initial number of sites, means that the prediction of soot surface growth rates in a diffusion flame environment remains highly uncertain. Further work is required to establish the factors influencing the initial site density and also to investigate the possibility that surface growth by addition of species of higher molecular weight might be significant.

16.1 Introduction

The rate of soot surface growth in premixed post-flame gases is generally well described by the kinetic expression of *Harris* et al. [16.1], and this expression has been used also with some success in diffusion flame environments also [16.2]:

$$\frac{\mathrm{d}f_V}{\mathrm{d}t} = k \, A_\mathrm{s} \, p_{C_2H_2} \qquad (16.1)$$

where A_s is the soot aerosol surface area, (m^2/m^3). However, while equation (16.1) has the correct (gas-kinetic) form, there are aspects of this expression which require further investigation.

There has been little doubt about the identification of acetylene as the dominant growth species although the strict first order kinetic might not always be the case, cf. [16.20]. In addition, some cracking of methane and of diacetylene has also been suggested to occur.

The apparent rate constant k, in Eq. (16.1), is not a constant, but decreases with time from the inception of the particles - this phenomenon has bee ascribed to the loss of reactivity of the particles as they "age" or "temper", or to changes in the gaseous atmosphere in which the growth is occuring. *Haynes* et al. [16.3] correlated surface growth rates as

$$\frac{df_V}{dt} = k_{sg}(f_{V\infty} - f_V) \tag{16.2}$$

where $f_{V\infty}$ is the ultimate soot volume fraction achievable under the particular conditions; and k_{sg}^{-1} is an empirical growth time constant which has been shown over a wide range of growth conditions [16.3-5] to be dependent only on the flame temperature. If the quantity k_{sg} is identified as an apparent rate constant for the loss of surface growth reactivity, then the expressions of *Harris* et al. (16.1) and *Haynes* et al. (16.2) can be reconciled [16.6] but the reasons for the loss of reactivity remain unknown.

The dependence of growth rates on the aerosol surface area in equation (16.1) is the expected result but has been difficult to verify directly because the surface area of a coagulating soot area undergoing simultaneous surface growth remains remarkably constant from near the point of inception of the first particles. However, inhibition of the aerosol coagulation by ionising metal additives does change the soot surface area (< 3-fold increase) without significantly affecting the observed rates of soot growth or the composition and temperature of the post-flame gases [16.7-8].

In the modelling of the heterogeneous reactions of carbons, it is widely accepted that rates may not correlate with the total surface area. Rather, the concept of active surface area is employed, with reaction occurring at specific active sites. An active-site model for the pyrolytic decomposition of hydrocarbon at about 1000 K was proposed by *Hoffmann* et al. [16.9] who found carbon deposition rates to vary as the initial number of substrate sites active for propylene chemisorption. These rates remained unchanged, even after many layers of carbon had been laid down on the original sites, implying that the sites are reproduced during deposition.

The concept of soot surface growth at regenerating active sites has recently been applied to the kinetics of soot surface growth as a means of explaining the observation that soot growth rates remain unchanged when coagulation is inhibited and the physical surface area increases [16.10-11]. Now, the growth law becomes

$$\frac{\mathrm{d}f_V}{\mathrm{d}t} = k' \, A_{\text{act}} \, p_{\text{C}_2\text{H}_2} \tag{16.3}$$

where A_{act} is the active growth site area, $A_{\text{act}} = N_s a_s$, (N_s: number density of active sites, each of area a_s).

For free-molecule kinetics, the site-specific growth rate constant, k', can be expressed as

$$k' = \gamma \frac{\bar{c}}{4k_{\text{B}}T} \, V_{\text{G}} = \frac{\gamma V_{\text{G}}}{\sqrt{2\pi m_{\text{G}} k_{\text{B}}T}} \tag{16.4}$$

where γ is the sticking coefficient (reaction probability), and V_{G} the volumetric growth due to accretion of one growth molecule of mass m_{G}.

As discussed by *Haynes* et al. [16.3], if one molecule of the growth species contains m_{C} carbon atoms, then $V_{\text{G}} \approx 10^{-29} m_{\text{C}}$ m^3 and $m_{\text{G}} \approx 2 \cdot 10^{-26} m_{\text{C}}$ kg. Thus for C_2H_2 as the growth species at 1800 K, $k' \approx 2.5 \cdot 10^{-7}\gamma$ m Pa^{-1} s^{-1}. Taking $a_s \approx 14 \cdot 10^{-20}$ m^2, the growth law now becomes

$$\frac{\mathrm{d}f_V}{\mathrm{d}t} = K \, \gamma \, N_s \, p_{\text{C}_2\text{H}_2} \tag{16.5}$$

where $K = 3.5 \cdot 10^{-26}$ m^3 Pa^{-1} s^{-1}.

Sites are presumed to be conserved by the growth process and the decrease in soot surface growth rates with increasing age of the particles is described in terms of the loss of active sites by purely thermal processes:

$$N_s = N_{s0} \, \exp(-k_{\text{sg}}t) \tag{16.6}$$

In premixed flames at atmospheric pressure, the concentration of acetylene in the growth region changes only slowly. Therefore, we can combine the previous two expressions:

$$\frac{\mathrm{d}f_V}{\mathrm{d}t} = k_{\text{sg}}(f_{V\infty} - f_V) \tag{16.7}$$

where

$$f_{V\infty} = \frac{K \, \gamma \, N_{s0} \, \bar{p}_{\text{C}_2\text{H}_2}}{k_{\text{sg}}} \tag{16.8}$$

Equation (16.7) is clearly the empirical growth law of *Haynes* et al. [16.3] and, as shown by them, k_{sg} is a function only of the average growth temperature and does not depend on other factors such as pressure, composition, the nature of the parent fuel, nor the soot volume fraction.

An alternative approach to the factors determining the numbers of active sites present under particular circumstances has been adopted by Frenklach and Wang [16.12] and Howard [16.13] who proposed that active sites are created by abstraction of hydrogen from aromatic structures:

$$\text{Ar} - \text{H} + \text{H} \rightleftharpoons \text{Ar} \cdot + \text{H}_2$$

The radical sites thus created are then active for acetylene addition:

$$Ar \cdot + C_2H_2 \rightarrow AR' - H + H$$

Clearly, surface growth by these processes will depend on the the local conditions, especially the concentration of H-atoms. Such a model for locally determined rates is highly attractive because it should be transportable to other gaseous environments such as exist in diffusion flames.

In this paper, we assess the apparent behaviour of active sites in soot formation in the post-flame gases of a variety of premixed flames, with a view to understanding the factors determining that behaviour.

16.2 Experimental

Laminar, premixed ethylene/air (+ oxygen) flames were stabilised on a water-cooled Meker-type flat flame burner at atmospheric pressure. The burner head was cooled by a cross-flow of water through embedded capillaries (0.2 mm) to provide a uniform temperature across the flame. The primary flame had a circular section, 60 mm in diameter, and was surrounded by a 20 mm annular shield flame. The combined flame was stabilised by a stainless-steel mesh screen located 40 to 70 mm above the burner surface.

Gas flows were metered and controlled with mass-flow controllers. The ethylene was 99.5% minimum purity; ambient air was filtered and dried. Some flames were established with oxygen-enrichment of the air, and high-purity (>99.9%) O_2 was employed for this.

Concentration profiles for major species, hydrocarbons, and hydrogen atoms were measured in the post-flame regions of the sooting flames described in Table 16.1. Measurements of stable species concentrations were made by sampling the post-flame gases with a water-cooled quartz probe fitted with a silica-wool soot trap. The dried samples (Permapure Inc driers) were analysed by gas chromatography for N_2, CO_2, CO, and C_1-, C_2-, C_3-, and C_4- hydrocarbon species. Concentrations of H_2O and of H_2 were determined by mass balances on elemental H and O.

Optical techniques were employed for the measurement of temperature (sodium line- reversal and Kurlbaum methods, $\pm$ 15 K), H-atom concentration (Li/LiOH method, $\pm$ 15% absolute), and soot volume fraction and number density profiles (light scattering and extinction at 633 nm). The application of these standard techniques is described in detail elsewhere [16.14].

Table 16.1 summarises the conditions studied in the various flames. Peak flame temperatures were 1780 to 1920 K, with the temperature in the post-flame region falling by 100 to 200 K.

Table 16.1. Summary of sooting ethylene/air flame conditions [16.11]

Flame Number	C/O Ratio	$f_V\infty$	Temperature T/K
S1	0.79	$2.7 \cdot 10^7$	1810
S2	0.82	$5.0 \cdot 10^7$	1780
S3	0.75	$1.3 \cdot 10^7$	1800
S4	0.75	$1.9 \cdot 10^7$	1780
S5	0.87	$1.2 \cdot 10^7$	1920

16.3 Results and Discussion

Our results for the behaviour of the soot aerosol, in terms of N and f_V, are qualitatively the same as has been reported elsewhere by various investigators and will not be described here. Figures 16.1 to 16.3 report the results obtained for the mole fractions of C_2H_2 (Fig. 16.1), H (Fig. 16.2), and for the ratio of the H-atom mole fraction to its equilibrium value calculated for $2\,H \rightleftharpoons H_2$ (Fig. 16.3).

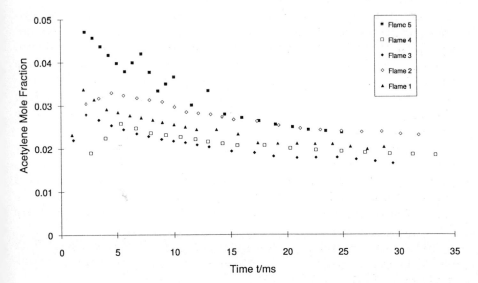

Fig. 16.1. Profiles of the mole fractions of acetylene in flames 1 to 5

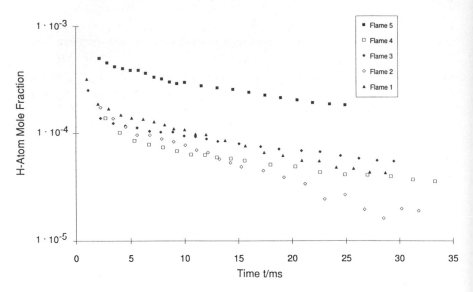

Fig. 16.2. Profiles of the hydrogen atom mole frations in flames 1 to 5

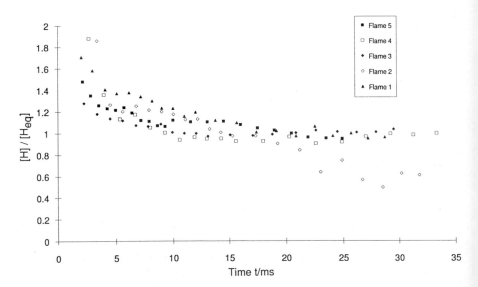

Fig. 16.3. Hydrogen atom concentrations relative to the equilibrium values based on $2\,H \rightleftharpoons H_2$

16.3.1 Surface-specific Growth Rate Constant

Figure 16.4 shows the values of the surface-specific mass growth rate constant, k, defined in Eq. 16.1, as a function of time in the various flames. In calculating these values, we have used the apparent surface area based on the soot particle number density and size determined from the light-scattering and extinction data by assuming a monodisperse soot aerosol. Different assumptions for the polydispersity of the aerosol lead to different values of A_s but the influence of polydispersity on the estimation of A_s is not great. For example, compared with the self-preserving free-molecule size distribution (SPSD), the monodisperse assumption underestimates the aerosol surface area by 12% [16.15].

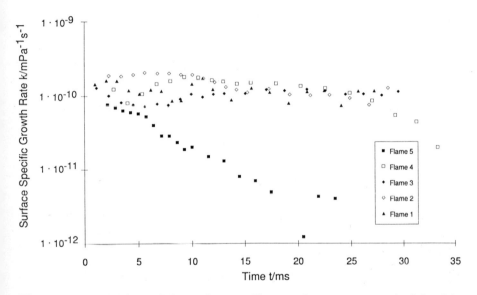

Fig. 16.4. Local values of the surface specific growth rate constant, k, defined in Eq. 16.1, for flames 1 to 5

The cooler flames (1-4) in Fig. 16.4 show a gradual decline in surface growth rate constant as the particles move away from the zone of their inception - the time constant for this decline is approximately 14 ms; $k_{sg} \approx 70$ s^{-1}. The hotter flame 5 shows a much more rapid decline in reactivity, $k_{sg} \approx 200$ s^{-1}. These values of k_{sg} agree closely with previous reports.

Figure 16.5 shows the data plotted as a function of the local gas temperature. As has previously been suggested [16.3], there is clearly no simple correlation or effective activation energy which can be discerned to describe the overall dependence of k on temperature. However, although the value of the surface growth rate constant, k, does vary in and among the various flames, its initial value, k_0, appears to be more nearly constant.

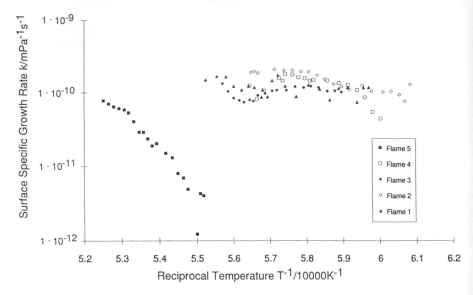

Fig. 16.5. Arrhenius plot of data from Fig. 16.4

The apparent initial value of the surface growth rate constants reported in the literature for a variety of flame conditions (temperatures in the vicinity of the reaction zone ranging from 1680 to 2000 K; pressures from 10 kPa to 100 kPa; methane, acetylene, ethylene, and propane fuels) are shown in Fig. 16.6 as a function of the (estimated) initial post-flame gas temperature [16.1,16.5,16.8,16.11,16.16-19]. It should be noted that no allowance for the various methods of determining the aerosol surface area has been made in preparing Fig. 16.6 - as discussed above, this effect is unlikely to contribute an uncertainty in excess of 10 to 15% in the calculation of k_0.

There is remarkably little variation in k_0 apparent for natural hydrocarbon/oxidant mixtures in Fig. 16.6, with an apparent activation energy of about 80 kJ mol^{-1} describing the trend of the data. Qualitatively, this correlation of k_0 indicates that there are common features in the soot growth characteristics of these flames; however, the anomalously low values of k_0 in the Cs-seeded flames of *Wieschnowsky* et al. [16.8] indicate that (16.1) is not a general description of these characteristics.

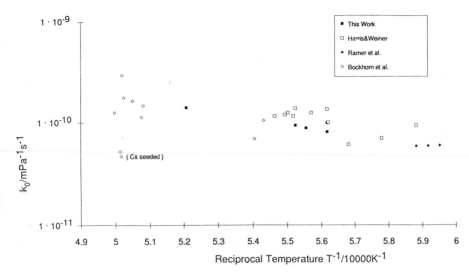

Fig. 16.6. Summary of values of the initial value of the surface specific growth rate constant obtained by various authors in a range of flames

16.3.2 Surface Growth at Active Sites

Figure 16.7 shows the effective active site density derived from the surface growth kinetics in our flames. These data were presented in a smoothed form previously [16.10] but a numerical error was present in the earlier presentation with the result that the effective number densities reported then were too low by a (variable) factor of aproximately 3.

Initial effective number densities of sites ($\gamma\, N_{s0}$) in Fig. 16.7 are of the order of 10^{17} m^{-3}, with the site density decreasing logarithmically with the age of the particles, the rate of loss of reactivity being considerably higher for flame 5 than for the other, cooler, flames. Phenomenologically, the time constant for this loss of sites is to be equated with the quantity k_{sg}^{-1} - there have been many studies of soot growth which have reported values for k_{sg}, and it is now well accepted that k_{sg} depends only on the flame temperature, and not on the fuel composition, pressure, stoichiometry, soot loading, particle size, or degree of coagulation.

It is tempting to equate k_{sg} with an intrinsic rate of destruction of active growth sites, perhaps as the carbon structure "tempers" or re-orders. However, recent approaches to the heterogeneous reactions of high-molecular-weight carbons and soot in flames have emphasised the importance of the gaseous atmosphere in determining the apparent overall reactivity.

Howard [16.13] suggested that the gas-phase hydrogen atoms create active sites for soot growth by abstracting H from the carbonaceous substrate.

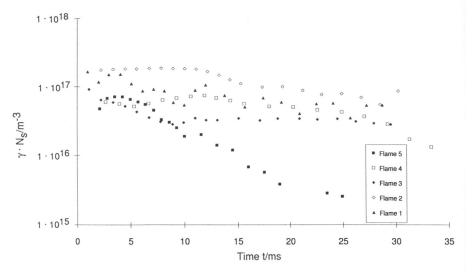

Fig. 16.7. Effective number density ($\gamma\ N_s$)

The presumed reversibility of this process leads to the result that the active site density should vary as

$$N_s = \frac{[\mathrm{H}]}{[\mathrm{H}_2]}\ \exp(-\frac{E_a}{RT}) \tag{16.9}$$

but, as shown in Fig. 16.8, there is no such correlation apparent among the various flames we studied. Indeed, the apparent temperature dependence in Figure 16.8 is reversed in going from the high-temperature Flame 5 to the lower-temperature Flames 1 to 4. It should also be noted that the temperature in the downstream regions of Flame 5 is close to that near the particle inception region in the other flames, but the factor relating N_s to $[\mathrm{H}]/[\mathrm{H}_2]$ differs by more than one order of magnitude between these conditions.

Frenklach and Wang [16.12] also suggested that H-abstraction reactions are responsible for the creation of active sites on the surface. They suggested that the apparent lack of dependence of acetylene surface growth rates on surface area observed by *Wieschnowsky* et al. [16.8] is a consequence of the high rate of destruction of the growth sites at the high temperatures (ca. 1950 K) in those flames; growth rates in lower temperature flames were predicted to be controlled by the rate of H-atom abstraction and to be formally proportional to the available surface area. The division between low and high-temperature flames found by *Frenklach* et al. [16.12] is however not supported by studies which cover the full temperature range - thus, *Haynes* et al. [16.7] found growth rates to be independent of surface area

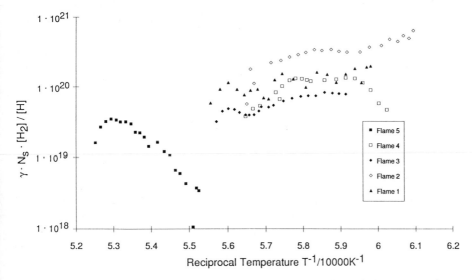

Fig. 16.8. Correlation of site number density with local conditions; an unique correlation is expected if the number of active sites is determined by the partially equilibrated abstraction of hydrogen atoms

perturbation by Cs additions over the temperature range from 1650 to 1900 K. Also, as shown in Fig. 16.6, initial surface-specific growth rates vary little over the temperature range 1650 - 2000 K and do not lead one to suspect that a dramatic shift in mechanism underlies them.

We therefore conclude that local conditions appear not to determine surface growth rates except for the influence of the growth species concentration and the mean flame temperature. This is unfortunate because it means that we are left with no obvious way of translating results from these premixed flame studies to a diffusion flame environment. In particular, the magnitude of the quantity N_{s0}, the number of growth sites present on the first particles, is used as an initial condition in the phenomenological description of (16.5) to (16.8) but cannot be predicted *a priori*.

As *Harris* [16.11] has pointed out, the fact that the initial value of the surface growth rate constant, k_0, hardly varies with conditions (see Fig. 16.6) may be taken as implying that the effective fraction of the initial aerosol surface area which is active , α_{eff}, is relatively invariant.

$$\alpha_{\text{eff}} = \gamma \cdot \alpha = \gamma \frac{N_{s0} a_s}{S_0}$$

For the variety of flames shown in Fig. 16.6, this fraction is $\alpha_{\text{eff}} = 0.0001$ within a factor of 2 - as discussed by *Woods* et al. [16.10], this in turn means that the collision efficiency, γ, must be at least as high as this as α cannot exceed 1. Equilibrium hydrogen abstraction, following *Howard* [16.13], may allow young soot particles to have 0.1 radical sites per carbon atom or, for a particle containing 100 carbon atoms, $\alpha \approx 0.3$ radical sites per surface carbon - in this case, $\gamma \approx 3 \cdot 10^{-4}$. A much lower value of $\gamma = 2 \cdot 10^{-5}$ was assumed by *Howard* [16.13], in which case C_2H_2 cannot be the major growth species - Howard suggested that this role was taken by "tar" but this remains conjecture at this stage.

The possibility that large molecules, "PAH", may be the dominant soot growth species has also been raised by Wang and Frenklach [16.12], who concluded that this route is responsible for most of the surface growth in the high-temperature, low-pressure flames of *Wieschnowsky* et al. [16.8]. However, *Frenklach* et al. [16.12] found these species not to be important, relative to C_2H_2, in the atmospheric-pressure flames of *Harris* et al. [16.1], in direct contrast to *Howard's* [16.13] conclusion that "tar" growth dominates in those flames.

Clearly, if large aromatic molecules do contribute significantly to surface growth rates, the kinetics and mechanisms involved have yet to be established. Both *Haynes* et al. [16.3] and *Harris* [16.11] have argued against a role for large aromatic molecules and the success of the correlations based on $[C_2H_2]$ has generally been taken as indicating growth by C_2H_2 addition. Of course, this evidence remains indirect and further work is required to resolve this issue.

16.4 Conclusions

The concept of soot growth at active sites provides a useful framework for understanding soot growth phenomena in premixed flames. The number of active sites initially present is determined by the flame conditions in a way which is not yet understood - however, the fraction of the soot aerosol area which is active early in the life of the particles is remarkably constant across a very wide range of conditions.

As particles age, the number of active sites present decreases, apparently by purely thermal processes - local conditions, especially the H radical concentrations, appear not to be important in determining the active site density.

The failure to establish a relationship between local conditions and the number of active sites makes it difficult to transfer the premixed flame results to diffusion flames in which a growing particle sees a vastly wider range of temperatures and compositions than can be studied in a premixed post-flame region. In particular, there is no way of estimating the initial active site density arising with the formation of new particles.

Current correlations based on acetylene growth at active sites appear very reasonable. Nevertheless, the possibility that larger hydrocarbon entities, "PAH" and "tar", may be responsible for some surface growth needs further investigation.

References

16.1   S.J Harris, A.M. Weiner: Combust. Sci. Technol. **38**, 75 (1984)
16.2   C.M. Megaridis, R.A. Dobbins: "Soot aerosol dynamics in a laminar ethylene diffusion flame", in Twenty-Second Symposium (International) on Combustion (The Combustion Institute, Pittsburgh 1988) p. 353
16.3   B.S. Haynes, H. Gg Wagner: Z. Phys. Chem. NF **133**, 201 (1982)
16.4   L. Baumgärtner, D. Hesse, H. Jander, H.Gg. Wagner: "Rate of soot growth in atmospheric premixed laminar flames", in Twentieth Symposium (International) on Combustion (The Combustion Institute, Pittsburgh 1984) p. 959
16.5   H. Bockhorn, F. Fetting, A. Heddrich, G. Wannemacher: "Investigation of the Surface Growth of Soot in Flat Low Pressure Hydrocarbon Oxygen Flames", in Twentieth Symposium (International) on Combustion (The Combustion Institute, Pittsburgh 1984) p. 979
16.6   C.J. Dasch: Combust. Flame **61**, 219 (1985)
16.7   B.S. Haynes, H. Jander, H.Gg. Wagner: "The effect of metal additives on the formation of soot in premixed flames", in Seventeenth Symposium (International) on Combustion (The Combustion Institute, Pittsburgh 1979) p. 1365
16.8   U. Wieschnowksy, H. Bockhorn, F. Fetting: "Some New Observations Concerning the Mass Growth of Soot in Premixed Hydrocarbon Oxygen Flames", in Twenty-Second Symposium (International) on Combustion (The Combustion Institute, Pittsburgh 1988) p. 343
16.9   W.P. Hoffmann, F.J. Vastola, P.L. Walker Jr.: Carbon **23**, 151 (1985)
16.10 I.T. Woods, B.S. Haynes: Combust. Flame **85**, 523 (1991)
16.11 S.J. Harris: Combust. Sci. Technol. **72**, 67 (1990)
16.12 M. Frenklach, H. Wang: "Detailed modelling of soot particle nucleation and growth", in Twenty-Third Symposium (International) on Combustion (The Combustion Institute, Pittsburgh 1990) p. 1559
16.13 J.B. Howard: "Carbon addition and oxidation reactions in heterogenous combustion and soot formation", in Twenty-Third Symposium (International) on Combustion (The Combustion Institute, Pittsburgh 1990) p. 1107
16.14 I.T. Woods: "Hydrocarbon reactions and soot growth in fuel-rich flames"; Ph.D. Thesis, University of Sydney (1988)
16.15 F.S. Lai, S.K. Friedländer, J. Pich, G.M. Hidy: J. Coll. Interface Sci. **39**, 395 (1972)
16.16 E.R. Ramer, J.F. Merklin, C.M. Sorensen, T.W. Taylor: Combust. Sci. Technol. **48**, 241 (1986)
16.17 H. Bockhorn, F. Fetting, G. Wannemacher, H.W. Wenz: "Optical Studies of Soot Particle Growth in Hydrocarbon Oxygen Flames", in Nineteenth Symposium (International) on Combustion (The Combustion Institute, Pittsburgh 1982) p. 1413

16.18 S.J. Harris, A.M. Weiner: "Some constraints on soot particle inception in premixed ethylene flames", in Twentieth Symposium (International) on Combustion (The Combustion Institute, Pittsburgh 1984) p. 969

16.19 S.J. Harris, A.M. Weiner: Combust. Sci. Technol. **38**, 75 (1984)

16.20 H. Bockhorn, Th. Schäfer: "Growth of Soot Particles by Surface Reactions", this volume, sect. 15

Discussion

Frenklach: What would be the difficulty in explaining your low temperature flames in terms of coalescence of PAHs? The arguments we are offering to explain the high temperature flames of Bockhorn may apply also to your flame. Or let me ask the question in the other way: have you measured PAH concentrations?

Haynes: That is a much nicer question. The answer of that is, I have not. I do not have the information to be able to answer the question, in fact.

Wagner: All these arguments are a kind of soft arguments, because we only can cover a very small range of conditions. However, there are a number of arguments independent of your observations, which support that active site concept. Bonne observed the tempering of soot particles, Homann did electron spin resonance measurements with aging particles. When measuring the hydrogen content of the particles in dependence on their age there is always a one to one relation between the hydrogen content and the reactivity of the particles no matter how big they are. There is one more argument in favour of the tempering of the particles under these conditions which again supports the active site approach. When you measure the C-C distance in the particles, which is possible for particles larger than 5 nm, it also correlates with the age of the particles. All these properties show the same temperature dependence as the corresponding "life time of the active particles".

Concerning a relation of surface growth with surface area: If we take the work of Tesner on the growth of graphite into consideration, in general, surface area will give the tendency but surface growth might not be directly proportional to surface area. Finally one remark on Bockhorn's results. I don't think you find a one over r dependence of surface growth rate. You find the surface growth rate decreasing with time, but you can have particles with 5 nm diameter which are as old as others with 500 nm diameter and do not grow anymore.

Bockhorn: Could I just comment on the last remark of Prof. Wagner? It was not a one over r dependence on the size of the soot particles. It was one over r dependence of size of the growing PAH-like structures which are on the surface of the soot particles. And they grow if they pick up acetylene or any growth components. And their size is somehow independent of the size of the soot particles or the external surface of the soot particles.

Santoro: I have two questions. The first one is, if the active sites are conserved in the growth process I would argue that inception is where those numbers are determined and, therefore, inception controls everything from that point on.

I have to ask my second question to John Kent to confirm my understanding. When he makes his flames very long, the growth process just keeps increasing with time. So even though the particles should be aging in that flame it seems whenever he increases the length the growth process just keeps going on. Now, those flames are cooler flames than the flames that you are talking about.

Haynes: As to the influence of particle inception. Yes, that must be very important, in fact. The model as it now stands - and it is a strictly empirical model - requires but one quantity and that is the initial number of active sites which leaves the primary reaction time. We can try to get a feeling for that number in terms of the total surface area. We can not have more active sites than there is total area. And basically, I think if we accept a sticking factor of somewhere of the order of 10^{-3}, which I think we have to accept as being realistic, then initially the total area of the first particles leaving the primary reaction zone is active and from then on lose activity. Approximately the physical area leaving the primary reaction is active.

On the second question: Of course there is a strong activation energy in this tempering or annealing time constant k_{sg} and that time constant at 1600 K is more than 20 ms. By the time in the tail of those flames there is no expectation that the activity should be decreasing.

Bockhorn: I did not understand your explanation for the surface growth reactions or appearence rate of soot being of first order in the acetylene partial pressure.

Haynes: The argument is simply based on taking the initial surface growth rates expressed in terms of Harris's growth constant. We are looking at data going in temperatures from 1600 K to 2000 K and pressures 101.33 kPa to 10.133 kPa and acetylene concentrations in all of these flames span more than one order of magnitude. So that is the best demonstration I can give you that the specific growth rate is independant of the acetylene concentration. Therefore, the growth itself is first order in acetylene.

Wagner: That dependence of the growth rate on acetylene concentration is a fairly special property of a very small group of fuels. It may not hold for other fuels. I think one should take the concentration of unsaturated species or something like that.

Haynes: I have no comment to make on that other than that I think it is acetylene in most atmospheric pressure flames. At higher pressures in particular there are many other things around that contribute to the surface growth.

Soot Precursor Particles in Flames

Richard A. Dobbins, Haran Subramaniasivam

Division of Engineering,
Brown University,
Providence, RI 02912, USA

Abstract: Thermophoretic sampling was conducted in buoyancy dominated diffusion flames fueled by ethene, methane and acetylene and in the laminar ethene flame. Small polydisperse singlet particles which are more transparent to the electron beam are found at intermediate temperatures on the fuel side of the flame front in these flames. The particles form agglomerates at the flame front or in the upper regions of the flame. The microparticles conform to the description that has been provided by Wersborg et al. of similar species found in the premixed acetylene flame at low pressures which suggests that they are a common particulate precursor to carbonaceous soot in hydrocarbon flames. It is concluded that the particles are likely to be composed of polyaromatic hydrocarbon compounds, which have been found by others to be present in hydrocarbon flames, and that the precursor particles undergo annealing to form a partially graphitic soot upon exposure to elevated temperatures at the flame front.

17.1 Thermophoretic Sampling in the Buoyancy Dominated Flames

The thermophoretic sampling technique has been applied to unsteady buoyancy dominated flames using ethene, methane and acetylene as fuels. For the three fuels, the volume flow rate $\dot{V}$ was 50 cm^{3}s^{-1}, the burner inner diameter is 11.1 mm, and a coannular air flow is provided. The Reynolds numbers of the gas flows for CH_4, C_2H_4, and C_2H_2 at the mouth of the burner were 350, 635 and 380 while heat release rates were 1.85, 2.72 and 2.94 kW, respectively. The flames from all three of these fuels display an 11 to 15 hertz oscillation capped by an unsteady roughly toroidal vortex. In these buoyancy dominated flames, which are well described in the literature [17.1-2], the vortex is formed near the height $h = 150$ mm above the burner. The vortex is shed aperiodically and breaks into a turbulent flame brush that extends to $h \approx 400$ mm. The flame geometry is similar for the three fuels but the ethene flame does display a stronger vortex action.

In thermophoretic sampling, a pneumatic drive system is employed to insert momentarily into the flame a small metal plate upon which are attached three electron microscope grids [17.3]. The metal strip is normally the bulk specimen carrier that is an accessory of the electron microscope, and which has been modified by a milled central slot as is shown in Fig. 17.1. Standard grids having an amorphous carbon substrate of 20 nm thickness were bonded with epoxy cement to the modified bulk specimen carrier. The motion into the flame of the grid assembly is controlled by the drive system [17.3] that results in the trajectory shown in Fig. 17.2. The transit time into and out of the flame is 10 ms while the controlled residence time used in the tests reported herein was 70 ms

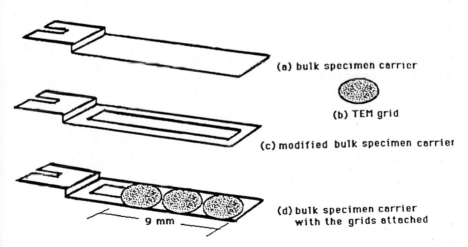

(a) bulk specimen carrier

(b) TEM grid

(c) modified bulk specimen carrier

(d) bulk specimen carrier with the grids attached

9 mm

Fig. 17.1. Thermophoretic probe assembly

After exposure to the particle bearing flame gases, the grid–carrier assembly is then inserted directly into the Philips 420 STEM using a grid holder that provides a 9 mm traverse within the electron microscope. A careful alignment procedure allows mapping from flame coordinates to position within the microscope that is accurate to within ± 0.5 mm. The TEM examination of the sample was conducted with a beam accelerating volt-

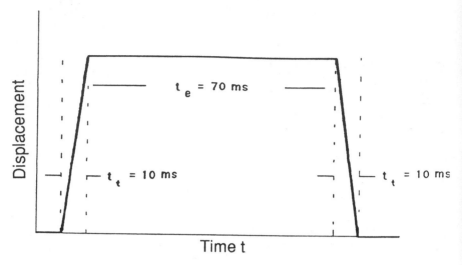

Fig. 17.2. Thermophoretic probe trajectory

age of 120 kV and with magnifications ranging from 24,000 to 105,000. The many micrographs displayed at the workshop cannot be shown in the space available here allowed. In Fig. 17.3. and 17.4. two micrographs of samples from the ethene flame that display highly contrasting morphologies are shown and are discussed in detail below.

Ethene: Aggregates show systematic growth over the first 150 mm of the height above the burner h where the vortex is located. At this height the primary particle diameter is approximately 35 nm. No further growth or oxidation of the aggregates is observed, and it is presumed that all soot generated is released to the surroundings. Fig. 17.3. shows an aggregate sampled at the top of the flame, $h = 330$ mm, where most aggregates are similar to those found at the vortex level. The aggregates at $h = 330$ mm are quite large consisting of chains of nearly monodisperse primary particles whose diameters are ca. 35 nm. A small fraction of the aggregates at the top of the flame have uniformly smaller primary particles. Low in the flame on the fuel side of the flame front, a substantial population of solitary, polydisperse particles is found as shown in Fig. 17.4. These particles often coexist with young agglutinated aggregates and are more transparent to the electron beam of the TEM. The characteristic features of the microparticle displayed in Fig. 17.4. are their greater transparency to the electron beam, their solitary non aggregate morphology, and the polydispersity of their sizes. This important result is discussed further below.

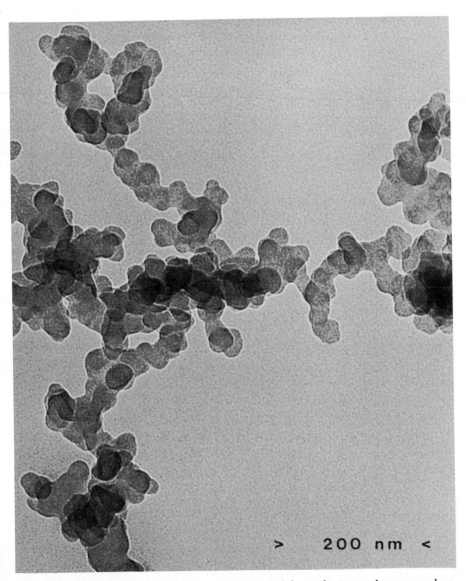

Fig. 17.3. Micrograph of a soot particle sampled from the upper buoyancy dominated diffusion flame, $h = 330$ mm

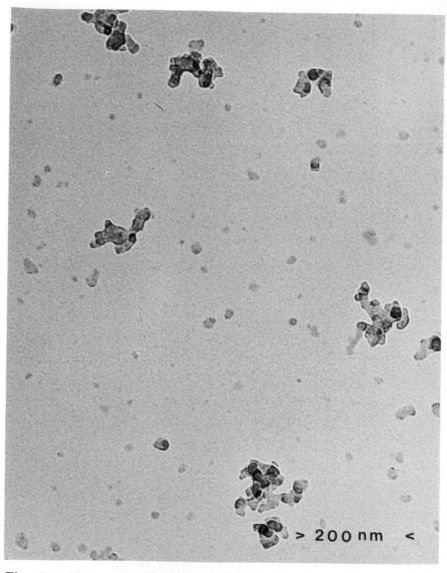

Fig. 17.4. Micrograph of particles sampled from the lower buoyancy dominated diffusion flame, $h = 10$ mm, $R = 5$ mm

Methane: In this lightly sooting flame, very small singlet particles are found near the burner mouth, and they develop into small aggregates with increasing height. The maximum diameter of the primary particles is observed to be about 35 nm. A moderate population of microparticles that precede the formation of carbonaceous soot is found again on the fuel side in the lower flame. Unlike the ethene flame, microparticles in the methane flame are found in the upper portion of the flame and some are probably released to the surroundings. The inception zone for methane is spatially extended into the upper portion of the flame.

Acetylene: Intense formation of aggregates is observed even below the height $h = 10$ mm. Primary particle diameters range up to 45 nm. Microparticles are prevalent on the fuel side of the lower portion of the diffusion flame. The vortex region is noticeably weaker in the acetylene flame, and some microparticles are observed in the upper portion of the flame.

17.2 The Nature of the Soot Precursor Particles

To learn more of the nature of the microparticles, we refer to some of the literature on the formation of soot particles in flames and to the literature on the laminar ethene diffusion flame [17.4-10]. Thermophoretic probing conducted on the laminar flame ($\dot{V} = 3.85$ cm^3 s^{-1}, nonsmoking, 88 mm high) along the radial direction at heights above the burner $h = 10$, 20, and 30 mm has revealed the existence of the microparticle field on the fuel side of the flame front as was found in the buoyancy dominated flames. The particle morphology along a given radial direction for prescribed height h displays strong radial gradients in which "transparent" microparticles are found on the fuel side, more opaque aggregates are found near the luminosity front, and small highly opaque singlets are found near the flame front. The optical observations by *Santoro* et al. [17.4] report particle number concentrations on the order of $5 \cdot 10^{11}$ cm^{-3} at $h = 15$ mm, $R = 2.4$ mm in this flame. At this concentration the interparticle collision frequency is very high and growth by particle coalescence is expected. The products of such collisions are singlet particles rather than chained aggregates that appear in higher temperature flame front regions of the flame.

The microparticles are found in the same regions of the laminar ethene flame where fluorescence was quantitatively measured [17.4]. Since fluorescence is attributed to PAH species, this finding constitutes evidence that the microparticles are composed one or more of the fluorescing PAH compounds.

Wersborg et al. [17.11] observed low opacity particles subject to coagulation in the lower portion of their low pressure premixed C_2H_2/O_2 flame. In these experiments the particulate and gaseous species were collected by means of a quartz supersonic nozzle. These authors specifically commented on the high contrast observed only in the case of the older particles and clusters that are found higher in the flame. The coagulating particles had volume mean diameters ranging from 1.5 to 15 nm. These observations are entirely consistent with our discovery of "transparent" coagulating microparticles on the fuel side of the flame front and more opaque aggregates in the higher temperature regions of the flame. *McKinnon* [17.12] and *Feitelberg* [17.13] have pointed to the role of the PAH species in the early formation of carbonaceous soot particles. Feitelberg has shown that the total tar component of the PAH's is quantitatively correlated with the magnitude of the fluorescence cross section observed at 514 nm in response to laser stimulation at 488 nm. Tar was defined as the dichloromethane soluble fraction of the solid material collected from the atmospheric pressure premixed ethene flame by a stainless steel probe. Chemical analyses employed by Feitelberg, who studied a C_2H_4 premixed atmospheric pressure flame, showed the dominant PAH species to be acenaphthalene $C_{12}H_8$, napthalene $C_{10}H_8$, and cyclopenta(cd)pyrene $C_{18}H_{10}$ which were present in the 5 to 10 ppm mole fraction range.

The fluorescence intensity observed by *Santoro* et al. [17.4] of $1.4 \cdot 10^{-8}$ per cm-str at $h = 10$ mm and $R = 2.4$ mm in the laminar nonsooting flame yields, employing Feitelberg's correlation, a PAH concentration in the ethene diffusion flame of 400 ng cm^{-3}. This value, which assumes that the bandwidths of the fluorescence emission and of the receiver systems employed in the two separate investigations [17.4, 17.11] are identical, can be regarded merely as an estimation. Nevertheless, it is in approximate agreement with 240 ng cm^{-3} that is derived from the scattering/extinction data reported [17.4] at the same flame location if one assumes the particle density to be 1000 kg m^{-3} as is characteristic of PAH compounds.

Frenklach [17.14] and *Howard* [17.15] have pointed to the role of coagulation in the growth of PAH species. *Howard* [17.15] has described the formation of reactive radical sites in PAH molecules as a result of H atom abstraction. He estimates that 10 to 30% of the peripheral C–H bonds in PAH species become reactive radical sites. These sites provide a chemical basis for reactive coagulation of PAH compounds with each other and with smaller radicals. Howard also speculates that the heavy PAH's contribute the early soot growth. Our observations indicate that the more mature soot particles are initially formed in the three flames observed above by the annealing at the flame front of the microparticles that are created in the fuel rich region.

17.3 Summary

The polydisperse, solitary microparticles that are found on the fuel side of the flame front in the buoyancy dominated diffusion flames of methane, acetylene and ethene and also in the laminar ethene diffusion flame conform to the description of particles found in the low pressure premixed acetylene flames [17.11]. These microparticles are present in sizes ranging up to 15 nm, are typically singlets or doublets, and appear highly transparent to the electron beam. The smallest observable microparticles are estimated to be 3 nm in diameter. The estimated size range of the PAH species is from 10^3 to 10^6 u if a material density of 1000 kg m^{-3} is assumed. Although the microparticle concentrations are sufficient to produce a high interparticle collision frequency, the particles appear as singlets and thus undergo a (liquid-like) coagulative growth. Nearer to the flame front the particles undergo an annealing action and form aggregates as they develop into the more familiar filous structures characteristic of soot aggregates. We believe these particles consist of mixtures of PAH fragments that are formed by reactive coagulation as described by *Howard* [17.15]. Microparticles are likely precursors to the carbonaceous soot that is formed in the higher temperature regions where annealing results in dehydrogenation and the formation of partially graphitic structures.

The progressive transformation of hydrocarbon fuel molecules to soot particles has been represented schematically by *Wagner* [17.16] in a graph of the particle hydrogen mole fraction vs. the logarithm of its mass, M, as measured in atomic mass units. In this representation the fuel molecules [$x_H = 0.5$, $M - O(25$ u$)$] are transformed to PAH molecules [$x_H = 0.4$ to 0.3, $M = O(200$ u$)$] which, in turn, are converted into soot primary particles [$x_H = 0.2$ to 0.1, $M \doteq 10^4$ to 10^7 u] and their aggregates. Our modification of this representation is shown in Fig. 17.5. and it includes, as an intermediate step, the presence of the microparticle stage as a particulate precursor to carbonaceous soot. The microparticles are envisioned to be composed of PAH species and accordingly have a material density near 1000 kg m^{-3} and $x_H = 0.4$ to 0.3. Their sizes range from single PAH molecules up to 15 nm spherules. When the microparticles are exposed to elevated temperatures near the flame front, the x_H decreases and the peripheral C–H bonds are replaced by graphitic crystallites. The coagulative growth associated with the reactive PAH species ceases and is supplanted by collisional growth by cluster–cluster aggregation CCA [17.3, 17.7-8]. Aggregative growth results in filous structures of fractal dimension ranging from 1.6 to 1.8 [17.8-10]. The soot primary particles undergo growth by surface reactions, so called monomer–cluster aggregation MCA, with small PAH or acetylene in the lower portion of the flame and are oxidized in the upper flame [17.9-10]. For higher volume flow rates the oxidation is incomplete and smoke is released to the surroundings.

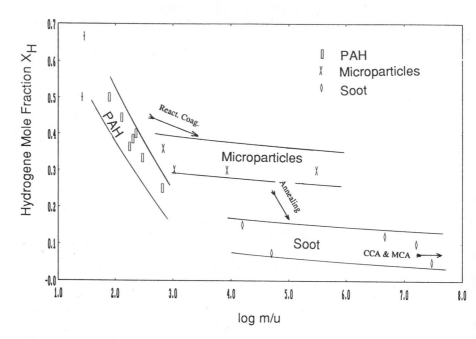

Fig. 17.5. The representation of the formation of carbonaceous soot from hydrocarbon fuels

17.4 Acknowledgements

The authors thank to Mr. Alan F. Schwartzman for assistance with electron microscopy and Mr. William D. Lilly for help with modifications of the test equipment and instrumentation. This research was sponsored by the U.S. Dept. of Commerce, National Institute for Standards and Technology, Center for Fire Research under Grant No. NANB1D1110.

References

17.1   H.A. Becker, D. Liang: Combust. Flame **52**, 247 (1983)
17.2   R.W. Davis, E.F. Moore, W.M. Roquemore, L.D. Chen, V. Vilimpoc, L.P. Goss: Combust. Flame **83**, 2663 (1991)
17.3   R.A. Dobbins, C.M. Megaridis: Langmuir **3**, 254 (1987)

17.4   R.J. Santoro, H.G. Semerjian, R.A. Dobbins: Combust. Flame **51**, 203 (1983)

17.5   R.J. Santoro, H.G. Semerjian: "Soot Formation In Diffusion Flames; Flow Rate, Fuel Species and Temperature Effects", in Twentieth Symposium (International) on Combustion (The Combustion Institute, Pittsburgh 1984) p. 997

17.6   R.J. Santoro, T.T. Yeh, J.J. Horvarth, H.G. Semerjian: Combust. Sci. Technol. **53**, 89 (1987)

17.7   C.M. Megaridis, R.A. Dobbins: "Soot Aerosol Dynamics in a Laminar Ethylene Diffusion Flame", in Twenty-Second Symposium (International) on Combustion (The Combustion Institute, Pittsburgh 1988) p. 353

17.8   C.M. Megaridis, R.A. Dobbins: Combust. Sci. Technol. **66**, 1 (1989)

17.9   R.A. Dobbins, R.J. Santoro, H.G. Semerjian: "Analysis of Light Scattering From Soot Using Optical Cross Sections for Aggregates", in Twenty-Third Symposium (International) on Combustion (The Combustion Institute, Pittsburgh 1990) p. 1525

17.10 R. Puri, T.F. Richardson, R.J. Santoro, R.A. Dobbins: in preparation

17.11 B.L. Wersborg, J.B. Howard, G. C. Williams: "Physical Mechanisms in Carbon Formation in Flames", in Fourteenth Symposium (International) on Combustion (The Combustion Institute, Pittsburgh 1973) p. 929

17.12 J.T. McKinnon: "Chemical and Physical Mechanisms of Soot Formation"; Ph.D. Thesis, M.I.T. (1989)

17.13 A.S. Feitelberg: "The Effects of Metal Additives on Soot Formation"; Ph. D. Thesis, M.I.T. (1990)

17.14 M. Frenklach, H. Wang: "Detailed Modeling of Soot Nuleation and Growth", in Twenty-Third Symposium (International) on Combustion (The Combustion Institute, Pittsburgh 1990) p. 1559

17.15 J.B. Howard: "Carbon Addition and Oxidation Reactions in Heterogeneous Combustion and Soot Formation", in Twenty-Third Symposium (International) on Combustion (The Combustion Institute, Pittsburgh 1990) p. 1107

17.16 H.Gg. Wagner: in Soot Formation in Combustion – an International Round Table Discussion, ed. by H.Jander, H.Gg. Wagner (Vandenhoek und Ruprecht, Göttingen 1990) p. 114

Discussion

Smyth: On the quantum yield business: I just calculated the value for OH which has a long radiative lifetime and the quantum yield in flames would be about 3/10 of a percent. When you go into polyclics, I think actually the low number sounds more reasonable to me than the three percent quantum yield you had your second analysis.

Dobbins: The numbers I have seen in the literature have varied considerably. But they are around a five percent number. As for the possibility of determining it from this experiment, it does not seem to be quite feasible because of the several assumption one has to make. Certainly one tries to do this data reduction to see if there is internal consistency and I think we are almost there, but not quite.

Frenklach: What kind of distribution function did you use for the analysis and have you tried different distribution functions? I refer to the reinterpretation of your results with respect to the complex index of refraction.

Dobbins: The entire dependence on size distribution function is through the ratio of the second moment to the square of the first moment, μ_1. So I don't have to assume a distribution function in particular. I just take a value of μ_1. The initial value I used in the original data reduction was 2.0 corresponding to a self preserving size distribution.

D'Alessio: I want to point out that the results by Prof. Dobbins seem to be quite consistent to what we have found and presented yesterday for premixed flames. I think it is exactly the same phenomenon as in diffusion flames. I was also very pleased to see that there is no coagulation of these particles. And that was probably the answer to Prof. Homann's question from yesterday because he found very large number densities of particles and the number densities of the particles that we found by optical methods did not decrease also. So it seems that they are not coagulating and Prof. Dobbins got some more evidence for this.

I do not think that these particle are composed of PAH. My feeling is, that they are composed by structures which should have aliphatic bonding and not so much aromatic. Otherwise it would be difficult to explain why they are transparent in the visible.

Dobbins: We do not have any information on chemical analysis at this time. That is certainly a most important area to fill in and there are several ways of doing that. One of them is to compare in an experiment in which you condition the various components in a temperature profile and do the high resolution microscopy and appropriate types of chemical analysis. If you are able to make bulk quantities of a material that has resemblance to what you see in the flame as manifested by a high resolution TEM attributes, then you can perform the indefinite number of the analyses of density, refractive index, chemical composition. I think there is a lot potential new work to be done in that area.

Homann: Your results remind me of some work by Anderson from the early sixties. He studied the thermal decomposition of various hydrocarbons in a flow reactor that had a hot part and a lower temperature part. He could find all kinds of particulate matter from droplets to more grey things up to this very black soot particles. And he stressed that he got the grey things and the droplets only in the lower temperature part of the reactor. So my question is, how sure are you that the micro particles are not formed by the action of the cold carrier that you are inserting into the flame.

Dobbins: This is a diffusion flame. It has a surrounding air stream but the fuel itself is pure acetylene or pure ethene. These particles were found on the fuel side of the diffusion flame.

Homann: What is the effect of the cooling of the surroundings by the carrier. I think it cools down the surroundings where you deposit your droplets or microparticles.

Dobbins: We hope the effect is to quench any possible chemical reactions and to maintain the state of the particles essentially unchanged. The best answer to that is very simply that thermophoretic sampling explains what you see in the optical (laser scattering/extinction) experiment. Therefore, I claim that thermophoretic sampling is to be used interactively with laser scattering and extinction. By using these two techniques you learn a great deal more than by either one of them individually.

D'Alessio: We did exactly the same thing in the reaction zone of premixed flames and we changed the C/O-ratio and we found roughly the same. Therefore, I can exclude that it is some quenching effect.

Santoro: An observation that always troubled me is that in regions where I can not see absorption with extinction measurements, I could see a very strong increase in light scattering. This is usually below the positions that Dick talked about. We could see three orders of magnitude increase in light scattering but not any absorption from the particles. We looked at this when Dick was on sabbatical at NBS and I had a notion that it could be an index of refraction change going on there from non-absorbing to absorbing material. There is some supporting evidence that very low in the flame there are large scatterers that do not absorb. So there is some consistency in what he is finding based on other independant measurements.

Haynes: Some ancient history: it was Steve Graham's original shock tube experiment which had precisely this model for the formation of the carbonaceous material from his various hydrocarbons. He formed initially a non-absorbing micro droplet which would carbonize and gradually become absorbing. It was in this way he explained the results that you obtained.

General Discussion on Soot in Flames: Mass Growth and Related Phenomena

chaired by

Klaus-H. Homann, Jaques Lahaye

Lahaye: An interesting point is this notion of active sites. It might be quite nice if you could give a physical meaning to this active sites in order to avoid the use of an adjustable parameter for the equation. In his work thirty years ago Walker tried to explain the reactivitiy of carbon using the notion of active sites. The active sites were correlated to defects of graphitic material and progressively this notion was used for pregraphitic materials. It came out that the concept of active sites is extremely useful to understand absorption, thermo-desorption and eventually reactivity. Listening to the different presentations we had this morning I am wondering whether you could define the active sites more precisely. Are they correlated to the active sites described by Walker or do they correspond to something else, e.g. to large hydrocarbons. Are there some comments on that particular point. An important property is that the number of sites is kept during the process.

Wagner: We should go back to the concept of Volmer about crystal grow. There we have the different crystal planes and the sites. These sites have different character and give different growth rates. Besides this we get a lot of dislocations and it is well known from oxidation and other kind of reactions, that this dislocations are active sites for catalytic and for gas phase surface reactions. We have to add polymerisation. We definitely do have a branching step when we are adding highly unsaturated species. We are in a balance between producing more active sites and less active sites and that is just the Flory concept. If you add e.g. acetylene you get a radical which is still an unsaturated one. This can form a C-C bond which is still active or it can split off H_2 and it can be built into a graphite lattice plane.

Santoro: The Walker model is a surface adsorption model and then migration on the surface to the reactive species. If you can saturate the surface then you can prevent reactions from happening if the desorption process is in balance. I think the concept here of finding a reactive site with the idea that the edges of these molecules are more reactive than the planes themselves goes back to some work Steve Stein did a number of years ago. He tried to estimate how he could continue to make large particles. So, I think that the quality of the structure comes in and this may link up with what H. Bockhorn was talking about where he was saying the particles are porous. May be they are not porous but you have these agglomerate structures that do preserve a lot of active sites. Now, B. Haynes brought up some objections to that point of view. But we have to remember the built in error when we calculate surface area from light scattering meassurements such as Harries did, because we assume that the particles are spherical. This error is about a factor of two. I think Dobbins has better numbers on that. The concept should focus on these edge planes versus single points. I think M. Frenklach has said something about that in his 23'rd Symposium paper. There has been some discussions on that subject but it is necessary to pay more attention here because those sites eventually do close up as time goes on, not through annealing but through reaction.

Sarofim: I really do not have a clearer idea of how soot agglomerates. If you examine aerosol dynamics with soot particles or any coagulation process freezing at a point of contact, you form a very open structure. Prof. Bockhorn's point would be that the area open structure is available for reactivitiy. The counter argument of Prof. Dobbins would be that particles coalesce as they collide and therefore would lose their reactivity. Which of those models makes more sense and when.

Frenklach: Why do you think those active sites must disappear by reaction and not by annealing?

Santoro: I do think that annealing is one aspect, but it seems to me from Steve Steins work that he showed those active edge sites in proportion to the surface area decreased as the particle grew in his model.

Homann: I would like to ask whether we have to think of these active sites more like radical sites or more like hydrogen rich regions on the soot particles. Because if you increase the temperature one increases the number of radical sites on the molecule. If you think of hydrocarbon molecules and you raise the temperature they will lose hydrogen. The very young soot particles are also just aggregates of carbon and hydrogen. I think that the concentration of hydrogen on the surface is more important than the concentration of radical sites since Dr. Haynes suggested that the number of active sites has increased by addition of acetylene. The process of mass growth by acetylene decomposition prevents the decrease of active sites which takes place by itself just as a thermal process.

Haynes: That was not my understanding of the role of surface growth. Referring back to your opening comments, I do think that the number of active sites is a fitting parameter at this point. The requirement of the model is that the number of active sites in fact is conserved whether you have growth or not. The only thing which effects that number is the annealing which depends only on temperature.

Frenklach: We should remember that when we have a reaction between gas and solid it doesn't proceed the same way as between gaseous species. Even though this is what assumed in our models, its not entirely correct. What we do now is a simple fix to be able to do some modeling. We consider the C-H bonds to be the active sites. So these are H-atoms which are bonded to carbons. Now, not all of these H-atoms are active sites because not all of them can make the system grow, neither by acetylene or other species. If you consider a zigzag there is one C-atom between two which can't grow. You cannot add something to it. So you have a certain number of sites which will not be active.

Lahaye: I completely agree. When a carbon surface is saturated with hydrogen the number of the active sites is vanishing. You have to outgas the material to bring out the hydrogens, to regenerate these bonds, and then you have active sites. It means that in the description of Walker the active sites are not hydrogen or C-H bonds but they are carbon having some vacancy because some hydrogen or oxygen left.

Frenklach: But even among those potential vacancies not all of them can be considered to be active sites. Some of the vacancies that you create cannot grow. We have to distinguish between potential active sites, C-H bonds, which cannot in principle sustain growth and those which can when you pull H-atoms and create vacancies.

Wagner: I think we should be careful with comparing the reactivity of graphite and soot particles. The reactivity of graphite has been investigated in great detail (see Stransky, Volmer, Walker et al.). If you compare these numbers with the growth rate of soot particles you see that they are different. What you say is true, but it is still a different process. We are at the beginning closer to a polymerisation situation than to a solid state surface reaction. One other point I wanted to mention is the growth between colliding particles. If you have small particles we have a high hydrogen content and the thermal mobility is very high. The second point is that we do see that the number density of the particles goes down and at the very beginning the particles seem to stay spherical. We examined young soot particles with very high resolution electron mirographs. We could differentiate between this small growing particles which have an initial hydrogen content of about 0.5. We saw the particles sticking together and the space in between being filled. It may well be that the space inbetween is really a very reactive zone.

Lahaye: I think this was a good summary for the moment. We don't know really if we have to refer to the solid state description given by Walker for graphitic material or whether we have to refer to radical polymerisation of material. We are not quite sure how to describe these aspect of active sites. We know it is useful but we certainly have to go much further to be capable to give description as good as the one given of catalysis or of carbon science.

Soot in Flames
Formation, Dependence, Practical Systems

Practical systems and flames at different burning conditions are the challenges for any model of soot formation
and oxidation. Therefore, this part contains articles on soot
formation, oxidation, and inhibition by additives in practical systems. Practical systems range from flames with fuel
additives, counter flow partially premixed diffusion flames,
shock tube devices, plug flow reactors, and diesel engines.
Modelling as well as experimental investigation are topics
of the discussion in part IV.

Metallic Additives in Soot Formation and Post-Oxidation

Jacques Lahaye, Serge Boehm, Pierre Ehrburger

Centre de Recherches sur la Physico-Chimie des Surfaces Solides,
Université de Haute-Alsace,
68200 Mulhouse, France

Abstract: A survey on the effect of metallic additives on the formation and oxidation of soot in hydrocarbon combustion is given. Possible chemical mechanisms for retarding or enhancing soot formation are discussed as well as catalytic actions of metals in soot oxidation.

19.1 General View

Some decades ago the influence of metallic additives on the formation of particulate carbon was essentially studied for controlling the structure of carbon black and to reduce smoke formation in fire from polymers. The main interest today is related to soot emission by diesel engine. Metallic additives may either decrease soot formation or increase the rate of soot oxidation [19.1-12]. Metallic additives can operate inside the engine or during post-oxidation of the particles collected on filters at the outlet.

19.2 Possible Influence of Metallic Additives Inside the Engine

A very large variety of cations has been added to the fuel, e.g. Li, Na, K, Cs, Mg, Ca, Ba, Sr, V, Cr, Mo, Co, Ni, Mn, Fe, Cu, Hg, Sn, Pb, In, La, Ce, Pr, Sm and Yb [19.13-19] Three mechanisms seem to be operating in flames [19.19]:

Mechanism I: An ionic mechanism occurring with metals of low potentials of ionisation such as Na, K, Cs, Ba (ionisation potential 5.14, 4.34, 3.29, and

5.21 eV, respectively). Such ions are expected to increase the nucleation rate and to lead to smaller individual particles for a given carbon yield and therefore to a higher chemical rate of oxidation [19.12, 19.15, 19.16, 19.18-24]. *Haynes* et al. [19.15] studied the influence of low concentrations of alkaline and alkaline-earth metals on soot particle size and number density in atmospheric pressure premixed flat flames. It was shown that the addition of low concentration (< 1ppm) of the alkaline metal slightly reduces the amount of soot. However, the particle number density is significantly increased apparently because these metals promote the charging of incipient soot particles. These charged particles resist to coagulation and preserve their individuality to a better extent, leading to larger number of smaller particles.

It must be kept in mind that carbon black producers have been using for decades potassium ions to reduce the degree of aggregation of carbon black without modifying significantly the specific surface area, i.e. the size of the individual particles.

Feugier et al. [19.25] have also studied the effect of alkaline metals on the amount of soot emitted by premixed hydrocarbon flames (rich ethylen-oxygen-nitrogen flames). It was concluded that two opposite effects are observed:

 (i) an increase of the rate of nucleation and of the amount of soot produced (promotors) and
 (ii) a decrease of agglomeration which increases its oxidatibility (inhibitors).

Mechanism II: Alkaline-earth additives (Ba, Ca, Sr) undergo a homogeneous reaction with flame gases to produce hydroxyl radicals which rapidly remove soot or gaseous hydrocarbon soot precursors [19.12, 19.13, 19.15-19, 19.26-27]. Few mechanisms can generate hydroxyl radicals. The following sequence is particularly relevant to flame conditions [19.28]:

$$A + H_2O \longrightarrow A - OH + H \tag{19.1}$$

$$H + H_2O \longrightarrow OH + H_2 \tag{19.2}$$

Mechanism III, (Mn, Fe, Co, Ni): As pointed out by *Howard* et al. [19.19], this mechanism, which only occurs to appreciable extents late in the flame, is the acceleration of the oxidation rate, possibly by occlusion of the metal within soot particle. It is worth to note that a part of the metal must be at the surface of the particles in order to begin to catalyse soot oxidation. As the oxidation proceeds the metal occluded inside the particle is progressively released and contributes to the carbon oxidation.

19.3 Oxidation of Soot Deposited on a Filter

The metallic additives may increase the rate of soot oxidation or/and decrease the ignition temperature. After ignition, however, the kinetics may be controlled by the diffusion of O_2 to the particle so that the catalytic effect is expected not to change the oxidation rate. The additives can be deposited on the filter before collecting soot. In that case the surface of interaction between the additive and soot is extremely small so that a negligible catalysis of oxidation of the solid is expected. However the catalyst may increase the rate of conversion of CO into CO_2 so that the temperature increases and can reach 500 to 550°C, temperature required for the ignition of soot particles (without additive). The most efficient process is expected to be the premixing of additives in the fuel. The main part of the additive is recovered on (and in) soot. Additives like Cu, Ba can decrease the ignition temperature of about 100°C [19.1,19.4,19.8,19.29-31].

Miyamoto et al. [19.29] studied by thermogravimetric analysis the oxidation of soot containing Ca, Ba, Fe and Ni with O_2/N_2 mixtures. These soot samples were obtained in a single cylinder diesel engine operating with fuel doped with metallic naphtenates. Figure 19.1 is a typical example of soot mass change with time at $T > 500$°C.

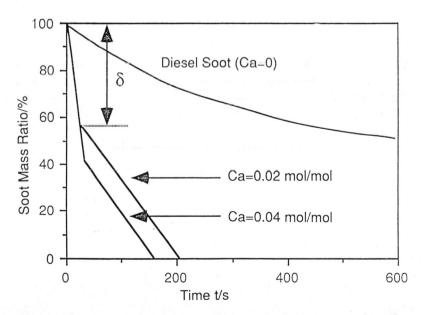

Fig. 19.1. Soot mass change by oxidation in a thermobalance [19.29]

For a given temperature, soot without additive exhibits a smooth decrease in mass, while soot with calcium additive is rapidly oxidized in two steps. The amount of soot oxidized in the first step is affected by the calcium concentration. The rate in the second step is independent of the calcium concentration. No clear interpretation is given by the authors though the second step is expected to be an oxidation in the diffusion controlled regime. As for the first step, it is difficult to explain the very high oxidation rates obtained. For the same soot samples, the ignition temperature was plotted versus Ca content. A 140°C decrease is observed from soot without Ca to soot with 0,06 mole Ca/mole C (Fig. 19.2).

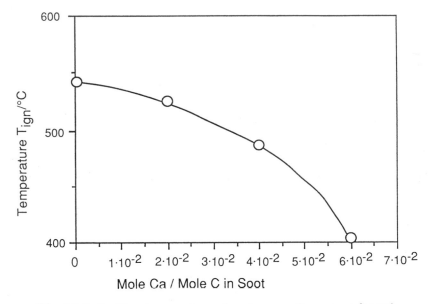

Fig. 19.2. Ignition temperature of soot versus Ca content [19.29]

We have studied by thermogravimetry the oxidation of a soot sample on which cerium was deposited by impregnation using cerium octoate. The results are presented in Fig. 19.3 for a constant heating rate equal to 2 K/min. It was found that the ignition temperature (intercept of the linear part of the oxidation rate with the x-axis) is decreasing with the Ce content.

Two mechanisms are usually proposed to explain the influence of metallic derivates on the gasification of soot:

(i) the oxygen-transfer mechanism [19.30-34] and

(ii) the electron-transfer mechanism [19.31,19.33].

In the first one, the metallic catalyst is considered as a oxygen carrier involved in the oxidation-reduction cycle. The metallic compound is oxidized by the oxidizing gases present in the system; it is reduced by the carbon substrate.

$$(-C) + (-MeO) \longrightarrow (-CO) + (-Me-) \qquad (19.3)$$

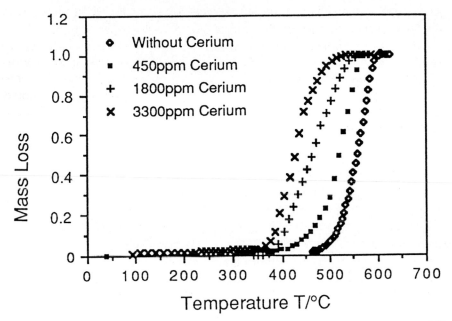

Fig. 19.3. Oxidation of Ce impregnated carbon black with air; heating rate: 2 K min^{-1}

$$(-CO) + (-MeO) \longrightarrow CO_2 + (-Me-) + \text{free carbon site} \quad (19.4)$$

or

$$(-C) + (-MeO) \longrightarrow CO + (-Me-) + \text{free carbon site} \quad (19.5)$$

$$CO + (-MeO) \longrightarrow CO_2 + (-Me-) \quad (19.6)$$

The catalytic reaction occurs at the interface soot/metallic oxide so that a good dispersion must be achieved for the oxidation to be efficient.

In the second mechanism, the electron transfer from the carbon matrix to metallic additives induces a reorganisation of electrons in carbon which weakens the carbon/carbon bonds and therefore decreases the energy required for an oxygen to pull out a carbon. From literature, it appears that oxygen transfer is more probable than electron transfer (catalytic oxidation occurs only at the interface carbon/oxide). Recent experiments carried out by *de Soete* [19.34] strongly suggest an oxygen-donor mechanism. Soot produced by pyrolysis of aromatic hydrocarbons were mixed with metallic oxides and diluted in pulverized quartz. The authors plotted (Fig. 19.4) at 800 K the combustion rate r_c obtained with O_2 (21 kPa) diluted in nitrogen versus the reduction rate of the catalyst r_{red} defined as the mass of oxygen stripped from oxide per unit time and unit mass of oxygen. For catalysts such as CeO_2, PbO, MnO_2 and CuO a fairly good experimental correlation exists between r_c and r_{red}, suggesting the catalytic efficiency of oxides to be related to its ability to act as oxygen-donor, rather than an electron transfer.

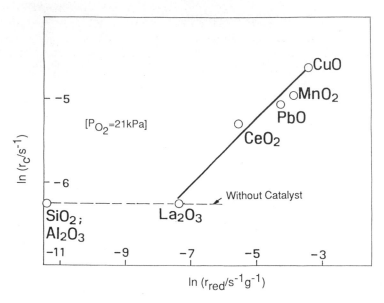

Fig. 19.4. Correlation between combustion rate and reduction rate of catalyst by soot; $T = 800$ K, number of metal atoms per surface carbon atoms of the initial sample, n = 3

References

19.1   M.R. Montierth: "Fuel Additive Effect upon Diesel Particulate Filters", SAE Paper No. 840072 (1984)

19.2   R. Noirot: "Etude Expérimentale et Paramétrique de la Combustion de Suies sur Filtre à Particule"; Ph.D. Thesis, University of South Alsace (1990)

19.3   D.W. Golothan: "Diesel engine Exhaust Smoke. The Influence of Fuel Properties and the Effects of Using Barium-Containing Fuel Additive", SAE Paper No. 670092 (1967)

19.4   N. Myamoto, Z. Hou, A. Harada, H. Ogawa, T. Murayama: "Characteristic of Diesel Soot Suppression with Soluble Fuel Additives", SAE Paper No. 871612 (1987)

19.5   T. Saito, M. Nabetani: "Surveying Test of Diesel Smoke Suppression with Fuel Additives", SAE Paper No. 730170 (1973)

19.6   C.O. Miller: "Diesel Smoke Suppression by Fuel Additive Treatment", SAE Paper No. 670093 (1967)

19.7   G.M. Simon, T.L. Stark:" Diesel Particulate Trap Regeneration Using Ceramic Wall-flow Traps, Fuel Additives, and Supplemental Electrical Igniters", SAE Paper No 850016 (1986)

19.8   B. Wiedemann, K.H. Neumann: "Vehicular Experience with Additives for Regeneration of Ceramic Diesel Filters", SAE Paper No. 850017 (1986)

19.9   O.A. Ludecke, K.B. Bly: "Diesel Exhaust Particulate Control by Monolith Trap and Fuel Additive Regeneration", SAE Paper No. 840077 (1985)

19.10   W.R. Wade, J.E. White, J.J. Florek: "Diesel Particulate Trap Regeneration Techniques" SAE Paper No. 810118 (1982)

19.11   B. Wiedemann, U. Doerges, W. Engeler, B. Poettner: "Application of Particulate Traps and Fuel Additives for Reduction of Exhaust Emissions", SAE Paper No. 840078 (1985)

19.12   G. de Soete, A. Feugier, J.P. Gyss, G. Mercier: Mécanismes de Formation et Moyens de Réduction des Polluants dus à la Combustion (Edit Technip, Paris 1973) pp. 15-105

19.13   J. Lemaire: Purification of Diesel Engine Exhaust. Contribution of Cerium-Based Additives (Japan Society of Lubrication Engine, Tokyo 1989)

19.14   A.Feugier: Adv. Chem. Sci. 12 (No. 166), 178 (1978)

19.15   B.S. Haynes, H. Jander, H.Gg. Wagner: "The Effect of Metal Additives on the Formation of Soot in Premixed Flames", in Seventeenth Symposium (International) on Combustion (The Combustion Institute, Pittsburgh 1979) p. 1365

19.16   E.M. Bulewicz, D.G. Evans, P.J. Padley: "Effect of Metallic Additives on Soot Formation Processes in Flames", in Fifteenth Symposium (International) on Combustion (The Combustion Institute, Pittsburgh 1975) p. 1461

19.17   D.H. Cotton, N.J. Friswell, D.R. Jenkins: Combust. Flame 17, 87 (1971)

19.18   J. Lahaye, G. Prado: Chemistry and Physics of Carbon 14, 167 (1978)

19.19   J.B. Howard, W.J. Kausch Jr.: Prog. Energy Combust. Sci. 6, 263 (1980)

19.20   E.R. Place, F.J. Weinberg: "The Nucleation of Flame Carbon by Ions and the Effect of Electric Fields", in Eleventh (International) Symposium on Combustion (The Combustion Institute, Pittsburgh 1967) p. 245

19.21   J.B. Howard: "On the Mechanism of Carbon Formation in Flames", in Twelfth (International) Symposium on Combustion (The Combustion Institute, Pittsburgh 1969) p. 877

19.22   J.L. Delfau, P. Michaud, A. Barassin: Combust. Sci. Technol. 20, 165 (1979)

19.23   K.-H. Homan: Ber. Bunsenges. Phys. Chem 83, 738 (1979)

19.24   D.B. Olson, H.F. Calcote: "Ions in Fuel-Rich and Sooting Acetylene and Benzene Flames", in Eighteenth Symposium (International) on Combustion (The Combustion Institute, Pittsburgh 1981) p. 453

19.25   A. Feugier, F. Mauss: "Efficacité des Additifs de Combustion, Mécanismes d'Action", in Journées d'Etudes sur l' Evolution de l'Utilisation des Fuels Lourds dans l'Industrie (AFTP 1979) p. 157

19.26   A.G. Gaydon, H.G. Wolfhard: Flames: Their Structure, Radiation and Temperature, (Chapman and Hall, New York 1978)

19.27   R.S. Sapienza, T. Butcher, C. Krishna, J. Gaffney: "Soot Reduction in Diesel Engines by Catalytic Effects", in Symposium on Chemistry of Engine Combustion Deposits (ACS, Atlanta 1981)

19.28 W.E. Kaskan: "The Reaction of Alkali Atoms in Lean Flames", in Tenth (International) Symposium on Combustion (The Combustion Institute, Pittsburgh 1955) p. 41

19.29 N. Miyamoto, Z. Hou H Ogawa: "Catalytic Effects of Metallic Fuel Additives on Oxidation Characteristics of Trapped Diesel Soot", SAE Paper No. 881224 (1988)

19.30 D.W. McKee, D. Chatterji: Carbon **13**, 381 (1975)

19.31 D.W. McKee: Chem. Phys. Carbon **16**, 1 (1981)

19.32 D.W. McKee: Carbon **8**, 623 (1970)

19.33 H. Marsh: Introduction to Carbon Science (Butterworths, London 1989)

19.34 G. de Soete: "Catalysis of Soot Combustion by Metal Oxides", in Western States Section Meeting of the Combustion Institut (The Combustion Institute, Salt Lake City 1988)

Discussion

Lepperhoff: I think the mechanims of oxidation of soot promoted by metal catalysts are different for soot in a diesel engine and for soot collected in a trap. Soot collected in a trap is old soot and soot particles are covered with hydrocarbons that are adsorbed on the surface. The amount of the adsorbed volatiles influences the ignition temperature of soot. With ferrocen as oxidation promoter the ignition temperature of trapped soot may be lowered down to about 100°C. For soot with little volatile content the ignition temperature is much higher. It is in the range of 410 to 430°C. Soot within the diesel engine resembles soot with low volatile contents. My question is what kind of technique do you use to determine the location of cerium within the particle?

Lahaye: In our work, the kinetic of oxidation at 450°C is not affected by PAH. Soot oxidized as collected and soot extracted with CH_2Cl_2 exhibit the same rates of oxidation.

For determining the cerium content of the particles, two techniques are mainly used: X-fluorescence and emission spectroscopy. These techniques, however, give no information about the distribution of cerium within the particles. Such a distribution might be obtained in oxidizing progressively the soot sample and trying to extract the cerium oxide of the sample partly oxidized e.g. by nitric acid and to continue this procedure step-wise.

Lepperhoff: We did not find any change in the activation energy for the oxidation of soot when iron oxide is mechanically mixed with soot. However, when introducing iron oxide into the soot through the fuel we found a decrease in the activation energy. Therefore, the kind of dispersion of the catalyst in the soot particle seems to influence the oxidation rates.

Lahaye: In the catalysis of soot oxidation by metallic additive, the oxygen-transfere mechanism requires, indeed, a good dispersion of the catalyst. Your results fit with such a mechanism. It must be kept in mind that the catalysis of oxidation may not modify the acitvation energy but the preexponential factor of the Arrhenius equation.

Sarofim: Clearly as you mentioned you would like to have an intimate contact between your catalyst and the carbon. Feitelberg whose work was referred to earlier found that with iron additive the iron condensed late and therefore coated the carbon and got mixed in that way. For the cerium what strategies do you propose to use?

Lahaye: In our experiments the cerium is introduced with a spray nozzle into a heated tube. About half of the cerium is collected with soot but, at this stage of our research work, we do not know whether cerium is into or on the particles.

Santoro: I think you have answered this question but I just want to raise it because Jack Howard mentioned in his review two precautions concerning the use of metal additives. One was the health effect of the metals getting into the environment. He thought this might be worse than the health problem of the soot. And the other one was that in gas turbine engines they had seen coating of the metals onto critical components. I think you answered that question possibly from your bus data. However, I was wondering if you have any further comment on those particular points.

Lahaye: Actually the cerium oxide is not considered as a toxic derivate. In the experiments with busses in Athena, the presence of cerium did not affect the working of the engines.

Critical Temperatures of Soot Formation

Irvin Glassman, Osami Nishida, George Sidebotham

Department of Mechanical and Aerospace Engineering,
Princeton University,
Princeton, NJ 08544-5263, USA

Abstract: A review of critical temperatures for soot formation considering new experimental results from Princeton University and other sources is given. From this the conclusion is drawn that incipient particle formation controls the total mass of soot formed in any process. Approaches to preventing soot formation and growth should be to prevent any soot particles from forming.

20.1 Scope of this Paper

During the last decade an extensive amount of research on soot formation and destruction has been performed. A great deal has been learned, but yet very little of the findings has found itself applied to solving practical problems. What the development engineer demands is interpretation of the results that they can be applied to industrial challenges. Thus at Princeton the continued work on soot processes has been dedicated to this task and focuses on the critical temperatures which are believed to control soot formation process and the soot burn-up process.

In this regard this presentation does not differ too greatly in context from what was presented previously [20.1]. Review of new experimental material from Princeton and the question period which followed the discussion reported in [20.1] leads this writer to state again that incipient particle formation controls the total mass of soot formed in any process. Thus the approach to preventing soot formation and growth should be to prevent any soot particles from forming.

20.2 Evaluation of Critical Temperatures of Soot Formation

Earlier work at Princeton estimated the temperature at which particles of soot are first observed in a diffusion flame by observing the position of the first particles along the centerline of a laminar diffusion flame and then measuring the temperature at that point. Since it is difficult to determine whether a particle was formed at the centerline or was brought there by thermophoresis, another procedure for determining the temperature of incipient particle formation was sought. The method chosen was, for a fixed flame height for every fuel, to dilute the fuel stream with an inert (primarily nitrogen) until all the observed soot luminosity just disappeared, and, then to measure the centerline temperature through the pyrolysis zone to the observed flame height. This temperature was assumed to be the incipient particle formation temperature. This procedure was followed for seven fuels: acetylene, allene, ethene benzene, 1-3 butadiene, 2-butene and toluene. Shown in Fig. 20.1 are the temperature profiles when zero luminosity is attained for allene and benzene flames. These curves are typical of those for the other fuels as well. Table 20.1 lists the corresponding flame temperatures for the fuels evaluated. A variation in the incipient temperatures of only 130 K exists. But, interestingly, the flame which requires the least dilution gives the highest particle formation temperature, i.e. these temperatures exactly correlate inversely with the amount of dilution required for observance of no luminosity. Table 20.2 lists other estimates including those of other investigators of incipient particle formation temperature both for diffusion and premixed flames. Although there appears to be a major variance in the data from various investigators, general arguments can be made that some of the variance is due to the experimental procedures.

Table 20.1. Soot inception temperatures, this study

Fuel	Inception Temperature T/K
Acetylene	1665
Allene	1585
Ethene	1700
Benzene	1580
1,3-Butadiene	1650
2-Butene	1684
Toluene	1570

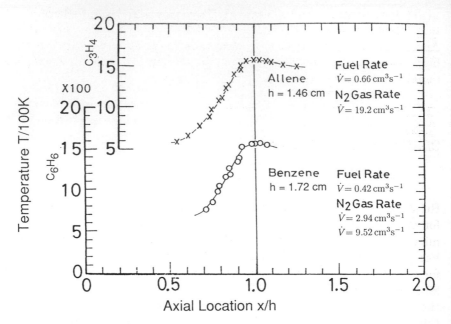

Fig. 20.1. Temperature profiles along centerline of laminar fuel jet diluted with N_2 to point of no luminosity

Table 20.2. Soot inception temperatures in K, other work

Fuel	1	2	3	4	5	6
Methane	1913					
Propane	1783					
Ethene	1753	1700		1750	1460	
n-Butane	1723					
i-Butane	1643					
Acetylene		1665	1345	1600		
Butadiene	1623	1650	1353			
Allene	1613	1586				
Benzene		1580	1332		1430	1750
Toluene		1570				

Remarks

(1): Princeton University, centerline temperature at flame height of oxygen diffusion flames (Smith);
(2): Princeton University, same as 1 (Nishida);
(3): Princeton University, centerline temperature of sooting diluted air diffusion flames in which particles first found (Gomez);
(4): from ref. [20.2];
(5): from ref. [20.3];
(6): from ref. [20.4]

Thus this work would indicate that the actual incipient particle formation temperature was the same for all fuels and would be about 1600K. And, it is concluded, as before, that there must exist a critical, perhaps high activation, step(s) that controls the system. An important challenge remaining to soot investigators is to determine what is this step(s).

20.3 Discussion

Referring to the discussion reported in ref [16.5], the argument is still made for the validity of Milliken's conceptual idea that soot formation in pre-mixed flames is the result of competition between the rate of soot precursor formation and the rate of oxidative attack on these precursors. Consider, for example, the data on pre-mixed flames from ref [20.3] shown in Fig. 20.2. The critical C/O-ratio declines as the temperature is lowered, then there is a sharp rise in the critical C/O sooting ratio. The decline in C/O to this point would indicate that formation processes are faster than precursor oxidation processes. The temperature at which the sharp rise occurs can only indicate that the soot formation process has essentially stopped. The abrupt termination of the formation in this case must indicate a controlling high activation temperature process.

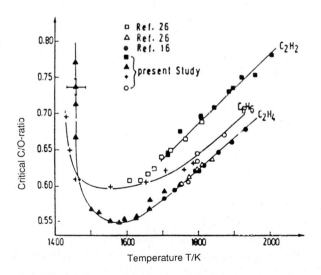

Fig. 20.2. Critical Sooting C/O -ratio of premixed flames versus controlled flame temperature from ref. [20.3]

Questions still arise as to whether the effect of diluents in diffusion flame soot experiments is due to either thermal or fuel concentration considerations. In order to resolve this question, the Princeton experiments performed to determine the incipient particle formation temperature were repeated with four fuels. However, three diluents were used; namely, argon, nitrogen and carbon dioxide. Obviously the amount of each required to obtain the same temperature varies significantly. Yet, as observed in Fig. 20.3, the same incipient formation temperature is obtained regardless of diluent. The conclusion is reached that concentration effects are negligible.

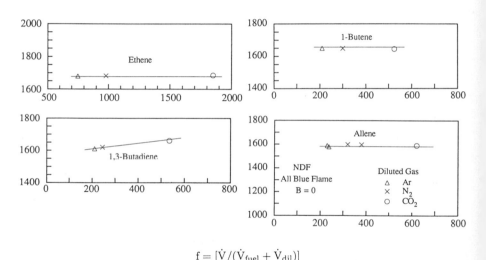

$$f = [\dot{V}/(\dot{V}_{fuel} + \dot{V}_{dil})]$$

Fig. 20.3. Incipient soot formation temperature as function of diluent; $f = \left[\dot{V}_{fuel}/\left(\dot{V}_{fuel} + \dot{V}_{dil}\right)\right] \cdot 100$

Finally it is within the approach of this paper to refer to *Kent* et al. [20.6]. These investigators, in a study of soot formation in diffusion flames, came to the conclusion that the smoke height, as noted in Fig. 20.4, was reached when the soot particle temperature dropped below 1300 K. Reasoning, then, that indeed the particle burning process is very slow at 1300 K and below, one can accept 1300 K as another limiting temperature. Thus, for most practical concerns, it is concluded that soot formation can be prevented in diffusion flames if the stoichiometric flame temperature can be kept below 1600 K, and, if soot formation is inevitable, that particle burnup can be achieved if temperatures are kept above 1300 K in an oxidizing atmosphere.

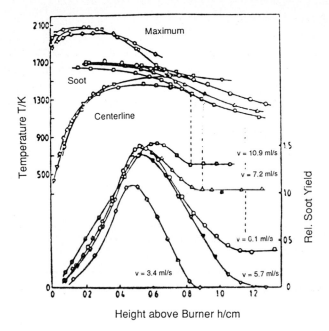

Fig. 20.4. Temperature and relative soot yield as function of flame height from ref. [20.5]

20.4 Acknowledgements

The author's work was supported by the Air Force Office of Scientific Research Under Contract No. F49620-86-C-0006.

References

20.1   H. Jander, II.Gg. Wagner: <u>Soot Formation in Combustion - an International Round Table Discussion</u> (Vandenhoeck and Ruprecht, Göttingen 1990), p. 67
20.2   L.R. Boedecker, G.M. Dobbs: "Soot distribution and CARS temperature measurements in axisymmetric laminar diffusion flames with several fuels", in <u>Twenty-First Symposium (International) on Combustion</u> (The Combustion Institute, Pittsburgh 1986) p. 1097
20.3   D. Böhm, D. Hesse, H. Jander, B. Lüers, J. Pietscher, H.Gg. Wagner, M. Weiss: "The influence of pressure and temperature on soot formation in premixed flames", in <u>Twenty-Second Symposium (International) on Combustion</u> (The Combustion Institute, Pittsburgh 1988) p. 403
20.4   D.B. Olson, S. Madronvich, Combust. Flame **60**, 203 (1985)
20.5   J.H. Kent, H.Gg. Wagner: "Temperature and fuel effects in sooting diffusion flames", in <u>Twentieth Symposium (International) on Combustion</u> (The Combustion Institute, Pittsburgh 1984) p.1007

Discussion

Wagner: I would just like to add some more information to your table.

Glassman: Are you going to support me?

Wagner: I am going to support you completely.

Glassman: Okay, good.

Wagner: You can add n-decane and you can add hexadecane and you can add polystyrene and even polyethylene to your table and they show the same inception temperatures.

My second point is: you can buy a "blue flame burner" in this country since about five years. This burner is operating exactly in the way you argued.

Glassman: Who has the patent here and in U.S.?

Wagner: I think it is DLR in Lampoldshausen.

Glassman: But Prof. Wagner, I must admit that this was more than five years ago that I was approached by this gentleman in the United States.

Wagner: They started to work twenty years ago and they sell it since about five years.

Glassman: Well, of course I have to say that is very pleasing to hear that. I think a number of environmental problems particularly with diesels and new restrictions being implaced in California on home furnaces could make this a very, very crucial problem.

Kent: In the data I showed this morning for the soot formation rates in our various flames you could observe soot formation right starting at different temperatures depending on the size of the flame. So there is a range of temperatures between about 1300 to 1500 K and I think this has to do with the fluid mechanics and where the trajectories of the particles sweep into the flame and what the temperature happens to be there. There may be a minimum temperature but I do not think that it is as high as what you said.

Glassman: I agree that your data are influenced by the effects you just mentioned. But you had another graph where you showed the maximum soot formation at indeed 1600 K. So I feel that this is a little confirmation of the process I have been discussing.

Frenklach: I understood, correct me if I am wrong, that you were saying that there is a physical process you are preferring rather than a process with an activation energy that stops soot formation or starts soot formation. If we look at the shock tube experiments that have been done in several laboratories we see a soot bell which means that you have a starting temperature and you have a temperature where soot formation stops. The soot bell is shifted depending on fuel, depending on pressure, depending on additives. For acetylene soot does not appear until 2200 K. If you add a small amount of oxygen the soot bell is shifted down by about 700 K. The question then is if it is a fundamental chemical process which shuts off or shuts on soot fformation we should clearly see it in shock tubes first before in any flame.

Glassman: It is very obvious that in any chemical kinetic process there is always a temperature-time factor, even if you have ignition phenomena. We say that there is an ignition temperature. Well, that is artificial. If I wait long enough I can ignite at a lower temperature. You have to realize that the temperature time histories in shock tubes are vastly different from the temperature-time histories which take place in flames. That is why you have to go to higher temperatures. If you could expose the wave longer so that the pressure and temperature behind the wave last longer, indeed your temperature would drop. The temperature time histories you get are out of phase with those you get within flames. If you go back to the early work on prompt NO there was the work of Just and that work of Bowman and there was a fundamental disagreement whether a prompt NO existed or not. Of course it did. The work of Just and the earlier work of Fenimore was in flames, but the work of Bowman was in shock tubes and basically had a completely different temperature-time history. So the prompt NO was overruled by the Zeldovich NO and they just could not observe it. There were two good experiments, both were right, but derived their arguments from experiments with different types of apparatus.

Santoro: I just drew a picture of what stream lines looks like in a diffusion flame. What we observed in our measurements of temperature and where the soot formed is that when you are near the outer edge where you are near the flame front you find that the soot begins to form at a higher temperature than when you are looking along the centerline. Comparing this with the results of Wagner's group we find that there is both a lower limit and an upper limit similar to what Michael is saying about the shock tube results. And it depends on the temperature-time history where you see those limits. So I am not sure that there is a single upper temperature, there might be a range of temperatures.

The second point I wanted to make is in fairness to the concentration argument. What I showed this morning was that you have to be very careful when you dilute flames because there is so much diffusion going on. You are not sure in your CO_2, Argon and N_2 experiment that you really change the concentrations. So you should measure the concentration.

Glassman: If you fundamentally assume that in a coannular diffusion flame there is diffusion around the rim, there is also a possibility of multi diffusion penetration through the flame zone of an oxidizer. In consistency to what I said about premixed flames, when you start to form soot near the flame boundary the effect of oxygen getting through the boundary can indeed make the flame temperature period different than what we observed. That was one of the reasons I argued we are to concentrate on the centerline where we get away from the extent of diffusion of any oxidizing species into the overall pyrolysis zone which exists in diffusion flames. George Sidebotham noticed that in his work on diffusion flames. It is somewhat experiment dependent. But if you look at Dr. Jander's data in premixed flames and you look at the stuff we are getting, there is a temperature limit. There is something controlling, Michael.

Frenklach: It may be true that you have some kind of constant temperature, whatever it is. But my point is that it can not be one physical fundamental process. It has to be a combination of several factors which somehow always lead to this temperature. A miracle.

Glassman: Michael I said steps. But I give you a challenge. You are doing a tremendous amount of modeling. I am getting my type of results, Dr. Jander is getting her type of results, both types of results are consistent with others and ignition measurements. Your theory has to tell me why.

Soot Formation in Partially Premixed Diffusion Flames at Atmospheric Pressure

Fabian Mauss, Bernhard Trilken, Hermann Breitbach,

Norbert Peters

Institut für Technische Mechanik,
RWTH Aachen,
52062 Aachen, Fed. Rep. of Germany

Abstract: The formation, growth and oxidation of soot has been studied in a laminar counterflow configuration. Soot inception occurs in a fuel rich burner stabilized premixed acetylene-air flame at the bottom of the system. Oxidation of soot takes place downstream in a diffusion flame, generated by unburnt hydrocarbons, carbonmonoxid and hydrogen in the burnt gas of the premixed flame and a counterflow of pure oxygen. In the region between the two flames soot particles grow due to surface reactions and coagulation. The residence time of the particles in this region is long, because the counterflow flame is placed near the stagnation point of the system. Mean particle diameter up to 100 nm are observed.

Concentration profiles of various stable species up to benzene were measured by gas chromatography. Thermocouples were used for temperature measurements. The soot volume fraction was determined by laser light extinction and Abel's inversion. Particle sizes measured with the combined scattering/extinction measurement technique, with the dispersion quotient method and with dynamic light scattering technique are compared.

Numerical calculations including a chemical model for soot formation have been performed for this system. The calculated profiles for temperature, concentrations, soot volume fraction and particle size were compared with the experimental data. The predicted profiles of soot volume fractions and particle sizes are found to be of the same accuracy as the predicted profiles of temperature and gas phase concentrations.

21.1 Introduction

In complex combustion systems soot particles are formed in fuel rich regions and can be oxidized in fuel lean regions. To control the emission of soot from real combustion systems the understanding of both formation and oxidation of soot is necessary. During the last years, the formation and growth of soot particles in premixed hydrocarbon flames have been investigated in dependence on temperature, C/O-ratio of the mixture, fuel and pressure [21.1-5]. At the same time detailed chemical models predicting the appearance of soot in premixed flames were developed [21.6-7]. But less is known about the oxidation of soot particles shortly after their formation.

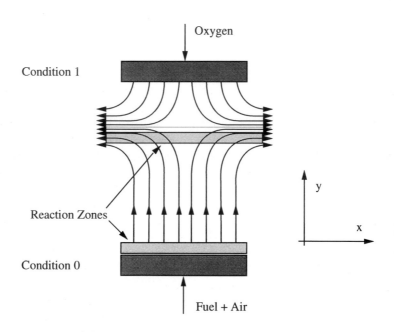

Fig. 21.1. Schematic illustration of the flame configuration

In order to study the formation and oxidation of soot particles, laminar flames in a counterflow geometry have been investigated experimentally and numerically (Fig. 21.1). A fuel rich mixture of C_2H_2 and air enters from the bottom, while pure oxygen enters from the top from two opposed tubes. Soot is generated in the postflame gases of the fuel rich premixed flame stabilized in the lower part on a cooled sinter metal plate. The soot is oxidized downstream in a diffusion flame, which is formed by the oxygen counterflow. The distance between the upper and lower tube is variable.

A surrounding annular shroud flow of nitrogen prevents a diffusion flame between the unburnt fuel in the postflame zone of the premixed flame with ambient air and stabilizes the flame configuration.

21.2 Experimental

Experimental measurements at atmospheric pressure have been performed for five different flames:

Number of flame	1	2	3	4	5
C/O-ratio	0.85	0.85	0.85	0.75	0.70
v_{pre}	4 cm/s	8 cm/s	12 cm/s	4 cm/s	4 cm/s
v_{oxygen}	2 cm/s	4 cm/s	6 cm/s	2 cm/s	2 cm/s

The flow velocity on the centerline of the flame configuration has been measured with Laser Doppler Velocimetry. The equipment was a standard single-component detection system working in the forward scattering mode without frequency shifting at a wavelength of 514.5 nm. ZrO_2-particles are introduced with a small flow of nitrogen through a horizontal ceramic tube with an outer diameter of 1 mm near the burner surfaces. The temperature at atmospheric pressure has been determined by fine coated thermocouples (Pt-6% Rh vs. Pt-30% Rh wires of 0.1 mm diameter). For measurements along the centerline in the sooting region the thermocouple was quickly moved into the flame and stopped at the desired location while the signal was recorded. The time trace allowed the initial clean thermocouple signal to be determined. The values were corrected for convective and radiative heat transfer. To determine particle sizes, different methods have been applied. Dynamic light scattering was used with scattering angles between 3 and 10 degrees. Sizes were also determined from combined measurements of scattered light intensity and extinction. Finally, size evaluation was obtained using the dispersion quotient method measuring light extinction at wavelengths of 350 nm, 488 nm and 632.8 nm.

In Fig. 21.2 the LDV- and thermocouple-measurements on the centerline of the flame 1 are shown. Close to the bottom the temperature rises rapidly to approximately 1650 K in the rich premixed flame. Downstream the slowly decreasing temperature in the sooting postflame gases of the premixed flame due to radiative losses can be observed. The temperature increases again in the counterflow diffusion flame. LDV measurements could be performed with the soot particles if their number density is small enough and the particle diameters are large enough. The measured values show good agreement with those determined with ZrO_2-particles. The stagnation point is located at 20 mm above the burner surface.

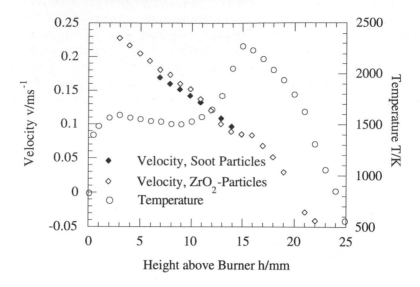

Fig. 21.2. LDV and thermocouple measurements in flame 1 (fuel: acetylene)

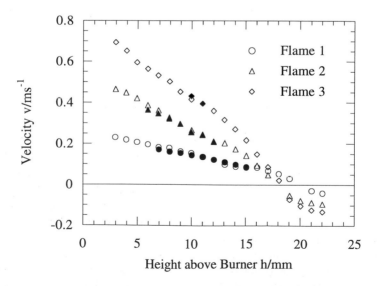

Fig. 21.3. LDV-measurement for different flames

Figure 21.3 shows the LDV results for different flow velocities. Because particle sizes decrease with increasing flow velocity, fewer LDV measurements using soot as tracer particles could be made for the flames with higher flow velocities. In Fig. 21.4 temperature profiles are shown for the same flames. With increasing velocities the temperature level of the premixed flame increases, due to shorter residence times of the radiating particles. This also affects the temperature level of the counterflow diffusion flame.

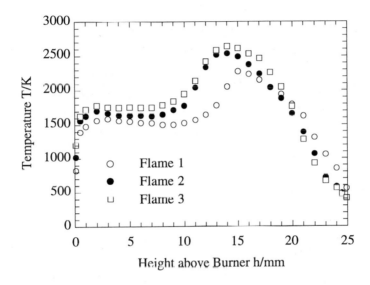

Fig. 21.4. Influence of velocity on the particle size

The soot particle sizes measured with the combined light scattering/extinction method are shown in Fig. 21.5 for different velocities. The values along the axis of symmetry were calculated from an Abel inversion of the extinction measurements. With increasing velocities the maximum particle size decreases because of decreasing residence times; simultaneously the beginning of the sooting zone moves towards the burner surface. The influence of the C/O-ratio is shown in Fig. 21.6.

A comparison of the results of the different particle sizing techniques is shown in Fig. 21.7 for flame 1. The values obtained with the combined light scattering/extinction measurement technique and with the dispersion quotient method are in good agreement. While the differences to the dynamic light scattering in the lower regime up to the height of 6 mm are relatively small, they increase with height above burner up to about 100 %. Because the dynamic light scattering technique is relatively independent of

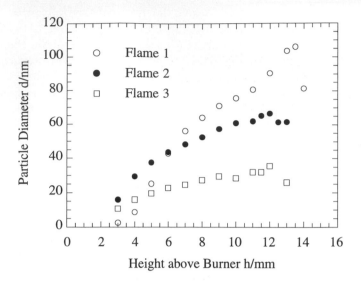

Fig. 21.5. Influence of velocity on the particle size

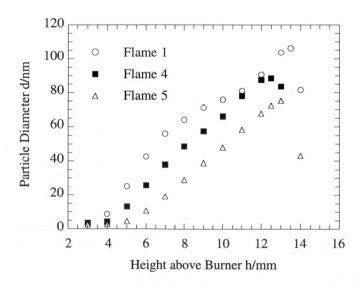

Fig. 21.6. Influence of C/O-ratio on the particle size

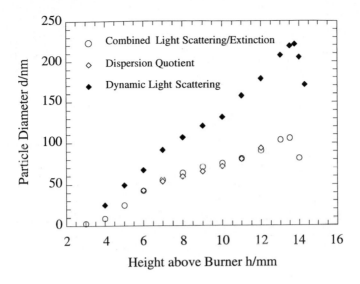

Fig. 21.7. Comparison of the different particle sizing techniques (flame 1)

the particle shape, a reason for this deviation could be a better detection of particle agglomerates.

21.3 Governing Equations

A planar, axis-symmetric counterflow between two tubes in a finite distance Δh is considered (Fig. 21.1). A fuel rich ethyne-air mixture enters the system through a porous-plug burner (condition 0, $y = 0$), and a sooting, laminar flame is stabilized on the lower burner. Unburnt hydrocarbons, CO, H_2 and soot are burnt in a downstream diffusion flame, established by an oxygen counterflow from the upper tube (condition 1, $y = \Delta h$).

A similarity solution exists for this flow field. Introducing $u = G \cdot x$ and $\partial p/\partial x = H \cdot x$ one obtains a set of one-dimensional equations in y-direction.

Continuity:
$$(j+1)\rho G + \frac{\partial(\rho v)}{\partial y} = 0 \tag{21.1}$$

Momentum:
$$\rho G^2 + \rho v \frac{\partial G}{\partial y} = -H + \frac{\partial}{\partial y}\left(\mu \frac{\partial G}{\partial y}\right) \tag{21.2}$$

Energy:

$$\rho v c_p \frac{\partial T}{\partial y} = \frac{\partial}{\partial y}\left(\lambda \frac{\partial T}{\partial y}\right) - \frac{\partial T}{\partial y} \sum_{i=1}^{N} c_{p,i}\rho Y_i v_i - \sum_{i=1}^{N} h_i w_i + q_R \qquad (21.3)$$

Species:

$$\rho v \frac{\partial Y_i}{\partial y} = \frac{\partial}{\partial y}\left(\rho Y_i v_i\right) + w_i\,, \quad i = 1,\dots,N_s \qquad (21.4)$$

Particles (free molecular regime):

$$\rho v \frac{\partial N_i/\rho}{\partial y} = \frac{\partial}{\partial y}\left(\rho D_{p,i}\frac{\partial}{\partial y}\left(\frac{N_i}{\rho}\right)\right) + \frac{\partial}{\partial y}\left(0.55\nu\frac{1}{T}\frac{\partial T}{\partial y}N_i\right) + \dot{N}_i\,, \; i = 1,\dots,\infty$$

$$(21.5)$$

Here $j = 0$ applies to the planar and $j = 1$ to the axis-symmetric configuration. The energy equation is solved including radiative heat losses in the limit of the thin gas approximation. In Eq. (21.5) the particle size class i is defined by $V_i = i \cdot V_1$, where V_1 is the volume of the smallest molecular structure occuring in soot particles, which is assumed to be $2.46^2 \cdot 3.51$ Å^3, where 2.46 Å is the width of the benzene ring and 3.51Å is the distance between PAH layers in soot. Following the HACA-mechanism, two carbon atoms expand the aromatic structure by one aromatic ring and with the assumption above by the volume V_1. The density of the soot particles can then be calculated to be $\rho_s = 1.87$ g/cm^3. The diffusion coefficient D_p varies in the free molecular regime with d_p^{-2} [21.8], where d is the particle diameter and we write:

$$D_{p,i} = i^{-2/3}D_{p,1}\,, \quad i = 1,\dots,\infty\,. \qquad (21.6)$$

To solve the infinite set of Eqns. (21.5) we use the method of moments [21.9]. With the definition of the statistical moment

$$\mu_r = \sum_{i=1}^{\infty} i^r N_i\,, \quad r = 0,\dots,\infty \qquad (21.7)$$

and with Eq. (21.6) we rewrite equation (21.5)

$$\rho v \frac{\partial \mu_r/\rho}{\partial y} = \frac{\partial}{\partial y}\left(\rho D_{p,1}\frac{\partial}{\partial y}\left(\frac{\mu_{(r-\frac{2}{3})}}{\rho}\right)\right) + \frac{\partial}{\partial y}\left(0.55\nu\frac{1}{T}\frac{\partial T}{\partial y}\mu_r\right) + \dot{\mu}_r\,. \quad (21.8)$$

To calculate the fractional moments $\mu_{(r-\frac{2}{3})}$ we use the interpolation technique outlined by *Frenklach* in [21.9]. In the following we truncate the infinite system of equations (21.8) after the third moment. The set of Eqns. (21.1–4 and 21.8) is completed by the ideal gas law and the boundary conditions.

We define the convective-diffusive operator for the statistical moments

$$L(\mu_r) = \rho v \frac{\partial \mu_r/\rho}{\partial y} - \frac{\partial}{\partial y}\left(\rho D_{p,1}\frac{\partial}{\partial y}\left(\frac{\mu_{(r-\frac{2}{3})}}{\rho}\right)\right) - \frac{\partial}{\partial y}\left(0.55\nu\frac{1}{T}\frac{\partial T}{\partial y}\mu_r\right)\,,$$

$$(21.9)$$

and rewrite Eq. 21.8:

$$
\begin{aligned}
L(\mu_r) = {}& \dot{\mu}_{\text{particle inception}} \\
&+ \dot{\mu}_{\text{condensation}} \\
&+ \dot{\mu}_{\text{coagulation}} \\
&+ \dot{\mu}_{\text{surface growth}} \\
&+ \dot{\mu}_{\text{oxidation}} \,,
\end{aligned}
\qquad (21.10)
$$

$$r = 0, \dots, \infty \,,$$

where the right hand side is given by the soot model outlined below.

21.4 Model for Soot Formation

The detailed chemistry soot model illustrated in figure 21.8 is essentially subdivided into three parts—gas phase chemistry, polymerisation of PAH and formation and growth of soot particles.

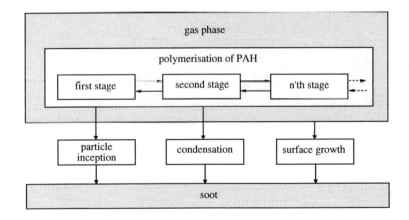

Fig. 21.8. Schematic illustration of the soot model

The gas phase chemistry up to the first aromatic ring is calculated using a detailed mechanism provided by *Warnatz* [21.10], consisting of about 250 elementary reactions between 52 chemical species. The first aromatic ring, viz. benzene, is formed via the reaction $C_3H_3 + C_3H_3 \rightleftharpoons c - C_6H_6$. The calculation of the growth reactions of benzene to small PAH follows the mechanism suggested by *Frenklach* and *Warnatz* [21.11]. Following *Frenklach* [21.12], the further growth of PAH is assumed to be a polymerisation

process. Differently from the approach describing the growth process by the
technique of linear (or chemical) lumping (see section 10.5) we assume that
the polymerisation of the PAH and the subsequent inception of soot particles is fast compared with convection and diffusion of the PAH. It follows,
that each PAH is in quasi steady state and that the concentrations of the
PAH can be calculated from an algebraic system of equations. The growth
and oxidation of the soot particles by heterogeneous reactions is discribed in
terms of the HACA mechanism. The coagulation of soot particles is calculated from Smoluchowski's coagulation equation in the free molecular regime
[21.9].

21.4.1 Fast Polymerisation of PAH

The individual polymers considered in each polymerisation stage are shown
in Fig. 21.9. Within each polymerisation stage two aromatic rings are added
to a given PAH. Each step from PAH to PAH involves several chemical
reactions. Additional oxidation reactions with OH and O_2 are not considered
in Fig. 21.9. The detailed growth and oxidation mechanism used for the
calculations presented here is given below.

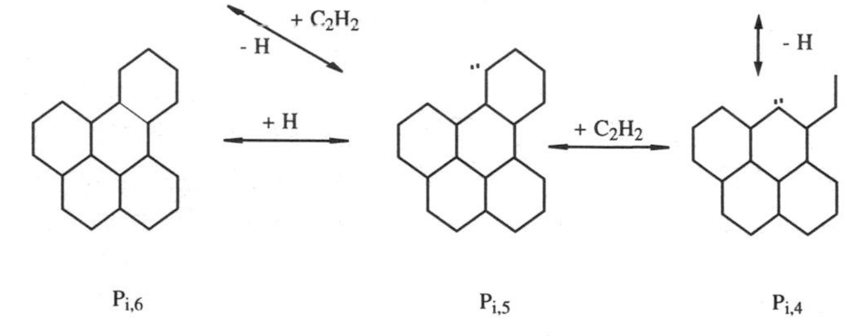

Fig. 21.9. Schematic illustration of the PAH growth within one polymerisation
stage

1.a	$P_{i,1}$	$+$ H	$\overset{k_{1a,P}}{\rightleftharpoons}$	$P_{i,2}$	$+$ H_2
1.b	$P_{i,1}$	$+$ OH	$\overset{k_{1b,P}}{\rightleftharpoons}$	$P_{i,2}$	$+$ H_2O
1.c	$P_{i,1}$	$+$ O	$\overset{k_{1c,P}}{\rightleftharpoons}$	$P_{i,2}$	$+$ OH
2.	$P_{i,2}$	$+$ H	$\overset{k_{2,P}}{\longrightarrow}$	$P_{i,1}$	
3.	$P_{i,2}$	$+$ C_2H_2	$\overset{k_{3,P}}{\longrightarrow}$	$P_{i,3}$	$+$ H
4.a	$P_{i,3}$	$+$ H	$\overset{k_{4a,P}}{\rightleftharpoons}$	$P_{i,4}$	$+$ H_2
5.	$P_{i,4}$	$+$ H	$\overset{k_{5,P}}{\longrightarrow}$	$P_{i,3}$	
6.	$P_{i,4}$	$+$ C_2H_2	$\overset{k_{6,P}}{\rightleftharpoons}$	$P_{i,5}$	
7.	$P_{i,5}$	$+$ C_2H_2	$\overset{k_{7,P}}{\rightleftharpoons}$	$P_{i+1,1}$	$+$ H
8.a	$P_{i,6}$	$+$ H	$\overset{k_{8a,P}}{\rightleftharpoons}$	$P_{i,5}$	$+$ H_2
9.	$P_{i,5}$	$+$ H	$\overset{k_{9,P}}{\longrightarrow}$	$P_{i,6}$	
10.a	$P_{i,6}$	$+$ OH	$\overset{k_{10a,P}}{\longrightarrow}$	$P_{i,3}$	$+$ CHO $+$ CH_2
10.b	$P_{i,3}$	$+$ OH	$\overset{k_{10b,P}}{\longrightarrow}$	$P_{i,1}$	$+$ CHCO
10.c	$P_{i,1}$	$+$ OH	$\overset{k_{10c,P}}{\longrightarrow}$	$P_{i-1,6}$	$+$ CHCO
11.a	$P_{i,5}$	$+$ O_2	$\overset{k_{11a,P}}{\longrightarrow}$	$P_{i,3}$	$+$ 2 CO
11.b	$P_{i,4}$	$+$ O_2	$\overset{k_{11b,P}}{\longrightarrow}$	$P_{i,2}$	$+$ 2 CO
11.c	$P_{i,2}$	$+$ O_2	$\overset{k_{11c,P}}{\longrightarrow}$	$P_{i-1,5}$	$+$ 2 CO

The growth mechanism is initiated by an H-abstraction step due to the attack of H, O and OH via reactions 1a–1c. The radical site of $P_{i,2}$ may be deactivated by H-addition (reaction 2) or by the revers of reactions 1a-1c. Ethyne is taken up at the radicalic site of $P_{i,2}$ via reaction 3. The repetition of this reaction sequence (reactions 4-6) leads to the closure of an aromatic ring. Because $P_{i,5}$ has a radicalic site, ethyne may be added again and an other ring closure occurs (reaction 7). This second ring closure leads to the next polymerisation stage, where the reaction sequence is repeated. In analogy to the radical sites of $P_{i,2}$ and $P_{i,4}$ the radical site of $P_{i,5}$ can be deactivated by H-addition (reaction 9) or by the revers of reactions 8a-8c. PAH may be oxidized by reactions with OH (10.a-c) and PAH with a radicalic site may be oxidized by reactions with O_2 (11.a-c).

Differently from the approach of Frenklach the growth of the PAH is assumed to be a fast polymerisation process. It follows that each PAH is in quasi steady state and that the concentrations of PAH can be calculated from an algebraic system of equations. In the following analysis we neglect reactions (10.a-c) in order to limit the size of the resulting system of equations.

$$[P_{1,1}] = \frac{r_{g,1} + K_{2,1}[P_{1,2}]}{K_{1,g} + K_{1,2} + K_{\text{soot}}}$$

$$[P_{1,2}] = \frac{K_{1,2}[P_{1,1}] + K_{3,2}[P_{1,3}] + K_{4,2}[P_{1,4}]}{K_{2,g} + K_{2,1} + K_{2,3} + K_{\text{soot}}}$$

$$[P_{1,3}] = \frac{K_{2,3}[P_{1,2}] + K_{4,3}[P_{1,4}] + K_{5,3}[P_{1,5}]}{K_{3,2} + K_{3,4} + K_{\text{soot}}}$$

$$[P_{1,4}] = \frac{K_{3,4}[P_{1,3}] + K_{5,4}[P_{1,5}]}{K_{4,3} + K_{4,5} + K_{4,2} + K_{\text{soot}}}$$

$$[P_{1,5}] = \frac{K_{4,5}[P_{1,4}] + K_{6,5}[P_{1,6}] + K_{1,5}[P_{2,1}] + K_{2,5}[P_{2,2}]}{K_{5,4} + K_{5,6} + K_{5,1} + K_{5,3} + K_{\text{soot}}}$$

$$[P_{1,6}] = \frac{K_{5,6}[P_{1,5}]}{K_{6,5} + K_{\text{soot}}}$$

$$[P_{2,1}] = \frac{K_{5,1}[P_{1,5}] + K_{2,1}[P_{2,2}]}{K_{1,5} + K_{1,2} + K_{\text{soot}}}$$

$$\cdots$$

$$[P_{i,1}] = \frac{K_{5,1}[P_{i-1,5}] + K_{2,1}[P_{i,2}]}{K_{1,5} + K_{1,2} + K_{\text{soot}}} \quad , i = 1, \cdots, \infty \qquad (21.11)$$

In Eqns.(21.11) we collected in $K_{j,k}$ all rate coefficients of the reactions leading from $P_{i,j}$ to $P_{i,k}$ multiplied with the respective gas phase compound, for example:

$$K_{1,2} = [H]k_{1a,p} + [OH]k_{1b,p} + [O]k_{1c,p} . \qquad (21.12)$$

$K_{i,g}$ collects oxidation and fragmentation reactions interrupting the polymerisation process. K_{soot} accounts for soot particle inception, calculated from PAH-PAH coagulation and for the condensation of the PAH on the soot particle surface, calculated from PAH-soot particle coagulation. Both processes are assumed to be independent of the size of the PAH. For the further analysis we define the fraction $F_{i,j}$ by:

$$F_{i,j} = \frac{K_{i,j}}{\sum_k K_{i,k} + K_{i,\text{soot}}}. \qquad (21.13)$$

For example $F_{1,2}$ is defined by:

$$F_{1,2} = \frac{K_{1,2}}{K_{1,5} + K_{1,2} + K_{1,\text{soot}}}. \qquad (21.14)$$

From this we find the fraction of growth from one to the next polymerisation stage

$$F_{\text{f}} = F_{1,2}F_{2,3}F_{3,4}F_{4,5}F_{5,1}, \qquad (21.15)$$

and the fraction of shrinkage

$$F_{\text{b}} = ((F_{5,4}F_{4,3} + F_{5,3})F_{3,2} + (F_{5,3}F_{3,4} + F_{5,4})F_{4,2})(F_{2,1}F_{1,5} + F_{2,5}) . \qquad (21.16)$$

Considering reactions (10.a-c) would involve additional terms in Eq.(21.16), such as $(F_{1,6}F_{6,3}F_{3,1})$ accounting for the reaction sequence $((10.c) \rightarrow (10.a) \rightarrow (10.b))$. With these definitions the elimination of PAH concentrations on the right hand side of the system of equations (21.11) leads to equations such as:

$$[P_{i,5}] = \frac{F_{\mathrm{f}}}{F_{1,5}Z_1 + F_{1,2}F_{2,5}Z_2}[\widehat{P}_{i,12}]$$
$$+ \frac{Z_{5k}}{F_{1,5}Z_1 + F_{1,2}F_{2,5}Z_2}[\widehat{P}_{i+1,12}] \quad , i = 2, \cdots, \infty \qquad (21.17)$$

In Eq. (21.17) we defined the joined concentration

$$[\widehat{P}_{i,12}] = \left[\frac{K_{1,5}}{K_{5,1}}[P_{i,1}] + \frac{K_{2,5}}{K_{5,1}}[P_{i,2}]\right], \qquad (21.18)$$

and calculate from (21.11)

$$[\widehat{P}_{1,12}] = \frac{F_{1,5}Z_1 + F_{1,2}F_{2,5}Z_2}{N}\frac{r_0}{K_{5,1}} + \frac{F_{\mathrm{b}}}{N}[\widehat{P}_{2,12}] \qquad (21.19)$$
$$[\widehat{P}_{i,12}] = \frac{F_{1,5}Z_1 + F_{1,2}F_{2,5}Z_2}{N}[P_{i-1,5}] + \frac{F_{\mathrm{b}}}{N}[\widehat{P}_{i+1,12}] \quad , i = 2, \cdots, \infty,$$

with

$$N = Z_1 - Z_2 F_{1,2}F_{2,1}$$
$$Z_1 = Z_2 - Z_3 F_{2,3}F_{3,2} - Z_4 F_{2,3}F_{3,4}F_{4,2}$$
$$Z_2 = Z_3 - Z_4 F_{3,4}F_{4,3} - F_{3,4}F_{4,5}F_{5,3}$$
$$Z_3 = Z_4 - F_{4,5}F_{5,4}$$
$$Z_4 = 1 - F_{5,6}F_{6,5} \qquad (21.20)$$

and

$$Z_{5k} - Z_{5i}F_{1,5}F_{5,1} + Z_{5j}F_{1,2}F_{2,5}F_{5,1}$$
$$Z_{5i} = Z_{5j} - F_{2,3}F_{3,2} - F_{2,3}F_{3,4}F_{4,2}$$
$$Z_{5j} = 1 - F_{3,4}F_{4,3} . \qquad (21.21)$$

The elimination of $[P_{i-1,5}]$ in Eq. (21.19) with the help of Eq. (21.17) leads to

$$[\widehat{P}_{i,12}] = \frac{F_{\mathrm{f}}}{N^*}[\widehat{P}_{i-1,12}] + \frac{F_{\mathrm{b}}}{N^*}[\widehat{P}_{i+1,12}], \qquad (21.22)$$

with

$$N^* = N - Z_{5k} . \qquad (21.23)$$

The analysis is completed by the recursiv elimination of $[\widehat{P}_{i+1,12}]$ in Eq. (21.22) and of $[\widehat{P}_{i+1,12}]$ in Eq. (21.19).

$$[\widehat{P}_{1,12}] = F_\infty \frac{F_{1,5}Z_1 + F_{1,2}F_{2,5}Z_2}{N}\frac{r_0}{K_{5,1}}$$
$$[\widehat{P}_{i,12}] = F_\infty \frac{F_{\mathrm{f}}}{N^*}[\widehat{P}_{i-1,12}] \quad , i = 2, \cdots, \infty \qquad (21.24)$$

where F_∞ is given by the limit of the recursion

$$F_k = \frac{1}{1 - F_{k-1}\frac{F_\text{f}}{N^*}\frac{F_\text{b}}{N^*}} \,. \tag{21.25}$$

This limit is found for the growth process of the PAH by:

$$F_\infty = \lim_{k \to \infty} F_k = \frac{1 - \sqrt{1 - \frac{F_\text{f}}{N^*}\frac{F_\text{b}}{N^*}}}{2\frac{F_\text{f}}{N^*}\frac{F_\text{b}}{N^*}} \,. \tag{21.26}$$

The second solution of the quadratic equation (21.26) describes a source of PAH at high molecular mass and a subsequent depolymerisation. It is evident from Eq. (21.24) that each PAH of the sizeclass i can be calculated from

$$[\widehat{P}_{i,12}] = \left(F_\infty \frac{F_\text{f}}{N^*}\right)^{i-1} [\widehat{P}_{1,12}] \quad , i = 2, \cdots, \infty. \tag{21.27}$$

The molar concentration of the PAH decreases exponentially with the size of the PAH because it can be shown, that

$$F_\infty \frac{F_\text{f}}{N^*} \leq 1 \,. \tag{21.28}$$

With Eqns. (21.27), (21.24) and the Equations of type (21.17) we calculate the moments of the PAH size distribution function and the source terms $\dot\mu_\text{particle inception}$ and $\dot\mu_\text{condensation}$ in Eq. (21.10).

21.4.2 Growth and Oxidation of Soot Particles

The heterogeneous surface growth and oxidation of soot particles follows the mechanism given below:

1.a	$C_{soot,i}H$	$+ H$	$\overset{k_{1a,s}}{\longleftrightarrow}$	$C^*_{soot,i}$	$+ H_2$	
1.b	$C_{soot,i}H$	$+ OH$	$\overset{k_{1b,s}}{\longleftrightarrow}$	$C^*_{soot,i}$	$+ H_2O$	
2.	$C^*_{soot,i}$	$+ H$	$\overset{k_{2,s}}{\longrightarrow}$	$C_{soot,i}H$		
3.a	$C^*_{soot,i}$	$+ C_2H_2$	$\overset{k_{3a,s}}{\longleftrightarrow}$	$C_{soot,i}C_2H_2$		
3.b	$C_{soot,i}C_2H_2$		$\overset{k_{3b,s}}{\longleftrightarrow}$	$C_{soot,i+1}H$	$+ H$	
4.a	$C^*_{soot,i}$	$+ O_2$	$\overset{k_{4a,s}}{\longrightarrow}$	$C^*_{soot,i-1}$	$+ 2\,CO$	
4.b	$C_{soot,i}C_2H_2$	$+ O_2$	$\overset{k_{4b,s}}{\longrightarrow}$	$C^*_{soot,i}$	$+ 2\,CHO$	
5.	$C_{soot,i}H$	$+ OH$	$\overset{k_{5,s}}{\longrightarrow}$	$C^*_{soot,i-1}$	$+ CH + CHO$	

The mechanism follows basically the ideas of the HACA mechanism, but has been modified compared with the one introduced by *Frenklach* [21.6]. The reverse of reaction 1.b accounts for the radical site consuming influence of

H_2O. Reaction 3 consists of two reactions—carbon addition, ring closure—, because it has been found recently that the reverse of reaction 3 accounts for the limitation of surface growth at high temperatures [21.13, 21.7]. The more detailed analysis here shows that the reverse of reaction 3.a is responsible for this effect, while reaction 3.b is approximately irreversible. In analogy to Frenklachs formulation and the planar PAH growth the total concentration of radical sites $[C_{soot}^*]$ and $[C_{soot}C_2H_2]$ are replaced by the assumption of quasi stationarity. We first introduce a factor accounting for the progress of soot growth,

$$ f_{3a} = \frac{k_{3b,f}}{k_{3b,f} + k_{3a,b} + k_{4b}[O_2]} \; . \tag{21.29} $$

If the the ring closure via reaction (3b,f) is fast against the limiting reaction (3a,b) and the oxidation reaction (4b) then $f_{3a} = 1$. If the $[C_2H_2]$-site consuming and oxidation reactions are fast then $f_{3a} = 0$. With this definition we write

$$ [C_{soot}^*] = \frac{k_{1a,f}[H] + k_{1b,f}[OH] + k_{3b,b}[H](1 - f_{3a}) + k_5[OH]}{k_{1a,b}[H_2] + k_{1b,b}[H_2O] + k_2[H] + k_{3a,f}[C_2H_2]f_{3a}} \cdot [C_{soot}] \tag{21.30} $$

$$ [C_{soot}C_2H_2] = \frac{k_{3a,f}[C_2H_2]}{k_{3b,f} + k_{3a,b} + k_{4b}[O_2]} \cdot [C_{soot}^*] $$

$$ + \frac{k_{3b,b}[H]}{k_{3b,f} + k_{3a,b} + k_{4b}[O_2]} \cdot [C_{soot}] \tag{21.31} $$

$$ \frac{df_v}{dt} = ((k_{3a,f}[C_2H_2]f_{3a} - k_{4a}[O_2]) \cdot [C_{soot}^*] $$

$$ - (k_{3b,b}[H](1 - f_{3a}) + k_5[OH]) \cdot [C_{soot}] \cdot V_1 N_A \; . \tag{21.32} $$

In the soot forming limit ($f_{3a} = 1$) Eq.(21.32) and the formulation in [21.6] are equivalent. With increasing temperature f_{3a} decreases because of the high activation energy of reaction (3a, b).

The concentration of sites on the surface S_i of the soot particles of all size classes i available for the HACA-mechanism is calculated from

$$ [C_{soot}] = \sum_{i=1}^{\infty} \chi S_i \frac{N_i}{N_A} , \tag{21.33} $$

where N_A is the Avogadro constant and χ is the number density of sites on the particle surface. We assume that each surface element of size S_1, this is one aromatic ring, has one site and then find for spherical particles:

$$ \chi S_i = \chi \pi^{1/3}(6V_i)^{2/3} = \chi \pi^{1/3}(6V_1)^{2/3}i^{2/3} = \chi S_1 i^{2/3} = i^{2/3} \; . \tag{21.34} $$

We calculate $\dot{\mu}_{surface\,growth}$ in Eq. (21.10) as the influence of reaction 3f on the particle size distribution [21.9] and find:

$$\dot{\mu}_{0,\mathrm{sg}} = -k_{3\mathrm{b,b}}[\mathrm{H}](1 - f_{3\mathrm{a}})N_x$$
$$\dot{\mu}_{1,\mathrm{sg}} = (k_{3\mathrm{a,f}}[\mathrm{C_2H_2}]f_{3\mathrm{a}}A - k_{3\mathrm{b,b}}[\mathrm{H}](1 - f_{3\mathrm{a}}))\mu_{2/3}$$
$$\dot{\mu}_{2,\mathrm{sg}} = k_{3\mathrm{a,f}}[\mathrm{C_2H_2}]f_{3\mathrm{a}}A(\mu_{2/3} + 2\mu_{5/3})$$
$$\qquad - k_{3\mathrm{b,b}}[\mathrm{H}](1 - f_{3\mathrm{a}})(-\mu_{2/3} + 2\mu_{5/3})$$
$$\dot{\mu}_{3,\mathrm{sg}} = k_{3\mathrm{a,f}}[\mathrm{C_2H_2}]f_{3\mathrm{a}}A(\mu_{2/3} + 3(\mu_{5/3} + \mu_{8/3}))$$
$$\qquad - k_{3\mathrm{b,b}}[\mathrm{H}](1 - f_{3\mathrm{a}})(\mu_{2/3} + 3(-\mu_{5/3} + \mu_{8/3})) \qquad (21.35)$$

where A is defined by Eq. (21.30):

$$A = \frac{k_{1\mathrm{a,f}}[\mathrm{H}] + k_{1\mathrm{b,f}}[\mathrm{OH}] + k_{3\mathrm{b,b}}[\mathrm{H}](1 - f_{3\mathrm{a}}) + k_5[\mathrm{OH}]}{k_{1\mathrm{a,b}}[\mathrm{H_2}] + k_{1\mathrm{b,b}}[\mathrm{H_2O}] + k_2[\mathrm{H}] + k_{3\mathrm{a,f}}[\mathrm{C_2H_2}]f_{3\mathrm{a}}} . \qquad (21.36)$$

The source of oxidation $\dot{\mu}_{\mathrm{oxidation}}$ is given by:

$$\dot{\mu}_{0,\mathrm{ox}} = -(k_{4\mathrm{a}}[\mathrm{O_2}]A + k_5[\mathrm{OH}])N_x$$
$$\dot{\mu}_{1,\mathrm{ox}} = -(k_{4\mathrm{a}}[\mathrm{O_2}]A + k_5[\mathrm{OH}])\mu_{2/3}$$
$$\dot{\mu}_{2,\mathrm{ox}} = -(k_{4\mathrm{a}}[\mathrm{O_2}]A + k_5[\mathrm{OH}])(-\mu_{2/3} + 2\mu_{5/3})$$
$$\dot{\mu}_{3,\mathrm{ox}} = -(k_{4\mathrm{a}}[\mathrm{O_2}]A + k_5[\mathrm{OH}])(\mu_{2/3} + 3(-\mu_{5/3} + \mu_{8/3})) . \qquad (21.37)$$

It can be seen from Eq. (21.35) and (21.37) that the system of equations is not closed, because the number density of the smallest physical size class x N_x is needed to calculate the total number of oxidized particles. The solution of the differential equation for N_x involves terms, dependent on N_{x+1}, as any differential equation of a single size class involves terms dependent on the next size class. If we assume that a constant fraction β of the total oxidized mass of soot is lost by the destruction of soot particles we find:

$$\dot{\mu}_{0,\mathrm{ox}} = -(k_{4\mathrm{a}}[\mathrm{O_2}]A + k_5[\mathrm{OH}])\beta\mu_{-1/3} . \qquad (21.38)$$

For the calculations shown in the next section $\beta = 0.01$ has been used. The coagulation of the soot particles is calculated with help of the method 2 in reference [21.9] (see also section 10).

21.5 Numerical Results

Figure 21.10 shows the profile of the flow velocity between the two burner surfaces for flame 1—3. The calculated values show good agreement with the experimental data obtained from LDV-measurements. Calculated velocities for flame 3 are slightly higher than the experimental data. The estimated stagnation point is located at about 18 mm above the lower burner surface.

The temperature profile in flame 1 is shown in Fig. 21.11. Soot particles are formed in the fuel rich premixed flame and burnt in the counterflow diffusion flame. In addition to the temperature measurements with thermo-couples (rapid insertion technique), the temperature of the soot particles

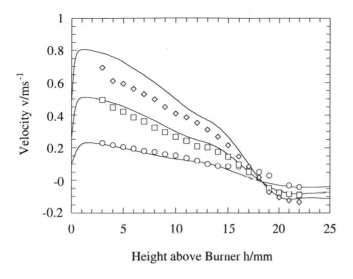

Fig. 21.10. Calculated centerline velocities (lines) in comparison with experimental data obtained from LDV-measurements for ○ flame 1, □ flame 2 and ◇ flame 3

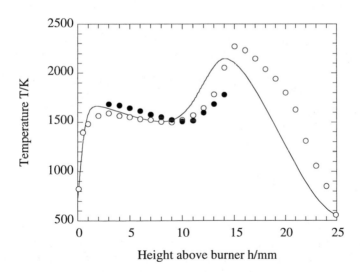

Fig. 21.11. Calculated temperature profile (line) for flame 1 in comparison with measurements with thermocouples ○ and with the Kurlbaum-method ●

has been measured with the Kurlbaum method. The calculated temperature profile in the premixed flame and the counterflow flame agrees quantitatively with the measurements from the Kurlbaum method. Near the premixed flame, at low heights above the burner, temperature measurements
with thermocouples are 200 K lower than the corresponding Kurlbaum temperatures. At 10 mm height above the lower burner–near to the diffusion
flame–both temperature measurements agree. Downstream the temperature
increases again in the counterflow diffusion flame. The predicted maximum
temperature of 2100 K is 200 K lower than the temperature from thermocouple measurements. Corresponding particle temperatures are not available,
because the soot particles are burnt in the diffusion flame.

Calculated and experimental temperature profiles of flame 2 ($v_{pre} =$
8 cm/s) and flame 3 ($v_{pre} = 12$ cm/s) are similar to flame 1 ($v_{pre} = 4$ cm/s).
The maximum temperature of the premixed flames increases with increasing flow velocity and with decreasing heat losses to the burner surface. The
predicted position of the counterflow diffusion flame in flame 2 and flame
3 does not correspond to the experimental findings. While the calculation
predicts an increasing distance between the premixed and the counterflow
flame with increasing flow velocity, a decreasing distance is found in the
experiment.

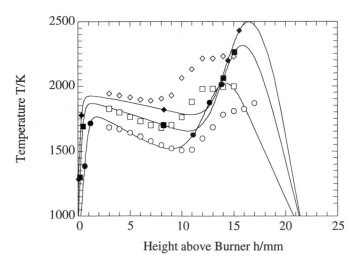

Fig. 21.12. Calculated temperature profiles (lines) in comparison with experimental data obtained with the Kurlbaum-method for ∘ flame 1, ▫ flame 2 and ◇
flame 3

From experiments in [21.3-5] it is well known that the final soot volume
fraction shows a bell shaped dependence on temperature. The maximum

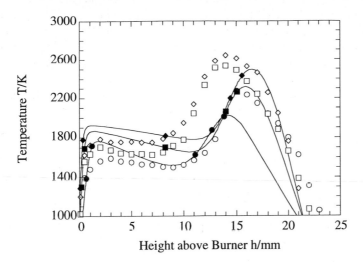

Fig. 21.13. Calculated temperature profiles (lines) in comparison with experimental data obtained from measurements with thermocouples for ○ flame 1, □ flame 2 and ◇ flame 3

amount of soot is obtained at a temperature of about 1650 K. Calculated and experimentally obtained soot volume fractions for flame 1—3 are shown in Fig. 21.14. Both, experiments and calculations show decreasing soot volume fractions with increasing temperatures in the premixed flames. The decreasing soot volume fraction can be explained with the increasing rate coefficient of the backward of reaction (3.a). This has been discussed in section 21.4.2.

The agreement between the measured and the calculated soot volume fraction is very good for flame 3, but is less good with decreasing temperature. The predicted soot volume fraction is less sensitive with respect to changes of the flame temperature. The calculated soot volume fraction in flame 1 is 50% lower than found in the experiment. However, the error in the calculated soot volume fraction of flame 1 is even small against the error in the calculated concentrations of acetylene and oxygen. This is shown in Fig. 21.15, where the calculated mole fractions of C_2H_2 and O_2 are compared with experimental data from gas chromatography for flame 1. It is found in the experiment that a remarkable part of oxygen remains in the burnt gas of the premixed flame, while the numerical calculation predicts a complete conversion of oxygen. This difference results in a considerable error in the calculated C_2H_2 profiles. The corresponding differences in the CO_2 profiles in flame 1 shown in Fig. 21.16 is also referred to the differences in O_2. A reason for the remaining O_2 could not be found from the kinetic mechanism.

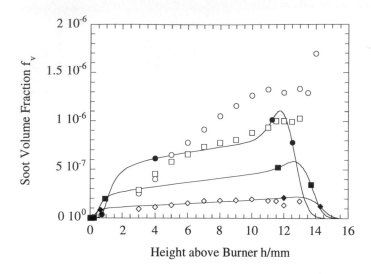

Fig. 21.14. Calculated profiles of the soot volume fraction for ○ flame 1, □ flame 2 and ◇ flame 3

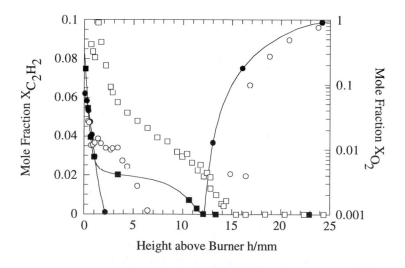

Fig. 21.15. Calculated mole fractions (lines) in comparison with experimental data for flame 1 (□ C_2H_2, ○ O_2)

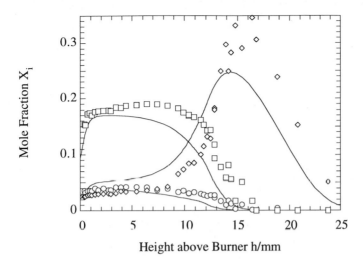

Fig. 21.16. Calculated mole fractions (lines) in comparison with experimental data for flame 1 (□ CO, ○ H_2, ◇ CO_2)

A comparison of the calculated profiles of benzene and diacetylene in flame 1 with experimental data is shown in Fig. 21.17. Considering the differences between the calculated profile of acetylene and the experimental data, the differences found in the benzene profiles are minor. It is the early increase of diacetylene in the premixed flame that indicates an error in the flame structure. Downstream of the premixed flame both profiles agree with the experimental data.

The calculated influence of the C/O-ratio of the premixed flame on the soot volume fraction is again weeker than found in the experiment. This is shown in Fig. 21.18 for flame 1 (C/O=0.85), flame 4 (C/O=0.75) and flame 5 (C/O=0.70). The agreement between the calculated soot volume fraction and the experimental data in flame 5 is good while flame 1 shows differences, as discussed above. It can be seen from Fig. 21.19 that the differences between the calculated and the measured C_2H_2 profiles in flame 5 are smaller than in flame 1. However, consumption of C_2H_2 and O_2 is found to be slower in the experiment than in the calculation. Calculated profiles of CO and CO_2 are shown in Fig. 21.20. The calculation predicts more CO to be burnt to CO_2 than found experimentally. The profiles of H_2 and H_2O, also shown in Fig. 21.20 are well predicted by the calculation.

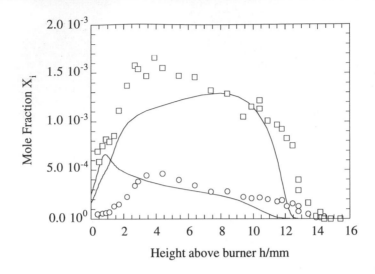

Fig. 21.17. Calculated mole fractions (lines) in comparison with experimental data for flame 1 ($\square\; c - C_6H_6$, $\circ\; C_4H_2$)

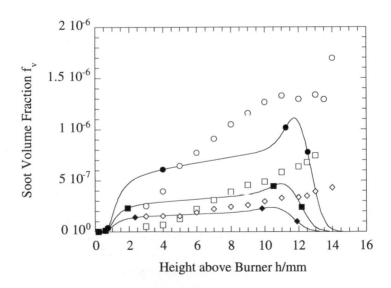

Fig. 21.18. Calculated profiles of the soot volume fraction for $\circ$ flame 1, $\square$ flame 4 and $\diamond$ flame 5

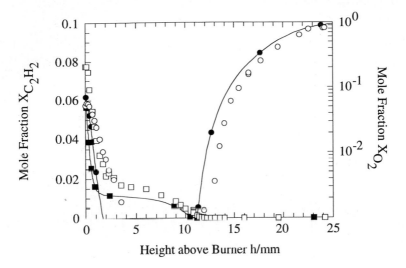

Fig. 21.19. Calculated mole fractions (lines) of the fuel and oxygen in comparison with experimental data for flame 5 ($\square$ C_2H_2, $\circ$ O_2)

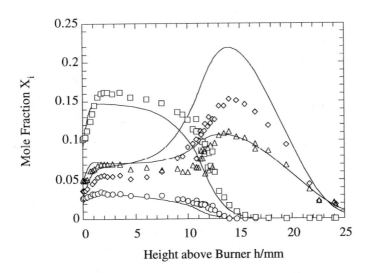

Fig. 21.20. Calculated mole fractions (lines) in comparison with experimental data for flame 5 ($\square$ CO, $\circ$ H_2, $\diamond$ CO_2, $\triangle$ H_2O)

21.6 Conclusion

A detailed chemical model for soot formation has been presented. The model has been embedded in a numerical code for the calculation of one dimensional laminar flames. The assumption of a fast polymerisation of PAH involves a set of algebraic equations describing the size distribution of the PAH. It is stated that the molar concentration of PAH decreases exponentially with the size of the PAH. The inception of soot particles is calculated in terms of the coagulation of PAH. The heterogenous surface growth is calculated from a hydrogen abstraction, carbon addition, ring closure mechanism. From experiments it is known that the final soot volume fraction shows a bell shaped dependence on temperature. The maximum amount of soot is obtained at a temperature of about 1650 K. It is found that at high temperatures the decrease of the final soot volume fraction with temperature can be explained by the reversibility of the carbon addition step.

Formation, growth and oxidation of soot particles in hydrocarbon flames at atmospheric pressure has been investigated experimentally and numerically in a configuration of a premixed and a counterflow flame. The temperature of the flames has been measured with thermocouples, and in regions where soot particles are available with the Kurlbaum method. Temperatures obtained from the Kurlbaum method are higher than temperatures measured with thermocouples. A better agreement of numerical calculation and experiment has been found for the temperature from the Kurlbaum method. The calculated profiles of the flow velocity show a good agreement with the experimental data obtained from LDV-measurements. This also indicates that the calculated temperature profiles are well predicted.

Concentration profiles of various stable species up to benzene were measured by gas chromatography. Significant differences of calculated and measured profiles of several species are reported for flame 1. This flame has a high C/O-ratio in the premixed flame (C/O=0.85) and low flow velocities (v_{pre}= 4 cm/s). It is found experimentally that only a small part of the fuel is burnt in the premixed flame. The remaining acetylene is availabe for the heterogenous surface growth of the soot particles that have been formed in the premixed flame. There is no decrease of surface growth rates found in the region between the premixed flame and the diffusion flame. The numerical calculation, however, predicts that a significant amount of fuel is burnt in the premixed flame. At the same time a marked decrease of the growth rate of soot is calculated downstream of the premixed flame. It is concluded that the differences between the calculated and the measured profile of the soot volume fraction can be explained in terms of the error in the calculation of the gasphase. A better description of the gasphase is the task of further investigations. For the other flames with lower C/O ratios or higher flow velocities the agreement between experiment and calculation is better.

21.7 Acknowledgement

This research was partially supported by the commission of the European Communities, the Swedish National Board for Technical Development (STU) and the Joint Research Committee of european car manufacturer (Fiat, Peugeot SA, Renault, Volkswagen and Volvo) within the IDEA program.

References

21.1   B.S. Haynes, H.Gg. Wagner: Prog. Energy Combust. Sci. **7**, 229 (1981)

21.2   I. Glassman: "Soot Formation in Combustion Processes", in Twenty-Second Symposium (International) on Combustion (The Combustion Institute, Pittsburgh 1988) p. 295

21.3   H. Böhm, D. Hesse, H. Jander, B. Lüers, J. Pietscher, H.Gg. Wagner, M. Weiss: "The Influence of Pressure and Temperature on Soot Formation in Premixed Flames", in Twenty-Second Symposium (International) on Combustion (The Combustion Institute, Pittsburgh 1988) p. 403

21.4   M. Bönig, Chr. Feldermann, H. Jander, B. Lüers, G. Rudolph, H.Gg. Wagner: "Soot Formation in Premixed C_2H_4 Flat Flames at Elevated Pressure", in Twenty-Third Symposium (International) on Combustion (The Combustion Institute, Pittsburgh 1990) p. 1581

21.5   H. Böhm, Chr. Feldermann, Th. Heidermann, H. Jander, G. Rudolph, H.Gg. Wagner: "Soot Formation in Premixed C_2H_4-Air Flames for Pressures up to 100 bar", in Twenty-Fourth Symposium (International) on Combustion (The Combustion Institute, Pittsburgh 1992) p. 991

21.6   M. Frenklach, H. Wang: "Detailed Modelling of Soot Particle Nucleation and Growth", in Twenty-Third Symposium (International) on Combustion (The Combustion Institute, Pittsburgh 1990) p. 1559

21.7   M.B. Colket, R.J. Hall: "Successes and Uncertainties in Modeling Soot Formation in Laminar, Premixed Flames", this volume, sect. 28

21.8   J.H. Seinfeld: Atmospheric Chemistry and Physics of Air Polution (J. Wiley & Sons, New York 1986)

21.9   M. Frenklach, S. Harris: J. Coll. Interface Sci., **Vol. 118**, No. 1, 252 (1987)

21.10 J. Warnatz, C. Chevalier: in Combustion Chemistry, 2nd edition, ed. by W.C. Gardiner Jr. (Springer Verlag, New York 1994)

21.11 M. Frenklach, J. Warnatz: Combust. Sci. Technol. **51**, 265 (1987)

21.12 M. Frenklach: Chem. Eng. Sci. **40**, 1843 (1985)

21.13 F. Mauss, N. Peters, H. Bockhorn: "A Detailed Chemical Model for Soot Formation in Premixed Acetylene- and Propan-Oxygen Flames", in Proceedings of the Anglo-German Combustion Symposium (The British Section of The Combustion Institute, Cambridge 1993) p. 470

Experimental Study on the Influence of Pressure on Soot Formation in a Shock Tube

Axel Müller, Sigmar Wittig

Institut für Thermische Strömungsmaschinen,
Unversität Karlsruhe,
76128 Karlsruhe, Fed. Rep. of Germany

Abstract: Soot formation in highly argon-diluted methane,- ethylene-, propane-, and acetylene- oxygen mixtures was studied at temperatures from 1600 to 2000 K and total pressures from 3 MPa up to about 10 MPa applying the shock tube technique. Time resolved measurements of soot particle sizes, number concentrations and soot volume concentrations were performed behind the reflected shock wave by monitoring the attenuation of two laser beams with the wavelengths of 488 nm and 633 nm. The influence of temperature, total pressure, and fuel partial pressure on soot formation was of major interest. According to flame experiments it was found that the soot formation is very sensitive to temperature and that the soot volume concentration increases with the fuel partial pressure proportional to p_{fuel}^2.

22.1 Introduction

The efforts to increase the efficiency of thermal power units lead to higher pressures during combustion. Pressures in gas turbine combusters attain values of about 3 MPa and in diesel engines maximum pressures of 20 MPa are reached. These high pressure conditions support the tendency of soot formation and thereby thermal radiation. For the design of modern combustors the knowledge of soot formation mechanisms therefore becomes more and more significant.

In recent years soot formation was examined extensively at low pressure conditions and only few high pressure experiments were reported. Flame experiments were performed by *Flower* et al. [22.1-3], *Mätzing* et al. [22.4], *Bönig* et al. [22.5] and *Luers-Jongen* et al. [22.6] who analized the soot formation at total pressures up to 7 MPa. Investigations at higher pressures are available only from shock tube experiments. *Fussey* et al. [22.7], *Mar'yasin* et al. [22.8], *Buckendahl* [22.9] and *Geck* [22.10] reported about shock tube experiments on soot induction times at elevated pressures. *Parker* et al. [22.11] performed experiments in a shock tube at high inert gas pressures (1 MPa - 3 MPa) and measured induction times as well as the soot yields at various pressures and temperatures. The aim of this work is to get more information about the influence of fuel partial pressure and total pressure on soot formation.

22.2 Experimental

The experiments were conducted in a conventional shock tube with an internal diameter of 31.4 mm, a total length of 4950 mm and a driven section beeing 2970 mm in length. Details are given in [22.12]. The test gas mixtures were manometrically prepared in a separate mixing tank using gases of high purities: He 99.996%, Ar 99.999%, O_2 99.995%, CH_4 99.995%, C_2H_4 99.950%, C_2H_2 99.600%, C_3H_8 99.950%. The shock tube was cleaned and evacuated before every run.

To get reproduccable results sets of mylar diaphragms of different thickness were used to reach bursting pressures which were only unessential higher as the aspired pressure in the driver section of the tube. The investigations were performed behind the reflected shock wave, 10 mm in front of the end wall of the driven section. The thermodynamic states of the gas in the measuring plane were calculated with the conservation equations and the ideal gas equation of state using the incident shock velocity measured with piezoelectric pressure transducers. According to calculations, heat losses caused by radiation of the produced soot particles are below 10 K $msec^{-1}$ and therefore were neglected.

The appearance of soot and the growth of the soot particles was observed by the dispersion quotient (DQ) method. This optical measuring technique, which has been discussed frequently in previous work [22.13-15] is based on the extinction of two laser beams with different wavelengths. The optical arrangement for the dispersion quotient method adapted to high pressure shock tube experiments is shown in Fig. 22.1.

The intensities of the original and attenuated laser beams (488 nm, 633 nm) are measured with photodiodes stored by a transient recorder and processed in a microcomputer. Errors, caused by scattered light and light emissions are minimized using narrow apertures, interference filters and running the lasers at maximum power.

The dispersion quotient DQ can be determined with the aid of the measured time-resolved intensity traces, compare Fig. 22.2. On the other hand the DQ can be calculated by employing the Lorenz-Mie [22.21] theory:

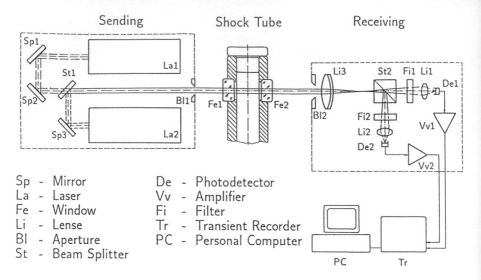

Fig. 22.1. Optical arrangement

$$DQ = \frac{\ln\left(\frac{I}{I_0}\right)_{\lambda_1}}{\ln\left(\frac{I}{I_0}\right)_{\lambda_2}} = \frac{\int_0^\infty P(r) \cdot r^2 \cdot Q_{\text{ext}}(r, \lambda_1, m) \cdot dr}{\int_0^\infty P(r) \cdot r^2 \cdot Q_{\text{ext}}(r, \lambda_2, m) \cdot dr} \qquad (22.1)$$

For a given complex index of refraction $m = n - ik$, given wavelengths λ_1, λ_2 and a known particle size distribution $P(r)$ the dispersion quotient DQ is only a function of the mean particle radius r. Figure 22.2. shows typical intensity traces of the laser beams during the combustion of a methane mixture. Beside the short attenuations, caused by a deflection of the laser beams due to the primary and reflected shock wave, the onset of soot formation as well as the increase of the soot volume is obvious. The increase of the intensity traces at the end of the observed time interval comes from an expansion wave passing the measuring plane.

In correspondence to *Cundall* et al. [22.16] *Haynes* et al. [22.17] showed, that visible laser light is attenuated during soot formation not only by soot particles but also by absorbing gases and that the absorption decreases with increasing wavelength. On the assumption that the extinction at 633 nm is solely due to soot, see [22.17], the dispersion quotient can be calculated as illustrated in Fig. 22.2.

The intensity I_0 at 633 nm is measured just at starting attenuation by soot particles. The intensity I_0 at 488 nm is measured at the same time, the laser beam at this wavelength already beeing attenuated by absorbing gases. This procedure affects the measured dispersion quotient only at small attenuation of the laser beams. It leads to a lower DQ-maximum.

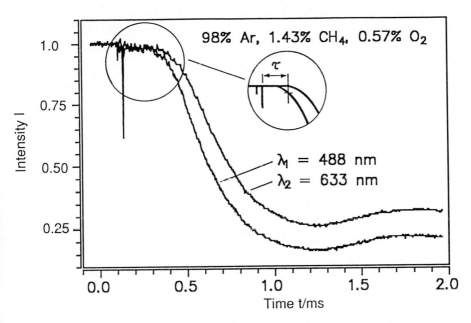

Fig. 22.2. Determination of the original laser intensity I_0

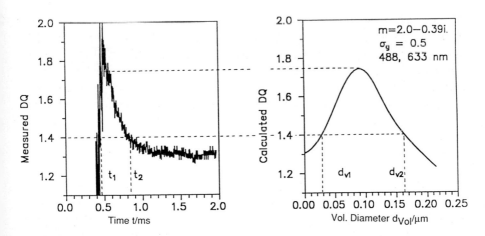

Fig. 22.3. Evaluation of the measured DQ-values

Fig. 22.3. elucidates the evaluation of the measured dispersion quotient. The right part of Fig. 22.3. displays the calculated DQ. It can be seen that the DQ passes a maximum with increasing particle diameter. A measured

DQ thus would correspond to two particle sizes. However, this ambiguity is resolved with the assumption that particles are growing with time.

All data were evaluated with calculated DQ values assuming a log-normal size distribution

$$P(r) = \frac{1}{\sigma_g \sqrt{2\pi}\, r} \cdot \exp\left[-\frac{(\ln r - \ln r_g)^2}{2\sigma_g^2}\right]$$
(22.2)

with a geometric standard deviation of $\sigma_g = 0.5$ and a refractive index for soot of $m = 2.0 - i0.39$ as proposed by *Mullins* et al. [22.18].

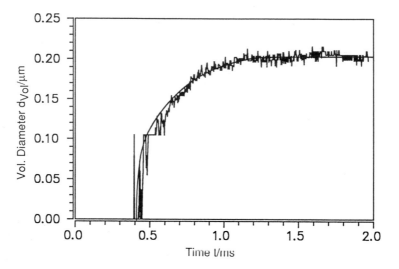

Fig. 22.4. Mean volume diameter versus time

Figure 22.4. shows a typical history of the mean volume diameter of soot particles. The induction time and the diameter corresponding to the maximum of the DQ-curve, which is in a wide range rather insensitive to any uncertainties in gas absorption, refractive index, and geometric standard deviation of the size distribution, can be determined without any significant error. For further calculations the measured d_{vol} versus time traces were approximated by a suitable function.

Knowing the volume mean diameter and the corresponding intensity it is possible to determine the particle number density and the soot volume fraction f_V using Lambert-Beer's law (Fig. 22.5.).

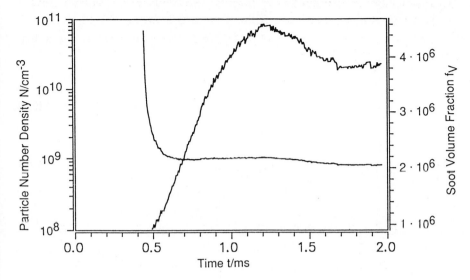

Fig. 22.5. Soot particle density and soot volume fraction versus time

22.3 Results and Discussion

With the time resolved intensity traces, which in first order are proportional to the produced soot volume, the influence of pressure on soot formation can be demonstrated qualitatively. Figure 22.6. shows the attenuation of the 633 nm laser beam during the combustion of a mixture consisting of 98% Ar, 1.25% C_2H_4, and 0.75% O_2 at a temperature of about 2000 K for several pressures.

With increasing pressure the attenuation increases and therefore the produced soot volume.

More complex is the influence of temperature on soot formation. For identical mixtures and a pressure of 4 MPa intensity traces are recorded for different temperatures (Fig. 22.7.).

With increasing temperatures soot formation starts earlier and the rate of soot production just at the beginning increases. The final produced soot volume increases with temperature for temperatures below 1697 K and decreases with temperature for temperatures higher than 1808 K.

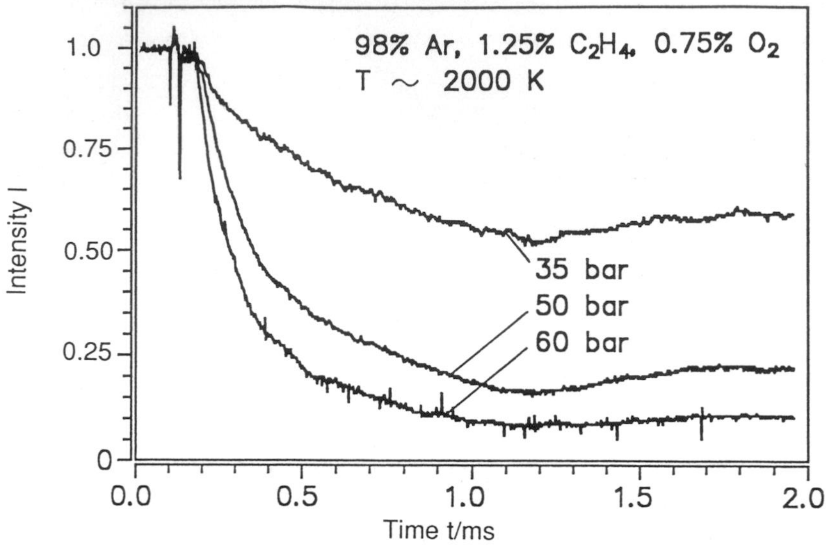

Fig. 22.6. Pressure dependent light attenuation

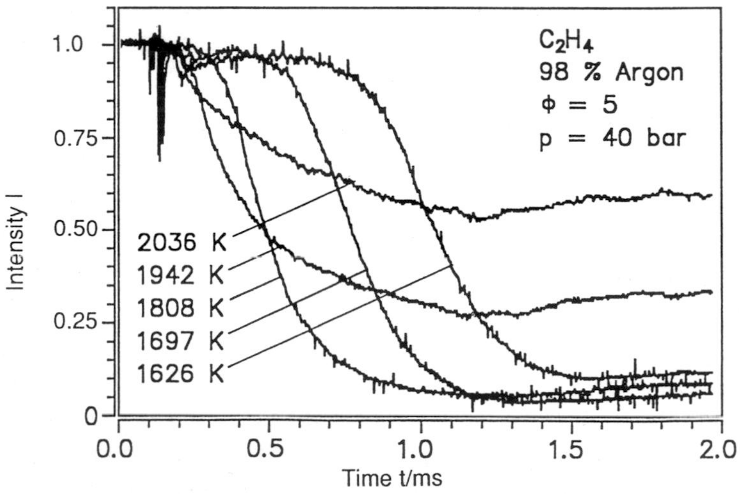

Fig. 22.7. Temperature dependent light attenuation

22.3.1 Soot Induction Time

One of the most important quantities to describe the soot formation kinetics is the soot induction time τ. During this time first solid soot particles are formed from gaseous hydrocarbons. To determine the soot induction time the delay from the onset of shock heating to the onset of detectable light attenuation of the 633 nm laser beam was measured, compare Fig. 22.2.

The soot induction time strongly depends on temperature and fuel as can be seen in Fig. 22.8. The logarithms of the induction times are plotted versus T^{-1} for methane, acetylene, propane, and ethylene. With increasing C/H-ratio and increasing temperatures the soot formation starts earlier.

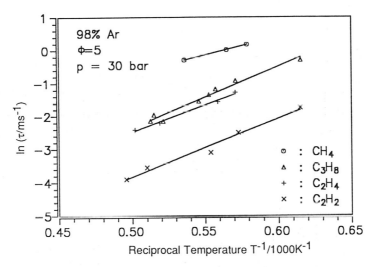

Fig. 22.8. Soot induction times for small hydrocarbons

The fuels under consideration behave qualitatively similar. Therefore, the influence of temperature, pressure, fuel partial pressure, and argon concentration in the test gas mixture on soot formation is demonstrated in the following only for selected fuels.

In Fig. 22.9. the soot induction time for methane combustion at 1950 K with an equivalence ration of $\phi = 5$ is plotted versus total pressure for various Ar-concentrations in the test gas mixture.

With increasing pressure and decreasing Ar-concentrations the soot induction time decreases. The gradient of all curves increases towards lower total pressures. This indicates that for low fuel partial pressures the induction time is very sensitive to changes in pressure. In Fig. 22.10. this dependence is demonstrated in more detail. All measured points of Fig. 22.9. are plotted versus the fuel partial pressure.

With increasing fuel partial pressure the induction time decreases. The measured data for different Ar-concentrations scatter around one curve, indicating that the influence of total pressure on the soot induction time is negligible.

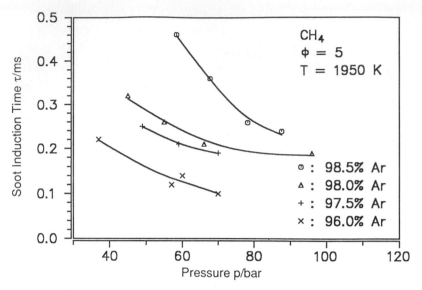

Fig. 22.9. Influence of total pressure and diluent-concentration on the soot induction time

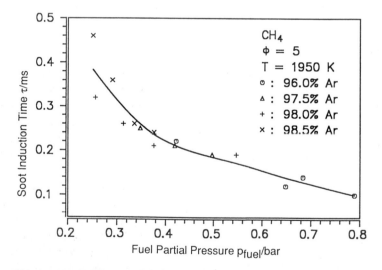

Fig. 22.10. Influence of fuel partial pressure on induction time

To show the effect of the equivalence ratio on soot induction time, experiments were performed with ethylene mixtures at 4 MPa. Equivalence ratios were varied between 3 and 5 (this results in changes of the fuel partial pressures of less than 25%) at temperatures in the range from 1600 up to 2000 K.

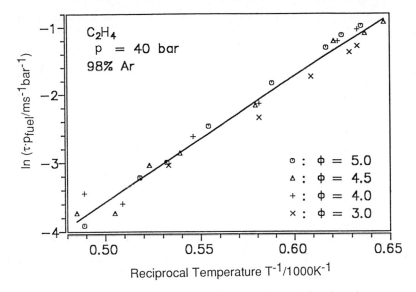

Fig. 22.11. Influence of equivalence ratio on induction time

Figure 22.11. shows the results of these tests. There is no measurable influence of the equivalence ratio on the induction time.

22.3.2 Soot Volume Fraction

In Addition to soot induction times the soot volume concentrations were measured. At the end of the induction time first solid soot particles are formed which then begin to grow by coagulation, surface growth and agglomeration as shown in Fig. 22.5. The soot volume fractions f_V, which indicates the produced soot quantity, were calculated with the Lambert-Beer-law using the particle number density, the mean volume diameter of the soot particles, and the measured intensities.

Figure 22.12. illustrates the influence of the fuel on the produced soot volume. Three mixtures of methane, propane, and ethylene respectively with Ar-concentrations of 98% and equivalence ratios of 5 are compared over a wide range of temperatures. To consider the different induction times the soot volume fractions were measured 0.5 msec after the appearance of first solid soot particles. Furthermore, to compare of the produced soot volumes from experiments at different total pressures and Ar-concentrations at constant temperatures, the soot volume concentrations were divided by the corresponding fuel partial pressure $(f_V p_{fuel}^{-1})$.

The three fuels show a similar temperature dependence of the normalized soot volume fractions. However, they differ in their abolute values. The maximum of the normalized soot volume fraction increases from methane to propane and ethylene. The soot volume produced by acetylene under the

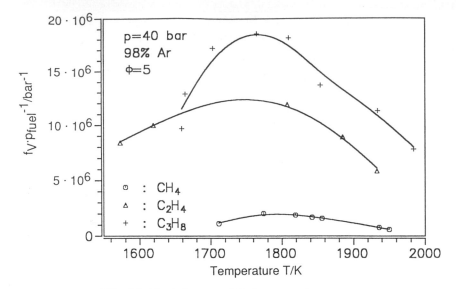

Fig. 22.12. Influence of fuel on soot formation

same conditions is too high to be measured with the dispersion quotient method and therefore is not plotted. The bell shaped curves for the temperature dependence of the soot formation are comparable with results of other shock tube experiments (for example [22.19]).

In the following the influence of pressure on soot formation is presented for tests with methane.

In order to demonstrate the influence of total pressure on soot formation the combustion of a mixture, consisting of 98% Ar, 1.43% CH_4 and 0.57% O_2, was observed at temperatures between 1700 up to 2100 K and pressures ranging from 4.5 MPa to 10 MPa. Typically bell-shaped curves were obtained with their maxima near 1800 K (see Fig. 22.13.). With increasing pressure the normalized soot volume fraction increases linearly.

This is shown in more detail in Fig. 22.14., where for three selected temperatures the normalized soot volume fractions are plotted versus total pressure.

The pressure dependence of the normalized soot volume fraction at constant temperature can be approximated by a straight line which means that the influence of pressure on the produced soot volume is proportional p^2. At low temperatures the effect of pressure is more important what is supported by the higher gradient of the corresponding lines.

With increasing total pressure the fuel partial pressure inreases too - if there is no change in the mixture composition. In order to change the fuel partial pressure at constant total pressure several experiments were performed with mixtures of different Ar-concentration.

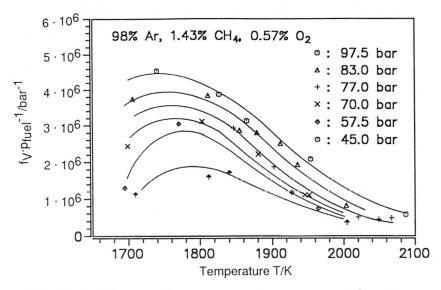

Fig. 22.13. Influence of temperature and pressure on soot formation

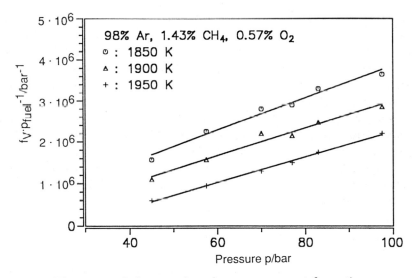

Fig. 22.14. Influence of total pressure on soot formation

In Fig. 22.15. the normalized soot volume fractions at 1950 K and $\phi = 5$ for different Ar-concentrations are compared. The experiments show that there is a linear relation between the normalized soot volume fraction and the total pressure at constant Ar-concentrations. As expected the produced soot volume fraction as well as the slope of the linear relation decrease with increasing Ar-concentrations.

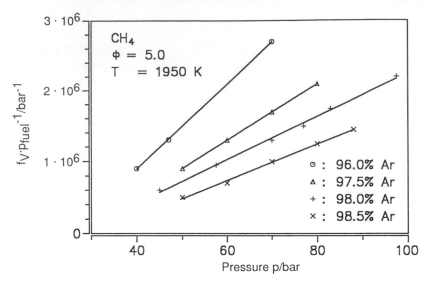

Fig. 22.15. Influence of *Ar*-concentration on soot formation

The slope of the curve for the mixture with an Ar-concentration of 96% has twice the gradient of that with an Ar-concentration of 98%. At the same total pressure and temperature the mixture with 96% Ar produces four times the soot volume as the mixture with 98% Ar which can be seen in the twofold value of the normalized soot volume fraction. This is due to the twofold fuel partial pressure of the mixture with 96% Ar.

From Fig. 22.16. can be seen that more soot is produced at constant fuel partial pressure when increasing the total pressure. Here the measured normalized soot volume fractions of Fig. 22.15. are plotted versus the fuel partial pressure.

Figure 22.16. shows that the normalized soot volume fraction is a linear function of the fuel partial pressure. With increasing total pressure the slopes of the lines increase. If the straight lines are extrapolated, they intersect in the origin: no fuel – no soot.

The results of numerous experiments performed at various total pressures, fuel partial pressures and Ar-concentrations in the test gas mixture are summarized in Fig. 22.17. The normalized soot volume fraction $\left(f_V p_{\text{fuel}}^{-1}\right)$ obtained from the combustion of methane with oxygen in argon $(\Phi = 5)$ at 1950 K is plotted versus pressure and fuel partial pressure.

This three dimensional plot, which is a synthesis of Fig. 22.15. and Fig. 22.16., shows in a more comprehensive manner the results discussed in the previous figures.

It turned out, that lines of constant Ar-concentration are straight lines in this graph. Furthermore, the produced soot volume increases with increasing

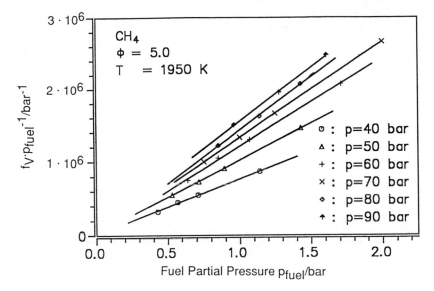

Fig. 22.16. Influence of fuel partial pressure on soot formation

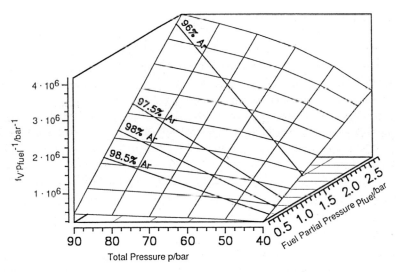

Fig. 22.17. Influence of pressure, fuel partial pressure and *Ar*-concentration on soot formation

total pressure although the Ar-concentration increases and the fraction of methane in the test gas mixture decreases. This indicates, that there are pressure dependent processes in soot formation. The effect of total pressure diminishes with increasing total pressure.

Another possibility to change the fuel partial pressure at constant pressure is to vary the equivalence ratio at constant Ar-concentration. Results are presented for experiments performed with ethylene and equivalence ratios ranging from 3 to 5. The normalized soot volume fractions obtained from experiments with mixtures containing 98% Ar at pressures of 4 MPa are shown in Fig. 22.18.

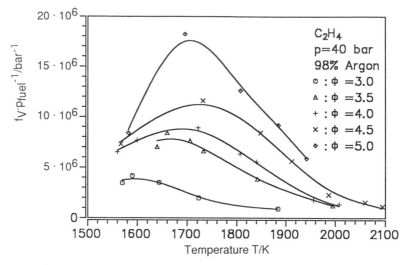

Fig. 22.18. Influence of temperature and equivalence ratio on soot formation

The measured normalized soot volume fractions decrease with decreasing equivalence ratios over the whole range of temperatures. The maxima are shifted towards lower temperatures for smaller equivalence ratios. This corresponds with results of *Frenklach* et al. [22.19] who examined acetylen, toluene, benzene and 1,3-butadiene at pressures smaller then 0.32 MPa.

If the measured values from Fig. 22.18. for four selected temperatures are plotted versus the equivalence ratios straight lines are obtained (see Fig. 22.19.).

This is in agreement with the effect of the fuel partial pressure on soot formation. For ethylene combustion the increase of soot volume fraction in the range under consideration is approximately proportional to the equivalence ratio. It should be mentioned that the slopes of the straight lines have their maximum at temperatures of about 1800 K. With decreasing temperatures the extrapolated straight lines intersect the x-axsis at lower equivalence ratios, which are compareable with the critical equivalence ratios. This corresponds with results of *Böhm* et al. [22.20] who found, that in a temperature range from 1500 to 2000 K the critical equivalence ratio increases with increasing temperatures.

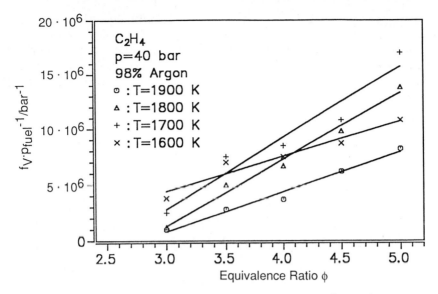

Fig. 22.19. Influence of equivalence ratio on soot formation

22.3.3 Particle Size and Density

For soot growth modelling not only the soot induction time and the soot volume fraction are of interest, but especially the soot particle size and density. Figure 22.4. and Fig. 22.5. show the time dependent behavior of these quantities. In the following the effect of pressure and temperature will be discussed.

In Fig. 22.20. the mean volume diameters of soot particles, generated during propane combustion at different pressures and measured 0.5 msec after the soot induction time are plotted versus temperature. The soot particle size is approximately independent of the pressure but increases with increasing temperature.

This behavior is due to a higher mobility of the particles and therefore due to a higher collision rate.

The consequence is, that the particle density is lower for higher temperatures as can be seen in Fig. 22.21., where the corresponding normalized particle density $(N\rho^{-1})$ is plotted versus temperature.

The influence of pressure is more significant in the temperature range lower than about 1800 K. At high pressures more soot particles are formed than at lower ones.

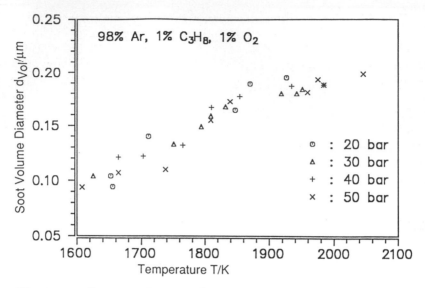

Fig. 22.20. Soot particle size as function of temperature and pressure

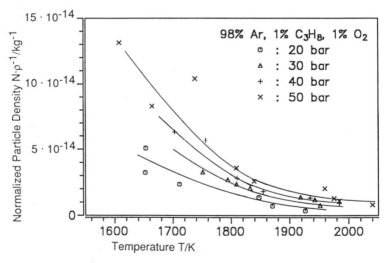

Fig. 22.21. Normalized particle density as function of temperature and pressure

22.4 Conclusions

To investigate the effect of pressure and fuel partial pressure on soot formation under high pressure conditions systematic experiments with methane, acetylene, propane, and ethylene in a shock tube were performed. Soot volume fractions and soot induction times were measured at temperatures from 1600 to 2000 K and total pressures ranging from 4 MPa up to about 10 MPa.

All fuels under consideration showed qualitatively similar sooting tendencies. With increasing C/H-ratio of the fuel the soot induction time decreases. In accordance with previous work in the literature it could be shown that the temperature is the dominant parameter for soot formation even at high pressures. Maximum soot formation was observed at temperatures about 1800 K. It was found that the produced soot volume fraction is proportional to the square of the fuel partial pressure ($\sim p_{\text{fuel}}^2$). Furthermore the experiments showed, that with increasing temperature the particle size increases and the particle density decreases.

22.5 Acknowledgements

The authors gratefully acknowledge the support by the Arbeitsgemeinschaft TECFLAM - technische Flammen.

References

22.1   W.L. Flower, C.T. Bowman: "Soot Production in Axisymmetric Laminar Diffusion Flames at Pressures From One to Ten Atmospheres", in Twenty-First Symposium (International) on Combustion (The Combustion Institute, Pittsburgh 1986) p. 1115

22.2   W.L. Flower: Sandia Report, SAND 88-8627 (1988)

22.3   W.L. Flower: Combust. Flame **77**, 279 (1989)

22.4   H. Mätzing, H.Gg. Wagner: "Measurements About the Influence of Pressure on Carbon Formation in Premixed Laminar C_2H_4-Air Flames", in Twenty-First Symposium (International) on Combustion (The Combustion Institute, Pittsburgh 1986) p. 1047

22.5   M. Bönig, Chr. Feldermann, H. Jander, B. Lüers, G. Rudolph, H.Gg. Wagner: "Soot Formation in Premixed C_2H_4 Flat Flames at Elevated Pressure", in Twenty-Third Symposium (International) on Combustion (The Combustion Institute, Pittsburgh 1990) p. 1581

22.6   B. Luers-Jongen, G. Rudolph, H. Jander: VDI Berichte **922**, 191 (1991)

22.7   D.E. Fussey, A.J. Gosling, D. Lampard: Combust. Flame **32**, 181 (1978)

22.8   L. Mar'yasin, Z.A. Nabutovskii: "Investigation of the Kinetics of Black Carbon Formation During the Thermal Pyrolysis of Benzene and Acetylene in a Shock Wave", in 3. Kinetika, Vol. 14 (Kataliz, 1973) p. 175

22.9   W. Buckendahl: "Untersuchung der Bildungsgeschwindigkeit von Ruß bei der Pyrolyse von Äthylen hinter reflektierten Stoßwellen"; Ph.D. Thesis, Universität Göttingen (1970)

22.10 C.C. Geck: "Untersuchung der Bildungsgeschwindigkeit von Ruß bei der Pyrolyse von Äthylen und Acetylen in Stoßwellen"; Ph.D. Thesis, Universität Göttingen (1975)

22.11 T.E. Parker, R.R. Foutter, W.T. Rawlins: "Optical Shock Tube Investigations of Soot Formation at High Inert Gas Pressures", in Seventeenth Symposium (International) on Shock Tubes and Shock Waves (1989) p. 481

22.12 S. Wittig, A. Müller, T.W. Lester: "Time-Resolved Soot Particle Growth in Shock Induced High Pressure Methane Combustion", in Seventeenth Symposium (International) on Shock Tubes and Shock Waves (1989) p. 468

22.13 T.W. Lester, S.L.K. Wittig: "Particle Growth and Concentration Measurements in Sooting Homogeneous Hydrocarbon Combustion Systems", in   Tenth Symposium (International) on Shock Tubes and Shock Waves (1975) p. 632

22.14 K. Bro, S. Wittig, D.W. Sweeney: "In Situ Optical Measurements of Particulate Growth in Sooting Acetylene Combustion", in Twelfth Symposium (International) on Shock Tubes and Shock Waves (1979) p. 429

22.15 S. Wittig, H.-J. Feld, A. Müller, W. Samenfink, A. Tremmel: "Application of the Dispersion-Quotient-Method under Technical System Conditions", in   Proceedings of the Second International Congress on Optical Particle Sizing (1990) p. 335

22.16 R.B. Cundall, D.E. Fussey, A.J. Harrison, D. Lampard: "An Investigation of the Precursors to Carbon Particles During the Pyrolysis of Acetylene and Ethylene",   in   Eleventh Symposium (International) on Shock Tubes and Shock Waves (1977) p. 375

22.17 B.S. Haynes, H. Jander, H.Gg. Wagner: Ber. Bunsenges. Phys. Chem. **84**, 585 (1980)

22.18 J. Mullins, A. Williams: Fuel **66**, 277 (1987)

22.19 M. Frenklach, M.K. Ramachandra, R.A. Matula: "Soot Formation in Shock-Tube Oxidation of Hydrocarbons" in Twentieth Symposium (International) on Combustion (The Combustion Institute, Pittsburgh 1984) p. 871

22.20 H. Böhm, D. Hesse, H. Jander, B. Lüers, J. Pietscher, H.Gg. Wagner, M. Weiss: "The Influence of Pressure And Temperature on Soot Formation in Premixed Flames" in Twenty-Second Symposium (International) on Combustion (The Combustion Institute, Pittsburgh 1988) p. 403

22.21 G. Mie: Anna. Phys. (Leipzig) **25**, 377 (1908)

Discussion

Lepperhoff: I have a question regarding the mean particle size. You reported particle sizes in the range from 150 to 200 nm. This seems to me quite big particles. You measured very young soot quite after the induction time. So there is no time for any coagualtion process.

Müller: The thing is that we did our experiments at very high pressures where the initial concentrations of nuclei are very high.

Lepperhoff: Just a small comment on these particle sizes: we measured in the cylinder of a diesel engine particle sizes in the range of less than 10 nm. The size we measured in the cylinder under high pressure conditions is in the range that has been reported today.

Müller: I think you cannot compare our experiments with diesel engine combustion. We have homogeneous conditions behind the reflected shock. In the diesel engine there are also mixing processes which you have to take into account.

Bockhorn: In consistency with the high pressure results from Aachen and Göttingen you found that the coagulation coefficients change with increasing pressure. Did you some calculations for coagulation at higher pressures and can you explain these effects? What quantity changes with pressure under these conditions?

Müller: We did some calculations but we measured higher coagulation rates than we can calculate. We have to consider that we are in the transition regime at our experimental conditions where we have to take into account very low Knudsen numbers. However, there must be a very high sticking co-efficient. May be that there is also a contribution of an ionic mechanism from ions of different charges. At the moment I have no consistent explanation for this.

Wagner: If I remember correctly there is for a given fuel no influence of pressure on particle size provided the carbon density is the same.

Müller: I have to mention that we see with our technique the agglomerates and what we measure is a mean volume diameter.

Wagner: I think the method overestimates the large particles.

Müller: Yes, this is correct.

Santoro: I have a comment on the coagulation rates. I believe that Kennedy and Harris wrote a paper a couple of years ago about van der Waals inter-actions at high pressure and their effects on coagulation. They calculated enhancement factors at high pressures and it might be useful to look at their results.

The question I have is on the assumptions about the index of refraction of the particles and the log-normal distribution. I do not remember exactly your definition of the log-normal distribution. However, the σ_g can have very different meanings. Your size distributions seem to be a very wide distribution compared to what other people have seen with σ_g about 0.3 or so. I would like your comments on those two numbers that were in your viewgraphs.

Müller: There is also some work from Wersborg on coagulation rates of soot particles in flames. He reported some influence of pressure and versus higher pressures he suggested an increase of the coagulation rates by a factor of 30.

The log-normal distribution we used is given by Eq. (20.2). We derived the value of $\sigma_g = 0.5$ for the log-normal distribution from the analyses of electron micrographs of the soot particles. The refractive index of very young soot particles is a serious problem. This morning we heard from Prof. D'Alessio that young soot contains a large part of material that is nearly transparent. That means that the imaginary part of the refractive index has to be very low. The value for the refractive index we used was proposed by Mullins and Williams for soot in methane combustion. The uncertainty in the refractive index affects the dispersion quotient at its maximum value and in the regime of higher particle sizes. If we take the particle sizes from the DQ and calculate the particle number density from Lambert-Beers law then the volume fraction is a function of particle diameter and Q_{ext} or DQ. At our measured values there is only little dependence of the calculated volume fractions on the dispersion quotient and, hence, on the imaginary and real part of the refractive index.

Feldermann: Don't you think that the two wavelengths that you used are too close together to get reliable values for the dispersion quotient?

I have another question regarding the soot volume fraction. You measured the soot volume fractions quite at the beginning of soot formation. These are not the final soot volume fractions. In flames we have also a dependence of the soot volume fractions at incipient soot formation from the square of the pressure.

Müller: This is right. We have seen this morning that the dependence changes when the acetylene concentration decreases rapidly so that there is no more surface growth by the addition of acetylene to the soot particles. We made our measurements 0.5 ms after induction time. That means we dont measure the $f_{V\infty}$.

Haynes: There was the question relating to the particle size determination and the coagulation coefficient. The dispersion quotient method may depend to some extend on the fact that the particles are not spherical and I think that is a very difficult area to assess. The second point is that the hard sphere diameter which presumably is what you are interpreting for the coagulation coefficient is not an appropriate collision diameter for these web-like structures and how one deals with that is a very difficult question.

Müller: As I said previously at the moment I can not answer the questions concerning the coagulation rates of these agglomerates but they will surely rise with increasing mean soot particle size.

Soot Formation from Hydrocarbons in a Plug Flow Reactor: Influence of Temperature

Michael Huth, Wolfgang Leuckel

Engler-Bunte-Institut, Universität Karlsruhe,
76128 Karlsruhe, Fed. Rep. of Germany

Abstract: Soot formation in a plug flow reactor has been investigated. For this a secondary fuel (propane) is injected into the hot flue gas of a primary natural gas-air flame. Parameters varied have been temperature of the hot gas, secondary fuel concentration and residence time. From the dependence of soot volume fractions on temperature and secondary fuel concentration formal relations are derived. These relations may be used to predict the soot volume fractions and the formation rates under the conditions investigated.

23.1 Introduction

The soot content of technical flames and gasification reactors is of practical relevance for two basic reasons. The presence of soot within a flame increases its emissivity, which influences its radiative charateristics, and hence, its heat transfer potential. Furthermore, unburnt soot is an undesired flue gas emission component. Reliable and universal concepts for the prediction of soot concentrations in technical flames are still not available.

The aim of this work is to make a contribution to the prediction of soot formation in technical diffusion flames. For this purpose, the variety of reaction conditions present in technical flames, i.e. conditions of temperature, flame gas stoichiometry and residence time, is being simulated within a turbulent plug-flow reactor.

23.2 Experimental Set-up

Natural gas is burnt inside a pre-reactor at the bottom end of the reactor tube in order to produce a hot flue-gas flow (Fig. 20.1). In continuing and completing our previously reported experiments [23.5] the temperature of the hot flue gas flow entering the reactor has been varied now in a much wider range between 1000°C and 1500°C. The soot forming hydrocarbon is being injected radially into the main flow through a multihole nozzle placed at the end of an injection tube inserted into the lower part of the reactor tube. Propane was used as the soot generating hydrocarbon. In our previous work we reported about a correlation of the soot formation rate with the initial propane concentration and the oxygen content of the incoming exhaust gas. The temperature dependence was not systematically investigated. The main purpose of the present work is the investigation of this temperature dependence.

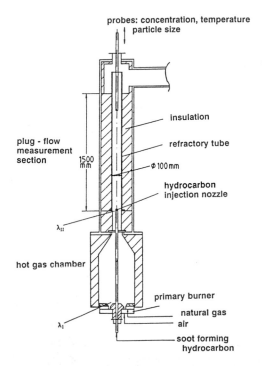

Fig. 23.1. Schematic drawing of the plug flow reactor

By introducing different suction probes, soot concentration, soot particle size distribution, gas temperature and gas phase species concentrations were measured at different distances downstream of the injector nozzle. Soot mass concentration was determined by gravimetry, soot particle size distribution by measuring the mobility of the particles in an electrical field. Gas temperature was measured by suction pyrometry, the main gas phase species by infrared absorbtion (CO, CO_2, CH_4), by determination of thermal conductivity (H_2) and by determination of the paramagnetic properties (O_2). For the hydrocarbons CH_4, C_2H_6, C_2H_4, C_2H_6, and C_3H_8 gaschromatography was applied.

23.3 Results and Discussion

Experiments with residual oxygen mole fraction $X_{O_2}^0$ in the primary flue gas between 9.4 % and 0.8 % were performed corresponding to a primary fuel air ratio λ_1 between 1.04 and 1.92. Different amounts of propane were added to each flue gas flow resulting in initial propane mole fraction concentrations $X_{C_3H_8}^0$ up to 3.2%. Table 23.1 gives an overview of the 19 different experimental runs. The table also presents data of the flow rate of the primary air $\dot{V}_{Air}$, the temperature of the incoming flue gas flow T_I and the temperature resulting after mixing and reacting with the propane T_{II} (measured at a height of about 30 cm above the injection level) and the critical C/O$_{II}$-ratio (see below) for soot formation. At $\lambda_I = 1.37$ T_I was varied between 1460°C

Table 23.1. Summary of experimental conditions (flowrates are given at s.t.p.)

Run	$\dot{V}_{Air}$/m^{-3}s^{-1}	λ_I	$X_{O_2}^0$/%	T_I/°C	$X_{C_3H_8}^0$/%	T_{II}/°C	C/O$_{II\,crit}$
1	20.8·10^{-4}	1.92	9.34	1018	2.6..3.2	1540..1480	0.58
2	20.4·10^{-4}	1.56	6.95	1162	2.4..3.2	1520..1380	0.61
3	21.2·10^{-4}	1.37	5.20	1458	1.6..3.0	1690..1490	0.73
4	21.2·10^{-4}	1.37	5.21	1385	1.6..3.0	1630..1430	0.72
5	21.2·10^{-4}	1.37	5.21	1290	1.6..3.0	1560..1370	0.71
6	20.4·10^{-4}	1.37	5.21	1173	1.6..3.0	1380..1210	0.73
7	17.8·10^{-4}	1.37	5.19	1090	2.0..3.0	1290..1190	0.74
8	15.8·10^{-4}	1.37	5.18	1065	1.8..3.0	1280..1160	0.72
9	14.1·10^{-4}	1.37	5.18	1030	2.0..3.0	1200..1130	0.73
10	11.0·10^{-4}	1.37	5.18	975	2.0..3.0	1190..1100	0.74
11	21.2·10^{-4}	1.28	4.22	1200	1.4..3.0	1390..1190	0.78
12	19.4·10^{-4}	1.26	3.96	1368	1.8..3.2	1440..1320	0.84
13	20.9·10^{-4}	1.21	3.28	1240	1.2..2.6	1340..1180	0.87
14	20.8·10^{-4}	1.14	2.25	1290	1.2..2.6	1300..1170	1.11
15	19.9·10^{-4}	1.13	2.23	1465	1.8..3.2	1400..1280	1.32
16	15.9·10^{-4}	1.13	2.22	1195	1.8..3.2	1140..1060	1.15
17	16.7·10^{-4}	1.10	1.66	1285	1.2..3.2	1210..1090	1.20
18	19.2·10^{-4}	1.05	0.81	1495	1.2..3.2	1400..1220	2.49
19	16.3·10^{-4}	1.04	0.80	1316	0.6..3.2	1240..1070	2.13

and 980°C by preheating the air, using a smaller flue gas chamber and varying the flow rates through the reactor.

Figure 23.2 shows the measured soot concentrations as function of time for the runs 4 to 10 with $\lambda_I = 1.37$ for a constant initial propane mole fraction $X^0_{C_3H_8}$ of about 3% and different temperatures. For rather low temperatures the soot growth rate is approximately independent of time. It increases with temperature. For reaction temperatures T_{II} above 1440° C the soot growth is slowing down with time in the observable residence time of 30 to 40 ms. This high temperature behaviour of the soot growth rate is similar to the behaviour of soot growth in flat premixed laminar flames, where a decline of the soot growth has been observed as well, cf. e.g. ref. [23.1-2, 23.4].

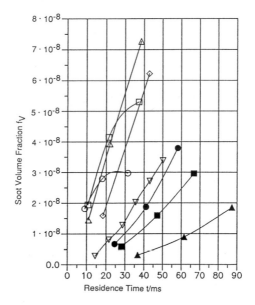

Fig. 23.2. Soot volume fraction f_V in dependence on residence time; $\lambda_I = 1.37$; $2.92 \leq X^0_{C_3H_8} \leq 3.08$; ○: run 3, $T_{II} = 1490..1420°C$; □: run 4, $T_{II} = 1430..1360°C$; $\triangle$ = run 5, $T_{II} = 1370..1310$ °C; ◇ = run 6, $T_{II} = 1200..1180°C$; ∇ = run 7, $T_{II} = 1190..1170°C$; ● = run 8, $T_{II} = 1150..1130°C$; ■ = run 9, $T_{II} = 1130..1100°C$; filled triangles = run 10, $T_{II} = 1100..1070°C$

Soot is formed in premixed flames only beyond a critical C/O ratio, which is about 0.57 for propane [23.6]. A similar parameter for the plug flow reactor experiments could be the ratio of the added amount of C from propane and the available free oxygen in the flue gas:

$$C/O_{II} = \frac{3 \, \dot{V}_{C_3H_8}}{2 \, X^0_{O_2} \, \dot{V}_I} \sim \frac{3 \, X^0_{C_3H_8}}{2 \, X^0_{O_2}} \qquad (23.1)$$

(the subscript I is designating the incoming exhaust gas flow, the subscript II the secondary combustion of the added propane with the available free oxygen).

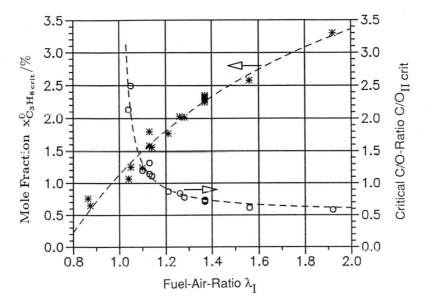

Fig. 23.3. Critical initial propane mole fraction, $X^0_{C_3H_8}$, and C/O$_{II \, crit}$ as a function of the primary fuel to air ratio, λ_I

Figure 23.3 shows the experimentally determined critical C/O$_{II \, crit}$-ratio and the corresponding critical initial propane mole fraction $X^0_{C_3H_8 \, crit}$ as function of the stoichiometric fuel air ratio of the incoming flue gas λ_I. The critical values were defined as the values at which the soot growth rate df_V/dt exceeds 10^{-10} ms^{-1}, corresponding to a soot volume fraction of about $4 \cdot 10^{-9}$ at 40 ms. At high λ_I, the critical C/O- ratio approaches the value known from premixed propane flames of 0.57. For λ_I approaching unity, C/O$_{II \, crit}$ is increasing strongly, hence, the critical propane concentration does not vanish for $\lambda_I = 1$. An explanation for this rather unexpected behaviour could be that not only the free oxygen is available for the oxidation of the secondary fuel, but also a certain portion of the bound oxygen present as H_2O and CO_2 will be consumed. For higher λ_I the critical C/O-ratios or the critical $X^0_{C_3H_8 \, crit}$, respectively, is only slightly influenced by temperature (cf. Tab. 23.1 for the values for $\lambda_I = 1.37$). At lower fuel air ratios there is possibly an influence of temperature, but the scatter of the data is too large to deduce such a dependence.

Referring to those critical values of $X^0_{C_3H_8}$, a correlation based on the description of soot growth in laminar premixed flames is being proposed

[23.4]:

$$f_V = f_{V\infty}[1 - \exp(-k_{sg}\, t)] \qquad (23.2)$$

$$\frac{df_V}{dt} = k_{sg}\, f_{V\infty}\, \exp(-k_{sg}\, t) \qquad (23.3)$$

with the phenomenological soot growth rate k_{sg} depending only on temperature:

$$k_{sg} = c_k \exp(-E_a/RT) \qquad (23.4)$$

and the ultimate soot volume fraction $f_{V\infty}$ as function of temperature and the surplus of the initial propane mole fraction beyond the critical propane mole fraction value $X^0_{C_3H_8} - X^0_{C_3H_8\text{ crit}}$:

$$f_{V\infty} = c_\infty \exp[-\frac{1}{2}(\frac{T - c_{T\text{ max}}}{c_{\sigma T}})^2][(X^0_{C_3H_8} - X^0_{C_3H_8\text{ crit}}) \cdot 100]^n \qquad (23.5)$$

In none of the experiments the ultimate soot volume fraction $f_{V\infty}$ was finally attained, so that $f_{V\infty}$ could not be determined directly from the experiments. *Bönig* et al. reported [23.3] a bell shaped temperature dependence for the final soot volume for premixed laminar C_2H_4 flames with a temperature maximum at about 1400°C and a decay of about 1.5 orders of magnitude in soot volume fraction towards 1600°C. Therefore, a similar temperature dependence was encorporated in Eq. (23.5).

The six model parameters c_k, E_a, c_∞, $c_{T\text{ max}}$, $c_{\sigma T}$ and n were determined from a least square fit to the measured soot volume fractions in two steps. In the first step, the apparent activation energy E_a and the constant c_k were calculated from fits of Eq. (23.2) for soot volume fractions-time-curves at higher temperatures with a decline in the soot growth rate. In the second step, the remaining parameters c_∞, $c_{T\text{ max}}$, $c_{\sigma T}$ and n were determined from all the data by using the values of c_k, E_a found in the first step. Table 23.2 presents the parameters resulting from applying this procedure.

Table 23.2. Model parameters determined by fitting eq. 23.2 to 23.5 to the experimental results

c_k/ms^{-1}	$E_a/\text{kJ mol}^{-1}$	c_∞	$c_{T\text{ max}}/°C$	$c_{\sigma T}/°C$	n
$1.244 \cdot 10^6$	231	$3.39 \cdot 10^{-7}$	1175	146	1.44

Figure 23.4 shows the temperature dependence of k_{sg} together with the data points used for the fit. Also plotted are the values for k_{sg} from Ref. [23.1], which are lower by a factor of 2 to 3 compared with the results of this work. The activation energy E_a of 231 kJ/mol (cf. Tab. 23.2) is about 50 kJ/mol higher than the value given in [23.1]. It seems that the soot growth rate is leveling off more rapidly in our exhaust gas fuel mixtures than in

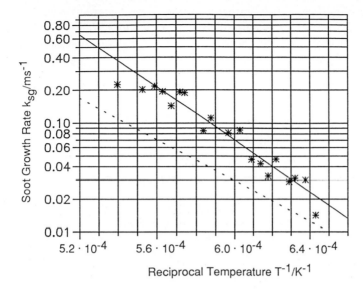

Fig. 23.4. Soot growth rate k_{sg} as a function of temperature with data points from the measurements; dotted line: values of k_{sg} taken from Ref. [23.1]

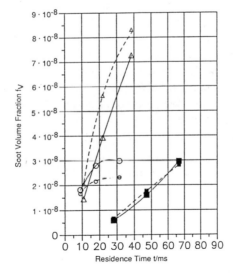

Fig. 23.5. Soot volume fraction from measurement and calculated (model parameters see table 23.2) as a function of residence time for $\lambda_I = 1.37$, $X^0_{C_3H_8} \approx 3\%$, o: run 3, $T_{II} = 1490..1420°C$; $\triangle$ = run 5, $T_{II} = 1370..1310 \,°C$; ■ = run 9, $T_{II} = 1130..1100°C$; - - -: calculated

premixed flames. Further experiments at still higher temperatures have to be carried out to verify this trend.

The soot volume fraction-time-curves, as calculated by the model equations and from the measurements, are compared in Fig. 23.5 (λ_I =1.37; run 3, 5, 9; $X^0_{C_3H_8} \approx 3\%$; cf. Fig. 23.2). There is some discrepancy in the absolute values, but the general trend is predicted with reasonable accuracy.

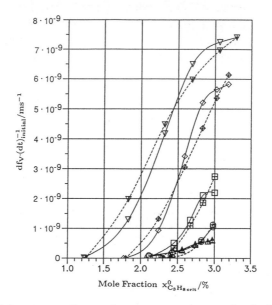

Fig. 23.6. Initial soot growth rate from measurement and calculated (model parameters see table 23.2) as a function of initial propane mole fraction $X^0_{C_3H_8}$; λ_I = 1.37: ○: run 3, T_{II} = 1620..1490°C; □ = run 5, T_{II} = 1480..1370°C; △ = run 9, T_{II} = 1200..1130°C; λ_I = 1.13: ◇: run 15, T_{II} = 1400..1280°C; λ_I = 1.05: ▽ = run 18, T_{II} = 1400..1220°C; - - -: calculated

Figure 23.6 shows initial soot growth rates ($t - t_0 \approx 15$ ms) as a function of the initial propane mole fraction for three runs with different temperature but equal λ_I, and also for two runs with different λ_I.

The influence of temperature upon the prediction from the model equations is demonstrated in Fig. 23.7. The values of k_{sg}, $k_{sg} \cdot \exp(-k_{sg}\, t)$, $f_{V\infty}$ and $f_{V\infty}\, k_{sg} \cdot \exp(-k_{sg}\, t)$ are calculated for the temperatures of the data points and for the model parameters indicated in Tab. 23.2. For the residence time and the propane surplus, constant values were assumed: $t - t_0$ = 15 ms, $X^0_{C_3H_8} - X^0_{C_3H_8\ crit}$ = 1%. The temperature dependence of $f_{V\infty}$ shows nearly no decline for low temperatures in contrary to premixed flame data. The decline at 1600°C is comparable with C_2H_4 flames. A lower limit for soot formation was not found. The model predicts a decreasing growth rate for temperatures lower than 1350°C, and a limit would have been reached presumably below 1000°C. An explanation for this difference compared to other premixed flame experiments, where a low temperature limit for soot formation between 1300°C and 1400°C has been measured, may possibly be deduced from the gravimetric method used in this work for the determination of soot concentrations. A certain amount of tar, which exists in the vapour phase under flame environment, but condenses when flame gases are cooled down, is being collected as "soot" on the filters. If the amount of tar collected increases for lower temperatures, a gravimetric method will find more soot than an optical method which really detects only particles present in the flame. The premixed flame measurements taken as a reference were carried out by using the extinction technique.

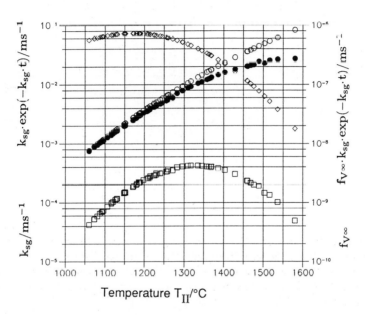

Fig. 23.7. Temperature dependence of $k_{sg}(\circ)$, $k_{sg} \cdot \exp(-k_{sg}\, t)(\bullet)$, $f_{V\infty}(\diamond)$ and $f_{V\infty}\, k_{sg} \cdot \exp(-k_{sg}\, t)(\square)$ for temperatures from the data points; for $X^0_{C_3H_8} - X^0_{C_3H_8\,crit}$ arbitraryly 1 mol % is choosen, for the residence time $t - t_0$ a constant value of 15 ms is used

References

23.1 L.Baumgärtner, D. Hesse, H. Jander, H.Gg. Wagner: "Rate of Soot Growth in Atmospheric Premixed Laminar Flames" ,in Twentieth Symposium (International) on Combustion (The Combustion Institute, Pittsburgh 1985) p. 959

23.2 H. Bockhorn, F. Fetting, A. Heddrich, G. Wannemacher: "Investigation of the Surface Growth of Soot in Flat Low Pressure Hydrocarbon Oxygen Flames", in Twentieth Symposium (International) on Combustion (The Combustion Institute, Pittsburgh 1985) p. 979

23.3   M. Bönig, Chr. Feldermann, H. Jander, B. Lüers, G. Rudolph, H.Gg. Wagner: "Soot Formation in Premixed C_2H_4 Flat Flames at Elevated Pressure", in Twenty-Third Symposium (International) on Combustion (The Combustion Institute, Pittsburgh 1990) p. 1581

23.4   B.S. Haynes, H.Gg. Wagner: Z. Phys. Chem. NF **133**, 201 (1982)

23.5   M. Huth, W. Leuckel: "Experiments on Soot Formation from Propane under Partial Oxidation Conditions in a Turbulent Plug-Flow Reator", in Twenty-Third Symposium (International) on Combustion (The Combustion Institute, Pittsburgh 1990) p. 1493

23.6   D.B. Olson, J.C. Pickens: Combust. Flame **57**, 199 (1984)

Discussion

D'Alessio: You had a wide range of temperatures in your experiments. In this range you find a quite different behaviour. So my question is, why do you try to put all these different things together. Probably the chemistry is different for the formation of tar like material at lower temperatures and the formation of carbonaceous material at higher temperature. Why don't you treat the formation of these two classes of compounds kinetically different and why do you try to put everything together?

Huth: Our results for the plug flow reactor resemble closely to that from premixed flames. We found the bell shaped temperature dependence and a similar dependence of soot volume fractions on fuel concentration. Therefore, we treated our system analogously to premixed flames and the correlations that we derived from our experiments show the same numbers as those derived from premixed flames.

Wagner: May I ask a simpler question? Could you tell us a little bit about the fraction of the added propane that went into the watergas components and into other compounds like hydrocarbons or solid material at the different conditions?

Huth: As a general rule we can say that most of the carbon added as secondary fuel is converted into carbon monoxide. However, we also find a decrease in carbon dioxide concentration in the plug flow reactor. So, some of the carbon dioxide formed in the primary flame is also converted into carbon monoxide.

Glassman: What is the possibility of the disproportionation of carbon dioxide in this particular system. Is it still the Boudouard reaction by which two carbon monoxide go into carbon plus carbon dioxide? The residence times are relatively long in this system. And my second question is, did you ever attempt to measure whether there is any oxygen in the system at the point of injection of the propane.

Huth: We have some carbon at the walls of the reactor. However, this is not very much and I do not think that it is formed in the Boudouard reaction.

Concerning your second question, we have a primary fuel to air ratio of 1.37 what means that the hot gas at the point of propane injection has an oxygen content of 5.2 %. At the first measuring point at about 20 cm above the injection level we found no more oxygen, even at 10 cm above the injection plane all the oxygen is consumed.

Homann: I did not understand the meaning of your critical carbon to oxygen ratio. Is it related with the fate of the carbon from propane? If most of the carbon goes into the water gas components you should expect a value below one and if most of it goes into other hydrocarbons, you should expect a value larger than one. Do you have any correlation for that?

Huth: The meaning of the carbon to oxygen ratio is simply the carbon to oxygen ratio at the point of injection calculated from the free oxygen concentration and the propane concentration at that point. It should characterize the soot forming tendency in the plug flow reactor.

Soot Formation and Oxidation in Diesel Engines

Franz Pischinger, Gerhard Lepperhoff, Michael Houben

Lehrstuhl für Technische Thermodynamik,
RWTH Aachen,
52062 Aachen, Fed. Rep. of Germany

Abstract: The formation of soot in diesel engine combustion is discussed. Different mechanisms of formation of soot corresponding to the different phases of diesel engine combustion, viz. soot formation in the fuel rich zones of the inhomogeneous premixed combustion, soot formation from the fuel injected into the flames, soot formation from the fuel injected into the burnt gases and, finally, soot oxidation, are identified. Some phenomena of soot formation in diesel engines are investigated experimentally by means of rapid sampling techniques and other experiments under diesel engine combustion conditions. From these experiments the need of further experiments for the investigation of single processes for soot formation under diesel engine combustion are concluded. Possible boundary conditions for the experiments are identified.

24.1 Introduction

Soot emission from diesel engines is subjected to increasing lower levels. With engine modifications development engineers are able to improve the combustion process regarding low soot and NO_x emission as well as low fuel consumption. For further improvements in reducing soot emission it is necessary to look deeper into the details of soot formation and oxidation processes. During diesel combustion different soot formation processes occur simultaneously and interacting. They are influenced by pressure and local temperature changes in very short times.

This paper deals with the soot formation and oxidation processes in diesel engines. With the aid of the in-cylinder pressure history and fuel ignition timing the different phases of diesel combustion will be elucidated. Results of measured soot concentrations in the cylinder of a DI diesel engine are presented that demonstrate the soot formation and oxidation processes.

In addition soot formation from fuel injected into burnt gas under diesel engine conditions is examined. Finally the boundary conditions for further detailed investigations of soot formation and oxidation under diesel engine conditions will be summarized.

24.2 The Diesel Combustion Process

For engine development engineers the normal way for in-cylinder combustion analysis is to measure the in-cylinder pressure versus crank angle or versus time. For a direct injected charged diesel engine these pressure curves for full and part load are demonstrated in Fig. 24.1.

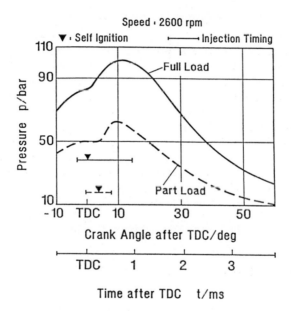

Fig. 24.1. Cylinder pressure history of a charged DI diesel engine

For truck engines the engine speed of 2600 rpm is a maximum value, for passenger cars it is in the medium range. The duration of burning and post flame reaction processes in the engine varies between approximately 28 ms for 1000 rpm and 5.7 ms for 5000 rpm. In the example from Fig. 24.1. the fuel injection time is approximately 1.1 ms at full load and 0.6 ms at part load. The ignition delay (time between start of injection and start of combustion) is about 0.22 ms in both cases. The start of combustion is defined by engineers by a discontinuity in the slope of the pressure curve.

With the aid of in-cylinder pressure and the gas mass in the cylinder the mean gas temperature can be calculated as a function of time, see Fig. 24.2.

The mean gas temperature calculated in this way is a good approximation for the real gas temperature during the expansion phase due to the mixing of the load with air. However, during combustion the local gas temperatures vary between the adiabatic combustion temperatures and the temperature due to adiabatic compression of air.

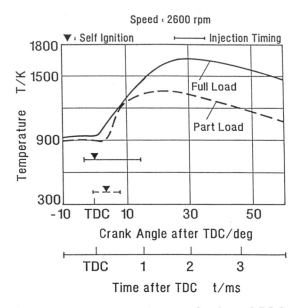

Fig. 24.2. Mean gas temperature history of a charged DI diesel engine

From a more detailed view with respect to soot formation and oxidation the combustion process can be devided in four different phases [24.1], see Fig. 24.3.:

Phase 1: Soot is formed in the fuel rich zones of the inhomogeneous pre-mixed combustion.
Phase 2: Soot is formed from the fuel injected into the flames.
Phase 3: Soot is formed from the fuel injected into the burnt gases.
Phase 4: Soot that has been formed is oxidized in the presence of oxygen.

In the example given in Fig. 24.3. the duration of phase 1 is approximately 0.4 ms, the duration of the phases 2 and 3 approximately 1.3 ms and that of phase 4 about 4.0 ms.

The processes occuring in the four phases are interacting. Finally, the remaining soot is measured in the exhaust gas.

In addition soot is formed from the lubricant burned or pyrolysed near the cylinder wall in the expansion phase [24.2].

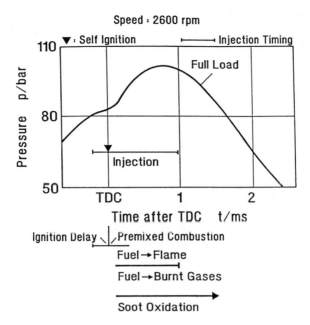

Fig. 24.3. History of diesel engine combustion

24.3 Soot Concentrations in the Cylinder

After start of injection and just before self-ignition the injected fuel is mixed
with air. For a single fuel spray a map of the C/O-ratios and the location of
self-ignition are demonstrated in Fig. 24.4. [24.3]. The results given in Fig.
24.4. are derived from spontaneous raman scattering measurements. In this
example the spray is not influenced by air motion (squish or swirl) or wall
interactions.

Self-ignition takes place in fuel rich mixtures with C/O-ratios from 0.56
to 0.42 ($0.6 \leq \lambda \leq 0.8$). It can be assumed that soot is formed at incident
combustion.

With the aid of a fast gas and soot sampling valve probes are taken from
the cylinder in dependence on time [24.4-5]. The position of the sampling
valve is located within one spray of a five hole nozzle about 13 mm from
the nozzle tip. Figure 24.5. shows the experimental set-up used for these
measurements, indicating the axis of one spray of the five hole nozzle, the
injection nozzle tip, the piston bowl, the cylinder head and the valve tip.
The tip diameter of the valve is small enough to avoid influence of the
emissions from this cylinder at part load. The disadvantage of gas sampling
is averaging during the opening duration of 1.2 ms. The advantage is that
all measured components are sampled from the identical probe volume.

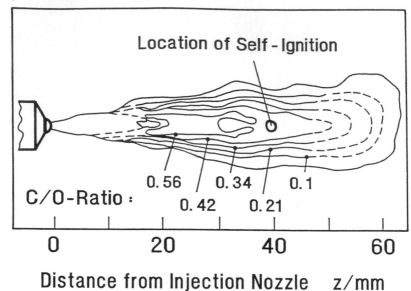

Fig. 24.4. Map of C/O-ratios in a spray at self-ignition

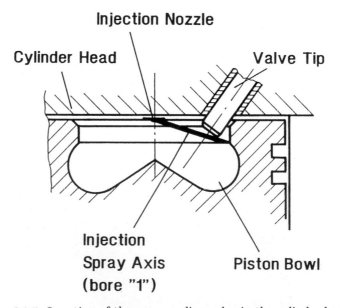

Fig. 24.5. Location of the gas sampling valve in the cylinder head

The gas sampling valve is mounted in a cylinder of a DI Diesel engine for passenger cars. Results are given in Fig. 24.6. for engine operation point at 2000 rpm and part load. The injection starts 0.13 ms before TDC, the injection duration is about 0.58 ms and the ignition delay is 0.42 ms. Most of the fuel injected is premixed before self-ignition occurs.

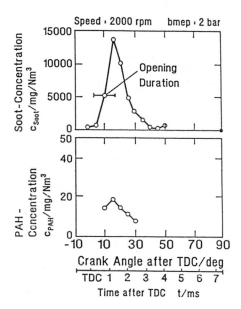

Fig. 24.6. Time resolved gas sampling in the cylinder of a DI diesel engine

The maximum of soot concentration in the mostly premixed flame is measured at 0.8 ms after ignition. For the interpretation of the results the opening duration of the sampling valve of about 1.2 ms has to be taken into account. The time dependence of the measured PAH concentrations is similar to time dependence of soot.

Figure 24.7. shows again the soot concentration as well as the calculated soot yield as a function of time. In the lower part of Fig. 24.7. the C/O-ratio for each sample is given. Correlating the C/O-ratios of the mixtures with the measured soot concentrations one finds that for a soot yield of 20% the C/O-ratio is 0.43 ($\lambda = 0.83$) and for a soot yield of 18% the C/O-ratio is 0.36 ($\lambda = 0.96$). After this time the soot oxidation process seems to dominate because the mixture is lean (C/O = 0.27; $\lambda = 1.22$) and the soot yield decreases rapidly.

It surprises that only a soot yield of 12% is obtained in a comparatively rich mixture. Probably for the rich mixture the local temperature is to low to form a high amount of soot in the short time of the engine combustion process. Clarification of this question will be a point of further investigations.

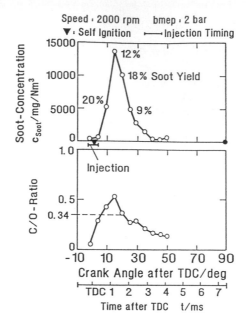

Fig. 24.7. Time resolved gas sampling in the cylinder of a DI diesel engine

With the same sampling technique the particle sizes in the cylinder are measured using an electrical aerosol analyser (EAA), [24.6], compare Fig. 24.8. Just after ignition and therefore at the beginning of soot formation nearly all particles are smaller than 10 nm (80% are in the range between 3 and 5 nm, 20% between 6 and 9 nm). During the combustion and expansion process of about 4.5 ms (to an end of in-cylinder condition of about 1 MPa and 1000 K) the particles grow. Nearly all particles are smaller than 100 nm. Only 30% are in the range from 3 to 5 nm and 65% in the range 6 to 9 nm. The remaining part shows sizes larger than 10 nm. It seems to be important to get more information regarding particle oxidation and particle growth under high pressure and temperature conditions in the time scale of engine combustion.

24.4 Soot Formation in the Burnt Gas

As mentioned above fuel is injected into the flame and into burnt gases during diesel engine combustion. The formation of soot from fuel injected into the burnt gases under diesel engine conditions has been investigated using a particular experimental arrangement [24.7]. The equipment is given in Fig. 24.9. The burnt gases are produced by burning natural gas under fuel rich conditions, so that no oxygen could be measured in the exhaust. The fuel is injected and mixed with the hot gas. The fuel-exhaust gas mixture is

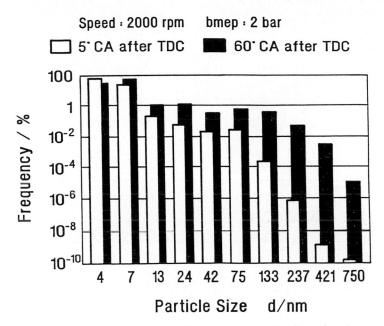

Fig. 24.8. Particle size distribution during diesel combustion

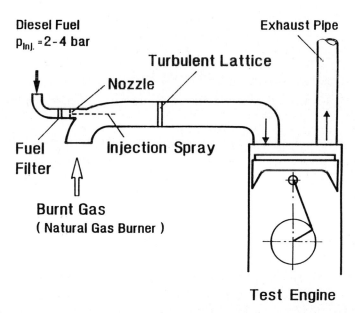

Fig. 24.9. Experimental arrangement to examine soot formation in burnt gas mixtures

compressed in the engine. The soot formed at high pressure and temperature is measured in the exhaust pipe [24.7].

In Fig. 24.10. the in-cylinder gas temperature versus time or crank angle of this engine is shown. To attain a peak temperature of 1700 K the fuel/exhaust gas inlet temperature were set to about 900 K. The time at which the fuel/gas temperature is above 1400 K amounts to approximately 5.5 ms.

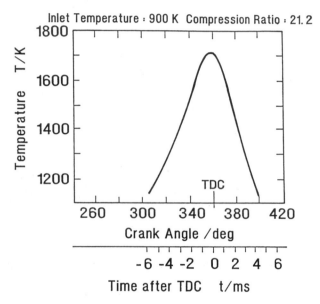

Fig. 24.10. Gas temperature history during compression of hot burnt gas

The measured soot yield for two different amounts of fuel in the exhaust gas (12% and 2%) and two air fuel ratios of the natural gas flame (0.83 and 0.97) versus the maximum compression temperature is demonstrated in Fig. 24.11. For both air fuel ratios no oxygen is present in the burnt gases. But it can be expected that a higher OH concentration exists in the near stoichiometric mixture. For an air fuel ratio of 0.83 an influence of the amount of fuel injected on soot formation can be observed. Higher fuel concentrations increase the soot yield. The influence of fuel concentration vanishes at an air fuel ratio of 0.97. For low fuel concentrations the temperature dependence of the soot yield is similar for both air fuel ratios of the natural gas flame. Soot is formed in the compressed "inert" exhaust gas at temperatures higher than 1450 K [24.7].

Comparing these results with data measured in shock tubes, compare Fig. 24.12., a much higher minimum temperature limit of soot formation is observed for shock tube experiments [24.8].

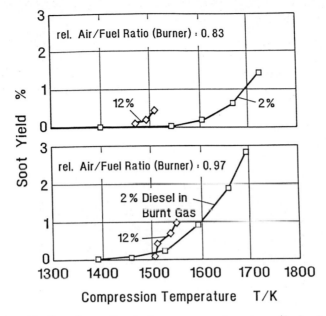

Fig. 24.11. Soot formation in homogeneous burnt gas/fuel mixtures

In Fig. 24.12. the soot yield measured in this special engine is compared with those measured in shock tubes. In a shock tube the reacting mixtures is exposed to high temperature for short times (1 ms). Therefore, the induction times for soot formation have to be shorter than 1 ms to measure soot [24.8]. In the special engine as well as in commercial diesel engines at low speeds the time for keeping the high temperature level is longer than in shock tubes and, therefore, longer than the induction time at lower temperatures. Therefore, soot formation in the engine is observed at lower temperatur than in shock tubes.

24.5 Boundary Conditions for Detailed Investigations of Soot Formation and Oxidation in Diesel Engines

To improve the diesel combustion process regarding no or low soot emission and low NO_x emission detailed knowledge of the process of soot formation and oxidation under diesel engine conditions is necessary. It is demonstrated in this paper that four different phases of soot formation or of soot oxidation processes take place, which have to be investigated under diesel engine conditions separatly. In the following the boundary conditions for such investigations are given.

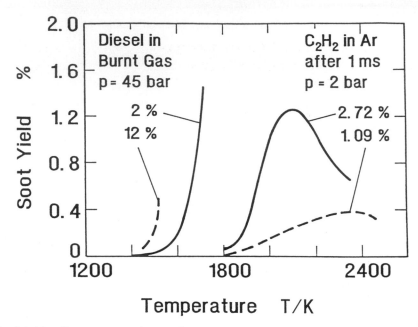

Fig. 24.12. Comparison of soot formation from shock tubes and from burnt gas/fuel mixtures

It seems that results of high pressure premixed flames [24.9] can be transfered to the premixed combustion phase in diesel engines. Pressures of 5 MPa to 8 MPa and temperatures of 1300 K to 2200 K occur during this phase of diesel combustion. The time for soot formation at these high temperatures is lower than 1 ms.

To investigate soot formation during injection of fuel into the flame the influence of radicals on soot formation and oxidation should be inspected in detail. Here the pressure and temperature is higher than for premixed combustion (up to 11 MPa and 2600 K).

For soot formed by fuel mixed with burnt gas first results are given in this paper. More data are necessary regarding the influence of fuel and radical or oxygen concentrations. The boundary conditions during diesel combustion are similar to those of fuel injection into the flame.

The soot oxidation process in diesel engines is very important. Beside pressure, temperature and oxygen the particle size will influence the combustion of soot in the engine. In addition radical and oxygen concentrations and hydrocarbons are influencing parameters.

The following boundary conditions must be kept in mind for soot oxidation in diesel engines:

Pressure: 1 MPa - 11 MPa

Temperature: 1000 K - 2200 K

Oxygen content: 0

Particle size: 4 nm - 100 nm

It would be very helpful for diesel engine developers if basic research work in the field of soot formation and oxidation can be done in the future under boundary conditions similar to those in diesel engines.

References

24.1 F. Pischinger, H. Schulte, J. Jansen: "Grundlagen und Entwicklungstendenzen der dieselmotorischen Brennverfahren", in VDI- Bericht Nr. 714 (VDI-Verlag, Düsseldorf 1988) p. 61

24.2 M. Maurer, G. Lepperhoff, F. Pischinger: "The Influence of the Engine Lubricant on Particulate Emissions of Diesel Engines", in Le Moteur Diesel D'Application: Automobile, vehicles industriels, vehicles speciaux,   S.I.A. Congress, Lyon, Juni 1984

24.3 E. Scheid, U. Reuter, H. Xu: "Orte der Selbstzündung bei der dieselmotorischen Verbrennung", in Forschungsbericht SFB 224 (RWTH, Aachen 1988) p. 363

24.4 G. Lepperhoff, J. Hansen: "Soot Formation in Diesel Engines", Autotech, Seminar Paper, Integrated Diesel European Action (IDEA) II No. C399/9, Birmingham 1989

24.5 G. Lepperhoff, M. Houben: "Particulate Emission and Soot Formation Process in Diesel Engines", IMechE Congress, Chester, September 1990

24.6 TSI: Electrical Aerosol Size Analyser, Manual of Model 3030

24.7 G. Lepperhoff, M. Houben:"Soot Formation in Local Zones of Fuel/Burned Gas Mixture in Diesel Engines", CIMAC Congress, D18, Florence, 1991

24.8 M. Frenklach: "Shock Tube Study of the Fuel Structure on the Chemical Mechanisms Responsible for Soot Formation"; NASA Report CR - 174661, May 1985

24.9 M. Bönig, C. Feldermann, H. Jander, B. Lüers, G. Rudolph, H. Gg. Wagner: "Soot Formation in Premixed C_2H_4 Flat Flames at Elevated Pressure", in Twenty-Third Symposium (International) on Combustion (The Combustion Institute, Pittsburgh 1990) p. 1581

Discussion

Sarofim: I have just a minor question on particle sizes. First let me say that it is a very impressive study. You said that the size distribution was measured by a mobility analyzer. This gives you a number distribution. The mass distribution would be very different.

Lepperhoff: This is correct, it is a number distribution.

Sarofim: But what would be the mass mean particle size. That would be much more interesting.

Lepperhoff: We can calculate the mass mean particle size and this value lies at higher particle sizes. There is a difference between young soot particles and older ones. For older particles we have more particles in the upper size range of 500 nm to 700 nm.

Müller: I have a little comment on the difference between the number density weighted average diameter and the mass weighted average diameter. If you take your number distribution and weight it to get the mass distribution you will get similar results as I presented from our shock tube experiments. The mean volume diameter or mean mass diameter is higher than the mean number diameter.

Lepperhoff: That is correct.

Glassman: I don't want to talk about soot as much as I want to talk about your concept of the combustion process as you inject the fuel into burnt gases. I find it very difficult to accept the fact that in a diesel engine all of the fuel is evaporated and we have basicly just a distribution of oxygen and fuel. I have seen some pictures of a diesel engine with an optical window which show that some liquid particles must remain in the core of the jet after the injection is completed. I find it very difficult to analyse any soot problem in a diesel engine by supposing every thing is premixed. Do you have evidence for this assumption?

Lepperhoff: We have measurements for diesel fuel injection concerning droplet sizes and the ratio of the liquid part to the evaporated part of the fuel. We know that the droplets are very small and most of them are in the range of less than let say 50 μm. Comparing the liquid and the evaporated part of the fuel we could see from these measurements, depending on the induction time or the ignition delay and depending on the temperature, that there is high amount of fuel evaporated and mixed with the gases. Therefore, we are sure that in the first stage of combustion we have a premixed flame. Particularly, if you look at engines that produce swirl or at charged engines where temperatures of the loading are much higher. For these engines there is an improvement of the mixing process so that we have a large part of the fuel burnt in a premixed flame.

Frenklach: I want to refer to the comparison of your data to shock tube data at 0.2 MPa. We did our measurements with acetylene and acetylene is very unique in that it starts to soot at very high temperatures. All other hydrocarbons formed soot at lower temperatures, at that pressure at about 1600 K. I would not expect that you have only acetylene in the engine. You have a mixture of different hydrocarbons. So if you compare you should take the data from the same paper for butadiene or other hydrocarbons. To me it seems not to be the difference in time from 1 ms to 2 ms or to few ms. If you want I can give you 2 ms data.

Lepperhoff: You can't compare acetylene with diesel fuel, I agree. However, what we wanted to demonstrate by this comparison is that it is important to look at the time scales. It is important to look what is happening at which time, when the mixing process dominates and when the soot formation starts. That was the reason why I compared your data with our data.

D'Alessio: Did you try to evaluate the oxydation time of soot particles in the engine for example by means of the Nagel-Strickland-Constable-equation. Is the rate of oxidation comparable to literature data or is it faster or slower. Can you try to make this kind of evaluation?

Lepperhoff: We did not evaluate the measurements from the diesel engine. We plan experiments with the experimental set-up I showed concerning the oxidation of soot particles during the expansion phase.

D'Alessio: Do you just take the soot away from your engine and then oxidize it later?

Lepperhoff: We want to use the soot from a burner. We mix it with air and then we go to the intake port of the engine compress it and expand it. After this we measure the soot. The distance between the burner and the intake port of the engine is very small in our concept.

D'Alessio: The reason why I am asking this is because yesterday we heard that in soot formation at high pressures the mass growth of soot is limited by the decay of growth compounds. Under these conditions soot acts as a "cleaner" for hydrocarbons. So, the point is in this case the particle should be more reactive and, therefore, oxidation rates should become very fast.

Dobbins: I believe your soot particles are aggregates made up of primary particles. Your size distribution are for aggregates of various chain length. Is that correct? What about the primary particle sizes. Could you comment on that?

Lepperhoff: I have no information about the structure of the soot particles because we did not take electron micrographs. We have only information about the total size of the particles. I suppose that these are agglomerates because of their large sizes.

Dobbins: Another question about the composition of the soot. Would it be reasonable to assume – and there is some evidence by the observation of quite a few people – that there is a large oily content to the diesel soot. And I wonder what you know about that and especially if there is a variation in the oily fraction according to load conditions or other variables in the operation of a diesel engine.

Lepperhoff: In our measurements soot consists of all particulate matter collected on a filter at a temperature of 360°C. In the exhaust soot particles may adsorb high boiling hydrocarbons. In our measurements we kept temperatures so high that we can exclude any adsorption process. Therefore, I suppose that the soot I have been talking about contains a negligible amount of hydrocarbons.

General Discussion on Soot in Flames: Formation, Dependence, Practical systems

chaired by

Paul Roth, Robert J. Santoro

Smyth: I have a comment and a question for Dr. Müller. The comment is about the critical temperatures that you have discussed. They came up in the Göttingen Workshop as well. I think it' s of interest to look at the Sandia experiments that were published in the 23rd Symposium on a range of turbulent flames. They measure shot by shot OH concentrations and temperature from Rayleigh measurements. In the temperature region from 1400 to 1600K the radical pool collapses. It doesn't go down just a little bit, it collapses. This would explain both inception shutting off and oxidation shutting off below this temperature. Did I understand that the coagulation rate increases with pressure?

Müller: The coagulation rate increases with pressure and with temperature but the temperature has the major influence.

Smyth: But both are small.

Müller: You know we are in the transition regime with Knudsen numbers which are very low and therefore the coagulation rate has to be very high.

Feldermann: Coagulation constant has to go down with higher pressure because the Knudsen number is smaller.

Müller: But we are in the transition regime, where for constant pressure, there is a maximum of the coagulation rate.

Feldermann: You are not in transition regime at your pressures. You have a Knudsen number of 1.0 and then you are not inside the transition regime. You are in slip flow regime and then at higher pressures the constant goes down.

Müller: That is not correct. In our experiments the increase of particle diameter is so large that you can not calculate it neither in the transition regime nor in the continuum regime nor in the free molecule regime. There must be some other influence.

Homann: We have heard a lot about the formation of soot at very high pressure, 7 MPa to 10 MPa. Can anyone tell me whether this kind of soot has been characterized as far as the carbon black in industry has been characterized. What is the nature of this soot under this extreme conditions.

Lepperhoff: I can not characterize this soot. If you measure the soot directly behind the engine you will find two different kinds of soot.

Smyth: Could I ask the leading question of the Wagner group who has made measurements. You have collected soot and you are using optical techniques to measure it. You might have some feeling for how the properties are changing as we go into this pressure range.

Wagner: We investigated some of the properties of soot up to a few hundred, E.U. Franck up about 200 MPa. In that case we do not know the exact temperatures. But there are a few points which are interesting. The soot does not have a special appearance. May be that high pressure soot looks a little bit greyer than the other one and if you take it in between your fingers it is not just like the soft material which you get out of a Christmas candle. The carbon to hydrogen ratio goes to similar values as you obtain in normal pressure flames. The size of the spherical soot balls is limited. A kind of growth process must be interrupted for some reasons earlier than in the other case. We have not seen very big soot balls neither did Franck in the 200 MPa diffusion flame experiment. X-ray experiments which have been performed show a structure you find in soot also at lower pressure. We have not used electron-spin-resonance signals yet. The density of the soot which we get at 10 MPa is about 1.96 and the density of the soot which we have got from the 20 MPa experiments is also in that range. So it is a little bit higher probably than what you get at normal pressure.

Schindler: I have some numbers about the diesel particles and I must say these are diesel particles from VW-engines. So the composition of those particles are 88 % carbon then about 5 % oxygen, 3 % hydrogen, 2.5 % sulfur and 0.5 % nitrogen.

Gösberg: Could you tell us what the operating conditions of that engine were?

Schindler: These are medium results from a test cycle.

Santoro: One of the hopes of diesel engine designers is to go through this magic temperature regime where they can avoid the formation of soot by keeping the temperature in right range such that combustion is going on and mixing occurs at the optimum rate to maintain the combustion and to minimize the formation of soot. I think that the question of this temperature time variation has been brought up at this meeting and it was brought up at a recent diesel workshop. This community can make a contribution in terms of how we tailor these processes. Dr. Lepperhoff was saying they have very good atomization procedures now, but I do not know what your pressure was in that nozzle.

Lepperhoff: This was a medium pressure of about 40-60 MPa.

Santoro: So these atomization processes tend to produce small droplets around rather quickly. I think that there are only two answers to soot control question, either people in this room really think that magic path can be found or we are really dealing with enhancing burn-out. Either you avoid the formation or you have to burn the soot. The approaches to achieve that in the engine are different. This is an important question to answer for this community.

Schindler: In a diesel engine you will not find droplets coming out of a nozzle. There you have nearly an optical dense part and this optical dense part is like a cheese with holes in it. It is mainly consisting of liquid surrounded by some sort of layers and then droplets. So this picture of fine droplets is true for gasoline engines but not for diesel engines.

Peters: I don' t think that this concept of a magic path to find this temperature window will be the one that solves the problems. This critical temperature may exist in premixed combustion. What we find in a diesel engine is a non-premixed situation where you have ignition in the stoichiometric part and then flame propagation into the rich part towards the fuel. A lot of the soot is formed during the propagation of the flame into the rich part. So I think the mixing is one of the major problems in the diesel engine not finding the right temperature.

Moss: I would like to make an observation on the whole of the day really. This morning we had people looking at models of processes that would describe diffusion flames and they characterized these flames in terms of mixture fraction. Then we had a session where people talked about these flames in premixed terms using carbon oxygen ratio. Besides this we had practical devices which one might traditionally have thought, certainly in the burning regime, were non-premixed. We had Prof. Peters describing a burner that was partially premixed. In circumstances where we clearly do not have an unifying description of the soot formation process and perhaps do not have one for the immediately foreseeable future, I wondered where should we all be going. Perhaps this is a question at the end of a long day and it is inappropriate to ask it. I just wondered how everybody thought we were moving in some coherent direction.

Lepperhoff: I believe that everybody is moving in different directions at the moment. There are some results which are quite good. We can compare for example these results in an engine with the results Dr. Jander and Prof. Wagner got for high pressure flames. So there are some connections and I believe that there are different processes. It is possible to look at these different processes in a practical way.

Glassman: I like to response to Norbert Peters comments. We had a paper at the Seattle Combustion Symposium where we did something similar to what Prof. Peters was doing. We basically took on opposed jet diffusion flame and on the fuel side we began to add oxygen. So, we went from a mixture ratio of infinity to the lean flammability limit. I would like to generalize what I think happens. We only dealt with two fuels so I can' t make a complete generalization, the two fuels where ethylene and propane. As you add oxygen to the ethylene fuel starting from infinity you increase the rate of pyrolysis and the extinction coefficient starts to rise. We get to a certain point where the actual energy releases even though it is still acting more like a diffusion flame and the extinction coefficient rises very rapidly. As you add more oxygen it begins like a premixed flame to drop off. I agree with your final conclusion if you going to stop soot formation in a diesel engine you have to achieve such good mixing that it acts like a lean premixed flame because the system is overall lean. How you do that? This is a typical fluid dynamic problem. I' m still intrigued by the Kent / Wagner data that you can burn up these particles regardless of the overall content.

Schindler: Perhaps a last remark from my side. I would like to make it a little bit more complicated. If an engineer tries to optimize an engine with respect to soot he will run into another problem. The NO_x production and this is a real trade off.

Lepperhoff: We know how to get a process without soot formation and NO_x formation, that is to keep this process below 1400 K. But in the case of an engine you have to need for 100 km about 13 or 14 liter diesel. You have no efficiency that means we have high CO emissions. Therefore we need a higher temperature than 1400 K. This is the problem that engineers have.

Glassman: I don' t know whether we generally accept the condition. I always thought that the rule of thermal NO_x formation was around 1700 K for the Zeldovich mechanism and I think there is a range you can go higher than 1400 K and keep the NO_x down. There is a window where you can both keep soot from forming and still keep NO_x down and that is in the area of 1500 to 1700 K.

Santoro: I have to say something about the soot / NO_x trade-off which is a well known problem. The engine companies have very good correlations that show an universal curve. How much soot they have and how much NO_x they produce. They can make very low soot emission engines right now, but they produce too much NO_x. They can make very low NO_x emission engines but they produce too much soot. They have tried to trade off between the two emission species without complete success. In fact it's exactly that problem that this community can lend help to the engineering community. I think there is a lot of understanding of premixed and diffusion flames which can contribute to practical problems, for example much of the spray vaporization process is like a diffusion flame. We can possibly say something about diesel engines where we always have two parts to the combustion process: the premixed part and the diffusion controlled part. I don't think this community can back away from making a contribution to this problem.

Models for Soot Formation
in Laminar Flames

Models for soot formation in laminar flames inevitably are prerequisites for models of soot formation in practical systems such as turbulent flames. The discussion of successes and uncertainties in modelling of soot formation in laminar flames is one important topic of part V of this book. Models that have been referred to several times in other parts of the book are evaluated and rated. Alternative routes - e.g. ionic mechanisms - from gaseous fuel molecules containing some carbon atoms to large soot particles containing some millions of carbon atoms are discussed as well as some essential parts of detailed chemistry models for soot formation.

Soot Deposition
from Ethylene/Air Flames and the Role
of Aromatic Intermediates

Joanne M. Smedley, Alan Williams,

Auemphorn Mutshimwong

Department of Fuel and Energy,
Leeds University,
Leeds, LS2 9JT, UK

Abstract: An understanding of the mechanism of soot formation is of importance not only because it is a pollutant but also because it forms deposits in combustion chambers. In the present work soot deposition has been studied using a flat flame, water-cooled, premixed burner burning two rich ethylene/air flames ($\phi = 2.52$, $\phi = 2.76$). Rates of soot deposition for both cooled copper and uncooled stainless steel surfaces were investigated and soot samples from these experiments were analysed for PAH. A quartz microprobe sampling system was used to determine the concentration profiles of aromatic and polyaromatic species in the deposition region. Experimental results support the theory that soot deposition occurs by a thermophoretic mechanism when cooled surfaces are used. It was found that soot deposition rates and associated PAH varied in the flame zone and with the temperature of the cooled plate. If uncooled high temperature surfaces are used then direct surface deposition can take place.

26.1 Introduction

The formation and deposition of soot or carbon deposits in combustion systems is usually undesirable. Such deposits can have adverse effects on heat transfer characteristics or combustion behaviour which can cause performance or failure problems in a range of systems from rocket engines to diesels. The emission of soot particles from combustion chambers into the atmosphere is also of environmental concern due to the fact that soot particles can contain significant concentrations of PAH. There is thus a need to accurately predict under what conditions soot formation will occur and to quantify the amount of soot deposited in any part of a combustion system.

Over recent years more progress has been made in understanding the chemical route of carbon formation but work in this area is difficult due to the complex interaction of aromatic species in sooting flames [26.1-5]. Such aromatic species may form soot particles which by thermophoresis or convective flow effects become deposited as spherio-soot deposits; alternatively the aromatic species may, under suitable conditions, form pyrocarbon which exists as lamellae.

Most recent research has been concerned with spherio-particulate soot because of its environmental implications. Species such as C_2H_2 have been accurately modelled by *Miller* et al. [26.1] and by *Harris* et al. [26.2] since reliable data are available for these species. *Harris* et al. [26.2, 6] have also attempted to model single ring aromatic species but have only partially succeeded because of the complexity of the growth steps. Not only must most of the rate constant data be estimated but there is uncertainty about actual reaction steps. Even though these models exist, many gaps remain in the mechanism of the subsequent steps leading to soot inception and growth. Whilst acetylene is recognised as a major growth species there is uncertainty about the level of participation of the PAH species. *Harris* et al. [26.7] investigated soot particle inception and suggested that PAH species were the main contributors to soot formation early in the flame. They decided that C_2H_2 was only involved when the PAH species were depleated. Soot deposition onto a cooled surface has been recently investigated by *Makel* et al. [26.8] using a laser diagnostic technique to measure soot deposit thickness and free stream soot concentrations. A numerical model to make predictions of soot deposition rate was also developed by them. In the present work this model has been applied to soot deposition on cooled and uncooled plates. In addition, information has been obtained on the role of PAH.

26.2 Experimental Methods

Two rich ethylene/air flames ($\phi = 2.52$ and $\phi = 2.76$) were studied using a flat flame, water-cooled, premixed burner (McKenna Industries). The gas flows in l min^{-1} for the $\phi = 2.52$ flame were $O_2 = 2.22$, $N_2 = 8.36$, C_2H_4 = 1.79 and the flows for the $\phi = 2.76$ flame were $O_2 = 2.36$, $N_2 = 8.90$, $C_2H_4 = 2.09$. To investigate soot deposition rates an uncooled stainless steel plate and a cooled copper plate were used to support stainless steel and copper sample squares respectively within the flame. The sample plate dimensions were 10 mm times 10 mm times 2 mm. Samples were taken at four flame heights, namely 5, 10, 15 and 20 mm above the burner surface. At each height soot deposit rates were determined gravimetrically using plates inserted into the flame for intervals of 15 s (up to 75 s).

Free stream soot which was collected on a Whatman glass microfilter GF/C paper from both flames was analysed for PAH using pyrolysis (desorption) chromatography. Deposited soot from both the metal plates was also analysed using this method. A Perkin Elmer 8320 gas chromatograph fitted with a Quadrax 007' series fused silica capillary column was coupled to a CDS Pyroprobe to desorb hydrocarbons from the soot. Approximately 2.5 mg of soot is heated at a rate of 0.1°C per millisecond up to 600°C. A quartz microprobe with a 6 mm outside diameter similar to that used by *Harris* et al. [26.3] was used to obtain gas samples at various heights above the burner surface and a syringe method was used to transfer samples into a gas chromatograph (Perkin Elmer 8700 gas chromatograph fitted with a J & W megabore GS-Q column).

26.3 Results and Discussion

26.3.1 Soot Deposition

Figures 26.1. and 26.2. give typical experimental data for the rate of soot deposition for the uncooled and cooled plates. In these each line represents different sampling heights above the burner. As expected the deposition rate of soot increases on an uncooled stainless steel plate as the plate is moved vertically away from the burner surface. However the deposition rate of soot at 20 mm above the burner surface is less than the deposition rate at 15 mm above the burner surface when the water-cooled copper plate is used. This was found to occur in both flames ($\phi = 2.52$ and $\phi = 2.76$). It has been observed that the ultimate soot load on sample squares is approximately 0.7 mg regardless which plate is used. For the uncooled plate a 0.7 mg soot load occurs at 20 mm but when using the water-cooled plate a soot loading of 0.7 mg is achieved at 15 mm when both had 75 s exposure in the flame. Above 0.7 mg the soot load is too great and the soot breaks off and is dispersed back into the flame and this is why a reduced soot deposition rate is seen for the 20 mm height samples when using the water-cooled plate. In general, for both plates the soot deposition rates are greater for the $\phi = 2.76$ flame at all heights than for the $\phi = 2.52$ flame. The exception for the metal plate is at 20 mm. Similar rates are experienced for both flames at this height and this again suggests that there is a limit to the amount of soot that can be deposited under these conditions.

When comparing the results from the uncooled plate and water-cooled plate for $\phi = 2.52$ flame it can be seen the soot deposition rate at 10 mm above the burner surface for the uncooled plate is much less than the soot deposition at the same height for the cooled plate. For example, at 75 s a soot load of under 0.1 mg is recorded for the uncooled plate. Whereas on the water-cooled plate a soot load of approximately 0.4 mg is recorded at 75 s. The same is noticed at 15 mm above the burner surface at 75 s. The uncooled plate has a soot load of 0.4 mg while the cooled plate has a soot load of 0.55 mg. This is consistent with the thermophoretic transport of soot particles. From the results it also seems that thermophoresis is more prominent at lower regions in the flame. This may be because the soot particles are smaller earlier in the flame and as they become larger later thermophoresis has less influence on the movement of the soot particles. *Makel* et al. [26.8] assumed that thermophoresis was the primary transport process for soot particles although no allowance was made for variation in particle size. Particle sizes on the uncooled plate were determined at various heights using both scanning and transmission electron microscopes, particle sizes ranged from 12 – 38 nm. They also experienced the process of resuspension when the soot load increases to a certain limit. This was incorporated into their model to try and determine the final soot loading.

The soot mass flux to the surface can be expressed [26.8] as

$$\dot{m}_s'' = \rho_s \, v_d \, f_V \qquad (26.1)$$

where ρ_s is the particle density (1900 kg m^{-3}), f_V is the soot volume fraction at the edge of the diffusion sublayer and the deposit velocity, v_d, is equal to the thermophoretic drift velocity, v_t, given by

$$v_t = -0.55 \, \nu \, \frac{\partial \ln T}{\partial y} \qquad (26.2)$$

where ν is the viscosity and which is largely determined by the temperature gradient.

Experimental deposition rates were found to be $1 \cdot 10^{-5}$ g cm^{-2}s^{-1} at 15 mm above the burner using the water-cooled plate and $0.75 \cdot 10^{-5}$ g cm^{-2}s^{-1} using the uncooled metal plate. This is consistent with *Makel* and *Kennedy's* experimental data although they used a water-cooled cylinder to collect their samples. The calculated deposition value for our experiments using their theory is $1.2 \cdot 10^{-5}$ g cm^{-2}s^{-1} for the water-cooled plate and $0.6 \cdot 10^{-5}$ g cm^{-2}s^{-1} for the uncooled plates. The agreement is excellent.

26.3.2 PAH Content of the Soots

Deposited soot samples from both the cooled and uncooled plates were collected from both flames after 75 s exposure within the flame. These soot samples were analysed in the GC-Pyroprobe apparatus and typical analyses are shown in Fig. 26.3. and 26.4. Most of the peaks in the chromatograph were identified using a standard PAH mixture or by using retention indices [26.9-10]. The major peaks are identified in Fig. 26.3. The gas chromatograph results for both the flames studied were very similar. Generally each PAH component was found in greater quantities in the richer $\phi = 2.76$ flame. The $\phi = 2.52$ flame contained much more volatile material which is expelled from the soot very early and is the first peak. Soot collected from

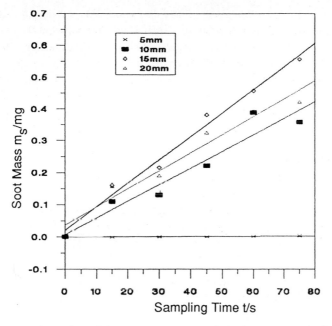

Fig. 26.1. Soot deposition rates onto a cooled plate; $\phi = 2.52$ flame

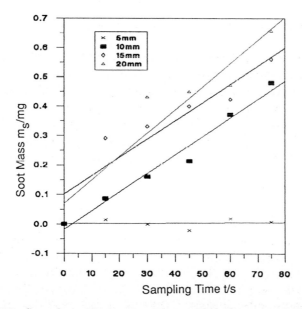

Fig. 26.2. Soot deposition rates onto a cooled plate; $\phi = 2.76$ flame

the cooled plate has a significantly higher concentration of PAH. The soot collected from the uncooled metal plate contains the smallest concentration of PAH and the free stream soot lies between. The temperature at which the soot is collected seems to have an important effect on the composition of the soot. From these results it could be interpreted that the cooled plate creates a low temperature region within the flame.

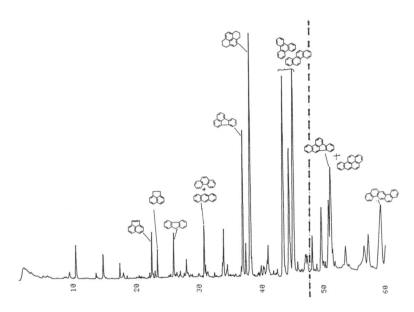

Fig. 26.3. Gas chromatograph sample for uncooled metal plate with major peaks labelled; $\phi = 2.76$

Free stream soot samples were collected on to Whatman glass microfibre filters were also analysed using the Pyroprobe system. Results from the ϕ = 2.76 flame are shown in Fig. 26.4a. The collecting conditions for the free stream soot are identical to those quoted above for the deposited soot samples.

26.3.3 Soot Formation Steps

Figure 26.5. shows the gas composition profiles obtained using the quartz probe sampling system. The reaction zone (based on O_2 decay) is quite extended. Figure 26.6. shows profiles of some aromatic species and their precursors. The three most abundant PAH species which were Pyrene (202), Chrysene (228) and Benzo(a)pyrene (252) are labelled. The results show that as combustion takes place C_2H_4 and O_2 are depleted. In Fig. 26.5. it can be seen that C_2H_2 peaks at about 5 mm above the burner after which it levels off to a fairly constant value between 10 and 20 mm. C_6H_6 rises gradually at about 6 mm and begins to fall at 12 mm. Initially there was more C_6H_6 in the $\phi = 2.52$ flame but at 20 mm there was more in the $\phi =$

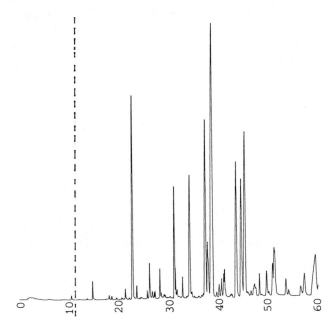

Fig. 26.4a. Gas chromatograph of free stream soot at 15 mm above burner; $\phi -$ 2.76

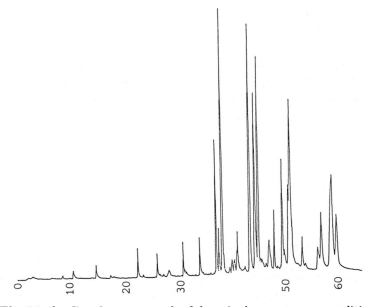

Fig. 26.4b. Gas chromatograph of deposited soot at same conditions

2.76 flame. Laser beam attenuation studies at 670 nm indicated that soot formation began at 6 mm above the burner. From the results it can be seen that as C_2H_2 is consumed C_6H_6 is produced as well as soot; this is consistent with the fact that C_2H_2 is involved in C_6H_6 formation and it is generally considered that C_2H_2 is an important precursor to soot formation. C_2H_2 and C_6H_6 fall as soot is formed and other larger hydrocarbons are produced in the flame. In Fig. 26.5. both CO and H_2 increase gradually as the probe is moved up from the burner surface. It was found that H_2 was in greater concentration in the $\phi = 2.52$ flame and CO was in a higher concentration initially. O_2 was only found in small amounts in the $\phi = 2.52$ flame. These increase as combustion takes place and are products of oxidation reactions. These trends were seen by *Harris* et al. [26.3] but they only probed to about 3 mm above burner so they did not observe the fall in C_2H_2 and C_6H_6 in the later part of the flame. *Miller* et al. [26.1] however undertook experiments up to 22.5 mm and also found similar profiles for CO, O_2 and H_2.

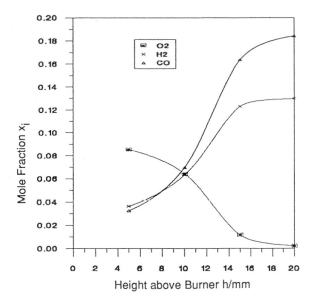

Fig. 26.5. Small molecule species profiles for $\phi = 2.76$ flame

26.3.4 Mechanism of Soot and PAH Deposition

The mechanism of soot formation from an ethylene flame involves the formation of acetylene and its polymerisation to single and then multi-ring species. The subsequent growth of the initial soot particles involves surface growth involving acetylene and polyaromatic species the relative extent of them being subject to different interpretation, eg. [26.5,26.8-12].

The experimental deposition rate of soot onto both the cooled and uncooled plates is consistent with predicted rates based on the thermophoresis model. It is interesting to note that the soot deposition rates ($\approx$ $1 \cdot 10^{-5} \mathrm{g\,cm}^{-2}\mathrm{s}^{-1}$) are approximately 10% of soot growth rate $\approx 1 \cdot 10^{-4}\mathrm{g}$ $\mathrm{cm}^{-2}\mathrm{s}^{-1}$ [26.7]. The PAH contents of the deposited soots are significantly different depending on the temperature of the plate used to collect it and the sampled position in the flame. This is demonstrated in Tab. 26.1. The value for anthracene from the uncooled soot deposited on the plate of $0.0057 \cdot 10^{-3}$ is taken as 1 and used to calculate the relative amounts of other PAH species which are present. It is evident from Tab. 26.1. that the PAH content of the soot deposited on the cooled plate is greatly enhanced at low temperatures. This is highlighted when comparing the columns for the cooled and uncooled plates for example the relative amount of pyrene in the cooled plate sample is over one hundred times greater than that recorded for the uncooled plate sample. The concentration of PAH in the flame gases were deduced from the difference between the free stream values and the hot deposited values. In the free stream soot samples taken during our experiments, the product involves soot particles, surface adsorbed PAH and gas phase PAH. However the deposited samples can only contain PAH associated with the soot particles as adsorbed or growth species. The ratios given in Tab. 26.1. reflect the wide range of PAH concentrations seen in the flames.

Table 26.1. Amounts of PAH produced in the $\phi = 2.52$ flame 10 mm above the burner normalized with $0.0057 \cdot 10^{-3}$

PAH	Free Stream Soot	Soot Deposited on Uncooled Plate	Soot Deposited on Cooled Plate
Benzene	318	-	-
Naphthalene	14	4	29
Phenanthrene	9	4	741
Anthracene	2	1	65
Pyrene	305	12	1323

In Fig. 26.7. the predicted ratios of $(C_{2n}H_{2m}/C_2H_2)$ are plotted against temperature in the way adopted by *Lam* et al. [26.4]. Using measured acetylene and the deduced PAH concentrations we also found that in the flame

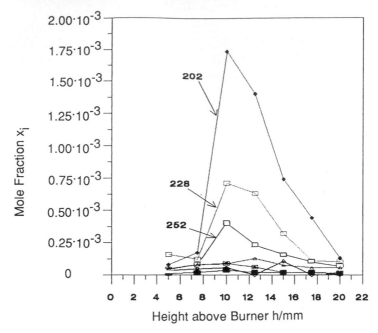

Fig. 26.6a. Aromatic hydrocarbon species profiles for free stream soot; $\phi = 2.52$ flame

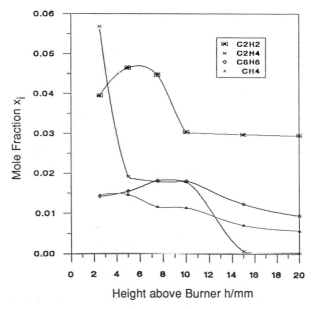

Fig. 26.6b. Hydrocarbon species profiles for free stream soot; $\phi = 2.52$ flame

gases the smaller molecules were in thermodynamic equilibrium whilst the larger species were increasingly below their equilibrium values in the same way as found by Lam et al [26.4]. Hence the abundance of pyrene in the flame which dominates the soot samples examined.

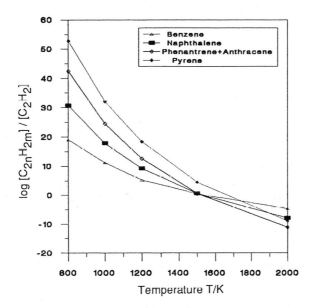

Fig. 26.7. Predicted $\log(C_{2n}H_{2m}/C_2H_2)$ results plotted against temperature for $\phi = 2.52$ flame at 10 mm above the burner

The PAH samples taken by the uncooled plate do not show such a wide variation. This can be explained on the basis that the PAH is adsorbed onto the soot particle surface. Further, by assuming reasonable dimensions for the PAH molecule, it is possible to calculate the ratio of PAH/soot (see Tab. 26.2.) and to show that the fraction of the surface covered in the case of pyrene is $1 \cdot 10^{-4}$.

Table 26.2. Calculated ratio of PAH/soot on the basis of maximum particle surface coverage

PAH	Ratio
Naphthalene	0.2
Phenanthrene	0.4
Anthracene	0.6
Pyrne	0.8

If soot growth is assumed to involve C_2H_2 and the dominant PAH, pyrene then we can estimate the reaction probability using experimentally determined soot growth rates and the acetylene/pyrene concentrations. At 10 mm above the burner and $\phi = 2.52$ we obtain a reaction probability of $1.2 \cdot 10^{-4}$ for C_2H_2 and $0.4 \cdot 10^{-4}$ for pyrene.

As the temperature within the flame falls for example when the cooled plate is inserted the surface concentration increases and so does the reaction probability factor (as $T^{-1/2}$). This accounts (approximately) for the greater PAH content recorded in the section collected from the cooled plate.

The suggestion must be implicit in these findings that the acetylene grows on the soot surface generating multi-ring compounds there which ultimately become part of the soot particle. During pyrolysis GC these compounds are desorbed as shown in the experimental results.

26.4 Conclusions

The deposition rate of soot increases with height above the burner surface (up to 20 mm) and sampling time (up to 75 s). More soot is deposited as ϕ is increased, but there seems to be a limit in the amount of soot that can be deposited regardless of the method of deposition because of soot break off at relatively low loadings.

There is evidence that thermophoresis is involved in soot deposition on to a cooled plate and that it is more dominant earlier in the flame when soot particles are smaller. The measured rates are consistent with the values calculated using the model proposed by *Makel* and *Kennedy*.

Soot samples from the uncooled and cooled metal plates contain large aromatic compounds with 4 and 5 and larger rings. The $\phi = 2.52$ had a higher concentration of these compounds.

Free stream soot samples contain a higher concentration of aromatics than deposited soot on the uncooled plate. The cooled plate contained the largest amount of aromatics. This implies that these compounds are implicated in the soot growth mechanism and that the temperature at which soot is collected caused marked differences in the PAH content.

26.5 Acknowledgements

One of us (JMS) wishes to thank the Esso Research for an SERC CASE Studentship. We also wish to acknowledge assistance from Dr. K.D. Bartle, Mr. J.M. Taylor and Miss A.L. Thomas.

References

26.1   J.A. Miller, J.V. Volponi, J.L. Durant Jr., J.E.M. Goldsmith, G.A. Fisk, J.R. Kee: "The Structure And Reaction Mechanism of Rich, Non Sooting $C_2H_2/O_2/Ar$ Flames", in Twenty-Third Symposium (International) on Combustion (The Combustion Institute, Pittsburgh 1990) p. 187

26.2   S.J. Harris, A.M. Weiner, R.J. Blint: "Concentration Profiles in Rich Sooting Ethylene Flames", in Twenty-First Symposium (International) on Combustion (The Combustion Institute, Pittsburgh 1986) p. 1033

26.3   S.J. Harris, A.M. Weiner, R.J. Blint: Combust. Flame **72**, 91 (1988)

26.4   F.W. Lam, J.B. Howard, J.P. Longwell: "The Behaviour of Polycyclic Aromatic Hydrocarbons During the Early Stages of Soot Formation", in Twenty-Second Symposium (International) on Combustion (The Combustion Institute, Pittsburgh 1988) p. 323

26.5   J.B. Howard: "Carbon Addition and Oxidation Reaction in Heterogeneous Combustion and Soot Formation", in Twenty-Third Symposium (International) on Combustion (The Combustion Institute, Pittsburgh 1990) p. 1107

26.6   J.H. Miller: "The Kinetics of Polynuclear Aromatic Hydrocarbon Agglomeration in Flames", in Twenty-Third Symposium (International) on Combustion (The Combustion Institute, Pittsburgh 1990) p. 91

26.7   S.J. Harris, A.M. Weiner: "A Picture of Soot Particle Inception", in Twenty-Second Symposium (International) on Combustion (The Combustion Institute, Pittsburgh 1988) p. 333

26.8   D.B. Makel, I.M. Kennedy: "Experimental and Numerical Investigation of Soot Deposition in Laminar Stagnation Point Boundary Layers", in Twenty-Third Symposium (International) on Combustion (The Combustion Institute, Pittsburgh 1990) p. 1551

26.9   K.D. Bartle: Handbook of Polycyclic Aromatic Hydrocarbons, Vol. 2, ed. by Bjorscht and Randahl (M. Dekker, 1985) p. 193

26.10   C.E. Rostad, W.E. Pereira: J. of High Resolution Chromatography and Chromatography Communicators, 328 (1986)

26.11   M. Frenklach, D.W. Clary, W.C. Gardiner Jr., S.E. Stein: "Detailed Kinetic Modeling of Soot Formation in Shock Tube Pyrolysis of Acetylene", in Twentieth Symposium (International) on Combustion (The Combustion Institute, Pittsburgh 1984) p. 887

26.12   M. Frenklach, H. Wang: "Detailed Modeling of Soot Particle Nucleation and Growth", in Twenty-Third Symposium (International) on Combustion (The Combustion Institute, Pittsburgh 1990) p. 1559

Discussion

Sarofim: Did you mention what the temperature of your hot plate was?
Williams: 900 K.
Sarofim: It is interesting to see your PAH concentrations. We usually use filters at around 200°C to prevent PAH condensing on our soots. We can get fairly PAH free soot. However, I am not sure if it is down to the level you have. So that is a very interesting observation.
Kennedy: I should say how nice it is that you got agreement with what I did. Did you keep the velocity constant when you varied the temperature?
Williams: Yes, we did different things and we had different velocities. But within a system we have constant velocity.
Kennedy: There are two ways of changing the deposition rate. One is to change the temperature of the free stream and the other of course is to change the velocity. I wonder if you get the same results by changing velocity as you do by changing temperature.
Williams: We just have not done these measurements.
Moss: For those who are not familiar with the Kennedy model, can I ask what you actually measure? You measure the temperature profile with a thermocouple?
Williams: As I pointed out we don't measure the temperature profile. We make an assumtion about the boundary layer temperature. That is the weakness in the model. Of course we have some estimates of temperature from optical or if you like photographical experiments. But we can not put a thermocouple into that boundary layer.
Kennedy: A stagnation point flow is actually fairly easy to treat mathematically. So it would be quite straight forward to solve a one-dimensional ODE for the temperature field and this would probably be a better way to get estimates of temperature. If you want to have a look at the effect of different strain rates or wall shear stresses it might be nice to use wedges in the flow instead of the flat plate. You could have wedges of different angle. That might be an interesting experiment.

Simplified Soot Nucleation and Surface Growth Steps for Non-Premixed Flames

Peter R. Lindstedt

Department of Mechanical Engineering,
Imperial College of Science, Technology and Medicine,
London SW7 2BX, England

Abstract: Simplified reaction steps for the formation and growth of soot particles in laminar non-premixed flames are outlined. The resulting models are combined with detailed gas phase chemistry and incorporate simplified steps for nucleation, surface growth and particle agglomeration. The soot nucleation and surface growth reactions are linked to the gas phase chemistry by the simplifying assumptions that benzene and acetylene are indicative of the locations in the flame structure where nucleation and soot mass growth occurs. The reaction mechanisms are applied to a range of ethylene and propane counterflow diffusion flames and the sensitivity of soot predictions to different nucleation and surface growth formulations are investigated. The formation paths of benzene in flames of this type are also discussed due to the importance of aromatic species in soot nucleation. It is shown that good qualitative and quantitative agreement with measured data for soot volume fraction, particle growth and number density can be obtained using simplified reaction steps.

27.1 Introduction

The need for simplified soot models arises due to the turbulent nature of most practical combustion applications such as gas turbines, gas flares and internal combustion engines. The degree of simplification required is to some extent dependent on the turbulent combustion modelling approach adopted. For example, with the use of a modified laminar flamelet approach a more detailed description of the gas phase chemistry is possible while a transported *pdf* approach in principle permits a more accurate evaluation of turbulence-kinetic interactions. However, the latter approach currently places significant restrictions on the possible number of independent scalars used in reaction schemes.

Simplified models have in the past been proposed by among others, *Tesner* et al. [27.1], *Kennedy* et al. [27.2] and *Moss* et al. [27.3]. These models have a very simple description of the gas phase chemistry-soot interaction whereby the formation of soot is linked directly to the fuel concentration or mixture fraction. This approach has been found to work well for conditions close to those were the models were calibrated. However, the application of such models to appreciably different conditions may yield significant errors due to the direct link between the parent fuel concentrations and soot formation - an aspect which is not in agreement with experimental data.

From past experimental and theoretical work it can be noted that there is fairly broad agreement [27.1-5] on the basic steps required to model the formation of soot particulates within the framework of simplified models. The steps should include soot nucleation, surface growth, particle agglomeration and finally destruction via combustion. In view of the above comments concerning simplified models, a different approach was adopted by *Leung* et al. [27.5], who assumed that the soot formation process is dependent upon the fuel breakdown process. It was shown that with the approximation of acetylene as the indicative critical species in the soot formation process good agreement could be obtained for counterflow, ethylene and propane flames [27.5] as well as coflowing methane flames [27.6]. The choice of acetylene is strongly supported by many other studies (see [27.5] for a discussion) as the adsorbed species responsible for soot mass growth at high temperatures. In contrast the use of acetylene to indicate the location in the flame structure where soot nucleation occurs is a much less satisfactory assumption. The sensitivity of the overall model to this approximation is less then may be expected as many intermediate hydrocarbon profiles in diffusion flames tend to differ predominantly in magnitude but not in shape. However, for higher hydrocarbons, starting with propane, additional reaction paths leading to minor C_3 species and benzene formation appear. As soot nucleation is linked to the appearance of aromatic species these reaction paths are likely to affect the sooting characteristics of flames with higher hydrocarbon fuels. This is naturally even more important for practical fuels with an aromatic content such as kerosene. Consequently, the use of benzene as a more realistic indicative species for incipient particle formation is likely to significantly improve the generality of any simplified model. The major disadvantage of this approach as far as the prediction of *laminar* flames are concerned is the increased uncertainty in the gas phase chemistry. However, for *turbulent* flame predictions the use of the first or any subsequent aromatic species to indicate the onset of soot nucleation currently limits the modelling approach to a modified laminar flamelet approach. This is due to the large number of independent scalars required to describe the gas phase chemistry.

The present paper discusses (i) the formation of benzene in diffusion flames, (ii) the sensitivity of soot predictions to the nucleation step, (iii) the formulation of idealised models for soot mass growth and (iv) a simple approximation describing soot agglomeration. Comparisons of soot predictions are made with data obtained from counterflow diffusion flames by *Vandsburger* et al. [27.7]. This data set is particularly attractive as it represent a consistent set of measurements of soot volume fraction, particle size and soot number densities for a large variation in conditions. Predictions of benzene formation in methane-air diffusion flames are compared in mixture fraction space with data obtained by *Smyth* et al. [27.8].

27.2 Basic Equations

The equations used in the present study are the customary transformed boundary layer equation describing the counterflow geometry. The essential mechanics of the actual transformation has been discussed by many authors, e.g. [27.9], for the steady flow case and a corresponding transformation of the time dependent equations gives,

$$\frac{\partial V}{\partial \eta} + \Phi' = 0 \tag{27.1}$$

$$\frac{1}{a}\frac{\partial \Phi'}{\partial t} + V\frac{\partial \Phi'}{\partial \eta} - \frac{\partial}{\partial \eta}\left\{\mu'\frac{\partial \Phi'}{\partial \eta}\right\} + \left\{\frac{1}{\rho'} - \Phi'^2\right\} \tag{27.2}$$

$$\frac{1}{a}\frac{\partial w_k}{\partial t} + V\frac{\partial w_k}{\partial \eta} = -\frac{\partial J_k}{\partial \eta} + \frac{R_k M_k}{\rho a} \tag{27.3}$$

$$\frac{1}{a}\frac{\partial n}{\partial t} + V\frac{\partial n}{\partial \eta} = -\frac{\partial J_n}{\partial \eta} + \frac{R_n}{\rho a} \tag{27.4}$$

$$\frac{1}{a}\frac{\partial h}{\partial t}n + V\frac{\partial h}{\partial \eta} = \frac{\partial}{\partial \eta}\left\{\frac{\mu'}{\sigma_{\mathrm{Pr}}}\frac{\partial h}{\partial \eta}\right\} + \frac{\partial}{\partial \eta}\left\{\sum_{k=1}^{n_{\mathrm{sp}}} h_k\left\{-J_k - \frac{\mu'}{\sigma_{\mathrm{Pr}}}\frac{\partial w_k}{\partial \eta}\right\}\right\} \tag{27.5}$$

where,

$$R_k = \sum_{j=1}^{n_{\mathrm{reac}}} \Xi_{jk}\left\{k_j^f \prod_{l=1}^{n_{\mathrm{sp}}} \phi_l^{\xi_{jk}} - k_j^r \prod_{l=1}^{n_{\mathrm{sp}}} \phi_l^{\xi_{jk}}\right\} \tag{27.6}$$

$$\Phi' = \frac{v_x}{v_{ex}} \qquad \mu' = \frac{\rho\mu}{\rho_e\mu_e} \qquad \rho' = \frac{\rho}{\rho_e} \quad V = \frac{\rho v_y}{\sqrt{\rho_e\mu_e a}} \qquad \eta = \sqrt{\frac{a}{\rho_e\mu_e}}\int_0^w \rho\, dw$$

and where v_{ex} and v_y are the components of velocity in the x and w directions, ρ the fluid density, μ the fluid viscosity, w a mass fraction and M the molecular weight. The enthalpy of the gas mixture is denoted h and a is the strain rate. The subscript k refers to a particular species and subscript e to the state at the edge of the boundary layer.

The diffusive fluxes for gaseous species are evaluated in the same way as previously used by *Jones* et al. [27.10],

$$J_k = -\frac{\mu'}{\sigma_{\mathrm{Sc}}} \left\{ \frac{\partial w_k}{\partial \eta} - w_k \frac{1}{n} \frac{\partial n}{\partial \eta} \right\} - \frac{v_c}{v_y} V w_k \tag{27.7}$$

For soot particles an extra term accounting for thermophoretical transport has been added,

$$V_t = -0.55\mu' \frac{1}{T} \frac{\partial T}{\partial \eta} \tag{27.8}$$

The source terms in the species transport equations were computed using a Newton linearisation,

$$R'_k = \sum_{j=1}^{n_{\mathrm{reac}}} \Xi_{jk} \left\{ k_j^f \prod_{l=1}^{n_{\mathrm{sp}}} \phi_l^{\xi_{jk}} - k_j^r \prod_{l=1}^{n_{\mathrm{sp}}} \phi_l^{\xi_{jk}} \right\} \tag{27.6}$$

$$R_k^{\nu+1} = R_k^{\nu} + \sum_{l=1}^{n_{\mathrm{sp}}} \frac{\partial R'_k}{\partial \phi_l} \frac{\partial \phi_l}{\partial Y_l} \left\{ Y_l^{\nu+1} - Y_l^{\nu} \right\} \tag{27.9}$$

The solution technique adopted for the above equation set has been outlined elsewhere [27.10]. Mesh distributions in the flames were set to ensure a large number of mesh nodes in the portion of the flames where the maximum gradients occur. This was achieved by using distributed mesh spacing with a minimum of 110 nodes and further increases in grid resolution were found not to alter the computed profiles.

The boundary conditions applied consisted of an adiabatic wall with a prescribed mass flux at the cylinder surface and zero gradient conditions at the free stream boundary for all variables except V. The flames were in all cases computed using the strain rates estimated by the experimental investigators.

27.3 Soot Nucleation

A simplifying assumption in the present work is that acetylene or benzene is used to indicate locations in the flame structure where soot nucleation occurs. The former approximation clearly becomes poorer with the use of more complex fuels. However, surprisingly good agreement can be obtained with the use of acetylene as the indicative species as shown for ethylene flames [27.5] and methane flames [27.6]. However, in propane flames, it was noted [27.5] that errors up to 50 % in soot volume fractions forced changes in reaction rate constants for optimum agreement. There may be many contributing factors to this discrepancy. These include errors in the gas phase predictions of acetylene, a very simple treatment of the interaction between the gas phase and the soot surface and soot nucleation occurring in different locations in the flame structure for the two fuels. Due to an extensive validation of the gas phase kinetics the two latter effects are more plausible and are investigated as part of the present study.

The use of acetylene as the indicative species remains attractive for reaction schemes intended to be used in conjunction with transported pdf methods for the predictions of soot formation in turbulent flames as the number of scalars are reduced. It has been shown [27.6] that with this assumption global chemistry can be formulated which gives excellent agreement with laminar flame measurements with as few as five or six independent scalars. Results obtained with acetylene as the indicative species for soot nucleation are therefore included.

The nucleation steps used in the present work can be written as,

$$C_2H_2 \rightleftharpoons 2C_s + H_2 \qquad (27.10)$$
$$C_6H_6 \rightleftharpoons 6C_s + 3H_2 \qquad (27.11)$$

The reaction rates are formulated as first order in the indicative species giving,

$$r_{27.10} = k_{27.10}(T)[C_2H_2] \qquad (27.12)$$
$$r_{27.11} = k_{27.11}(T)[C_6H_6] \qquad (27.13)$$

The introduction of a nucleation rate dependent on benzene as the characteristic species is likely to significantly improve the generality of the model at the cost of increased complexity. There are also considerable uncertainties in the reaction paths leading to benzene formation. To investigate the different paths of benzene formation in counterflow methane, ethylene and propane flames, the detailed mechanism used by *Leung* et al. [27.5], for the prediction of the gas phase chemistry, was extended firstly to 47 species and 148 reversible reaction steps and subsequently to 60 species and 292 reversible reactions. In the latter case thorough consideration was given to isomerization reactions. The extension of the mechanism to include benzene was based on the work by *Miller* et al. [27.12], *Westmoreland* et al. [27.13] and the CEC Data Evaluation Group [27.14] among others. Full details are available elsewhere [27.11]. Thermodynamic data was obtained from the CHEMKIN [27.15] and *Burcat* [27.16] data bases.

Particular attention was given to the modelling of possible benzene formation paths via reactions of $C_2H_2, C_3H_2, C_3H_3, nC_4H_3$ and nC_4H_5. Reactions via iC_4H_3 and iC_4H_5 where not considered. The potentially most important benzene formation paths considered in the present study are as follows,

$$C_3H_3 + C_3H_2 \rightleftharpoons C_6H_5 \qquad (27.14)$$
$$C_3H_3 + C_3H_3 \rightleftharpoons C_6H_5 + H \qquad (27.15)$$
$$nC_4H_3 + C_2H_2 \rightleftharpoons C_6H_5 \qquad (27.16)$$
$$nC_4H_5 + C_2H_2 \rightleftharpoons C_6H_6 + H \qquad (27.17)$$

Reaction paths leading to C_3 species, such as the propargyl radical, has been extensively studied in alkane diffusion flames by *Leung* et al. [27.11] and in acetylene diffusion flames by *Lindstedt* et al. [27.17]. The potential importance of these species is obvious from reactions (27.14) and (27.15) above.

The differences in predictions of minor C_3 species for ethylene and propane flames can be clearly seen in Fig. 27.1. and Fig. 27.2. The most obvious difference is in the level of allene (AC_3H_4), but differences are also noticeable in C_3H_3 and C_3H_2 profiles. Regarding the C_4 route to benzene the predicted concentrations of the nC_4H_3 and iC_4H_3 isomers can be found in Fig. 27.3. As can be clearly seen concentrations of the nC_4H_3 isomer are strongly depleted by isomerization and benzene formation reactions. For both of these flames the dominant path for benzene formation with the current reaction mechanism passes via reaction (27.15). The reaction of C_3H_2 with C_3H_3 shows a complex behaviour with a large reverse rate in the high temperature flame zone where it is a major destruction path for the phenyl radical (C_6H_5). The most important benzene consumption path is via hydrogen abstraction.

$$C_6H_6 + H \rightleftharpoons C_6H_5 + H_2 \qquad (27.18)$$

Reactions (27.14), (27.15) and (27.18) largely control the benzene formation characteristics of the current diffusion flames as reactions (27.16) and (27.17) are significantly slower.

The effect of additional reaction paths for propargyl radical formation in the propane flames can be clearly seen from a comparison of acetylene and benzene profiles plotted in mixture fraction space in Fig. 27.4. where the results from two strongly oxygen enriched propane and ethylene flames are shown. While the acetylene profiles are similar and differ only in magnitude, the benzene profile for propane is extended in mixture fraction space towards the high temperature region of the flame. As soot nucleation is associated with a comparatively high activation temperature this feature is important for sooting characteristics.

To gain further confidence in the modelling of benzene formation a counterflow methane-air flame with a strain rate of 15 s^{-1} was computed. Two computations were performed using the reaction scheme by *Leung* et al. [27.11]. The first computation used the full scheme while the second computation excluded reactions (27.14) and (27.15) above. The latter computation was made to evaluate the relative importance within the scheme of alternative reaction paths to benzene formation. The results from these computations are here compared with the measurements by *Smyth* et al. [27.8] obtained 9 mm downstream in a coflowing methane-air flame on a Wolfhard-Parker burner. The benzene data obtained by *Smyth* et al. [27.8] has recently been re-evaluated [27.27] and the new values are used here. To enable a comparison with experimental data the computations were transformed to mixture fraction space. A similar comparison has been made by

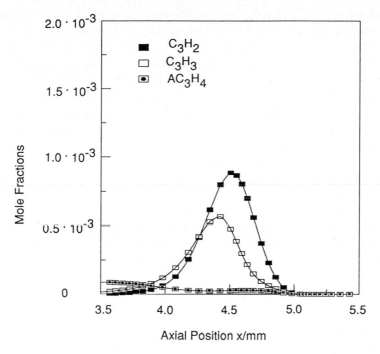

Fig. 27.1. Minor C_3 species in a counterflow C_2H_4 flame with an oxidant stream having an O_2/N_2 ratio of 28/72

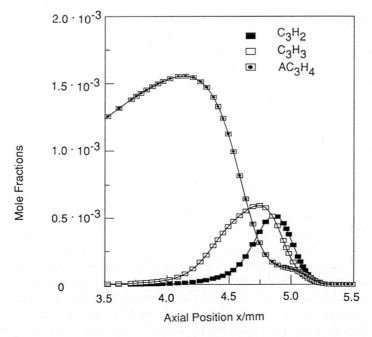

Fig. 27.2. Minor C_3 species in a counterflow C_3H_8 flame with an oxidant stream having an O_2/N_2 ratio of 28/72

Seshadri et al. [27.28] who used a different kinetic scheme. The results obtained with the present scheme are shown in Fig. 27.5. where they have been plotted in mixture fraction space using the definition proposed by *Bilger* [27.18]. The agreement must be regarded as very satisfactory for the case obtained with the full reaction scheme and adds confidence in the ability of the mechanism to provide quantitative predictions of benzene formation in diffusion flames. Also shown in Fig. 27.5. is the resulting benzene profile in the absence of the propargyl reaction paths via reactions (27.14) and (27.15) but including benzene formation via reactions (27.16) and (27.17). It can be clearly seen that for this latter case the obtained result is not in agreement with measurements. Furthermore, the discrepancies are sufficiently large to cause significant concern about to past predictions of soot formation in C_1 and C_2 flames based exclusively on these reaction paths. The predicted results are also significantly closer to those measured experimentally than the predictions by *Seshadri* et al. [27.28] who recorded peak mole fractions of around $9.5 \cdot 10^{-4}$ and a large shift in the profile towards richer mixtures. While it is recognised that there are currently considerable uncertainties concerning the formation paths of benzene in flames, the current reaction scheme is almost certainly sufficiently accurate to permit the intended sensitivity analysis of the nucleation step.

It should be noted that expressions (27.10) and (27.11) assume that the incipient soot particles do not contain hydrogen. This is a poor approximation and one of the problems encountered in the simplified modelling of nucleation processes is that newly formed particles appear to display a significantly higher reactivity than older particles [27.5, 7]. To partly solve this problem the reaction rate for the nucleation step has been chosen to represent both the processes associated with the formation of incipient soot particles as well as initial mass growth. The assumption has been made that particles arc formed containing a certain minimum number of carbon atoms $n_{C\,min}$. This assumption results in a source term in the number density equation which may be written as,

$$r_{27.10} = 2k_{27.10}(T)[C_2H_2]\frac{N_A}{n_{C\,min}} \qquad (27.19)$$

$$r_{27.11} = 6k_{27.11}(T)[C_6H_6]\frac{N_A}{n_{C\,min}} \qquad (27.20)$$

where N_A is Avogadros Number $6.022 \cdot 10^{23}$ mol^{-1}. In the present work it has been assumed that ($n_{C\,min}$) is equal to a C-60 shell which is the most abundant of the lower PAH ions [27.19]. This assumption gives an initial particle size of 1 nm. The approximation is not too serious as it can readily be shown that predictions in diffusion flames are relatively insensitive to the initial particle size provided it is kept below 10 nm. Many different activation energies for this process have been evaluated for a range of conditions [27.5] and it was found that good predictions could be obtained with a value

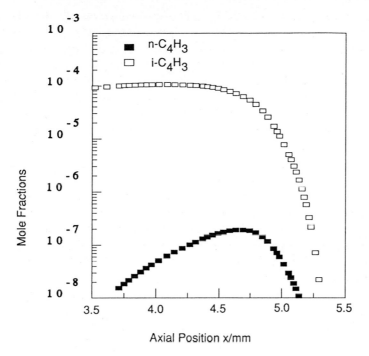

Fig. 27.3. C_4H_3 isomers in a counterflow C_2H_4 flame with an oxidant stream having an O_2/N_2 ratio of 28/72

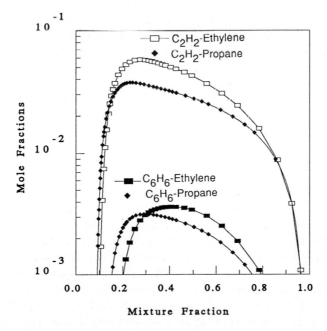

Fig. 27.4. C_2H_2 and C_6H_6 profiles in mixture fraction space in a counterflow C_2H_4 flame with an oxidant stream having an O_2/N_2 ratio of 28/72

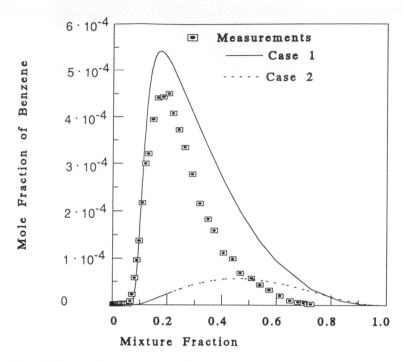

Fig. 27.5. C_6H_6 profiles in mixture fraction space for a coflow CH_4 - air flame; measurements from *Smyth* [27.27]; case 1 includes reactions (27.14) and (27.15) and case 2 excludes these reactions

of E/R of around 21000 K. A further discussion can be found in reference [27.5]. The pre-exponential factor was determined in both cases by the computation of an ethylene flame with an oxidant stream of 22% oxygen and 78% nitrogen.

27.4 Soot Mass Growth

The second reaction step, which is mainly responsible for the increase in soot mass, is assumed to be surface growth due to the adsorption of acetylene on the surface of soot particles. *Harris* et al. [27.21] e.g. have suggested, based on extensive measurements, that the soot mass growth is approximately first order in acetylene concentration. This approximation has been retained throughout this work. Directly relevant estimates of the activation energy have been made by *Vandsburger* et al. [27.7], who determined a value of the activation temperature (E/R) of around 12100 K in counterflow diffusion flames. Many other studies, e.g. *Bockhorn* et al. [27.23], have suggested similar values based on measurements in premixed flames. A large number of computations have been performed using this value and it has been found to describe the temperature dependence of the surface growth step well. The pre-exponential factor was determined by the solution of the complete equation set for the same flame used in the determination of the constant for the nucleation step as outlined above. Once the reaction rate constant had been determined in this way no further adjustments were made.

The dependence of the soot mass growth reaction step on the surface area of soot particles is less clear and many different suggestions have been made e.g. [27.2-5, 21, 22, 29]. To further investigate the behaviour in diffusion flames of some possible ideas a number of idealised cases are considered. These may be written in the following form,

$$C_2H_2 \rightleftharpoons 2C_s + H_2 \tag{27.21}$$

with the reaction rate,

$$r_{27.21} = k_{27.21}(T)f(A_s)[C_2H_2] \tag{27.22}$$

where $f(A_s)$ is a function of the total surface area (m^2/m^3-mixture). Particle related properties are obtained by the assumption of spherical particles. In the equations below the subscript s indicates soot, n the number of particles (particles/kg-mixture), w a mass fraction, ρ the density (kg m^{-3}), d_p the particle diameter (m) and a_p the surface area of an individual particle (m^2).

$$d_p = \left\{ \left[\frac{6}{\pi}\right] \left[\frac{\rho}{\rho_s}\right] \left[\frac{w_s}{\rho n}\right] \right\}^{\frac{1}{3}} \tag{27.23}$$

$$a_p = \pi d_p{}^2 \tag{27.24}$$

$$A_s = \pi d_p{}^2 [\rho n] \tag{27.25}$$

The density of soot is here assumed to be 2000 kg m^{-3}.

Four idealised approximations of the behaviour of the function $f(A_s)$ have been considered. Case (27.34) follows the suggestion by *Frenklach et al.* [27.4] that the reactivity is proportional to the available surface area and that the effects of surface chemistry can be accounted for by the use of a steady-state approximation for the hydrogen-abstraction/acetylene-addition (*HACA*) sequence. Following these authors the *HACA* sequence is here based on the properties of the first aromatic ring (benzene) and is defined as,

$$C_6H_6 + H \rightleftharpoons C_6H_5 + H_2 \tag{27.18}$$

$$C_6H_5 + H \leftarrow C_6H_6 \tag{27.26}$$

$$C_6H_5 + C_2H_2 \rightarrow C_8H_6 + H \tag{27.27}$$

$$C_6H_5 + O_2 \rightarrow products \tag{27.28}$$

where only reaction (27.18) is treated as reversible. The reaction rate constants used for all of the above gaseous reactions can be found in Tab. 27.1.

A truncated steady state approximation for the phenyl radical considering only these reactions leads directly to,

$$L([C_6H_5]) = w_{27.18f} - w_{27.18b} - w_{27.26b} - w_{27.27f} - w_{27.28f} \tag{27.29}$$

with the resulting expression,

$$[C_6H_5] = \frac{k_{27.18f}[C_6H_6][H]}{k_{27.18b}[H_2] + k_{27.26b}[H] + k_{27.27f}[C_2H_2] + k_{27.28f}[O_2]} \tag{27.30}$$

The above expression can now be inserted into reaction (27.27) to obtain a modified expression for acetylene adsorption on soot particles. However, Frenklach et al. [27.4] introduce an additional simplification by postulating a relationship between benzene and the number of active sites and further assumes that the latter is constant. The value of the number of active sites per unit area was chosen as $\chi_{s-h} = 2.32 \cdot 10^{19}$ sites m^{-2} [27.4]. For dimensional correctness these assumptions imply,

$$[C_6H_6] = \alpha \frac{\chi_{s-h}}{N_A} A_s \tag{27.31}$$

where α is a proportionality constant (particle/sites). Introducing these simplifications gives rise to a rate expression for case (27.34) which may be written as,

$$r_{27.34} = k_{27.27f} \chi_s A_s [C_2H_2] \tag{27.32}$$

where,

$$\chi_s = \frac{\alpha \, k_{27.18f}[H]}{k_{27.18b}[H_2] + k_{27.26b}[H] + k_{27.27f}[C_2H_2] + k_{27.28f}[O_2]} \frac{\chi_{s-h}}{N_A} \tag{27.33}$$

The adjustable constant α is here assigned the value 1.0. Despite the assumptions involved in the derivation, the above expression is interesting as it permits a first approximation to effects of surface chemistry.

Table 27.1. Gaseous reaction rate constants as $A T^b e^{-\frac{E}{RT}}$; units are K, kmol, m^3 and s

Reaction	A	b	E/R
(27.14)	$0.50 \cdot 10^{11}$	0	0
(27.15)	$0.10 \cdot 10^{11}$	0	0
(27.16)	0.28	2.9	705
(27.17)	0.28	2.9	705
(27.18)	$0.30 \cdot 10^5$	2.0	2510
(27.16)	$0.93 \cdot 10^{15}$	0	53347
(27.27)	$0.40 \cdot 10^{10}$	0	3019
(27.28)	$0.10 \cdot 10^{11}$	0	0

The second case (27.35) is identical to case (27.34) but the influence of the *HACA* sequence is here removed. This assumption therefore states that the reactivity is simply proportional to the local surface area and the acetylene concentrations. The third approximation (27.36) is that the reaction rate is proportional to the number of particles but independent of the surface area. This approximation has some experimental support from measurements in premixed flames e.g. *Bockhorn* et al. [27.29]. The final assumption (27.37) is that of a constant particle number density. The latter converts the function $f(A_s)$ to scaling factor constant throughout the flame. The four alternative expressions may be written as,

$$f(A_s) = A_s \chi_s \tag{27.34}$$

$$f(A_s) = A_s \tag{27.35}$$

$$f(A_s) = \frac{A_s}{a_p} = [\rho n] \tag{27.36}$$

$$f(A_s) = [\rho n] = \text{constant} \tag{27.37}$$

The final approximation also has the consequence of removing one scalar (n) from the equation set and is therefore attractive in situations where the number of scalars needs to be minimised. The corresponding reaction rate constants $k_{27.34}$ to $k_{27.35}$ can be found in Tab. 27.2. along with the reaction rate constants used for the nucleation steps.

Table 27.2. Soot reaction rate constants as $A T^b e^{-\frac{E}{RT}}$; units are K, kmol, m^3 and s

Reaction	A	b	E/R
(27.11)	$0.63 \cdot 10^4$	0	21000
(27.12)	$0.75 \cdot 10^5$	0	21000
(27.34)	$0.40 \cdot 10^{10}$	0	3019
(27.35)	$0.75 \cdot 10^3$	0	12100
(27.36)	$0.10 \cdot 10^{-11}$	0	12100
(27.37)	$0.50 \cdot 10^5$	0	12100
(27.38)	$0.12 \cdot 10^6$	0.5	19800

For all of the computations a simple soot oxidation step [27.5] based on the work by *Lee* et al. [27.24] was used. In this expression the burnout rate was adjusted to conform with the maximum surface specific oxidation rate observed by *Garo* et al. [27.25] in laminar co-flowing methane-air flames. The reaction step can be written as,

$$C_s + \frac{1}{2}O_2 \rightleftharpoons CO \tag{27.38}$$

with the reaction rate

$$r_{27.38} = k_{27.38}(T)A_s[O_2]$$

This step is clearly approximate in as much as oxidation via other species, such as the OH radical, have been neglected and as the overlap of different species, for example O_2 and OH, with the soot formation regions in flames are different. The reaction step is, however, considered sufficiently accurate for the present purposes as soot oxidation is of less importance in counterflow flames due to the resulting flame structure.

The soot nucleation step also gives rise to the source term in the number density equation as outlined above. The decrease in particle number density is simply assumed to occur according to particle agglomeration which is modelled using the normal square dependence used by many other investigators e.g. *Kent* et al. [27.26]. It is well known that this expression does not accurately represent the evolution of the soot number density throughout the flame [27.5]. However, it does provide a reasonable approximation of particle related properties.

$$r_{27.40} = k_{27.40}(T)[C_s]^{\frac{1}{6}}[\rho n]^{\frac{11}{6}} \tag{27.40}$$

$$k_{27.40} = 2C_a \left\{\frac{6M_s}{\pi \rho_s}\right\}^{\frac{1}{6}} \left\{\frac{6k_B T}{\rho_s}\right\}^{\frac{1}{2}} \tag{27.41}$$

In the above expression M_s is the molecular weight of soot (12.011 kg kmol^{-1}), C_a the agglomeration constant assigned a value 9 and k_B the Boltzmann constant ($1.38 \cdot 10^{-23}\,\mathrm{J\,K^{-1}}$).

27.5 Results

To enable comparisons with measurements it is essential to consider heat loss effects from flames of this type. *Kennedy* et al. [27.2] implemented a heat loss term based on the assumption of an optically thin medium and considered radiation only from the formed soot but not from gaseous species. For the moderately sooting flames investigated in the present work this is not a good approximation as radiation from the gas phase contributes significantly [27.20]. Therefore a simpler approach has been adopted here. This is based on matching the experimental temperature profile by the introduction of a heat loss factor as suggested by *Moss* et al. [27.3] and the following expression has been used,

$$T = T_{ad} \left\{1 - \beta \left\{\frac{T_{ad}}{T_{max}}\right\}^4\right\} \tag{27.42}$$

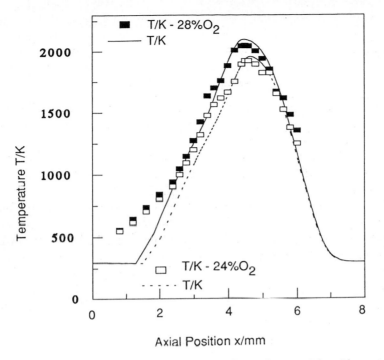

Fig. 27.6. Temperature profiles in counterflow C_3H_8 flames with oxidant streams having O_2/N_2 ratios of 24/76 and 28/72

where T_{ad} is the adiabatic temperature and T_{max} the maximum adiabatic temperature. A heat loss of 12% ($\beta = 0.12$) was found to give acceptable agreement with temperature profiles for all the ethylene flames measured by *Vandsburger* et al. [27.7]. For the propane flames a constant heat loss factor of 0.09 was used, resulting in acceptable agreement as can be seen in Fig. 27.6. The errors observed close to the burner indicate some degree of pre-heating of the fuel. The latter introduces an additional source of uncertainty in the predictions. To quantify the possible errors computations were performed with fuel temperatures up to 600 K. The results indicate that soot volume fractions can increase by up to 30% with this amount of pre-heat. However, as the intention of the present study is to assess the relative behaviour of simplified reaction steps this uncertainty is acceptable.

The counterflow flames investigated by *Vandsburger* et al. [27.7] provide a stern test as both peak temperatures and acetylene levels vary greatly with oxygen concentrations in the oxidant stream. For example the variations in peak temperatures are similar to, or in excess of, those encountered when initially unstrained counterflow alkane-air diffusion flames are strained to extinction.

The first set of computations were of three counterflow ethylene flames with oxidant streams consisting of oxygen/nitrogen ratios of 18/82, 22/78 and 28/72. The predictions obtained with the different modelling assumptions for soot mass growth, combined with a nucleation step based on acetylene, can be found in Tab. 27.3 where comparisons are made with measurements [27.7]. It can readily be seen that approximations based on the idealised expression for the total surface area (A_s) do not provide predictions which are in agreement with measured trends. Furthermore, the approximate account for surface chemistry effects by the use of the $HACA$ sequence does not improve agreement. In fact the predictions with this model gives errors larger than those of *any* other model and absolute errors in peak soot volume fractions are large. For example the predicted maximum soot volume fraction for the cooler flame is around $2.4 \cdot 10^{-8}$ compared to the measured value of $3.7 \cdot 10^{-7}$ and similar errors occur for other flames. Results of this kind are clearly unacceptable if the intention is to produce quantitative predictions of soot volume fraction for even modest variations in conditions. Predictions obtained without the $HACA$ modification are somewhat improved though still unacceptable. This indicates that the use of an idealised surface area in the rate expression for soot mass growth is unacceptable regardless of whether surface chemistry effects are taken into account or not. Recent studies of premixed flames [27.22] have also recorded similar findings for the expression where the effects of surface chemistry are not taken into account.

Proceeding with the cases outlined above it can be seen that the assumption that soot mass growth step is independent of surface area of the particles but dependent on the number of particles gives the best results of the present simple expressions. Errors are surprisingly small and typically less than 10-20%. These results are broadly in agreement with the trends obtained by *Bockhorn* et al. [27.29] who observed a similar behaviour in seeded premixed acetylene-air flames. The predicted soot volume fractions for case (27.36) can be seen in Fig. 27.7. for the three flames.

The introduction of a constant soot particle number density increases the maximum errors to around 40%. These errors are smaller in magnitude to those obtained with the models based on an idealised surface area ((27.34), (27.35)). This result is encouraging as it indicates that very simple soot models based only on the prediction of the evolution of the soot volume fraction can produce reasonable results. However, the range of applicability of such models will naturally be more restricted.

The fact that the approximate $HACA$ modification to expression (27.35) does not result in acceptable predictions is surprising in view of the apparent success in premixed flames [27.4]. However, the premixed flame predictions concerned very small soot particles with maximum sizes of around 5 nm. The present study considers significantly larger particles, up to around 70 nm, and correspondingly higher soot volume fractions. It is therefore quite possible that the flames studied by *Frenklach* et al. [27.4] are not dominated by soot mass growth to the same extent as the current flames. Some support for the latter is given below where it is shown that for a given fuel the sensitivity to the nucleation step is comparatively small for the current diffusion flames.

Table 27.3. Peak soot volume fractions obtained for a range of C_2H_4 flames with different soot mass growth steps; (27.34) proportional to surface area with *HACA* sequence modification; (27.35) proportional to surface area; (27.36) proportional to number of particles; (27.37) independent of particle characteristics; measurements from *Vandsburger* et al. [27.7]

Flame	(27.34)	(27.35)	(27.36)	(27.37)	[27.7]
18% O_2	$2.38 \cdot 10^{-8}$	$1.00 \cdot 10^{-7}$	$3.38 \cdot 10^{-7}$	$5.11 \cdot 10^{-7}$	$3.7 \cdot 10^{-7}$
22% O_2	$1.00 \cdot 10^{-6}$	$1.03 \cdot 10^{-6}$	$1.02 \cdot 10^{-6}$	$9.78 \cdot 10^{-7}$	$1.1 \cdot 10^{-6}$
28% O_2	$1.11 \cdot 10^{-5}$	$1.07 \cdot 10^{-5}$	$2.40 \cdot 10^{-6}$	$1.64 \cdot 10^{-6}$	$2.2 \cdot 10^{-6}$

It must also be pointed out that predictions of soot volume fractions are very sensitive to the value of the modelling constant α. For example a change in this constant by a factor two can result in changes in soot volume fraction predictions by a factor five or more. It would therefore appear essential to consider the effects of temperature and other parameters which may have an influence on α. The excellent results obtained with reaction step (27.34) for the ethylene flame with an oxygen/nitrogen ratio of 22/78 must therefore be viewed as quite fortuitous. The sensitivity of predictions to α also places quite unrealistic demands on the accuracy of the concept of relating soot particle characteristics to the phenyl radical and on the accuracy of the gas phase reaction rate constants used in the truncated steady state balance. The current rate expression used for reaction (27.27) is based on the value of the activation energy recommended by *Benson* [27.30] and the frequency factor was determined by ensuring that the proper rate, according to *Fahr* et al. [27.31], was obtained at 1100 K. If the unreasonable assumption is made that the reaction rate for this step is independent of temperature then improvements in prediction quality can be obtained. However, there is currently no kinetic justification for such an assumption.

The results obtained with reaction steps ((27.34), (27.35)) are not very encouraging as the errors in predicted soot volume fractions increase to in excess of an order of magnitude for the flames studied. However, it is possible that further refinements of the *HACA* sequence may yield improvements though it is arguably much more plausible that the core of the problem is in the simplistic assumption of soot mass growth occurring as a function of an idealised surface area based on the retention of a presumed spherical particle shape throughout the flame. If this is the case then the introduction of a size related shape factor may improve agreement significantly. Irrespective of the current difficulties the *HACA* sequence is an interesting idea which in principle is capable of incorporating surface chemistry effects.

The agreement between the model using a soot mass growth step independent of surface area (27.36) and the experimental data must be regarded as very good and the maximum errors in soot levels are generally low with the largest errors occurring for the most oxygen enriched flame. These discrepancies are sufficiently small to be attributable to uncertainties in the gas phase chemistry model and the simplified treatment of non-adiabaticity via a constant heat loss factor for all flames. Regarding the latter it may be more accurate to include radiation from the soot layer for the most highly oxygen enriched flames. However, if a nucleation step based on acetylene is used to compute propane flames with the most accurate surface growth model (27.36) then errors of around 50% in soot volume fractions occur. These errors are similar to those obtained by *Leung* et al. [27.5]. This is clearly not desirable and the problem appears difficult to remedy with improved soot mass growth descriptions alone. The most likely error is therefore in the very simplified description of soot nucleation.

The effects of introducing a nucleation step based on the formation of benzene is first assessed for the same three C_2H_4 flames discussed above. The results can be seen in Fig. 27.8. It is clear that while some further improvement in agreement is indeed obtained, the sensitivity to the choice of indicative species in the nucleation step is small for a particular fuel. However, this is not the case when additional reaction paths appear to benzene and PAH formation – as is the case when the fuel is changed beyond C_2 hydrocarbons. To evaluate the consequences of a change of the indicative species from acetylene to benzene computations were made also for the case of two oxygen enriched propane flames. These computations were made with the same set of reaction rate constants used for the ethylene flames. The propane flames have oxygen/nitrogen ratios in the oxidant streams of 24/76 and 28/72. The results can be found in Fig. 27.9.

It can be noted that the agreement between measurements and predictions is surprisingly satisfactory and the discrepancies are close to those observed experimentally. It is consequently clear that the introduction of a nucleation step based on benzene significantly improves the generality of the simplified soot models. This is not surprising. However, the level of generality obtained for these flames is quite extraordinary considering the simplicity of the soot formation steps.

It is evident from plots of soot particle sizes, see Fig. 27.10., that the simplified agglomeration expression used does not represent the process accurately over the entire range of conditions. However, these errors do not appear to result in large discrepancies in soot volume fractions. Some of the discrepancies between measured particle sizes and number densities and those computed may also be due to experimental uncertainties which are

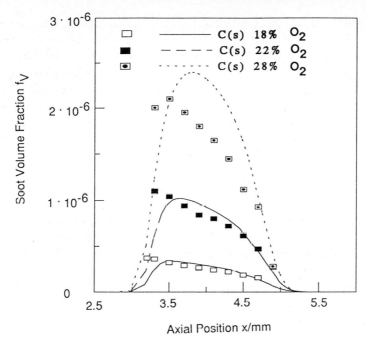

Fig. 27.7. Soot volume fractions in three counterflow C_2H_4 flames with oxidant stream ratios for O_2/N_2 of 18/82, 22/78 and 28/72; predictions based on nucleation step (27.10) and surface growth step (27.36); measurements from *Vandsburger* et al. [27.7]

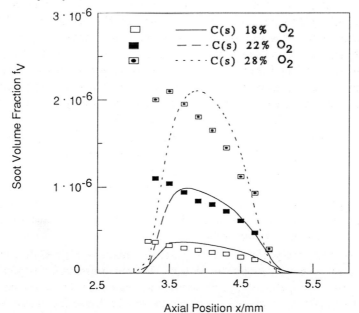

Fig. 27.8. Soot volume fractions in three counterflow C_2H_4 flames with oxidant stream ratios for O_2/N_2 of 18/82, 22/78 and 28/72; predictions based on nucleation step (27.11) and surface growth step (27.36); measurements from *Vandsburger* et al. [27.7]

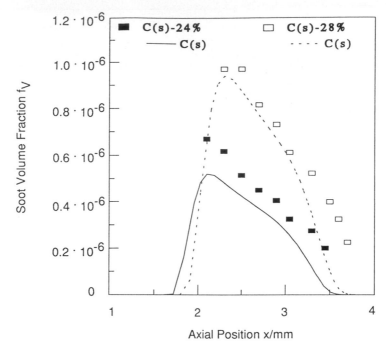

Fig. 27.9. Soot volume fractions in two counterflow C_3H_8 flames with oxidant stream ratios O_2/N_2 of 24/76 and 28/72; nucleation step based on C_6H_6; measurements from *Vandsburger* et al. [27.7]

considerable. The measurements indicate that freshly formed particles have a significantly higher rate of agglomeration than older particles and it is clear that older particles in the cold part of the flame essentially do not grow. However, the agglomeration model does yield acceptable results both qualitatively and quantitatively for particle sizes up to around 70 nm. At later stages the simplified rate expression appears to exaggerate the rate of growth, though it should be noted that this occurs close to the stagnation point and at flame temperatures below 900 K. A suitable modification to this simplified rate expression would be to take into account the 'stickiness' of the particles. This factor could in principle be related to the surface chemistry of the particle, though preliminary computation using the *HACA* sequence for this purpose are *not* encouraging. To solve this problem satisfactorily more rigorous methods such as the application of moment based methods [27.4] may be required. This is, however, outside the scope of the present work.

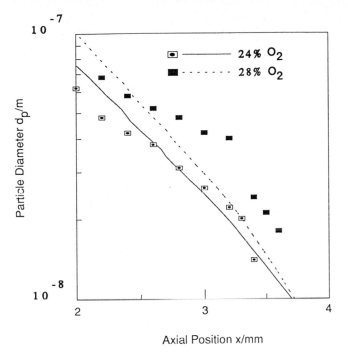

Fig. 27.10. Particle sizes in two counterflow C_3H_8 flames with oxidant stream ratios O_2/N_2 of 24/76 and 28/72; nucleation step based on C_2H_2; measurements from *Vandsburger* et al. [27.7]

27.6 Summary

Simplified reaction steps for soot formation have been proposed and tested for a wide range of counterflow ethylene and propane flames. The models are based on the use of characteristic pyrolysis products, benzene and acetylene, to link the gas phase chemistry to the soot formation steps. The soot formation models typically require solution of conservation equations for the soot mass-fraction and number density and are closed by the assumption of a spherical particle shape.

Benzene formation has been analysed in $C_1 - C_3$ diffusion flames and it has been shown that the dominant reaction path leading to benzene involves the propargyl radical and that reaction paths via C_4 species are unimportant by comparison. It has also been shown that in diffusion flames soot mass growth models based on the use of idealised surface area expressions do not reproduce measured trends. It would therefore be more appropriate to discuss mass addition to soot particles in terms of soot mass growth rather than surface growth.

The agreement obtained with a model based on benzene for the nucleation step and which has a soot mass growth step independent of surface area, but dependent on the number of particles, is very encouraging. The accuracy in predictions of soot volume fractions with this model is approaching the accuracy obtained for the prediction of gaseous species. Predictions of other properties such as particle number densities and aggregate sizes are less satisfactory, though experimental uncertainties in these properties are also considerable.

However, many issues remain which require attention. Among these are: (i) it has been shown that effects of surface chemistry even when accounted for in a very approximate manner can be large and this requires further attention. (ii) The current model is limited to comparatively high temperature flames where acetylene, not aromatic species, form the most likely growth species. (iii) The effects of soot oxidation must be accounted for more accurately and (iv) the expression used for the change in particle numbers throughout the flame has been shown to serve only as a first approximation. Finally, further validation work is essential for different flame geometries and combustion conditions.

27.7 Acknowledgement

The author wishes to gratefully acknowledge the financial support of the Ministry of Defence at RAE Pyestock and British Gas plc for parts of this work and Mr K.M. Leung for performing some of the computations. The data sets provided by Dr. K.C. Smyth on measurements in coflowing methane flames are also most gratefully acknowledged

References

27.1   P.A. Tesner, T.D. Snegiriova, V.G. Knorre: Combust. Flame **17**, 253 (1971)

27.2   I. Kennedy, W. Kollmann, Y. Chen: AIAA Journal **29 (9)**, 1452 (1991)

27.3   J.B. Moss, C.D. Stewart, K.J. Syed: "Flowfield Modelling of Soot Formation at Elevated Pressure", in Twenty-Second Symposium (International) on Combustion (The Combustion Intitute, Pittsburgh 1988) p. 413

27.4   M. Frenklach, H. Wang: "Detailed Modelling of Soot Particle Nucleation and Growth", in Twenty-Third Symposium (International) on Combustion (The Combustion Institute, Pittsburgh 1990) p. 1559

27.5   K.M. Leung, R.P. Lindstedt, W.P. Jones: Combust. Flame **87**, 289 (1991)

27.6   R.P. Lindstedt: "A Simple Reaction Mechanism for Soot Formation in Non-Premixed Flames", in Proceedings of IUTAM Symposium, Taipei, Taiwan, June 1991, ed. by R.S.L. Lee, J.H. Whitelaw and T.S. Wang, (Springer-Verlag, New York 1992) p. 145

27.7   U. Vandsburger, I. Kennedy, I. Glassman: Combust. Sci. Technol. **39**, 263 (1984)

27.8   K.C. Smyth, J.H. Miller, R.C. Dorfmann, W.G. Mallard, R.J. Santoro: Combust. Flame **62**, 157 (1985)

27.9   N. Peters, R.J. Kee: Combust. Flame **68**, 17 (1987)

27.10 W.P. Jones, R.P. Lindstedt: Combust. Sci. Technol. **61**, 31 (1988)

27.11 K.M. Leung, R.P. Lindstedt: Combust. Flame., in press

27.12 J.A. Miller, C.F Melius: Combust. Flame **91**, 21 (1992)

27.13 P.R. Westmoreland, A.M. Dean, J.B. Howard, J.P. Longwell: J. Phys. Chem. **93**, 8171 (1989)

27.14 D.L. Baulch, C.J. Cobus, R.A. Cox, L. Esser, P. Frank, T. Just, J.A. Kerr, M.J. Pilling, J. Troe, R.W. Walter, J. Warnatz: J. Phys. Chem. Ref. Data **21**, 411 (1992)

27.15 R.J. Kee, R.J. Rupley, J.A. Miller: "The CHEMKIN Thermodynamic Data Base", Sandia Report SAND87-8215, April 1987

27.16 A. Burcat: "Thermochemical Data for Combustion Calculations", in <u>Combustion Chemistry</u>, ed. by W.C. Gardiner, (Springer-Verlag, New York 1984) p. 455

27.17 R.P. Lindstedt, F. Mauß: "Reduced Kinetic Mechanisms for Acetylene Diffusion Flames", in <u>Reduced Mechanisms for Application in Combustion Systems</u>, ed. by N. Peters and B. Rogg, (Springer-Verlag, Berlin 1992), p. 241

27.18 R. Bilger, S.H. Stårner, R.J. Kee: Combust. Flame **80**, 135 (1990)

27.19 S. Löffler, K.H. Homann: "Large Ions in Premixed Benzene-Oxygen Flames", in <u>Twenty-Third Symposium (International) on Combustion</u> (The Combustion Institute, Pittsburgh 1990) p. 355

27.20 M. Fairweather, W.P. Jones, R.P. Lindstedt: Combust. Flame **89**, 45 (1992)

27.21 S.J. Harris, A.M. Weiner: Combust. Sci. Technol. **32**, 267 (1983)

27.22 S.J. Harris: Comb. Sci. and Tech. **72**, 67 (1990)

27.23 H. Bockhorn, F. Fetting, V. Meyer, G. Reck, G. Wannemacher: "Measurement of the Soot Concentration and Soot Particle Size in Propane-Oxygen Flames", in <u>Eighteenth Symposium (International) on Combustion</u> (The Combustion Institute, Pittsburgh 1981) p. 1137

27.24 K.B. Lee, M.W. Thring, J.M. Beer: Combust. Flame **6**, 137 (1962)

27.25 A. Garo, G. Prado, J. Lahaye: Combust. Flame **79**, 226 (1990)

27.26 J.H. Kent, H.Gg. Wagner: Combust. Flame **47**, 53 (1982)

27.27 K.C. Smyth: private communication, October 1991

27.28 K. Seshadri, F. Mauss, F. Peters, J. Warnatz: "A Flamelet Calculation of Benzene Formation in Coflowing Laminar Diffusion Flames", in <u>Twenty-Third Symposium (International) on Combustion</u> (The Combustion Institute, Pittsburgh 1991) p. 559

27.29 U. Wieschnowsky, H. Bockhorn, F. Fetting: "Some New Observations Concerning the Mass Growth of Soot in Premixed Hydrocarbon-Oxygen Flames", in <u>Twenty-Second Symposium (International) on Combustion</u> (The Combustion Institute, Pittsburgh 1988) p. 343

27.30 S.W. Benson: Int. J. Chem. Kin. **21**, 233 (1989)

27.31 A. Fahr, S.E. Stein: "Reactions of Vinyl and Phenyl Radicals with Ethyne, Ethene and Benzene", in <u>Twenty-Second Symposium (International) on Combustion</u> (The Combustion Institute, Pittsburgh 1989) p. 1023

Discussion

Santoro: Do you investigate a counter flow diffusion flame and do you use a kinetic model that is going through the stagnation plane from the fuel side to the oxygen side of the flame?

Lindstedt: We consider a standard counter flow diffusion flame geometry. The detailed kinetic mechanism incorporates in the benzene case 47 to 60 species and up to 242 reaction steps. The flames are modelled with an energy equation.

Santoro: In the comparison in the table that you were showing us between the various models was that the rate of change of the volume fraction or was that a rate constant?

Lindstedt: I was comparing the peak soot volume fractions predicted using several simplified surface growth steps.

Santoro: It appears then to me that the critical thing in your model is the surface growth part, because you seem to be able to change the nucleation expression widely without affecting what was going on and you seem to be able to change the kinetics that lead to the first particle over a fair range. So I think that it is the surface growth part that is critical in getting the right numbers. What confidence level do you have in saying that you correctly modelled nucleation in the mechanism. It seems to me that you are almost insensitive to those parameters.

Lindstedt: Yes, it is true to some extent that the nucleation step in terms of mass growth is always the less important than the mass growth by C_2H_2 addition. But the nucleation step related to benzene becomes more general if you go to higher hydrocarbons say propene or something of that kind. If you do not use benzene as an indicative species of the first ring formation or some subsequent PAH then you will not predict the correct nucleation rate. Whereas for C_1 and C_2 hydrocarbons it is less important.

Kennedy: What did you mean by the correct nucleation rate because it is not something we can measure. Do you have coagulation occuring in there?

Lindstedt: Correct this is the wrong word. What I mean is that it should occur in the correct part of the flame structure and that the species should be indicative of changes in relevant concentrations in absolute terms.

Frenklach: Firstly, to what degree can you trust your conclusion if you don't use a physical nucleation rate and, secondly, what source for your data, thermodynamics and rate coefficients, did you use for the kinetics?

Lindstedt: The source of the kinetic data was predominantly the CEC data base and the thermodynamic data came predominantly from the CHEMKIN chemical data base.

Dobbins: You said the number concentration was proportional to the eleven sixth power for the coagulation rate. And that expression comes from a self preserving size distribution function which I don't understand being used in the nucleation or the particle inception stage. I wonder whether you have something more to justify the use of that expression at that point.

Lindstedt: The justification for this in the nucleation part of a flame is as we have pointed out obviously suspect. But I have to bear in mind that we are trying to work with a simplified model and this expression has been used in the past on a global basis. It gives reasonable agreement even in the early region although the foundation of it may be suspect.

Dobbins: I wish to raise a personal protest over the excessive use of the self preserving distribution results under circumstances when it clearly does not apply unless one adds some other detailed justification.

Kent: I think maybe in that configuration you fortunately may have got agreement which you wouldn't get in a more general case of a co-flow diffusion flame. When I tried simple types of mass growth functions based on K. Smith's measurements of acetylene and benzene in these flames, the result was that you get large amounts of soot growth in very early parts of the flame down below and you can never manage to sustain it in the upper parts of the flame.

Lindstedt: We have computed co-flow methane and ethylene flames and the agreement obtained is very reasonable indeed.

Successes and Uncertainties in Modeling Soot Formation in Laminar, Premixed Flames

Meredith B. Colket, Robert J. Hall

United Technologies Research Center,
E. Hartford,
CT 06108, USA

Abstract: A model for soot formation in laminar, premixed flames is presented. The analysis is based on a simplified inception model, detailed kinetic calculations of soot surface growth, and coalescing particle collisions. A sectional aerosol dynamics algorithm which involves solving a master equation set for the densities of different particle size classes provides an efficient solution scheme. The calculation of surface growth and coalescence sectional coefficients has been simplified and extended to the entire temperature range of interest in flame simulations. In order to test convergence properties, the former geometric limitation on the number of size classes has been relaxed. Convergence of the soot volume fraction typically requires only a few size classes and balance equations. Several possible soot surface growth models have been compared. The inception and surface growth models require profiles of temperature and important species like benzene, acetylene, and hydrogen atoms, and oxidizing species. Extensive comparisons have been made with well-characterized flame data by using experimental temperature profiles and calculating the concentrations of the important species with a burner code. The calculated species concentrations and surface growth/oxidation rates are input to the aerosol dynamics program, which calculates the evolution of various soot size and density parameters. While aspects of the model are highly simplified, on balance it appears to give agreement with experiment that is comparable to that obtained from more elaborate models. The calculated sensitivity of soot growth to temperature and the important inception and coalescence parameters is discussed.

28.1 Introduction

The importance of soot production on pollution is well documented. In addition, soot formation can dramatically effect flame phenomena. The most obvious and acknowledged flame process affected by soot is radiation which can lead to substantial fractions of energy lost from the immediate flame environment [28.1-2]. Twenty percent energy loss due to radiation is not atypical for a coflowing sooting diffusion flame [28.3]. This energy loss leads directly to a temperature reduction (and therefore affects kinetics of pollution formation), a change in density (and local gas velocity), as well as changes in flame length.

A secondary and often unrecognized phenomenon is the effect on flame thermochemistry and kinetics. The thermochemistry of soot formation and oxidation can dramatically affect the heat release profile in a flame. The conversion of alkanes, for example, to acetylene (a principal soot intermediate) is nearly one-fifth as endothermic as the oxidation process is exothermic. Thus, a strongly endothermic process can occur just inside the flame front to form soot. The soot may be transported to another region and oxidized. The oxidation rates of soot particles are dramatically different than those of gas-phase species. Local flame temperatures may be lowered in order to provide energy to drive the endothermic soot forming reactions. This process is also strongly dependent on fuel-type and often has not been considered by researchers addressing fuel-type effects on soot formation. The importance of the thermochemistry of fuel components is dramatically demonstrated in a recent analysis [28.4]. Their model indicates the existence of a region of negative heat release just below the apex of a coflowing diffusion flame! This effect is attributed principally to the strongly endothermic reaction, $2 \, CH_4 \rightleftharpoons C_2H_2 + 3 \, H_2$.

These flame effects are often so dominant that the modeling of practical (soot-containing) flames will be highly inaccurate without predictions of soot formation as part of the flame modeling process. Further support for modeling efforts comes from the more commonly recognized deleterious effects of soot formation, such as pollution, hardware lifetimes, and plume visibility.

Consequently, over the past several years a procedure for calculating soot production in one-dimensional laminar, premixed flames has been developed. This procedure has similarities with techniques developed at other laboratories [28.5-7]. It is hoped that this or similar procedure could be used for predicting soot in more complex flame systems. Although there are limitations and uncertainties with the developed code, the procedure is quite versatile and is quite successful in predicting soot production from several different laboratory flames. A detailed description for most of the calculation procedure is provided in this manuscript as well as a discussion of the principal assumptions, successes, and uncertainties. The formulation and modifications to the MAEROS code [28.8] as used in this soot model are described elsewhere [28.9]. Despite uncertainties, the code in its present form is a useful tool for evaluating the controlling soot formation processes in different flames and for providing valuable understanding to the many interrelated and competitive processes of soot formation. Examination of this soot model and its predictions leads to some increased understanding of the overall successes and uncertainties in soot formation modeling.

28.2 Brief Description of Model

A flow chart of the model used to describe soot formation processes is provided in Fig. 28.1. Dashed lines are used for portions of the code which have not yet been implemented. The model couples detailed chemical kinetics calculations of gas-phase processes with mechanisms for particle surface growth, oxidation, and agglomeration. MAEROS, a widely-used aerosol dynamics code [28.8], has been modified for the latter part of the analysis.

The soot growth/aerosol dynamics program is based on a sectional representation of the growth equations with provision for inception source terms, surface growth through condensible vapor deposition, and coagulation. The program has been modified in a number of ways for the soot growth problem. Its temperature range capabilities have been extended to the full range of interest in combustion problems by a reformulation of the surface growth and coalescence sectional coefficient calculations. Provision for oxidizing vapors (oxygen and hydroxyl radicals) has also been made. The simulations require as input profiles of temperature, aromatics (C_6H_6), condensible (C_2H_2) and oxidizing vapor concentrations (OH and O_2), and the concentrations of other flame species (such as H-atoms and H_2). The input of the aerosol dynamics program has been made compatible with the output of the Sandia premixed, laminar flame program [28.10-11]. The facility to exclude small mass spheroids from coagulative processes has also been added. The code also calculates the radiative energy loss from soot assuming optical thinness and neglecting soot refractive index dispersion.

In the present version of the model, the inception calculation makes use of the local benzene formation rate. Starting with benzene, the surface growth process alone is used to generate small mass, solid carbon spheroids. At a specified threshold mass, particle coalescence is allowed; the inception rate is taken to be the flux of particles through this threshold mass, beyond which it is assumed that one is describing soot particles. This model of inception is based on an extrapolation of particle growth dynamics and kinetics into the pre-particle regime; although benzene is not the inception species, the particle dynamics calculations begin at a particle mass equivalent to the carbon content of benzene, with a source rate derived from the benzene formation rate. This simple, provisional assumption about inception may be

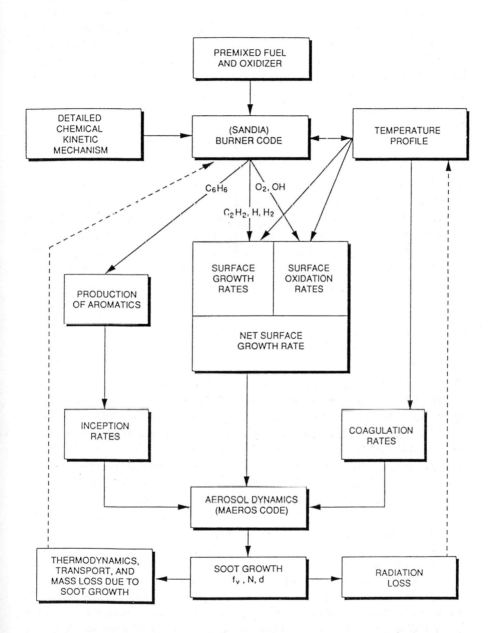

Fig. 28.1. A soot growth model for a premixed, laminar flame

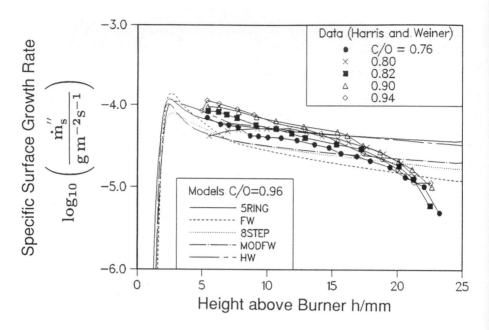

Fig. 28.2. Soot growth rate constant, ethylene flame, C/O=0.96

just as good as those presently obtainable from more elaborate calculations of the inception species, given the uncertainties presently associated with the latter. As additional information becomes available, this assumption may be modified as required. The sensitivity of the results to the inception species mass will be discussed later in the text.

Acetylene is assumed to be the surface growth species. The growth rate has been calculated using the Harris-Weiner [28.12] value, the Frenklach-Wang expression [28.5], as well as several other steady-state expressions. Local values of acetylene and hydrogen atom/molecule concentrations as well as estimates of rate coefficients for certain kinetic processes are required for these calculations. Oxidation is assumed to be given by the Nagle and Strickland-Constable expression [28.13], and oxidation by OH by a gas kinetic rate multiplied by a collision efficiency of 0.13 [28.14]. Recent studies have shown that soot mass also grows by the addition of polyaromatic hydrocarbons. These processes are included in the present model since low molecular weight polyaromatic hydrocarbons are included in the total soot mass (according to the model used in this study) and these species are allowed to grow by acetylene addition as well as by coagulation with other 'soot' particles.

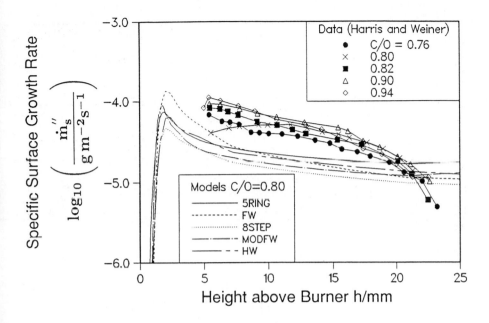

Fig. 28.3. Soot growth rate constant, ethylene flame, C/O=0.8

28.3 Modeling of Gas Phase Chemistry

In this investigation, extensive comparisons were made with the well-characterized, premixed, burner-stabilized flames examined by *Harris* et al. [28.12] and by *Bockhorn* et al. [28.15]. Concentrations of gas-phase species were calculated using detailed chemical kinetic models with CHEMKIN II [28.11] and the SANDIA premixed flame code [28.10]. The purpose of these calculations is to determine concentrations of gas-phase species as a function of height above the burner. These data in turn are used to calculate inception rates and specific growth rates. Thus far, experimentally determined temperature profiles have been used; in general, a code could calculate the temperature profile once radiation, thermodynamics of soot formation, and burner effects are included in the model. Results from the kinetic modeling efforts are described in the following paragraphs. Since benzene has been assumed to be the incepting particle in the present calculations, the discussion is focused on the predictions of benzene profiles.

For the atmospheric pressure, $C_2H_4/O_2/Ar$ flames examined by *Harris* et al. [28.12], the kinetic code of *Harris* et al. [28.16] was used with the addition of the reaction

$$C_3H_3 + C_3H_3 \leftrightarrow C_6H_5 + H \tag{28.1}$$

Recently, this and related reactions have been found to contribute significantly to benzene formation in rich flames. The rate constant for Reaction 1 was 10^{13} cm^{3}mole^{-1}sec^{-1} [28.17]. For the C/O = 0.92 flame, it contributes approximately 50% to the total formation of benzene.

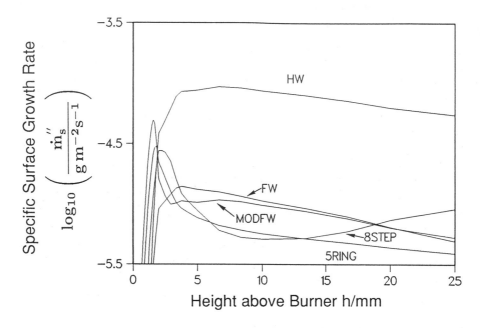

Fig. 28.4. Soot growth rate constant, acetylene flame

For the low pressure acetylene (12.159 kPa) and propane (15.20 kPa) flames studied [28.15], species profiles were calculated using a modified version of the kinetic code of *Miller* et al. [28.18] with propane kinetics from *Westbrook* et al. [28.19]. (Numerical problems were encountered with the Harris kinetics.) The same ring-forming reactions used to model the *Harris* et al. flames were also included in this kinetic sequence. A sensitivity analysis on benzene formation in the acetylene flame indicates that, besides chain branching and termination steps, its formation is principally dependent on reactions linked to the formation and destruction of C_3H_3. These results support the recent findings in several studies [28.17, 20-21] regarding the significance of C_3-species to benzene formation in flames. Our results which indicate the importance of Reaction 1 differ from the conclusions of *Frenklach* et al. [28.22]. It is likely that this difference is due in part to differences in the C_3-reaction set or related thermodynamics.

Calculated values for selected species are compared to experimental data in *Colket* et al. [28.9]. In general, agreement is reasonable except for a spatial misalignment with the *Bockhorn* et al. data [28.15]. As described in *Colket* et al. [28.9], the discrepancies are partly due to experimental uncertainties.

28.4 Simplified Inception Model: Justification and Motivation

The use of an incepting species derived from benzene, as explained previously, is justified by several facts. First of all, benzene has been found in this laboratory as well as by *Kern* et al. [28.23] to correlate directly with sooting tendencies for a variety of aliphatic hydrocarbons; secondly, while relatively small uncertainties exist in our ability to predict benzene concentrations (factor of two), large uncertainties exist in the prediction of multi-ringed aromatics (probably a factor of ten or more), so calculations of inception based on concentrations of these high molecular weight species are at the present time subject to large error; thirdly, results can be very sensitive to the selection of the incepting species and its selection appears to be somewhat arbitrary; and fourthly, the actual mechanism for inception is not yet known (although some good speculation is available). Consequently, we conclude that there are significant uncertainties in calculating the true inception rate and, therefore, starting our particle growth calculation with benzene and defining the inception species to correspond to that particle mass at which coalescence begins is as good as other uncertain alternatives. As will be seen, use of this assumption yields reasonable agreement with experimental data in many regards.

In addition to these justifications, there is a strong motivation for using a simplified inception process. Calculation times for solutions of flame systems increase dramatically as the number of flame species increase. This concern becomes even greater as multi-dimensional flames are considered. Already, the calculation time for solution of the gas-phase kinetics (many hours) by far dominates over the solution times of the aerosol dynamics (seconds to a couple of minutes on an IBM 486 pc).

28.5 Soot Growth Mechanisms

Soot growth rates were calculated using a variety of procedures, each of which exhibit different features. Two literature 'mechanisms' were used, i.e., the Harris and Weiner (HW) expression [28.12] with an activation energy of 31.8 kcal/mole and the Frenklach and Wang (FW) mechanism [28.5]. In addition, three other mechanisms (MODFW, 8STEP, 5RING) have been developed [28.5]. For the latter four mechanisms, steady-state assumptions are made for all intermediate 'species' and expressions for overall soot growth

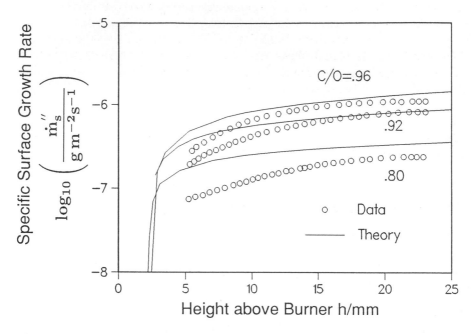

Fig. 28.5. Calculated soot volume fraction, Harris and Weiner flames, HW surface growth

rates were determined. In the latter three cases, rate constants were initially selected based on literature expressions but then adjusted (typically less than a factor of four) in order to provide better agreement with experimental data (that is, specific surface growth rates from *Harris* et al. [28.12] and the soot profiles from *Bockhorn* et al. [28.15]) . *Frenklach's* [28.5] approach of not altering rates from those which describe reactions for low molecular weight species is admirable. However, we justify our adjustments due to the fact that rate constants between low and high molecular weight species are not necessarily identical for similar processes (due to changes in molecular structures as well as reduced masses) and since a simplified sequence is used to describe what is probably a very complex process.

The soot growth rates do not employ the particle ageing equation [28.24] which reduces growth rates with increasing particle time. This process was not included in this study because of our initial acceptance of the arguments of *Frenklach* et al. [28.5] indicating that particle ageing was an artifact of the H-atom decay process. As described subsequently in this paper, this argument was found not to be consistent with experimental data. In future studies, equations for particle ageing will be included in the analysis; it can be expected that modifications to empirically derived rate constants used in this analysis will be necessary.

For simplification, only our modified Frenklach and Wang mechanism is described in this document. All mechanisms are described in *Colket* et al. [28.9]. As an alternative to the multiplying factor of 0.1 required to explain the high temperature flame data of Bockhorn and coworkers, we attempted (Frenklach and Wang also tried a similar approach) to include some reversibility in the acetylene addition process. The resulting mechanism (MODFW) is listed below.

Table 28.1. Modified version of the FW Soot Growth Mechanism (MODFW); activation energies are given in kcal mol^{-1}

	Reactions Considered	$\log_{10}(A_f)$	E_a^{for}	$\log_{10}(A_r)$	E_a^{rev}
1.	$H+C(s) \leftrightarrow \dot{C}(s)+H_2$	14.40	12	11.6	7.0
2.	$H+\dot{C}(s) \leftrightarrow C(s)$	14.34	-	17.3	109.0
3.	$\dot{C}(s) \rightarrow$ products $+ C_2H_2$	14.48	62	-	-
4.	$C_2H_2+\dot{C}(s) \leftrightarrow C(s)CH\dot{C}H$	12.30	4	13.7	38
5.	$C(s)CH\dot{C}H \rightarrow C'(s)+H$	10.70	-	-	-

The mechanism includes possible acetylene elimination from the soot radical (analogous to phenyl radical decomposition) and separates the acetylene addition process into a reversible formation of the radical adduct and a cyclization reaction. Assuming steady-state conditions for all intermediate species, the rate expression for soot mass growth is calculated to be

$$\frac{dm}{dt} = 2m_c \frac{(k_1[H] + k_{-2}[C_2H_2])k_4k_5\chi a_s}{(k_{-1}[H_2] + k_2[H] + k_3)k_{-4}k_5 + k_4k_5[C_2H_2]} \quad (28.2)$$

where m_c is the mass of a carbon atom, χ is a surface density of C_{soot}-H sites ($\approx 2.3 \times 10^{15}$ cm^{-2}, according to *Frenklach* et al. [28.5]) and a_s is the surface area. Rate constants used in this expression are listed in Table 28.1.

28.6 Predictions of Specific Surface Growth Rates

Specific surface growth rates, $\dot{m}_s''$, were calculated according to

$$\dot{m}_s'' = \frac{dm/dt}{a_s} \quad (28.3)$$

for each of the above expressions and by using the same value (2.3×10^{15} sites/cm^2) as derived by *Frenklach* et al. [28.5] for χ, the surface density of

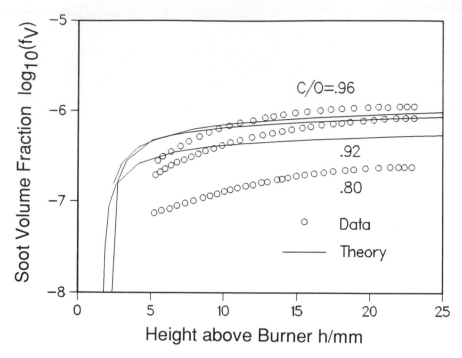

Fig. 28.6. Calculated soot volume fraction, Harris and Weiner flames, FW surface growth

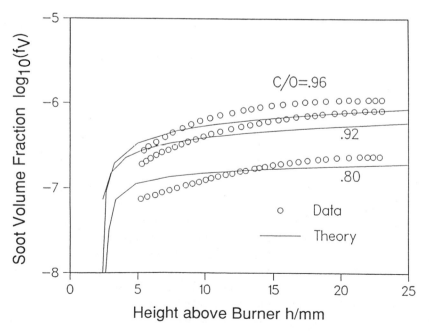

Fig. 28.7. Calculated soot volume fraction, Harris and Weiner flames, MODFW surface growth

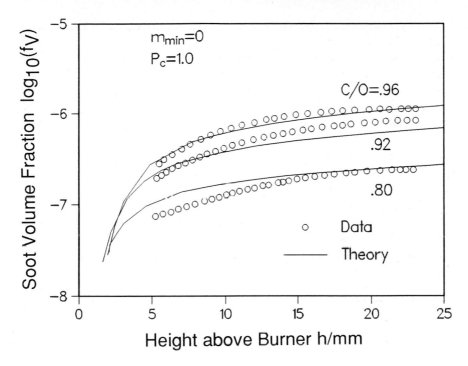

Fig. 28.8. Calculated soot volume fraction, Harris and Weiner flames, HW surface growth

C_{soot}-H sites. Comparisons of surface growth rates for three separate flames are shown in Figs. 28.2-4. The three flames include two of the (ethylene) *Harris* et al. flames [28.12] at differing stoichiometries and the acetylene flame examined by *Bockhorn* et al. [28.15]. For the acetylene flame, the Frenklach and Wang (FW) expression was calculated with the factor $\alpha = 0.1$. Net surface growth rates are reduced from the rates in these figures by the subtraction of oxidative terms. Oxidation has a negligible effect on the Harris and Weiner flames but affects the Bockhorn flame dramatically. The effect of oxidation will be discussed in more detail in the sensitivity section. *Harris* et al. measured soot growth rates for a variety of flames ($0.76 \leq$ C/O ≤ 0.94), and these data are included for comparison in Figs. 28.2 and 28.3. No data were presented for the C/O=0.96 flame, but we assume such data would lie slightly above the highest set of experimental data. Although the experimental data nearly collapses to a single curve, there is a noticeable stoichiometric dependence with the soot growth rates in the richer flames about a factor of two above those for the C/O=0.80 flame (except high in the post-flame zone where the data converge fairly well). The predicted curves tend to peak early and are all concave upwards, whereas the experimental data is concave downwards. The shape of these 'theoretical' curves is at least partially due to their dependence on H-atom concentrations and partially due to the decay in temperature. For the richer flames, all of the calculated soot growth rates predict the initial magnitude of the soot growth rate fairly well, although they all fall off too rapidly with increasing height above the burner. Furthermore, none of the models adequately describes the fall-off observed for 'older' soot particles. Consequently, we favor proposals which attribute decreasing soot growth rates with the particle ageing process. For the C/O=0.8 flame, the FW expression does not decrease by a factor of two from that for the C/O=0.96 flame (in fact the FW predictions for the two flames are nearly identical) as the experimental data indicates and leads to substantial overprediction of soot formation in this leaner flame. The failure in the mechanism is due to a low dependence on the acetylene concentration. The alternative mechanisms provide a slightly better description of the stoichiometric differences between flames. In fairness to the FW mechanism, the reader should be reminded that the rate constants used in the alternative mechanisms were 'fitted' in order to obtain the agreement.

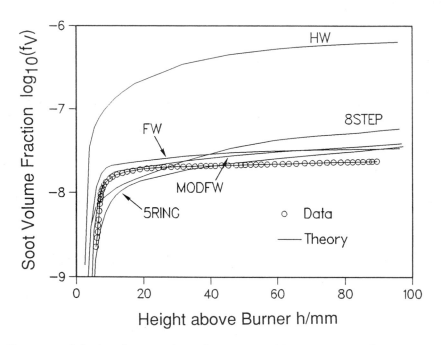

Fig. 28.9. Calculated soot volume fraction, Bockhorn acetylene flame, various growth models

Although relative stoichiometric dependence can be described, it is also obvious that none of the models predicts the absolute magnitude of the soot growth rate data. Yet, despite the low predicted values of soot growth rates, reasonable agreement between the experimental soot profiles and the modeling of soot production is obtained. The use of benzene production as a surrogate for the inception rate undoubtedly overstates the inception process. As a result, our present model favors the use of lower surface growth rates. In general, however, we do not feel that our model is far wrong because of its reasonable agreement with the results of Frenklach and Wang, who used a more elaborate scheme for simulating the inception process. In addition, uncertainties in χ as well as the rate constants, indicates that the predicted curves in Figs. 28.2-4 could easily shift up or down as much as a factor of two. We have not made such shifts in the present study since uncertainties still exist in the understanding of the inception process and since the fall-off in soot growth rates in the post-flame zone is not modeled by any of the mechanisms examined in this study. Without this fall-off, significant growth late in the post-flame zone would be predicted by our code (using any of the mechanisms described herein) if all curves were shifted upwards. As a result of these issues, we believe that it is not the absolute magnitude that is of concern when comparing these different mechanisms, but rather the shape of the soot growth profiles within a given flame as well as the comparison of the mechanisms for several different flames.

28.7 Oxidation Processes

The per particle net rate of growth due to surface mass addition and oxidation is assumed, as discussed, to be proportional to particle surface area (free molecule form)

$$\frac{dm_{\text{eff}}}{dt} = G'(t)a_s = G(t)m^{2/3} \tag{28.4}$$

where a_s is the particle surface area, and the specific growth rate has the overall form

$G'(t) =$ (Growth rate by acetylene or other growth species addition
     $-$ oxidation rate by O_2, OH)/unit surface area.

$$\tag{28.5}$$

or $G'(t) = R_G - R_{O_2} - R_{OH}$

The mass addition term R_G in Eqns. (28.3) and (28.5) has been derived for several kinetic models of the surface growth process, as discussed in the preceding section. Oxidation of soot by OH radicals is assumed to proceed at a gas kinetic collision frequency multiplied by a collision probability of 0.13 [28.14]. Thus, with N_{OH}, N_A, and m_{OH} representing the OH number density, Avogadro's number, and the OH radical mass, respectively, the OH oxidation is

$$R_{\text{OH}} = (0.13) \times N_{\text{OH}} \sqrt{\frac{KT}{2\pi m_{\text{OH}}}} \times \frac{12}{N_{\text{A}}}$$

$$= 16.7 \frac{p_{\text{OH}}}{\sqrt{T}} \tag{28.6}$$

where p_{OH} is the OH partial pressure in atmospheres, and the specific growth rate is in c.g.s. units. For oxidation by O_2, the Nagle & Strickland-Constable [28.13] expression is used.

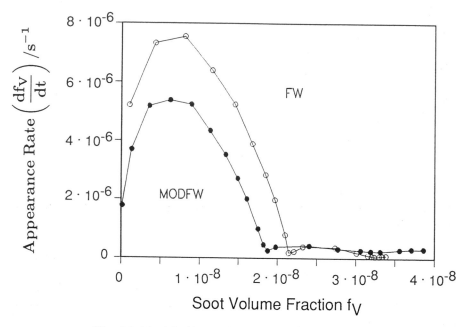

Fig. 28.10. df_V/dt vs. f_V, Bockhorn acetylene case

28.8 Soot Spheroid Growth Model

The growth of soot spheroids has been modeled as an aerosol dynamics problem, involving the division of the size range of interest into discrete intervals or classes, and then solving a master equation for the size class mass densities with terms representing inception, surface growth (or oxidation), and coagulation (coalescence). The spheroids are assumed to be comprised of the single component carbon only. The sectional analysis is discussed by *Gelbard* et al. [28.25] and *Gelbard* et al. [28.26], and the computer program we have developed is an outgrowth of the well-known MAEROS program [28.8].

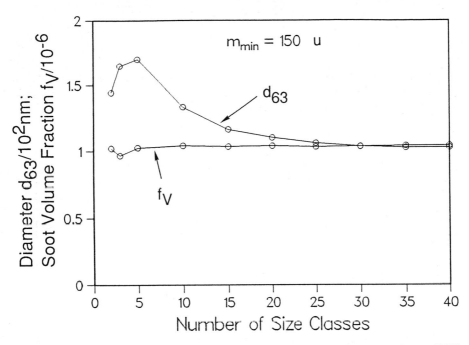

Fig. 28.11. Convergence properties of soot parameters, C/O=0.96 flame, HW growth rate

Because our application is specialized to the free molecule regime, we were able to simplify the calculation of growth and coagulation coefficients. The fundamental growth equations and numerical analysis algorithms are those of MAEROS, however. Soot spheroids vary in diameter from approximately one to 100 nanometers, representing a variation of six orders of magnitude in mass. In the sectional analysis, it is assumed that the boundaries of the sections vary linearly on a log scale. The important features of the aerosol dynamics analysis are discussed by *Colket* et al. [28.9].

28.9 Comparisons to Experimental Data

To compare theory with the soot growth data of *Harris* et al. [28.12] and *Bockhorn* et al. [28.15], profiles of temperature, benzene and acetylene concentrations, and net surface growth rate are provided to the aerosol code as a function of time or height above the burner surface. The net surface growth rate consists of the mass addition rate for acetylene vapor deposition, minus oxidative terms due to oxygen molecules and OH radicals. The latter typically are dominant low in the flame, and the starting procedure for the aerosol growth analysis is to advance in time or height to the point where the net growth rate first turns positive. As will be seen, the assumptions about the inception rate and the starting procedure yield predicted onsets of growth that agree reasonably well with experimental data in most cases. The aerosol code has provision for depletion of the acetylene vapor due to deposition, but this is typically on the order of 10%, and thus does not have a major influence on the results. Certain other assumptions have been made: a size class-independent coalescence sticking probability of 1.5 is assumed, and particles with masses below 150 u have been excluded from coalescence. As will be seen, the volume fraction tends to converge in relatively few sections, but the average particle diameter tends to require many more, so that the nominal number of sections assumed in the calculations was 25. The sensitivity of the calculations to assumptions such as these will be discussed later. Unless stated otherwise, all calculations to follow will employ the foregoing assumptions.

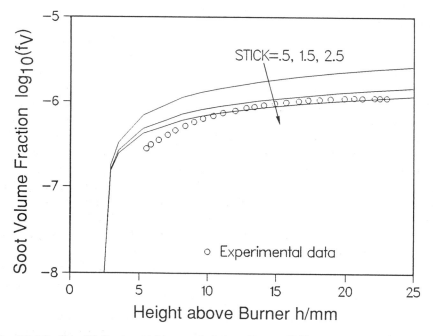

Fig. 28.12. Sensitivity to sticking probability, Harris C/O=0.96, HW growth rate

For comparison with the Harris flames, CHEMKIN simulations were carried out for C/O ratios of 0.8, 0.92, and 0.96 since temperature profiles were available for these flames. Experimental volume fraction data were available at values of 0.8, 0.84, 0.90, and 0.94. The comparisons of the Harris and Weiner (HW), Frenklach and Wang (FW), and modified Frenklach and Wang (MODFW) surface growth models with the data are shown in Figs.

5-7. With the exception of the Frenklach-Wang model, it can be seen that the stoichiometric dependence of the soot volume fraction is approximately satisfied. The FW model has little or no dependence on acetylene pressure (under the Harris and Weiner flame conditions), and cannot therefore reproduce the relative stoichiometric dependence with this inception model. It seems unlikely that the observed stoichiometric dependence of the volume fraction could be explained on the basis of inception alone, without a surface growth rate more strongly dependent on the acetylene concentration. The other surface growth models are seen to be reasonably consistent with the data. The FW growth model slightly overpredicts the data which is inconsistent with the presentation by *Frenklach* et al. [28.5]. This difference is due in part to the lower benzene profile (lower by about a factor of two) which were calculated by *Frenklach* et al. [28.5]. All the model calculations tend to suffer from overshoot at early times, particularly in the C/O=0.8 case. Suppression of coagulation involving the smallest particulates tends to raise the particle surface area. Figure 8 shows a simulation using the HW model in which all size classes are allowed to coalesce with unit sticking probability. This combination represents the best overall agreement obtained thus far, although allowing the smallest particles to undergo coalescence might well be dubious. Comparison of theory and experiment for the Bockhorn flame is shown in Fig. 28.9. Use of the HW rate with the 31.8 kcal activation energy gives poor agreement, serving as an indication of how uncertain the high temperature surface growth rates are. In fact, there is no reason to expect that surface growth rates can be extrapolated to the temperatures of the acetylene flame since available experimental data for specific soot growth rates is limited typically to below 1700K. Other models give more satisfactory agreement, as seen, with the MODFW model and the FW model (the latter with a 0.1 steric factor) giving the best overall agreement. If one accepts the use of this steric factor at high temperatures, the HW expression, which already gives excellent agreement with the Harris and Weiner soot data, also then describes the acetylene data very well.

This good agreement is not surprising considering that the HW specific surface growth rates (see Fig. 28.4) are similar in shape but a factor of 10 above the sterically corrected FW rates for the acetylene flame. It is interesting to note that the simple HW expression as modified with a steric factor provides as good agreement or better as any of the more complex mechanisms – and only acetylene, benzene, and temperature profiles are required. There is experimental evidence [28.27-29] and theoretical argument involving the surface density of "active sites" [28.24, 30-31] that the soot growth law is of the form

$$\frac{\mathrm{d}f_V}{\mathrm{d}t} \propto f_V - f_{V\infty} \qquad (28.7)$$

where $f_{V\infty}$ is an equilibrium, asymptotic value. A consequence of this is that a plot of rate of change of volume fraction versus volume fraction should be

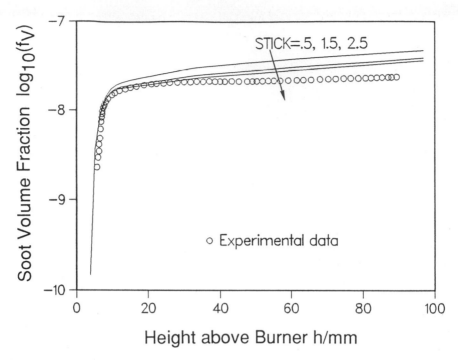

Fig. 28.13. Sensitivity to sticking probability, Bockhorn acetylene flame, MODFW growth rate

linear. A surface growth law proportional to total soot surface area can be consistent with (28.7) if the total surface area is constant and the surface reactivity has a simple exponential decay [28.24] therefore, we tested our simulations by making such plots from the best theory-experiment comparisons (see [28.9]). For the Harris 0.96 case, there is certainly a region of such linearity at the longer times where most of the experimental data exist, but in the 0.8 case it is hard to identify any such linear region. Similarly, the acetylene simulation seems somewhat ambiguous (Fig. 28.10), but the region of rapid growth between volume fractions of 10^{-8} and 2×10^{-8} could be said to conform to this law. Our results are not totally inconsistent with the growth law, (28.7), but do not yet confirm it, either. Even the experimental data [28.29] seems to have only a limited regime over which (28.7) is valid. In any case, Fig. 28.10 indicates that the MODFW mechanism is at least as good as the FW mechanism in providing linear plots.

28.10 Sensitivity Studies

Size class –
The sensitivity of the C/O = 0.96 case (ethylene) to the number of size classes is exhibited in Fig. 28.11. The number of classes is being varied, and we specify a mass threshold (m_{min}) such that particles of mass less than this value cannot coagulate. As seen, the volume fraction tends to converge quite rapidly, with 3-5 sections giving a good approximation; the surface area has similar convergence properties. The value of volume fraction with 40 size classes differs by no more than one per cent from the value obtained with 5 classes. The optical diameter converges more slowly, requiring 25-30 sections.

Sticking probability –
Some uncertainty is associated with the proper value of the sticking coefficient, and the sensitivity of the theoretical predictions is given in Figs. 28.12 and 28.13. Lowering the sticking probability below unity increases the surface area and contributes to much more predicted soot growth for the Harris 0.96 case. For the Bockhorn acetylene flame, however, the sensitivity is less, because the surface growth rate is relatively small except for a limited region low in the flame. The sensitivity of predicted soot volume fraction to coalescence sticking probabilities is tied to the magnitude of the local surface growth rates. Particle number densities and sizes can be expected to be much more sensitive.

Inception rate –
If the inception or nucleation source rate is multiplied by constant factors, some impression of the sensitivity of our predictions to source rate can be gained, as shown in Fig. 28.14. Very significant changes are predicted for the acetylene flame when the inception rate is increased or decreased by a factor of ten. A large increase in the inception rate over the base value (m_{min} = 72 u) is unphysical, however. Similar results were obtained for the *Harris* et al. flames, although the predictions were slightly less sensitive to decreases in the inception rate. The large sensitivity to inception rate is consistent with early models of soot formation but differs dramatically from recent suggestions. *Kennedy* et al. [28.2] argued that as soot concentrations in flames increase, the dependence on soot inception decreases and the quantity of soot produced is dominated by growth processes. The recent experiments by *Kent* et al. [28.32] support these arguments. The opposing result from our model under selected conditions, we believe, is due to the fact that as inception rates decrease, the losses in surface area due to coalescence are less effective, and relatively high soot growth rates result. These opposing trends highlight the complexity of the soot formation process and the fact that conclusions drawn from one study may not be necessarily applicable to another set of conditions.

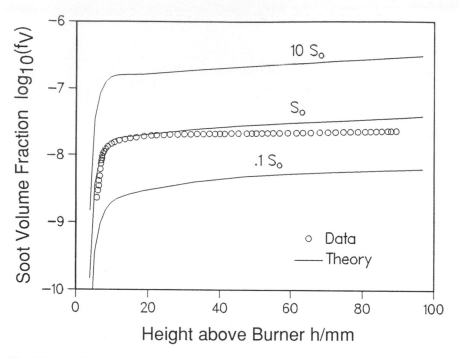

Fig. 28.14. Predicted sensitivity to source rate, Bockhorn acetylene flames, MODFW surface growth

Minimum size for coalescing spheres –
Figure 28.15 displays for the *Harris* et al. [28.16] 0.96 case the predicted sensitivity to the upper mass limit of small particles excluded from coalescence. This limit is the effective inception species mass in our model. The curves correspond to excluded masses of approximately 72, 150, 300, and 600 u respectively. As seen, the sensitivity is significant. In future efforts, we hope to include the recent results of *Miller* [28.33] regarding size dependent sticking rates for polyaromatic hydrocarbons. In the previous calculations, coalescence was found only to be important for particles whose mass exceeds 800 u, much larger than those assumed to coalesce in the present study. As stated previously, identification of a more realistic (lower) inception rate will enable the model to address more reasonable descriptions of the coalescence processes.

Temperature –
The importance of temperature to the soot formation process is well established [28.34]. The effect of temperature on benzene profiles (which in turn affects inception rates) has been shown to be significant in the report by *Colket* et al. [28.9]. Subsequent changes in inception rates lead to nearly linear changes in soot production for most of the flames examined here. Other

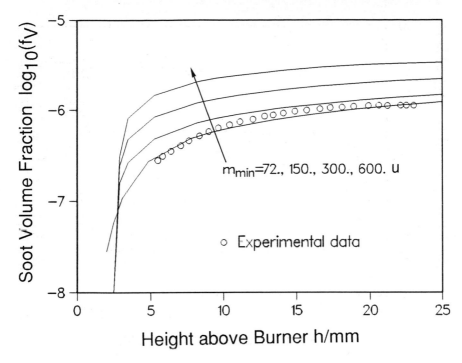

Fig. 28.15. Sensitivity to coagulation exclusion, Harris C/O=0.96, HW growth rate

species, particularly H-atoms which may affect specific surface growth rates, are also a strong function of temperature. In addition, temperature may affect rate constants used in the calculation of specific surface growth rates. To examine a portion of this complex dependence on temperature, we considered the effect of uncertainties in the temperature profile on predictions using just the HW mechanism, since this mechanism only depends on acetylene concentrations, which are weak functions of temperature. For the low pressure acetylene flames, and assuming a steric factor of 0.1 as well as the inception rates used for the base temperature case, soot profiles were calculated with temperature profiles shifted 100K above and 100K below the experimental values. These shifts resulted in about a 40% shift in the specific growth rates as well as a 30 to 50% shift in soot production. The changes in soot production due to shifts in temperature are shown in Fig. 28.16. Accompanying shifts in benzene profiles can be expected to enhance these differences.

Oxidation –

Oxidation plays an important role because of its competition with growth processes. As described previously, soot growth calculations are initiated at the point at which growth first begins to dominate over oxidation. The relative importance of oxygen and hydroxyl radicals changes as flame con-

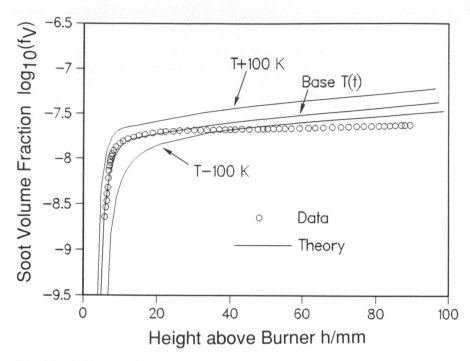

Fig. 28.16. Predicted temperature sensitivity of soot growth, acetylene flame, HW growth rate with steric factor=0.1

ditions are altered. For the lower temperature, atmospheric pressure flames of *Harris* et al. [28.12], oxygen is the dominant oxidizer. For the higher temperature, low pressure flames of *Bockhorn* et al. [28.15], hydroxyl radical concentrations are much higher and OH becomes the dominant oxidizer as molecular oxygen is depleted in the flames. To examine the sensitivity of soot formation to uncertainties in the calculated OH-radical profile, we doubled the OH concentration. A substantial shift (about 5 mm) was observed in the onset of soot formation, although the total volume fraction rapidly approached the soot profile without the OH modification. These results suggest that uncertainties in OH and oxygen concentrations and/or in rate constants and processes describing the oxidation lead to uncertainties in the point of soot onset and perhaps in the early growth rates, but they do not dramatically alter total soot production. Consideration of O_2 and OH profiles relative to these issues can provide some further insight for the effect of oxidation. Oxygen is depleted dramatically in the flame front as the oxygen is consumed by the fuel. Because its precipitous drop, uncertainties in this profile and in rates of oxidation by O_2 probably have a small effect on the location of soot onset. Therefore, lower temperatures, high pressures will be less sensitive to uncertainties in oxidation processes. Hydroxyl radicals, however, peak within the flame front and slowly decay downstream as the flame cools and the radical concentrations relax to equilibrium conditions. Since this profile is so gradual, uncertainties in absolute concentrations and in rates and mechanisms can lead to substantial errors in predicting the onset of soot formation. The difficulties in accurately predicting the soot profiles for the propane flame could in part be due to uncertainties in predicting the hydroxyl radical concentrations. This analysis indicates that at elevated temperatures and low pressures, when hydroxyl radical concentrations are large, soot predictions may be quite sensitive to uncertainties in oxidation processes.

28.11 Conclusions

An analytical model of soot formation in laminar, premixed flames has been presented which is based on coupling the output of flame chemical kinetics simulations with a sectional aerosol dynamics algorithm for spheroid growth. A provisional particle inception model based on benzene formation rates is employed. Justification for the use of this simplified model is provided and its use is motivated by a desire to develop a simple procedure which might be useful for predictions in more practical flames. Surface growth has been based on experimental measurements and ab initio calculations using various possible mechanisms for the surface chemistry. In the latter, the surface growth rate becomes a function of the local values of certain gas phase species concentrations and the gas temperature. Detailed comparisons have been made with various flame data by using experimental temperature profiles and calculating profiles of species concentrations needed for the inception rate and surface growth/oxidation calculations. Most of the models appear to overestimate soot production high in the flames. This overestimation is undoubtedly due to the fact that none of the mechanisms include effects due to particle ageing. Recent suggestions that decay of H-atoms is the cause of this 'ageing', although plausible, are inadequate because of differences between the spatial (or temporal) dependence of the decay in the H-atom profiles and the fall-off in specific growth rates observed in laboratory flames. Among the various models for the soot surface growth, best overall agreement is obtained with a modified form of the approach taken by Frenklach and Wang. The result is an analysis that is highly efficient; accurate soot volume fraction calculations can be obtained with only a few growth equations. While various aspects of this simple model can be challenged, it yields agreement with experiment that is comparable to that obtained using more elaborate models.

The modified Frenklach and Wang mechanism avoided the assumption of a high temperature steric factor, yet reproduced growth profiles while nearly providing a linear df_V/dt vs. f_V relationship. The FW mechanism was found to be deficient in that it does not properly describe the stoichiometric dependence observed in the Harris and Weiner flames; while the very simple HW expression could describe these flames as well as the acetylene flames (when the high temperature steric factor is used). Sensitivity analyses have been performed for parameters such as temperature, oxidation, number of size classes, sticking coefficients, inception rates, and coalescence of low molecular weight polyaromatic hydrocarbons. Each of these can have a significant effect on predictions of soot concentration depending on specific flame conditions.

28.12 Acknowledgements

This work has been sponsored in part by the Air Force Office of Scientific Research (AFSC), under Contracts No. F49620-88-C-0051 and F49620-85-C-0012. The United States Government is authorized to reproduce and distribute reprints for governmental purposes notwithstanding any copyright notation hereon. The authors are indebted to Dr. Julian Tishkoff, AFOSR contract monitor, for his support. The authors would like to express their appreciation to F. Gelbard for his advice and assistance on matters concerning the aerosol dynamics analysis. In addition, we would like to thank H. Bockhorn, M. Frenklach, I. Glassman, S. Harris, J. Howard, I. Kennedy, and R. Santoro for many fruitful discussions as well as for their contributions to the literature which helped to guide this research. The able assistance of K. Wicks and H. Hollick in the preparation of the manuscript and figures is gratefully acknowledged.

References

28.1   S. Bhattacharjee, W.L. Grosshandler: "Effect of Radiative Heat Transfer on Combustion Chamber Flows", in Proceeding on Western States Sectional Meeting of Combustion Institute (The Combustion Institute, Salt Lake City 1988)
28.2   I.M. Kennedy, W. Kollmann, J.Y. Chen: Combust. Flame **81**, 73 (1990)
28.3   R.J. Hall, P.A. Bonczyk: Appl. Opt. **29**, 4590 (1990)
28.4   M.D. Smooke, P. Lin, J.K. Lam, M.B. Long: "Computational and Experimental Study of a Laminar Axisymmetric Methane-Air Diffusion Flame" in Twenty-Third Symposium (International) on Combustion (The Combustion Institute, Pittsburgh 1990) p. 575
28.5   M. Frenklach, H. Wang: "Detailed Modeling of Soot Particle Nucleation and Growth" in Twenty-Third Symposium (International) on Combustion (The Combustion Institute, Pittsburgh 1990) p. 1559
28.6   J.T. McKinnon, Ph.D. Thesis, Massachusetts Institute of Technology (1989)

28.7   J.T. McKinnon, J.B. Howard: Combust. Sci. Technol. **74**, 175 (1990)

28.8   F. Gelbard: MAEROS User Manual, NUREG/CR-1391, (SAND80-0822) (1982)

28.9   M.B. Colket, R.J. Hall: "Description and Discussion of a Detailed Model for Soot Formation in Laminar, Premixed Flames", United Technologies Report No. UTRC91-20, August 9, 1991. Also Appendix A in M.B. Colket, R.J. Hall, J.J. Sangiovanni, D.J. Seery: "The Determination of Rate-Limiting Steps During Soot Formation", Final Report to AFOSR, under Contract No. F49620-88-C-0051, UTRC Report no. 91-21, August 14 (1991)

28.10 R.J. Kee, J.F. Gcar, M.D. Smooke, J.A. Miller: "A Fortran Program for Modeling Steady Laminar One-Dimensional Premixed Flames" in Sandia report, SAND85-8240 (1985)

28.11 R.J. Kee, R.M. Rupley, J.A. Miller: "CHEMKIN II: A Fortran Chemical Kinetics Package for the Analysis of Gas-Phase Chemical Kinetics" in Sandia Report, SAND89-8009-UC-401 (September 1989)

28.12 S.J. Harris, A.M. Weiner: Combust. Sci. Technol. **31**,155 (1983)

28.13 J. Nagle, R.F. Strickland-Constable: "Oxidation of carbon between 1000-2000°C" in: Proceedings of the Fifth Conference on Carbon (Pergamon Press, London 1962) p. 154

28.14 K.G. Neoh, J.B. Howard, A.F. Sarofim: "Soot Oxidation in Flames" in Particulate Carbon: Formation During Combustion, ed. by D.C. Siegla, G. W. Smith (Plenum Press, New York 1981) p. 261

28.15 H. Bockhorn, F. Fetting, H.W. Wenz: Ber. Bunsenges. Phys. Chem. **87**, 1067 (1983)

28.16 S.J. Harris, A.M. Weiner, R.J. Blint: Combust. Flame **72**, 91 (1988)

28.17 J.A. Miller, C.F. Melius: Combust. Flame **91**, 21 (1991)

28.18 J.A. Miller, C.T. Bowman: Prog. Energy Combust. Sci. **15**, 287 (1989)

28.19 C.K. Westbrook, F.D. Dryer: Prog. Energy Combust. Sci. **10**, 1 (1984)

28.20 F. Communal, S.D. Thomas, P.R. Westmoreland: "Kinetics of C_3 Routes to Aromatics Formation", Poster paper (P40) presented at the Twenty-Third Symposium on Combustion, Orleans, France (July 1990)

28.21 S.E. Stcin, J A. Walkcr, M. Suryan, A. Fahr: "A New Path to Benzene in Flames" in Twenty-Third Symposium (International) on Combustion (The Combustion Institute, Pittsburgh 1990) p. 85

28.22 M. Frenklach, H. Wang: "Aromatics Growth Beyond the First Ring and the Nucleation of Soot Particles", presented at the 202nd American Chemical Society, National Meeting, New York, August 25-30, 1991, also Prep. Div. Fuel Chem. **36(4)**, 1509 (1991)

28.23 R.D. Kern, C.H. Wu, J.N. Yong, K.M. Pamidimukkala, H.J. Singh: "The Correlation of Benzene Production with Soot Yield Determined from Fuel Pyrolysis", in Proceeding on Division of Fuel Chemistry (American Chemical Society, National Meeting, New Orleans 1987)

28.24 C.J. Dasch: Combust. Flame **61**, 219 (1985)

28.25 F. Gelbard, J.H. Seinfeld: J. Coll. Interface Sci. **78**, 485 (1980)

28.26 F. Gelbard, Y. Tambour, J.H. Seinfeld: J. Coll. Interface Sci. **76**, 54 (1980)

28.27 B.S. Haynes, H.G. Wagner: Z. Phys. Chem. NF **133**, 201 (1982)

28.28 U. Wieschnowsky, H. Bockhorn, F. Fetting: "Some New Observations Concerning the Mass Growth of Soot in Premixed Hydrocarbon-Oxygen Flames" in Twenty-Second Symposium (International) on Combustion, (The Combustion Institute, Pittsburgh 1988) p. 343

28.29 H. Bockhorn, F. Fetting, A. Heddrich, G. Wannemacher: "Investigation of the Surface Growth of Soot in flat low Pressure Hydrocarbon Oxygen Flames" in Twentieth Symposium (International) on Combustion (The Combustion Institute, Pittsburgh 1984) p. 979

28.30 S.J. Harris: Combust. Sci. Technol. **72**, 67 (1990)

28.31 I.T. Woods, B.S. Haynes: Combust. Flame **85**, 523 (1991)

28.32 J.H. Kent, D.R. Honnery: Combust. Sci. Technol. **75**, 167 (1991)

28.33 J.H. Miller, "The Kinetics of Polynuclear Aromatic Hydrocarbon Agglomeration in Flames" in Twenty-Third Symposium (International) on Combustion (The Combustion Institute, Pittsburgh 1990) p. 91

28.34 I. Glassman, "Soot Formation in Combustion Processes" in Twenty-Second Symposium (International) on Combustion (The Combustion Institute, Pittsburgh 1988) p. 295

28.13 Discussion

Haynes: Did your implementation of the Harris and Weiner growth kinetics incorporate a time dependent loss of reactivity. Did you have tempering in there?

Colket: We liked some features about the H-atom decay proposal which we hoped would describe ageing. So when we constructed the model we did not include tempering. Now, I think it' s probably appropriate to put in.

Haynes: I think when you get very poor agreement with the high temperature flames that would be the first place to look when using the HW kinetics because it is simply not appropriate to extended times such as more then 5 or 10 ms at those temperatures.

Colket: I have not checked the characteristic times for the ageing process under acetylene flame conditiones, but I guess it doesn' t make sense that particle reactivity could decay so rapidly. We shall take a look at it.

(note added in proof: We have since added the tempering effects and this feature improves the asymptotic agreement, but does not reduce the HW kinetics fast enough under the Bockhorn acetylene flame conditiones unless a steric factor is also used.)

Sarofim: With the MAEROS-code you have an extra dimension that you can carry on the particles and you could put in hydrogen content and therefore suggestions made earlier that the alpha or the sticking coefficient could be modelled depending upon the hydrogen content. You could carry an additional variable along.

Colket: I think you are probably right.

Warnatz: Where is the mechanism for the small molecules in your propane flame?

Colket: We used a derivative of a Westbrook and Dryer mechanism which I modified to some of the propane PSR data from Orlean.

Homann: In your flames the zone what you might call "soot nucleation zone" is very thin and very narrow, so if you take out just this zone from your model, don' t you get any more soot then or does it just happen a little later that the soot nucleation takes place. Has anyone tried this out? If you remove the very narrow zone of your flame where the nucleation of soot takes place do you get then zero soot in your models or is it just shifted to a little later time and you still get the same amount of soot.

Frenklach: If we look at Steve Harris' flame which several people try to model, we arrive at one problem and I think it is answering your question and it is also important in a very general sense. This is a measurement of H-atom profile for Steve's flame. It was done using multi–photon ionization spectroscopy by J.E.M. Goldsmith at the Sandia National Laboratories. The meassured and computed profiles agree reasonably well but slighthly shifted in distance with respect to each other. If we use a computed H-profile in the soot model we get for benzene a factor of two difference compared to the experimental profile which seems acceptable. If we do the same for 5 membered rings we get a difference of about eight orders of magnitude. The answer to your question is that if you don't model the inception zone right you can completely miss fluxes by orders of magnitude. And what I think happened to you since you probably don't use the correct H-profile in your model (nobody can model Steve's flame correctly) and you are using a wrong nucleation rate because of using benzene, which is four or five orders of magnitude higher than it should be, you get two errors which cancel each other out.

Colket: Firstly, I think that the inception process, the initial coagulation process is a magnitude larger or at least a factor of 5 to 10 higher in mass. We are talking about a 1000 or 2000 u in terms of the particles that are actually coagulating. When you consider the coagulation process in the inception it is really a dimerisation process which is dependent on the square of the concentrations of those species. So if we are talking about a difference of 5 or 10 in concentrations we shall have around two orders of magnitude difference in the inception process. So I think we still have to learn about how to do this. Secondly, I agree that the H-atoms have to be where you also have benzene production in order to see the growth processes. These things do have to occur simultaneously to get a rapid process going on. However typically in the calculations that we did the early growth process really starts at a point where the net growth becomes positive, prior to that we only observe oxidation. So that' s really the location in the flame where we begin to see the process starting up. I have some concern that if I postpone that by putting an artificially high OH or O_2 concentration I might still get growth downstream. I haven' t done this yet. There might be a concern because this is probably not very realistic. Finally, the HW kinetic set also adequately predicts soot profiles but does not require any H-atoms, expect as related to predictions of acetylene and benzene profiles. Thus, in this case, your concern should not apply.

Comparison of the Ionic Mechanism of Soot Formation with a Free Radical Mechanism

Haetwell F. Calcote, Robert J. Gill

AeroChem Research Laboratories, Inc.,
P. O. Box 12,
Princeton, NJ 08542, USA

Abstract: The possible mechanisms by which large hydrocarbon ions might be produced in a fuel rich sooty flame are examined and all of them are eliminated except growth through ion-molecule reactions starting from the chemiion, HCO^+. Large ions are not, as sometimes assumed, an indicator of neutral species. The ionic mechanism of soot nucleation is compared with a free radical mechanism by comparing the time each mechanism requires to add 10 carbon atoms to the growing species, ion or neutral. Experimental concentrations are combined with appropriate rate coefficients for the individual steps. The ionic mechanism appears to require less time, but the differences are within the accuracy of the data available.

29.1 Introduction

Ions may participate in the formation of soot in fuel rich flames at three stages: (1) nucleation; (2) coagulation of large aromatic ions; and (3) coagulation of soot particles. In the nucleation steps leading to large molecular species, equivalent to incipient soot particles, the ionic mechanism may compete with free radical mechanisms because the rate coefficients of ion-molecule reactions are greater than those for free radical reactions. The greater rate coefficients compensate for the lower concentrations of ions. The rate of coagulation of large ions or charged particles is enhanced over that of neutral species because of electrostatic effects. As molecules or soot particles become larger, their ionization potential or work function becomes smaller and neutral species may become charged by thermal ionization. Large molecules and particles can also become charged by electron attachment or by diffusive charging of ions, e.g., by low ionization potential metal ions. Electron attachment or diffusive charging of particles are accepted in the combustion community as playing a significant role in soot formation, especially when chemical additives are present. The contribution of ion-molecule reactions, i.e., ionic nucleation, to soot formation has not been generally recognized as competing with free radical mechanisms, although there is overwhelming evidence of the importance of ionic processes in soot nucleation [29.1-2]. Equivalent evidence is not present for the free radical mechanism!

Experimentally large concentrations of a great variety of ions are observed in hydrocarbon fuel rich flames [29.3-5]. Two fundamental questions arise; first, " what is the mechanism by which these large ions are produced ?" The second question, the answer to which may depend upon the answer to the first question, is, " do ions play a role in the nucleation step of soot formation?" Because of the parallel paths followed by free radical and ionic mechanisms, a decision on whether or not ions play a role requires a comparison between the two mechanisms for any given system. In this discussion we: (1) briefly consider the mechanisms by which ions might be produced in a fuel rich flame; and (2) compare the rate of growth of ion and neutral species in a soot producing flame.

29.2 Mechanism of Formation of Large Ions

A large variety of ions is observed in flames and the number of ionic species and size of the ions increase dramatically at the soot threshold [29.1, 29.6]. It is thus logical to assume that the large ions are a result of soot formation and not a cause. Soot is known to be ionized so that one may hypothesize that the molecular ions are produced from the ionized soot. The difficulty with this hypothesis is that the free energy per unit mass of a charged particle decreases with increasing size so there is no driving force to transfer a charge from a particle to a molecular species.

Thus the energy, E, to remove an electron from a particle of diameter d is [29.7-8]:

$$E = V + \frac{e^2}{4\pi\varepsilon_0 d} \tag{29.1}$$

where V = the work function of bulk material, e = the charge on an electron, and ε_0 = the dielectric constant of free space. As the particle diameter increases, E decreases. The same general trend is observed for ions, the larger the ion, the smaller the free energy per unit mass [29.9].

This same argument, the larger the ion the more stable it is, refutes any explanation of ionization in flames in which a large ion is assumed to produce a smaller ion, e.g., thermal ionization of large molecules. Thus ion-molecule reactions:

$$A^+ + B \rightarrow C^+ + D \tag{29.A}$$

where the ion or charged particle A^+ is larger than C^+ are rare. The reverse reaction is favored. Another proposed explanation of the appearance of a large range of ions in sooty flames is that the ions are in equilibrium with large neutral species [29.4]. It has, in fact, been assumed that the ions are good indicators of the neutral species [29.4, 29.9-11]. This, of course, is sometimes true because proton, hydride, and hydrogen atom shifts may be very rapid [29.12]. Ion profiles can, in fact, be used to measure some equilibrium constants [29.13] because ion-molecule reactions are very fast [29.12]. The source of large ions is, however, missing and there are too few reactions in equilibrium to account for all of the smaller ions by charge transfer from the larger ions [29.14]. Thermal ionization of the neutral species is generally eliminated because ionization potentials are too great and the concentrations of large neutral species are too small. As the molecule becomes larger, the ionization potential decreases, but so does the concentration of neutral species, see e.g., Fig. 29.1. If thermal ionization of large molecules did occur, there would still be no mechanism to produce the smaller ions, see above.

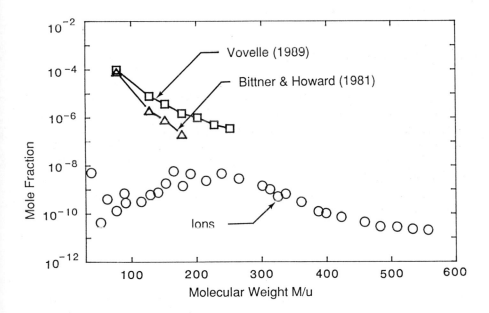

Fig. 29.1. Rate of decay of neutral and ionic species concentrations

There is further experimental evidence that the variety of ions observed in fuel rich flames cannot be the result of an equilibrium process. The rate of decay of total ions in a flame, where the temperatures are still high, is consistent with expected ion-electron or ion-ion recombination rates [29.3].

The initial ion concentration must be out of equilibrium for this to occur. There may be equilibrium between some ions and neutral species but the source of ions cannot be an equilibrium process for the observed recombination rates to be observed. Thus ions are not an indicator of neutral species behavior. Ions must be produced in supra equilibrium concentrations.

The accepted source of nonequilibrium flame ions is chemiionization via [29.8, 29.15]:

$$CH + O \rightarrow HCO^+ + e^- \ . \tag{29.B}$$

This also is apparently the prime source of ions in fuel rich and sooty flames [29.8, 29.10].

29.3 Comparison of the Rate of Growth of Large Ions and Large Neutral Species in Sooting Flames

A simple technique has been employed to compare the relative rates by which ionic and radical mechanisms might account for the growth of carbon containing species in the same sooting flame. It is accepted that soot is formed from small carbonaceous molecules by a mechanism in which they increase in size and carbon to hydrogen ratio until at some size they coagulate to produce even larger species, and eventually produce incipient soot particles. The least understood step in this process is the increase in carbon number from 2 to 100, a molecular weight of about 1,000 u. In this comparison, experimental data are combined with reaction rate coefficients to compute the time to add ten carbons by the two mechanisms for the same flux of carbon species moving through the system.

In modeling the ionic mechanism [29.16], the reactants and products used have been limited to those which have been observed experimentally and it has been required that the model calculate all observed ion species. The model for the free radical mechanism [29.17] has not been limited by either of these constraints. Thus that model includes cyclopentaphenanthrene which is not observed experimentally, and does not include phenanthrene and pyrene which are observed experimentally. Further, in the free radical model, only five calculated cyclic compound profiles are compared with experiment and the peak concentration of one of them, naphthalene, is computed to be 20 times smaller than measured. For this comparison we use only experimentally observed species and experimentally measured concentrations for both the ionic and the free radical mechanism.

We chose for comparison the well-studied acetylene/oxygen flat flame at 2.67 kPa, equivalence ratio = 3.0 and unburned gas velocity = 50 cm s^{-1}. For the experimental neutral species concentrations we use those of Vovelle [29.18] and Bittner and Howard [29.19]. Vovelle's data extend to mass 252 u; Bittner and Howard's data extend only to mass 178, $C_{14}H_{10}$. At mass

178, Bittner and Howard's concentrations are about an order of magnitude lower than those of Vovelle, Fig. 29.1, and seem to be approaching the concentration of ionic species for high masses. The burner and conditions in the Bittner and Howard experiments were duplicated in the AeroChem measurements of ion profiles [29.3] so that the profiles of neutral and ion species could be compared. Use of Vovelle's data enhances the growth rates calculated for the free radical mechanism and thus favors the free radical mechanism. Free radical concentrations have not been measured, so they are assumed equal to the concentration of the preceding neutral species from which they were formed. This again favors the free radical mechanism. The range of observed carbon numbers is very limited for neutral species because the concentrations rapidly diminish below detectability as the molecules become larger.

For the individual ionic species concentrations, we use those measured at AeroChem [29.3], Fig. 29.1, on a burner identical to that used by *Bittner* et al. [29.19]. The total concentrations have been confirmed by *Gerhardt* et al. [29.4] and are consistent with measurements of *Delfau* et al. [29.5].

For the free radical mechanism we use that of *Frenklach* et al.[29.17] and use the rate coefficients of *Frenklach* et al. [29.20] for the specific steps. Four elementary reaction types are employed by *Frenklach* et al. in the free radical mechanism; they are, with their rate coefficients:

$$MH + H\cdot \longrightarrow M\cdot + H_2 \qquad k = 4.5 \cdot 10^{12} cm^3 mol^{-1} s^{-1} \qquad (29.C)$$
$$M\cdot + C_2H_2 \longrightarrow MC_2H + H\cdot \quad k = 3.1 \cdot 10^{12} cm^3 mol^{-1} s^{-1} \qquad (29.D)$$
$$M\cdot + C_2H_2 \longrightarrow MC_2H_2\cdot \qquad k = 1.0 \cdot 10^{13} cm^3 mol^{-1} s^{-1} \qquad (29.E)$$
$$M\cdot + H\cdot \longrightarrow MH \qquad k = 1.3 \cdot 10^{13} cm^3 mol^{-1} s^{-1} \qquad (29.F)$$

where M is the large reactant species.

The ionic mechanism requires basically two elementary reactions:

$$M^+ + C_2H_2 \longrightarrow MC_2^+ + H_2 \quad k = 3.6 - 5.3 \cdot 10^{14} cm^3 mol^{-1} s^{-1} \quad (29.G)$$
$$M^+ + C_2H_2 \longrightarrow MC_2H_2^+ \qquad k = 5.6 \cdot 10^{14} cm^3 mol^{-1} s^{-1} \qquad (29.H)$$

The ion-molecule reaction rate coefficients are calculated by Langevin theory and are adjusted so that the reverse reaction rate calculated by thermodynamic equilibrium never exceeds the Langevin rate.

The times for the species to add a specified number of carbon atoms by the two mechanisms are compared by determining the time for each step and adding the individual step times to obtain a total time. Thus, for the reaction:

$$A + B \longrightarrow C + D \qquad (29.2)$$

the rate of reaction is:

$$r = \frac{d[C]}{dt} = k[A][B] \qquad (29.3)$$

The appropriate rate coefficient for k is used for either the free radical or the ionic mechanism. In the simplified free radical mechanism [29.17], A is a large stable species or a free radical and C is either a free radical with the same number of carbon atoms as A, or a stable species with two more carbon atoms than A; and B is a hydrogen atom or acetylene. In the corresponding ionic mechanism, A is a large ion and C is an ion with two more carbon atoms than A; and B is acetylene.

From (29.3):

$$d[C] = k[A][B]\, dt \tag{29.4}$$

At any distance from the burner, the concentration of each carbon number species is constant, in steady state. Hence, the concentration flux (molecule $cm^{-3}s^{-1}$) of molecules through any carbon number in a linear reaction scheme can be estimated from the production rate calculated from the local concentration of the preceding, smaller molecule or ion and the appropriate B species concentration and rate constant. Integration, assuming constant concentrations, yields a characteristic time required to produce a fixed concentration, n, of the molecule based on experimental reactant species concentrations:

$$\tau = \frac{n}{k[A][B]}. \tag{29.5}$$

Summing the times for each reaction required to carry n small species to n large species is then a measure of the overall growth time.

A more accurate account may be obtained by taking an average value of k, A, and B between the two positions over which we integrate in the flame, but for the approximate nature of the calculation toward which this analysis is leading, this would make no difference. For calculation purposes, for both ions and neutral species, we use the maximum (with respect to distance from the burner) experimental concentrations of A, and use the measured value of B at the position in the flame where the concentration of A is a maximum for the given step. The total time for n species per cm^3 to move through the reaction chain is simply the sum of the individual step times. We chose n equal to the maximum concentration of soot particles observed in the flame because this is the number concentration of growing species required to produce the soot. This neglects any loss of reactive species by oxidation, or in the ionic mechanism, by ion recombination. This assumption favors the ionic mechanism. For the subject flame, $n = 4 \cdot 10^{14} cm^{-3}$. The times required to add ten carbon atoms are displayed in Figs. 29.2, 29.3, and 29.4 for each step in the mechanism; total times are also shown.

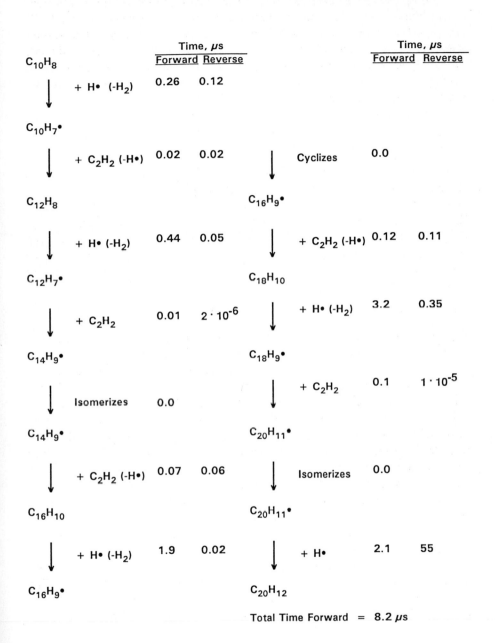

Fig. 29.2. Times to add ten carbon atoms by free radical mechanism, based on concentration measurements from Vovelle [29.18]

The reverse reaction times are also included in the figures for completeness. They are unimportant for the calculation of times because the reverse reactions are taken into account in the use of measured concentrations. Inspection of the figures leads to the observation that reverse reactions are much more significant for the free radical mechanism than for the ionic mechanism, i.e., the reverse reactions in the neutral mechanism are faster relative to the forward reactions than for the ionic mechanism and there are more steps in which this occurs. The total times to add ten carbon atoms by the two mechanisms are comparable: 8.2 μs for the free radical mechanism using Vovelle's data, Fig. 29.2, and 25 μs using Bittner and Howard's data, Fig. 29.3. The time for the ionic mechanism is 6.7 μs, Fig. 29.4. Use of the Bittner and Howard data is a fairer comparison because the two burners and conditions were exactly the same. Extrapolating the neutral species data of Bittner and Howard, Fig. 29.1, to higher mass also favors the ionic mechanism because the concentrations of neutrals and ions become equal; the ion reaction rate coefficients are greater than for the neutrals. The above analysis indicates that for modest size carbon species, 10 to 20 carbon atoms, the rate of growth is about the same or faster for the ionic than for the free radical mechanism. The greater concentration of neutral species is balanced by the greater reaction rate coefficients for ion-molecule reactions and the fewer number of steps involved in adding a specific number of carbon atoms to the growing species for the ionic mechanism: five steps compared to eleven steps for the free radical mechanism. Evaluation of the free radical mechanism suffers in the comparison because of lack of experimental data for large neutral molecules and for the free radicals involved.

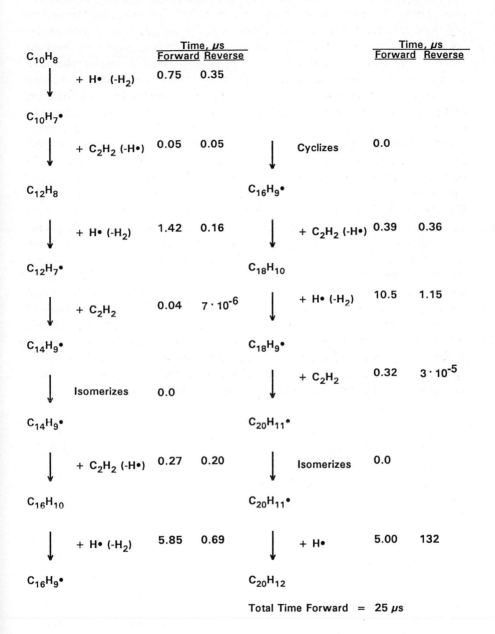

Fig. 29.3. Times to add ten carbon atoms by free radical mechanism, based on concentration measurements from Bittner and Howard [29.19]

		Time, μs	
		Forward	Reverse

$C_{11}H_9{}^+$

$\downarrow$ $+ C_2H_2\ (-H_2)$ 3.6   240

$C_{13}H_9{}^+$

$\downarrow$ $+ C_2H_2$ 0.51 $1.9 \cdot 10^{-4}$

$C_{15}H_{11}{}^+$

$\downarrow$ $+ C_2H_2\ (-H_2)$ 0.64   14

$C_{17}H_{11}{}^+$

$\downarrow$ $+ C_2H_2\ (-H_2)$ 1.3   160

$C_{19}H_{11}{}^+$

$\downarrow$ $+ C_2H_2\ (-H_2)$ 0.65   20

$C_{21}H_{11}{}^+$

Total Time Forward = 6.7 μs

Fig. 29.4. Times to add ten carbon atoms by the ionic mechanism

The reasonableness of the above approach can be evaluated by comparing the time to produce the maximum soot number density (Fig. 2 in Ref.

29.1), assuming the growth rate remains the same for the experimental time to reach the maximum soot concentration. The maximum soot concentration is reached at about 35 mm above the burner, or about 6.7 ms from the position in the flame at which the observed concentrations of large carbon containing molecules maximize. This is a good estimate of the time available, τ_s, for soot particles to be formed from molecular species. If we assume the time for the addition of one carbon atom to the growing species is τ_c, then the number of carbon atoms, n_c, that can be added to the growing nuclei is:

$$n_c = \frac{\tau_s}{\tau_c} = \frac{6.7 \cdot 10^{-3}}{7.4 \cdot 10^{-7}} \approx 9.000 \quad \text{carbon atoms} \qquad (29.6)$$

τ_c is taken as the average for the neutral (Vovelle data) and ion mechanisms. The calculated 9.000 carbon atoms corresponds to a molecular weight of about 110,000 u, equivalent to a particle diameter of about 4.5 or 3.0 nm, depending upon whether the particle is planar or spherical, respectively [29.8]. The experimentally observed particle diameters at 35 mm above the burner surface are 9-13 nm for neutral particles and 3-6 nm for charged particles [29.1]. The calculated diameter, assuming the equivalent of a fixed rate (fixed time) for adding carbon atoms to the growing species, neutral or ion, leads to a diameter of the carbon particle very close to that observed, within the accuracy of the calculation and the measurement. This confirms the reasonableness of the method.

Clearly a comparison of the two mechanisms requires extension to much larger species than can be included currently. This requires the measurement of concentrations of neutral species, including free radicals, and larger carbon containing stable species in the standard flame. Because of all the advantages given to the free radical mechanism in this calculation, it must be concluded, from this analysis, that the carbon flux through ions is greater than the carbon flux through neutral species. Thus it seems more likely that in this particular flame the ionic mechanism is more important in soot formation than the free radical mechanism.

29.4 Acknowledgements

This research was sponsored by the Air Force Office of Scientific Research (AFSC) under Contract F49620-91-C-0021. The United States Government is authorized to reproduce and distribute reprints for governmental purposes notwithstanding any copyright notation hereon. It is a pleasure to acknowledge stimulating discussions with Drs. William Felder and David G. Keil.

482   Haetwell F. Calcote, Robert J. Gill

References

29.1 H.F. Calcote, D.B. Olson, D.G. Keil: Energy Fuels **2**, 494 (1988)
29.2 H.F. Calcote, D.G. Keil: Pure Appl. Chem. **62**, 815 (1990)
29.3 H.F. Calcote, D.G. Keil: Combust. Flame **74**, 131 (1988)
29.4 P. Gerhardt, K.H. Homann: Combust. Flame **81**, 289 (1990)
29.5 J.L. Delfau, P. Michaud, A. Barassin: Combust. Sci. Technol. **20** 165 (1979)
29.6 D.B. Olson, H.F. Calcote: "Ions in fuel-rich and sooting acetylene and benzene flames" in <u>Eighteenth Symposium (International) on Combustion</u> (The Combustion Institute, Pittsburgh 1981) p. 453
29.7 F.T. Smith: J. Chem. Phys. **34**, 793 (1961)
29.8 H.F. Calcote: Combust. Flame **42**, 215 (1981)
29.9 S.E. Stein, S.A. Kafafi: "Thermochemistry of Soot Formation", in <u>Proceeding on Fall Technical Meeting, Eastern Section</u> (The Combustion Institute, NBS 1987)
29.10 A.N. Eraslan, R.C. Brown: Combust. Flame **74**, 191 (1988)
29.11 K.H. Homann: "Precursor Formation" in <u>"Soot Formation in Combustion"</u> -an International Round Table Diskussion, ed. by H. Jander, H.Gg. Wagner (Vandenhoeck and Ruprecht, Göttingen 1990) p. 127
29.12 S.E. Stein: Combust. Flame **51** 357 (1983)
29.13 P. Michaud, J.L. Delfau, A. Barassin: "The positive ion chemistry in the post-combustion zone of sooting premixed acetylene low pressure flat flames" in <u>Eighteenth Symposium (International) on Combustion</u> (The Combustion Institute, Pittsburgh 1981) p. 443
29.14 H.F. Calcote, R.J. Gill: in preparation
29.15 H.F. Calcote: in <u>Ion-Molecule Reactions, Vol. 2</u>, ed. by J.L. Franklin (Plenum Press, New York 1972) p. 673
29.16 H.F. Calcote, R.J. Gill: "Computer Modeling of Soot Formation Comparing Free Radical and Ionic Mechanisms" AeroChem TP-495 (April 1991)
29.17 M. Frenklach, J. Warnatz: Combust. Sci. Technol. **51**, 265 (1987)
29.18 C. Vovelle, personal communication (December 1989)
29.19 J.D. Bittner, J.B. Howard: "Preparticle chemistry in soot formation" in <u>Particulate Carbon: Formation During Combustion</u>, ed. by D.G. Siegla, G.W. Smith (Plenum Press, New York 1981) p. 109
29.20 M. Frenklach, H. Wang: "Detailed modeling of soot particle nucleation and growth" in <u>Twenty-Third Symposium (International) on Combustion</u> (The Combustion Institute, Pittsburgh 1991) p. 1559

Discussion

Smyth: Did you calculate carbon fluxes at just certain times?
Calcote: Fluxes are a number density going through. I just calculated the flame times.

Smyth: I think one can accept that an ion-molecule reaction of a given size, benzene size or acetylene size is faster than the corresponding radical reactions. If we start with one ionized benzene and then add acetylene it grows faster than if we start with phenyl-radical. But this is not the interesting question. If you could demonstrate that more carbon was added in going from benzene to naphthalene or benzene to anything through an ionic route than through a neutral route, in other words a real carbon flux, then this would be a much stronger argument for the mechanism.

Calcote: I have done essentially that. I assumed the flux is exactly the same in both cases. I kept the flux essentially constant and asked the question of what the time steps are. So the flux was the same in both cases.

Smyth: Why didn' t you start with the assumption that there are known phenyl radical concentrations in low pressure acetylene flames and you have measured the corresponding ionic concentrations, in other words different numbers at the starting point?

Calcote: No, I started with experimental values in both cases. All the concentrations used were experimental in the free radical mechanism and in the ionic mechanism. I started with an ion concentration which was very low, about a factor of 1000 lower than for the neutral species concentration. So this is a realistic comparison.

Homann: I would like to mention two points, one I agree with and one I disagree with. Let's start with the one I agree with. If you take this standard acetylene/oxygen flame, and measure the amount of charged and neutral soot at 20 mm above the burner where the particles have grown to a mass somewhere between 1500 and 100,000 u, you have 20% of the soot particle charged and I think that from then on a distinction between the ionic mechanism and neutral mechanism is obsolete.

Calcote: I agree.

Homann: Secondly, you used as your growth species ions like $C_{13}H_9^+$, $C_{19}H_{11}^+$, $C_{21}H_{11}^+$ ions that could be analysed in flames and we know their concentrations, but these ions are molecular ions and I think there is no difference in reactivity to normal neutral PAH-hydrocarbons other than they just carry a charge. They have no free electron, they have no single electron, they are molecular ions and I don' t think you should use the rate constant that the ion-molecule-people have used and found for radical ions. They might be orders of magnitude lower because these are rather unreactive species, for instance $C_{13}H_9^+$.

Calcote: The ion-molecule mechanism does not depend upon the ion being a reactive species as you seem to assume. It depends upon the Langevin equation which assumes the electrostatic charge reacts with the neutral particle. One of the problems in using the Langevin equation as we have done, is that the Langevin equation depends on a point charge source and a large molecular ion may not be a point charge source. This may very well slow down the ion-molecule rates. If it does, then one is left with a real major problem of how do you account for those large ions.

The Role of Biaryl Reactions in PAH and Soot Formation

Adel F. Sarofim, John P. Longwell, Mary J. Wornat,

Jaideep Mukherjee

Department of Chemical Engineering and Energy Laboratory,
Massachusetts Institute of Technology,
Cambridge, MA 02139, USA

Abstract: The pyrolysis of anthracene and pyrene for a residence time of about 0.7 seconds over a temperature range of 1200 to 1500 K has been studied in order to obtain insight on the contribution of aryl aryl reactions in the molecular weight growth of polycyclic aromatic hydrocarbons (PAH). The sequence of reactions observed are the formation of biaryls, cyclodehydrogenation of the biaryls to form fused PAH, the isomerization of the cyclodehydrogenation products, and their thermal decomposition. Soot is formed in amounts that increase progressively with increasing temperature. The concentrations of the biaryls, cyclodehydrogenation and thermal decomposition products however pass through maxima as the temperature is increased, as a consequence of the competition between the formation and destruction reactions. The relative concentration of the biaryls are found to be in the ratios expected from the probability of the reaction at the different sites, with deviations consistent with the effects of steric hindrance, and are not governed by the relative reactivity of the different sites on the aryl radicals. The aryl-aryl reactions are thought to be an overlooked path to PAH and soot formation that acts in parallel with the paths involving acetylene and its derivatives. The path involving the aryl radicals is believed to be particularly important in diffusion flames, where PAH and soot formation are favored on the fuel rich side of the flame front. Such reactions are difficult to identify at the lower temperatures for wood and coal pyrolysis because of the large number of compounds present. At high temperatures, however, only a few of the thermal decomposition products survive and these are found to be common for the pyrolysis of wood, coal, and model aromatic compounds.

30.1 Introduction

One of the products arising from the combustion of coal and other solid fuels is soot. Due to its physical properties, soot plays an important role in radiative heat transfer during combustion [30.1]. *Wornat* et al. [30.2] showed that soot particles as submicron aerosols also serve as surfaces for the condensation and like *Cautreels* et al. [30.3] discussed subsequent environmental transport of other combustion or pyrolysis products, particularly polycyclic aromatic compounds (PAH). Because some PAH are carcinogenic [30.4] and because soot particles lie within the respirable size range [30.5], soot from solid fuel combustion also presents a health concern.

PAH and soot come from the nonoxidative processes of solid fuel combustion – i.e., the thermal decomposition of the fuel and subsequent pyrolytic reactions. The chemical structure of the fuel, therefore, has a large influence on the distribution of products initially formed and the reactions they subsequently undergo.

An examination of the moieties comprising coal and wood underlines the features that distinguish them from other fuels. Bituminous coal is composed of aromatic units of one to four fused rings [30.6], many of which contain heteroatoms (N,S,O) within the rings and/or functional groups as substitutes for ring hydrogen [30.7]. Lower rank coals and wood contain fewer rings per unit but larger numbers of aliphatic groups and oxygen-containing functionalities. The aromatic units inherent in coal and wood thus provide larger building blocks for the formation of large PAH and soot, relative to light hydrocarbon fuels, whose pyrolysis yields small hydrocarbon fragments [30.8]. Heteroatoms within the solid fuels' aromatic moieties and side chains on their peripheries further differentiate the solid fuels from their light hydrocarbon counterparts, with respect to the pools of participants in the soot formation process.

Experiments of *Nenninger* et al. [30.9] and *Wornat* et al. [30.10] with a high volatile bituminous coal in a laminar flow, drop-tube reactor show that an increase in pyrolysis temperature brings about a decline in PAH yield that is accompanied by an equivalent rise in soot yield – indicating that coal-derived PAH are precursors to soot. Compositional analysis of the PAH reveals that PAH conversion to soot involves preferential loss of aromatic compounds having side chains [30.10] and/or heteroatoms within the rings [30.11]. Parallel to the formation of soot is the production of highly condensed unsubstituted PAH [30.12], formed at the expense of the other PAH. The presence of only small levels of acetylene in the coal pyrolysis environment, coupled with the absence of acetylene functionalities associated with the PAH [30.10], suggest a ring buildup mechanism that involves combinations of the aromatic units present, rather than addition of C_2 and C_4 species, as *Bittner* et al. [30.8] and *Harris* et al. [30.13] reported for light hydrocarbon flames.

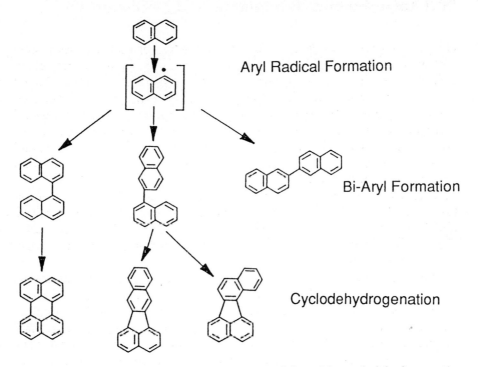

Fig. 30.1. The illustration of bi-aryl formation followed by cyclodehydrogenation

Consistent with the coal pyrolysis results is the *Badger* et al. mechanism, proposed for the pyrolysis of single aromatic compounds [30.14-20] (Fig. 30.1.). In this mechanism, an aromatic molecule loses a substituent group or an aryl hydrogen to form an aryl radical, which then combines with another aryl radical or molecule to form a biaryl molecule. When possible, the biaryl then undergoes further loss of hydrogen in a cyclodehydrogenation process that yields a condensed aromatic molecule of one or two more fused rings than the parent biaryl. *Badger* et al. [30.14, 30.15, 30.20] and *Bruinsma* et al. [30.21] show that conversion of PAH having alkyl substituents or ring heteroatoms [30.21] is faster than that of unsubstituted PAH since the formation of the initial aryl radical requires less energy in such cases.

The previous studies on biaryl formation have been mostly limited to temperatures below 1000°C and have not provided quantification of the product distribution. Advances in analytical chemistry now enable us to follow these reactions in greater detail and are used in this paper to study the biaryl reactions at temperatures of greater interest to combustion.

30.2 Experimental Equipment and Procedures

For our pyrolysis studies, the fuels or model compounds are ground and sieved for fluidized bed feeding into a laminar flow, drop-tube furnace described previously [30.10]. Pyrolysis occurs in argon at temperatures between 1 200 and 1 500 K, for an average gas residence time of approximately 0.75 seconds. Exiting the reaction zone, the pyrolysis products encounter a stream of argon quench gas, which causes condensation of the species of greater than two aromatic rings. These aromatic products, along with any soot produced, are collected on a Millipore Teflon filter (hole size, 0.2 μm), which is then sonicated in dichloromethane, as described earlier [30.10], for dissolution and further analysis of the PAH. The residue from the dichloromethane extractions is termed "soot," and the yields of PAH and soot are determined as before [30.10].

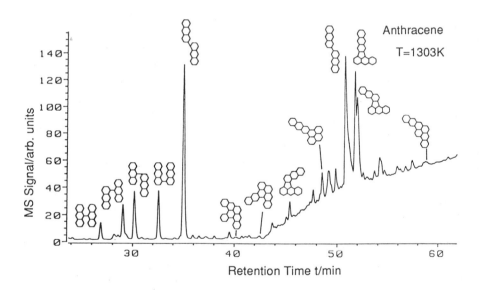

Fig. 30.2. Liquid chromatogram for the pyrolysis products of anthracene

The PAH product mixtures are analyzed by reverse phase HPLC on a Vydac 201TP octadecylsilica column. We employ a Hewlett-Packard Model 1040 chromatograph equipped with a 190-600 nm ultraviolet-visible diode-array detector. The product PAH are identified by matching the ultraviolet-visible absorption spectra of the chromatographically separated components with those of reference standards. Details of the analytical procedures are found elsewhere [30.22]. A liquid chromatogram for the pyrolysis products of anthracene is shown in Fig. 30.2, with the identification of six bianthryls, three cyclodehydrogenation products (benzonaphthofluoranthenes) formed from the bianthryls, and five seven-ring isomers of the cyclodehydrogenation products. The ramp in the chromatogram output at 43 minutes corresponds to the change in mobile phase from acetonitrile to tetrahydrofuran/dichlormethane.

30.3 Results and Discussion

The studies at MIT have covered the pyrolysis of a range of model compounds and selected coals and wood. The data below are chosen to illustrate the mechanisms of biaryl formation, cyclodehydrogenation, isomerization, and fragmentation that are found to determine the products of pyrolysis.

30.3.1 Biaryl Formation

As aryl radicals are produced from the parent molecules, they react with arene molecules or other aryl radicals to form biaryl molecules. For naphthalene (see Fig. 30.1) there are three biaryls, and for anthrance, six (see Fig. 30.2). For an aromatic molecule with n distinct substituent positions, there can be $n(n+1)/2$ biaryls. For a mixture of two aromatic compounds with n and m distinct substituent positions, the number of possible biaryls is $n(n+1)/2 + m(m+1)/2 + nm$. The rapid increase in the number of possible biaryls with the increased number of aromatic compounds makes it difficult to identify the formation of biaryls for complex mixtures although this has been achieved recently [30.23].

The use of model compounds permits the evaluation of the relative rates of formation of the different isomers. From the results, for anthracene [30.24] (Fig. 30.2) and pyrene [30.25] (Fig. 30.3), the relative abundance of the different isomers for the bianthryls is $1, 2' > 2, 2' > 1, 1' \cong 2, 9' \cong 1, 9' > 9, 9'$ and for the bipyrenes $1, 4' > 1, 2' \simeq 2, 4' > 1, 1' \simeq 4, 4' > 2, 2'$. If the probability of reaction between two molecules was determined entirely by the number of possible reaction sites, the relative abundance of the bianthryls would be $1, 2' : 2, 2' = 1, 1' = 2, 9' = 1, 9' : 9, 9' = 8 : 4 : 1$ and for the bipyrenes $1, 4' : 1, 2' = 2, 4' = 1, 1' = 4, 4' : 2, 2' = 8 : 4 : 1$. The relative abundances observed are generally in the order predicted from statistical considerations with deviations from the predictions being consistent with the expected effects of steric hindrance. It should be noted that the observed relative abundance is *not* consistent with the relative reactivity of the sites which for anthracene is $9 > 1 > 2$ and which for pyrene is estimated to be $1 = 4 > 2$. Preliminary calculations indicate that at temperatures about $1\,200\,K$ the bianthryl concentrations are those predicted from equilibrium.

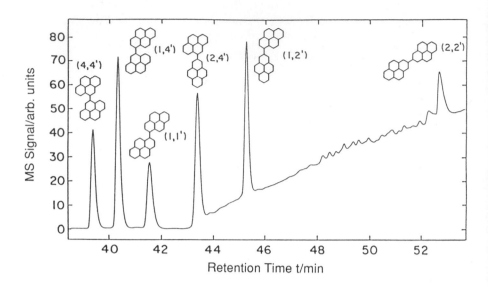

Fig. 30.3. Liquid chromatogram for the bipyrenes from pyrene pyrolysis

30.3.2 Cyclodehydrogenation

The steric factors are also important in the sequential cyclodehydrogenation reactions. The 2, 2' bianthryls and bipyrenes cannot form cyclodehydrogenation products. The products that are produced from the bipyrenes by cyclodehydrogenation involve the formation of either a five- or a six-membered ring, as shown in Fig. 30.4. The number of possible cyclodehydrogenation products is 3 for the binaphthyls, and 5 for the bianthryls and bipyrenyls. The relative amounts of the products are determined primarily by the concentration of their precursors modified slightly by the steric factors. Molecules, the structures of which have coves and fjords [30.26], will tend to have a higher steric hindrance and adopt nonplanar configurations reducing their reaction rates. For the bianthryls the only cyclodehydrogenation products observed are the three involving the formation of a five-membered ring (see Fig. 30.2); the six-membered ring equivalents were looked for and not found suggesting that their formation was precluded by steric factors. All five cyclodehydrogenation products of the bipyrenyls were observed.

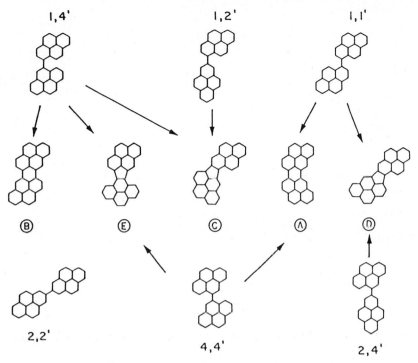

Fig. 30.4. Cyclodehydrogenation products produced from bipyrenes

30.3.3 Isomerization

A number of the pyrolysis products are found to be isomers of the cy-clodehydrogenation compounds. *Blumer* et al.[30.27] and *Scott* et al. [30.28] showed the importance of isomerization reactions. For example, *Scott* et al. [30.28] studied the interconversion of fluoranthene, aceanthrylene, and acephenanthrylene at temperatures of 1 223–1 378 K, showing how different startup mixtures approach equilibrium when heated. They proposed that the reactions occurred by means of 1, 2 carbon shifts (see Fig. 30.5).

A number of compounds are found in the products of pyrolysis of an-thracene that are isomers of the cyclodehydrogenation products of the bianthryls. Some of these seven-ring compunds have been identified on Fig. 30.2. This type of isomerization might be important in explaining the pathways for the carcinogens, dibenzo[a,h]pyrene and dibenzo[a,i]anthra-cene, which have been reported to account for a large fraction of the po-tency of these soot extracts when tested in mice [30.29]; a possible mecha-nism leading to the formation of the dibenzopyrenes from napthalene and anthracene is shown in Fig. 30.6 which draws on the types of reactions seen in our model compound studies.

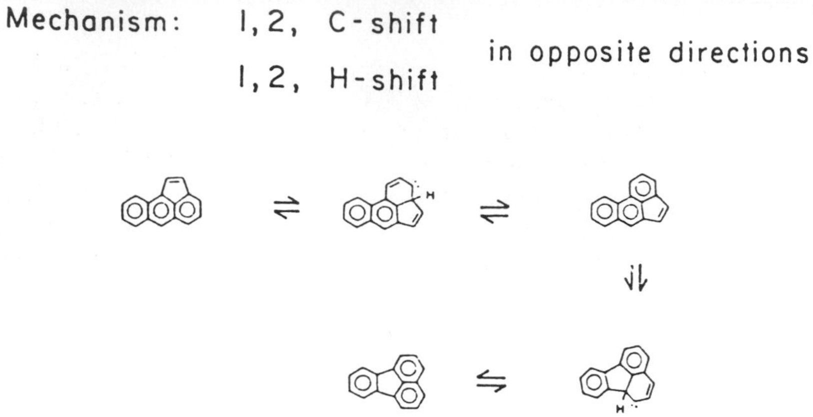

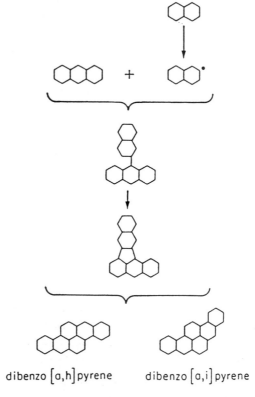

Mechanism: 1, 2, C-shift

1, 2, H-shift   in opposite directions

Fig. 30.5. Example of an isomerization of PAH [30.28]

dibenzo [a,h] pyrene   dibenzo [a,i] pyrene

Fig. 30.6. Postulated mechanism for formation of dibenzopyrenes

30.3.4 Fragmentation

As the temperature in the furnace is raised to about 1 350 K, the biaryls and their cyclodehydrogenation products decompose and are replaced by soot and polycyclic compounds that must have been produced by the fragmentation of the higher molecular-weight products. The results in Fig. 30.7 are for anthracene, but similar trends are observed with other model compounds. At high temperatures, a high conversion of the parent arene to soot is observed, consistent with our earlier studies showing high conversion to soot of bituminous coal tars [30.9].

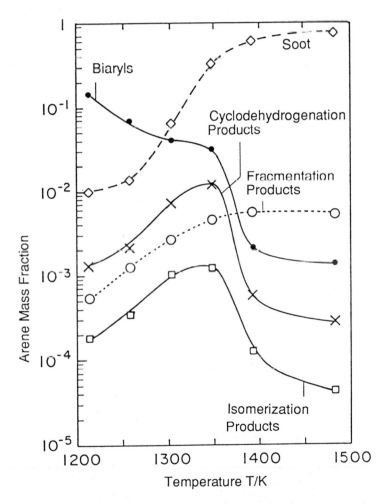

Fig. 30.7. The temperature dependence of the pyrolysis products of anthracene

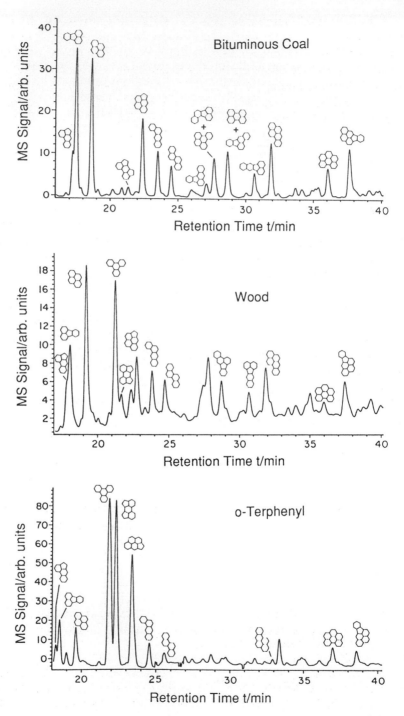

Fig. 30.8. The products of pyrolysis of a bituminous coal, a wood, and terphenyl

A relatively small number of four- to six-ring compounds persist at the higher temperatures, with very similar major components for a wide variety of fuels. This is evident from the liquid chromatograms of the products of pyrolysis at 1 475 K of o-terphenyl, a bituminous coal, and a hard wood (Fig. 30.8). Understanding of the kinetics of decomposition as well as formation is therefore critical in the development of tools for predicting the composition of polycyclic aromatic compounds (PAH).

30.4 Concluding Remarks

The present paper provides an assessment of the role of aryl-aryl reactions in the formation of PAH. The sequence of reactions involving the formation of biaryls, their cyclohydrogenation products, isomerization, and fragmentation have been quantified for several model compounds. The present paper presents an overview of data presented in greater detail elsewhere on anthracene [30.23] and pyrene [30.25]. This class of reactions should not be overlooked in modeling flames, particularly diffusion flames where soot and PAH are produced on the fuel-rich side of the flame front in temperature regimes that bridge those in this study. These reactions will be more important for fuels such as coal and wood that have an aromatic structure that readily produces aryl radicals.

Kinetics need to be obtained for the sequence of reactions so that their role in soot and PAH production may be evaluated in parallel to that of the molecular weight growth involving C_2 to C_4 radicals.

30.5 Acknowledgment

The authors are grateful to the National Institute of Environmental Health Sciences (Grant NIII-5PO1-ESO1640-14) for its financial support.

References

30.1   A.F. Sarofim: "Radiate Heat Transfer in Combustion: Friend or Foe", in Twenty-First Symposium (International) on Combustion (The Combustion Institute, Pittsburgh 1986) p. 1

30.2   M.J. Wornat, A.F. Sarofim: Aerosol Sci. Technol. **12**, 832 (1990)

30.3   W. Cautreels, K. van Cauwenberghe: Atmos. Environ. **12**, 1133 (1978)

30.4   R.C. Garner, C. Martin: Chemical Carcinogens, Second Edition, ACS Monograph 182, Vol. 4 (American Chemical Society, Washington, DC 1984) Chapter 1

30.5   R.D. Pierce, M. Katz: Environ. Sci. Technol. **9**, 347 (1975)

30.6 R. Hayatsu, R.G Scott, L.P. Moore, M.H. Studier: Nature **257**, 378 (1975)

30.7 A. Attar, G.G. Hendrickson: "Functional Groups and Heteroatoms in Coal", in <u>Coal Structure</u> (Academic Press, New York 1982) Chapter 5

30.8 J.D. Bittner, J.B. Howard: "Combustion Profiles and Reaction Mechanisms in a Near-Sooting Premixed Benzene/Oxygen/Argon Flame", in <u>Eighteenth Symposium (International) on Combustion</u> (The Combustion Institute, Pittsburgh 1981) p. 1105

30.9 R.D. Nenniger, J.B. Howard, A.F. Sarofim: <u>Proceedings of the International Conference on Coal Science</u> (Pittsburgh Energy Technology Center, Pittsburgh 1983) p.521

30.10 M.J. Wornat, A.F. Sarofim, J.P. Longwell: Energy Fuels **1**, 431 (1987)

30.11 M. J. Wornat, A.F. Sarofim, J.P. Longwell: "Pyrolysis-Induced Changes in the Ring Number Composition of Polycyclic Aromatic Compounds from a High Volatile Bituminous Coal", in <u>Twenty-Second Symposium (International) on Combustion</u> (The Combustion Institute, Pittsburgh 1988) p. 135

30.12 M.J. Wornat, A.F. Sarofim, J.P. Longwell, A.L. Lafleur: Energy Fuels **2**, 775 (1988)

30.13 S.J. Harris, A.M. Weiner: Combust. Sci. Technol. **31**, 155 (1983)

30.14 G.M. Badger, T.M. Spotswood: J. Chem. Soc. **Part IV**, 4420 (1960)

30.15 G.M. Badger, R.W.K. Kimber: J. Chem. Soc. **Part III**, 3407 (1961)

30.16 G.M. Badger, J. Novotny: J. Chem. Soc. Part III, 3400 (1961)

30.17 G.M. Badger, J.K. Donnelly, T.M. Spotswood: Aust. J. Chem. **17**, 1138 (1964)

30.18 G.M. Badger, J.K. Donnelly, T.M. Spotswood: Aust. J. Chem. **17**, 1147 (1964)

30.19 G.M. Badger, S.D. Jolad, T.M. Spotswood: Aust. J. Chem. **17**, 771 (1964)

30.20 G.M. Badger, R.W.K. Kimber, J. Novotny: Aust. J. Chem. **17**, 778 (1964)

30.21 A.S.L. Bruinsma, P.J.J. Tromp, H.J.J. de Sauvage Nolting, J.A. Moulijn: Fuel **67**, 334 (1988)

30.22 M.J. Wornat, A.L. Lafleur, A.F. Sarofim: Polycyclic Aromatic Compd. **3**, 149 (1993)

30.23 L.D. Pfefferle, G. Bermudez, J. Boyle: "Higher Hydrocarbon Formation During the Pyrolysis of Allene", this volume, sect. 2

30.24 M.J. Wornat, A.F. Sarofim, A.L. Lafleur: "The Pyrolysis of Anthracene as a Model Coal-Derived Aromatic Compound", in <u>Twenty-Fourth Symposium (International) on Combustion</u> (The Combustion Institute, Pittsburgh 1992) p. 955

30.25 J. Mukherjee, A.F. Sarofim, J.P. Longwell: Energy Fuels, in press

30.26 J. Dias: J. Chem. Inf. Computer Sci. **30**, 61 (1990)

30.27 G.-P. Blumer, K.-D. Gundermann, M. Zander: Chem. Ber. **110**, 2005 (1977)

30.28 L.T. Scott, J.H. Roelofs: J. Am. Chem. Soc. **109**, 5461 (1987)

30.29 W.F. Busby Jr.: personal communication, 1991

Discussion

D'Alessio: You examined the extract of tarry material from the pyrolysis. What was the proportion of the total material you where able to examine with the HPLC spectra technique and how much was left out?

Sarofim: We have a combination of unreacted material at lower temperatures which is anthracene plus dimerization products which account for the entire product spectrum; as we go up higher in temperature we begin to see less of the unreacted anthracene and some of the cyclodehydrogenation products around 1400 K; we begin to see compounds that were produced by secondary reactions and soot starting around 1300 K. We account for about 85 % of the mass of the tar.

D'Alessio: Do you get the same results starting either with model compounds or with wood?

Sarofim: All of the polycyclics you see represent 1 to 2 % of the wood weight. For coal you may get up to 15 % tar.

D'Alessio: How much PAH do you get from the tar of the wood.

Sarofim: About 30 to 40 %.

Moss: Based on what has been said about temperature limits it may be that the temperatures above 1400 K are the temperatures which are important in soot formation process and it might be worth saying a few more words about the secondary products.

Sarofim: For the coal for which we have the most extensive data we find invariably peaks in the ring number at the 5 to 6 membered rings and the maximum maybe at 30 to 40 carbon atoms. This seems to be consistent with the model and we believe that the higher PAH don' t exist because they are preferentially scavenged by the soot, suggesting that PAH are the building blocks for soot formation. If you do a collision analysis you don't have to have a very high collision efficiency in order to absorb all of the higher molecular PAH.

Moss: Do you see any light products?

Sarofim: We have aliphatics which we can break down in two groups. About 2/3 of the aliphatics come from decomposition products and 1/3 from the side chaines which are broken off.

Santoro: What were the residence times of these reactions?

Sarofim: About 0.6 to 0.7 seconds.

Santoro: In the early work of Lahaye and Prado one finds low temperature soot formation studies in flow reactors with benzene and other molecules where they had very long reaction times to produce soot particles. We didn't talk about that because in most of the combustion systems about which we are concerned the residence times are fairly short. How do you think your mechanisms, involving this purely aromatic coagulation and isomerization reactions, would compare with having a reactive species such as acetylene around to provide the "cement" and having a residence time of 0.1 second or even much less than that.

Sarofim: In Judy Wornat's thesis she had limited data where she varied residence time. She used conversion as a measure of reaction. At a given conversion the effect of residence time or temperature were interconvertible i.e. you could go to higher conversion by having increasing temperature or increasing residence time. We just got some results from Fred Merklin at Kansas State University. He has some shock tube studies on benzene where he observes biphenyls, some phenanthrene and higher ring members but has no naphthalene. If you think of naphthalene as being the stepping stone for C_2 addition than, at least for his shock tubes studies which went up to 1900 K, similar reactions might have been occuring.

Homann: How did you identify those very large PAH.

Sarofim: Judy Wornat has located all of the people around the world who have synthesized these compounds and has been in correspondence with them. We used absorbance over the 190 to 600 nm range for identification and we have been trying to track down as many of the reference compounds as possible.

Pfefferle: I want to make a remark on these mechanisms at short residence times (1-20 ms) in the 1600 K regime. We have studied the pyrolysis, oxidative pyrolysis of a large range of compounds from acetylene up to dicyclopentadiene. The temperature onset for 1 mg PAH appearance and the way they get there from a small species are different but the general features of high mass hydrocarbon distributions (<228 u) at long residence times and high temperatures are the same.

Wagner: Did you also check straw?

Sarofim: No, we have not looked at straw yet.

Wagner: You compared coal and wood. Coal is usually formed from wood under conditions which are not too much different from what you had here and you got quite a number of similarities. If you look more carefully at the coal then you find a lot of compounds which still contain heteroatoms. Is there also a similarity concerning the compounds which contain the heteroatoms from the products of the coal and the wood you investigated here?

Sarofim: We believe that similar reactions will occur. There is evidence from our wood that the oxygenated compounds are more resistant and therefore they persist and decay at lower rate. The heterocompounds we have looked at have been the nitrogen containing ones. The nitrogen content decreases with time. We find the nitrogen both in the soot and in the tars persisting up to fairly high temperatures but they are always decreasing.

Wagner: How far can you compare coal and wood?

Sarofim: Coal and wood look very similar once you get up to 1400 K and above. At lower temperatures we went to the model compounds because we get a continuum of peaks which give a big hump and it would be a Ph.D.

thesis in itself to resolve them all. That is why we went to the model compounds because we believe that these reactions are occuring. We thought we would study a large number of model compounds but the analytical work is too time consuming. We would like to look at heteroatoms containing compounds also. Looking at wood and coal at lower temperatures is complicated just by the enormous number of compounds present.

Colket: I just want to add a word about the benzene pyrolysis. We also have done shock tube studies of benzene pyrolysis and indeed biphenyl is a very important product and the modeling of the system indicates most of it is formed during the high temperature process not in a quenching way which seems to support very much the early modeling results of Prof. Frenklach and the importance of biphenyl in ringformation process as recombination steps. But we also see significant amounts of naphthalene.

Sarofim: I do not say this is the main pathway for production of the soot in conventionel fuels. But I think it should be looked at this as a possible parallel path.

Santoro: I' m getting the impression from what you said earlier that a key feature of the mechanism that you were talking about is that you do not need to have acetylene around to have the growth of these species. When you were showing us results from your work you were showing building blocks that had an acetylene addition to them which led to the same large compound. So to me there is a conceptional difference between those two models. Here you have a very low acetylene concentrations and still getting rapid build up by isomerization and the rearrangement of bonds in a possibly hydrogen rich environment.

Pfefferle: I believe both pathways are important in the lower mass range in our work. In the 300 u, 400 u range we were at first appearance observing very distinct peaks at large spacing which were not explanable by these acetylene groupings. And so I think it just depends on the conditions. Both mechanisms are competing but in that higher range for there reaction conditions we studied addition reactions of high molecular-weight species are important.

Models for Soot Formation in Turbulent Flames

The numerical effort for modelling turbulent flames is much larger than for one-dimensional laminar flames. From this soot models with detailed chemical mechanisms presently are not applicable to turbulent flames, unless the large system of chemical reactions can be reduced essentially. Consequently soot models for turbulent flames contain a limited number of variables to describe the process of soot formation. Due to the reduction or simplification parts of these models have to be "calibrated" for the system under consideration. Part VI contains different approaches to reduce the complex chemical description of soot formation and different approaches to link the complex chemistry to a turbulent flow.

Application of a Soot Model to a Turbulent Ethylene Diffusion Flame

Wolfgang Kollmann [1], Ian M. Kennedy [1]

Mario Metternich [1], J.-Y. Chen [2]

[1]Mechanical Engineering Department,
University of California,
Davis, CA 95616, USA
[2]Combustion Research Facility,
Sandia National Laboratories,
Livermore, CA 94551-0969, USA

Abstract: Modeling of a highly non-adiabatic turbulent ethylene diffusion flame has been tested with two mixing models in a pdf code. As a result of the radiative heat loss from the soot the relationships between the flow properties such as mixture fraction and enthalpy are much different than the more common adiabatic conditions. A comparison of the predictions with measurements of soot volume fraction that were obtained in a combustion wind tunnel indicate that satisfactory agreement is achieved. The statistics showed that most of the soot was at a temperature of about 1400K. The correlation of mixture fraction and soot volume fraction was small, indicating that state relationships are not appropriate for this particular flame.

31.1 Introduction

The combustion of fossil fuels in energy systems inevitably results in the production of soot and radiation within the combustor. A large fraction of the heat of combustion may be lost from the flame through the action of radiation, and it should be expected that the structure of these highly non-adiabatic flames is quite different from non-luminous or adiabatic flows. Because the calculation of radiative heat transfer from turbulent flames is difficult, particularly from luminous flames with substantial soot loadings that are not easily predicted, most computations of turbulent diffusion flames have neglected the coupling of radiation and flame structure.

Magnussen et al. [31.1] attempted to incorporate a model for soot formation into a k-ε model for a turbulent acetylene flame. The model for soot formation that they used was derived from earlier work by *Tesner* et al. [31.2]; it was not based fundamentally on our present understanding of soot formation. More recently, *Moss* et al. [31.3] have made some progress in formulating a soot model that is based on the essential physics of the phenomenon. They used a two equation model for soot production, one equation for particle nucleation and coagulation and a second equation for particle surface growth and oxidation. This model performed satisfactorily when the constants in the model were adjusted after comparison with experimental results in laminar flows. The usual questions arise in the application of the soot formation model in a turbulent flow with regard to the treatment of the non-linearities in the source and sink terms in the soot equations and the potential significance of correlations between the flow variables which must be neglected.

A probability density function (pdf) transport model of a reacting turbulent flow has the significant advantage that it can deal with chemical source terms exactly and it does not neglect or model correlations between the scalar variables [31.4-5]. However, the calculation of a turbulent flame with this approach is computationally intensive and requires a Monte Carlo technique. It is highly desirable to minimize the number of independent scalar variables which need to be solved. Therefore, some effort has been devoted to the development of a simple soot model which can adequately represent the physics and chemistry of soot formation and oxidation without incurring an enormous numerical penalty. The soot model was developed for a laminar diffusion flame [31.6-7] and it has been incorporated without any change into a pdf code for a turbulent ethylene flame.

31.2 Thermochemical Model and Soot Formation

The thermochemistry of the flame has been described in terms of a constrained equilibrium model. That is, chemical equilibrium is calculated for a given mixture fraction [31.8] subject to the constraint of a specified enthalpy. The enthalpy h

$$h = \sum_{i=1}^{n} \frac{\bar{h}_i}{M_i} W_i \tag{31.1}$$

is the sum of the enthalpies of the individual components and the molal enthalpy $\bar{h}_i$ is the sum of the sensible and chemical enthalpies

$$\bar{h}_i = \bar{h}_i^0 + \int_{T_0}^{T} dT' \bar{C}_p(T') \tag{31.2}$$

The low Mach number limit of the energy equation can be given in the form

$$\rho(\frac{\partial h}{\partial t} + v_\alpha \frac{\partial h}{\partial x_\alpha}) = -\frac{\partial q_\alpha}{\partial x_\alpha} \tag{31.3}$$

The energy flux consists of conductive-diffusive q_α^{cd} and radiative q_α^{r} contributions. The former can be expressed in terms of the enthalpy for Lewis number unity, and the divergence of the radiative part can be approximated by

$$\frac{\partial q_\alpha^{\mathrm{r}}}{\partial x_\alpha} = 4\sigma_{\mathrm{SB}} K_{\mathrm{abs}}^{\mathrm{Pl}}(T^4 - T_\infty^4) \tag{31.4}$$

The assumption is made that the soot loading in this flame permits the use of the optically thin approximation for radiation; this assumption is justified by reference to the experimental results of radiative heat fluxes measured by *Neill* et al. [31.9] for the same flame conditions. The divergence of the energy flux is therefore given by

$$-\frac{\partial q_\alpha}{\partial x_\alpha} = \frac{\partial}{\partial x_\alpha}(\rho \Gamma \frac{\partial h}{\partial x_\alpha}) - 4\sigma_{\mathrm{SB}} K_{\mathrm{abs}}^{\mathrm{Pl}}(T^4 - T_\infty^4) \tag{31.5}$$

where Γ denotes the thermal conductivity. Radiative heat loss is incorporated in the pdf method by including enthalpy as one of the independent variables for the pdf. The sink term in the energy equation accounts for black-body radiation from soot particles.

The chemical model for the combustion of ethylene with air was based on the assumption of constrained equilibrium which leads to local relationships for the thermodynamic properties as a function of mixture fraction and enthalpy. Temperature $T(f, h)$, is shown in Fig. 31.1 as a function of the mixture fraction and an enthalpy normalized by the maximum possible enthalpy (the adiabatic case) and an arbitrarily chosen lower limit that encompasses the expected variations in enthalpy that are likely to be encountered.

The soot model is identical to the one proposed by *Kennedy* et al. [31.6] for a laminar ethylene-air diffusion flame. Only one equation is used viz., a balance equation for the volume fraction of soot (the volume of soot per unit volume).

$$\rho \left[\frac{\partial f_{\mathrm{V}}}{\partial t} + (v_\alpha + v_\alpha^{\mathrm{th}}) \frac{\partial f_{\mathrm{V}}}{\partial x_\alpha} \right] = \frac{\partial}{\partial x_\alpha}(\rho \Gamma_{\mathrm{s}} \frac{\partial f_{\mathrm{V}}}{\partial x_\alpha}) + \rho R_{\mathrm{n}} + \rho R_{\mathrm{s}} - \rho R_{\mathrm{O}} \tag{31.6}$$

where Γ_{s} denotes the thermal conductivity of soot [31.7]. The thermophoretic velocity v_α^{th} is small compared to the fluid velocity for turbulent flows and can be neglected in this case. Three source and sink terms account for the formation and oxidation processes which are important in sooty flames. The first two terms R_{n}, R_{s} determine the amount of soot volume which is produced as a result of particle nucleation and surface growth.

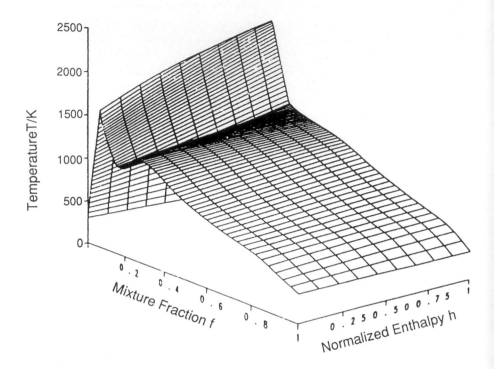

Fig. 31.1. Temperature as a function of mixture fraction f and normalized enthalpy h

Particles of soot are first formed as very small spherules of the order of 1 nm in diameter [31.10]. The number density of particles is very large initially but it is rapidly reduced through the mechanism of coagulation. Most sooty flames exhibit remarkably similar particle number densities. This observation suggested the possibility of using an average number density in the calculations so as to avoid the requirement of an additional variable (soot particle number density). Earlier calculations in a laminar flame showed that this approach leads to reasonable results [31.7]. The peak particle nucleation rate was estimated to be 10^{18} m^{-3}s^{-1} with a clipped Gaussian distribution over a narrow range of mixture fractions on the fuel rich side of stoichiometric. It was found that the results in laminar flames are not at all sensitive to the details of this part of the model. The model for nucleation is then

$$R_{n} = C_{n} \exp\left[-\frac{(f - f_{n})^{2}}{\sigma_{\text{SB n}}^{2}}\right] \tag{31.7}$$

where $C_{n} = 10^{18}$ m^{-3}s^{-1} is the peak formation rate and $f_{n} = 0.12$, $\sigma_{\text{SB}} = 0.02$ are empirical parameters as discussed by *Kennedy* et al. [31.6].

Most of the soot in a flame is produced as the result of surface growth, which refers to the addition of hydrocarbon species from the gas phase onto the particles. The rate of this process depends upon the concentrations of available carbon-bearing species in the neighborhood of the particles, on the temperature, and possibly on the reactivity of the particle surface. The first of these effects is accounted for through the use of an empirical correlation between mixture fraction and specific surface growth rates; the specific surface growth rate is the volume of soot added to a particle per unit aerosol surface area. The correlation that we used was derived from measurements in a laminar counterflow diffusion flame (see [31.6] for details). The temperature effect is estimated by assuming an Arrhenius expression for the growth rate with a suitable activation energy:

$$R_s = \pi^{1/3} 6^{2/3} N_{s\ avg}^{1/3} f_V^{2/3} \dot{m}''(f, T) \tag{31.8}$$

It has been found in our previous soot calculations in laminar diffusion flames that it is not necessary to include a particle "aging" effect; the effect of temperature appears to dominate any loss in particle reactivity in these flows.

Oxidation of soot particles occurs as a result of the attack on particles by OH radicals and by molecular oxygen; the latter mechanism is only significant on the lean side of the flame. Rates of OH oxidation were derived from the measurements of *Neoh* et al. [31.11]. The model of Nagle and Strickland-Constable [31.12] was used to account for the attack of molecular oxygen.

$$R_O = f_V^{2/3} \left[R_{OH}(f) + C_1 \exp(-\frac{C_2}{T}) \right] \tag{31.9}$$

where $R_{OH}(f)$ is the rate of soot destruction due to the OH-radical according to *Kennedy* et al. [31.6] and $C_1 = 4.74 \cdot 10^5$, $C_2 = 16000$ K are the parameters of the Nagle and Strickland-Constable formula [31.12] for the destruction rate due to O_2 in the low temperature limit.

31.3 PDF Method

The description of the local thermodynamic state requires three scalar variables $\psi_i, i = 1, 2, 3$. These variables were defined in the previous section as mixture fraction $\psi_1 \equiv f$, enthalpy $\psi_2 \equiv h$ and soot volume fraction $\psi_3 \equiv f_V$ and they are governed by transport equations of a general structure given by convection, diffusion, and sources.

$$\rho(\frac{\partial \psi_i}{\partial t} + v_\alpha \frac{\partial \psi_i}{\partial x_\alpha}) = \frac{\partial}{\partial x_\alpha}(\rho \Gamma_i \frac{\partial \psi_i}{\partial x_\alpha}) + \rho Q_i(\psi_1, .., \psi_3) \tag{31.10}$$

The source terms follow from the models for the combustion reactions and the soot formation discussed in the previous section. Mixture fraction is conserved leading to

$$Q_1 = 0 \qquad (31.11)$$

The energy equation for low Mach numbers was shown to be of the general form (31.10) and the source term represents the divergence of the radiative energy flux. Its form is according to the radiation model discussed above, with the explicit form of the absorption coefficient $K_{\text{abs}}^{\text{Pl}}$ according to *Kennedy* et al. [31.7] given by

$$Q_2(f, h, f_V) = -8\pi K_1 f_V^{2/3}[1 - (\frac{T_\infty}{T})^4] \int_0^\infty \frac{\mathrm{d}\lambda}{\lambda^6[\exp(\frac{C_2}{\lambda T}) - 1]} \qquad (31.12)$$

where $K_1 = 4.16808 \cdot 10^{14}$, $T_\infty = 300$ K and $C_2 = 14388$ μmK. The equation for the dynamics of the soot volume fraction contains a source term consisting of three distinct contributions.

$$Q_3(f, h, f_V) = R_n(f) + R_s(f, h, f_V) - R_O(f, h, f_V) \qquad (31.13)$$

They represent production of soot volume fraction due to nucleation and surface growth and the destruction due to the action of the OH radical and O_2, as discussed in the previous section. The transport equation for the pdf of the values of these three scalars at a single point $(\underline{x}, t)$ in the turbulent flow field can be obtained with standard methods [31.4, 31.5].

$$\langle \rho \rangle \left\{ \frac{\partial \tilde{P}_1}{\partial t} + \tilde{v}_\beta \frac{\partial \tilde{P}_1}{\partial x_\beta} + \sum_{j=1}^l \frac{\partial}{\partial \varphi_j} \left[Q_j(\varphi_1, \cdots, \varphi_l) \tilde{P}_1 \right] \right\} =$$

$$-\frac{\partial}{\partial x_\alpha} \left(\langle \rho \rangle \langle v_\alpha''|\psi_j = \varphi_j \rangle \tilde{P}_1 \right) - \langle \rho \rangle \sum_{j=1}^l \sum_{k=1}^l \frac{\partial^2}{\partial \varphi_j \partial \varphi_k} \left(\langle \varepsilon_{jk}|\psi_j = \varphi_j \rangle \tilde{P}_1 \right)' \qquad (31.14)$$

where $v_\alpha'' \equiv v_\alpha - \tilde{v}_\alpha$ are the velocity fluctuations and $\langle q|\psi_j = \varphi_j \rangle$ is the conditional expectation for variable q. The density-weighted pdf $\tilde{P}_1$ is defined by

$$\tilde{P}_1 \equiv \frac{\rho(\varphi_1, \cdots, \varphi_l)}{\langle \rho \rangle} P_1(\varphi_1, \cdots, \varphi_l; \underline{x}, t) \qquad (31.15)$$

and the scalar dissipation rates ε_{ij} are defined by

$$\varepsilon_{ij} \equiv \Gamma_{ij} \frac{\partial \psi_i}{\partial x_\alpha} \frac{\partial \psi_j}{\partial x_\alpha} \qquad (31.16)$$

The pdf equation contains two additional unknowns: the turbulent flux and the conditional expectations of the scalar dissipation rates. The closure model for the latter is called the mixing model. The present model is based on the notion of pairwise exchange of scalar properties within small fluid parcels [31.13, 31.14]. This idea can be carried over to two-phase flows if the solid phase moves randomly relative to the fluid and, therefore, diffuses relative to the gas phase fluid. The mixing model to be discussed briefly, allows control over the amount of mixing for each scalar variable individually. The mixing model is given by

$$-\langle\rho\rangle\sum_{j=1}^{l}\sum_{k=1}^{l}\frac{\partial^2}{\partial\varphi_j\partial\varphi_k}\left(\langle\varepsilon_{jk}|\psi_j=\varphi_j\rangle\tilde{P}_1\right)=$$

$$\langle\rho\rangle\frac{C_d}{\tau}\{\int_{\Re}\mathrm{d}\underline{\varphi}'\int_{\Re}\mathrm{d}\underline{\varphi}''\tilde{P}_1(\underline{\varphi}')\tilde{P}_1(\underline{\varphi}'')T(\underline{\varphi}',\underline{\varphi}''\rightarrow\underline{\varphi})-\tilde{P}_1(\underline{\varphi})\} \qquad (31.17)$$

where $\Re$ denotes the domain of definition in scalar space. The soot volume fraction is treated as a diffusive scalar variable if equal diffusivities are assumed for all scalars and the integration is carried out over the three-dimensional scalar space. However, if the integration is restricted to the $l-1$ dimensional scalar space excluding the soot volume fraction, then soot volume fraction is treated as a non-diffusive scalar. Both limiting cases will be considered in this paper. The properties of the mixing model are contained in the transition pdf T and the time scale τ. *Dopazo* [31.13] and *Janicka* et al. [31.14] showed that T must satisfy the following constraint

$$T(\underline{\varphi}',\underline{\varphi}''\rightarrow\underline{\varphi})=\begin{cases}|\underline{\varphi}''-\underline{\varphi}'|^{-1} & \text{for } \underline{\varphi}\in[\underline{\varphi}',\underline{\varphi}'']\\ 0 & \text{otherwise}\end{cases} \qquad (31.18)$$

This model has been applied to studies of gas-phase combusting flows [31.4, 31.15-16]. The time scale τ for the mixing model depends on the turbulence field and the scalar fields [31.15]. The present model relies on the assumptions of constant time scale ratios and is given by

$$\tau=C_d^{-1}\frac{\tilde{k}}{\tilde{\varepsilon}} \qquad (31.19)$$

with C_d denoting a constant of order unity. The turbulent flux in the pdf equation is closed with a gradient-flux model and the final form of the closed pdf equation is then given by

$$\langle\rho\rangle\left\{\frac{\partial\tilde{P}_1}{\partial t}+\tilde{v}_\beta\frac{\partial\tilde{P}_1}{\partial x_\beta}+\sum_{j=1}^{l}\frac{\partial}{\partial\varphi_j}(Q_j(\varphi_1,\cdots,\varphi_l)\tilde{P}_1)\right\}=\frac{\partial}{\partial x_\alpha}\left(C_s\langle\rho\rangle\frac{\tilde{k}}{\tilde{\varepsilon}}\widetilde{v_\alpha''v_\beta''}\frac{\partial\tilde{P}_1}{\partial x_\beta}\right)$$

$$+\frac{C_d}{\tau}\langle\rho\rangle\{\int_{\Re}\mathrm{d}\underline{\varphi}'\int_{\Re}\mathrm{d}\underline{\varphi}''\tilde{P}_1(\underline{\varphi}')\tilde{P}_1(\underline{\varphi}'')T(\underline{\varphi}',\underline{\varphi}''\rightarrow\underline{\varphi})-\tilde{P}_1(\underline{\varphi})\} \qquad (31.20)$$

Note that $\Re$ denotes either the full or the reduced scalar space. The mean values for thermodynamic properties such as density can be calculated by integration. The statistics of the velocity field are determined from a second order closure described by *Dibble* et al. [31.17]. The numerical solution of the pdf transport equation is carried out with a stochastic simulation technique [31.4, 31.16].

31.4 Application to Non-Premixed C_2H_4 Flames

Non-premixed round jet flames burning C_2H_4 in air were considered with a nozzle exit velocity of $v_0 = 44.5$ ms^{-1} and coflowing air streams at two velocities, $v_e = 5.0$ ms^{-1} and $v_e = 8.0$ ms^{-1}. The nozzle diameter was 5 mm. The present results refer to $v_e = 5.0$ ms^{-1} only. The Reynolds number based on nozzle conditions was $Re = 20000$. The velocity field is obtained through a Reynolds stress closure.

The soot model represented by the source terms Q_1, Q_2 and Q_3 was tested by comparison with measurements in buoyant, laminar, non-premixed flames. It was shown [31.6] that the essential properties of soot formation in a laminar diffusion flame could be predicted with this model. The present soot model is identical to the model of *Kennedy* et al. [31.6] except that the thermophoretic velocity was neglected in the turbulent flames. Figure 31.2. shows a comparison of the mean soot volume fraction along the centerline with the measurements of *Neill* et al. [31.9]. The full line corresponds to mixing in the restricted scalar space. This implies that the molecular diffusivity of soot was set to zero. This is the correct way to treat soot as long as the particles move like material points. The broken line is the result if soot is allowed to diffuse like the mixture fraction (mixing occurs in the full scalar space). The measurements were obtained by a laser extinction technique in which the attenuation of a light beam was recorded as it traversed the axis of the round jet flame. The computed results can be transformed into a soot volume fraction $\hat{f}_V(x)$ that is equivalent to the experimental line-of-sight measurement. The equivalent soot volume fraction is defined as

$$\hat{f}_V(x) \equiv \frac{1}{2L} \int\limits_{-L}^{L} dr \tilde{f}_V(x, r)$$

where $\tilde{f}_V(x, r)$ is the local mean soot volume fraction and $2L$ is the path length across the soot bearing zone of the flame. The integration was carried out for both lines in Fig. 31.2. It must be reiterated that the soot model was not changed in any manner when it was transferred from a laminar code to the present pdf calculation. No optimization at all has been attempted. Therefore, the agreement with the measurements is encouraging in view of the errors inherent in the line-of-sight extinction measurements and the approximations in the soot model. The use of the restricted scalar space for mixing results in a modest improvement of the predictions.

Figure 31.3 shows the mean profiles for mixture fraction along the centerline of the jet. The full line corresponds to mixing in the restricted scalar space and the broken line to mixing in the full scalar space. As might be expected, very little effect on the mixture fraction is observed. However, the situation for enthalpy in Fig. 31.4 is somewhat different. In the region of

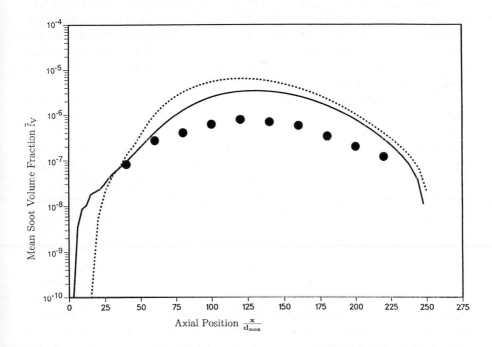

Fig. 31.2. Mean soot volume fraction $\tilde{f}_V$ along the axis compared to measurements by *Neill* et al. [31.9] for a turbulent C_2H_4 flame with zero pressure gradient; full line denotes mixing in the restricted scalar space (zero diffusivity for soot) and the broken line is the result for mixing in the full scalar space; profiles were smoothed using third order splines

the flame where radiation becomes important $(x/d_{noz} \geq 100)$ enthalpy is significantly reduced if soot mixes like the mixture fraction itself. This is a consequence of the higher soot levels in this case; they lead to higher radiative loss rates that are evident in the radial profile of enthalpy in Fig. 31.5. The impact of the mixing model on temperatures is, by contrast, rather modest. This may be seen in Fig. 31.6 which presents a radial profile of the mean temperatures at $x/d_{noz} = 100$. The comparison of mixture fraction with enthalpy demonstrates that, due to radiation heat loss, enthalpy is no longer a linear function of mixture fraction, as is the case up to about $x/d_{noz} \leq 50$. The profiles are not only different in shape but enthalpy is also negative, in contrast to the initial part of the jet where enthalpy is strictly positive (note that the standard enthalpy of formation of C_2H_4 is positive). This shift of the enthalpy to negative values is due to the radiative heat loss, as described by the sink term in the energy equation (Eq. 31.12).

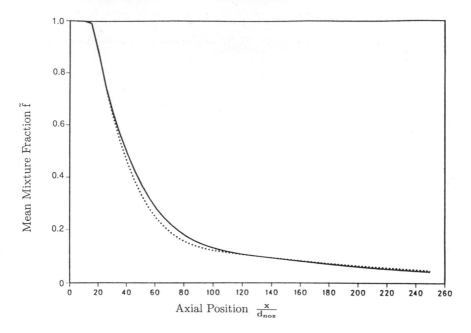

Fig. 31.3. Mean mixture fraction $\tilde{f}(x, o)$ on centerline; see legend for Fig. 31.2

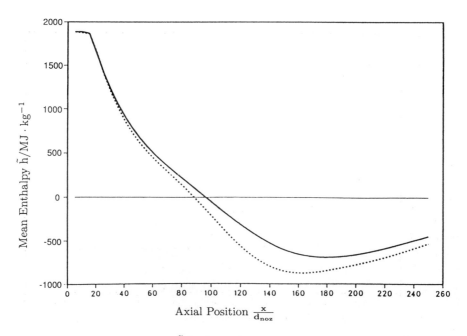

Fig. 31.4. Mean enthalpy $\tilde{h}(x, 0)$ on the centerline; see legend for Fig. 31.2

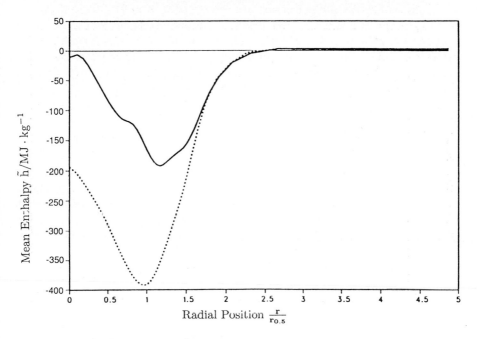

Fig. 31.5. Mean enthalpy $\tilde{h}(x,r)$ at $x/d_{\mathrm{noz}} = 100$; see legend for Fig. 31.2

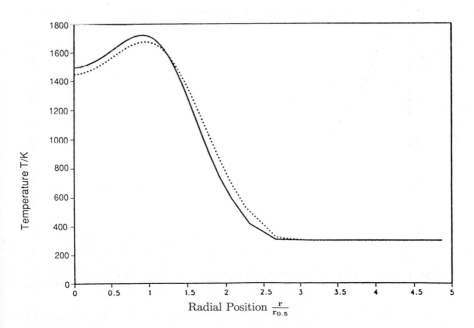

Fig. 31.6. Mean temperature $\tilde{T}(x,r)$ at $x/d_{\mathrm{noz}} = 100$; see legend for Fig. 31.2

Profiles of soot volume fraction are presented in Fig. 31.7. Allowing soot to mix in the full scalar space leads to higher levels of soot. The same point has been inferred indirectly from the results for enthalpy that were presented above. The properties of the soot volume fraction are presented in greater detail in terms of one-dimensional pdfs at $x/d_{noz} = 100$ in Figs. 31.8 $(r/r_{0.5} = 0.1)$ and 31.9 $(r/r_{0.5} = 0.74)$. The mixing model has a pronounced impact on the pdf for soot volume fraction as these figures clearly indicate. The full line is the pdf that corresponds to mixing in a restricted space i.e., soot does not diffuse and the broken line corresponds to mixing in the full scalar space. If the soot does not mix it is seen that the pdf has a long tail and a relative maximum at much smaller values of f_V than in the case where soot mixes like the other scalars (broken line). The pdfs without mixing bear a remarkable similarity to the light scattering pdfs of soot that were obtained by *Kennedy* [31.18] in a turbulent ethylene flame. Although it is not possible to relate laser light scattering directly to soot volume fraction it yields a qualitative indication of soot loadings at a point within the flame. The measured pdfs exhibited a spike at zero soot volume fraction and a very long tail similar to the results of the pdf calculation.

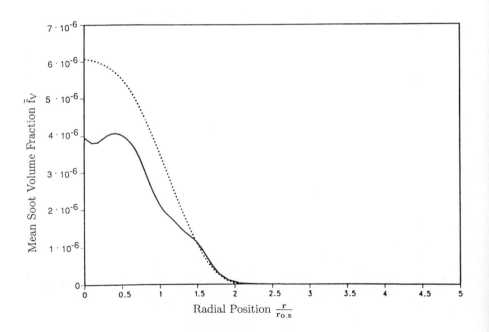

Fig. 31.7. Mean soot volume fraction $\tilde{f}_V(x, r)$ at $x/d_{noz} = 100$; see legend for Fig. 31.2

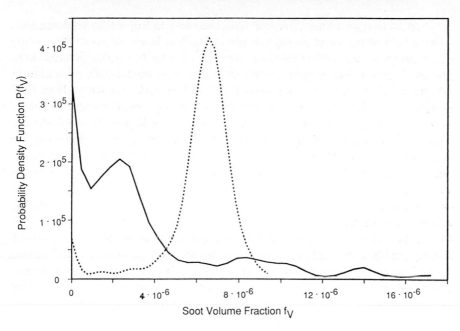

Fig. 31.8. Pdf of soot volume fraction $f(f_V, x, r)$ at $r/r_{0.5} = 0.1$; see legend for Fig. 31.2

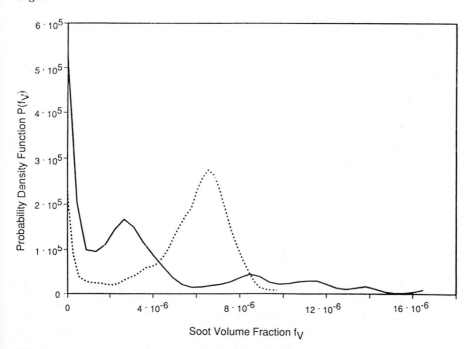

Fig. 31.9. Pdf of soot volume fraction $f(f_V, x, r)$ at $r/r_{0.5} = 0.74$; see legend for Fig. 31.2

A comparison of Figs. 31.8 and 31.9 demonstrates that the relative maxima persist in both cases across the flame. It should be noted that the change in the mean value is determined by the spike at zero f_V and the length of the tails. It is well known that the transport model in the second order closure for the velocity field leads to a variation of the relative maximum according to the mean profile. Such a variation is not observed in the present case and it follows that the relative maximum is created by the source terms, in particular surface growth, and not transport effects. These source terms are stongly dependent on temperature and mixture fraction. It follows that a better understanding of this phenomenon can be obtained from the temperature and mixture fraction statistics.

The pdfs for temperature at $x/d_{\mathrm{noz}} = 100$ are shown in Figs. 31.10 and 31.11 for which $r/r_{0.5} = 0.1$ and $r/r_{0.5} = 0.92$ respectively. At the inner station there is only one peak but as the outer edge is approached bi-modality is seen to develop. At the outer location there is a peak at about 1400K and another at around 2300K which is close to the adiabatic flame temperature. The origin of the lower mode may be found in an examination of both the pdf for mixture fraction and the equilibrium function of temperature that was presented in Fig. 31.1. Temperatures around 1400K are seen to occupy quite a large portion of the mixture fraction space. Bimodality of the pdf is enhanced by mixing in the restricted scalar space as is evident in Fig. 31.11.

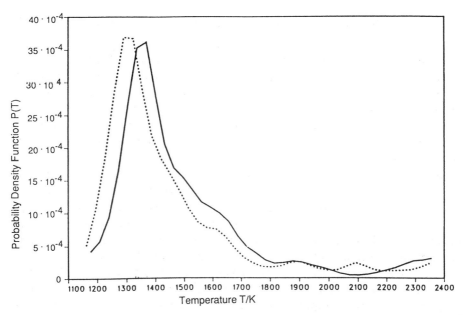

Fig. 31.10. Pdf of temperature $f(T, x, r)$ at $r/r_{0.5} = 0.1$; see legend for Fig. 31.2

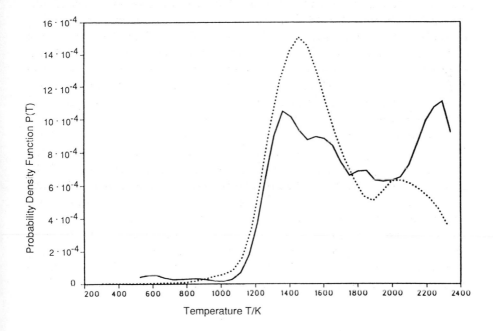

Fig. 31.11. Pdf of temperature $f(T, x, r)$ at $r/r_{0.5} = 0.92$; see legend for Fig. 31.2

The joint statistics of temperature and soot volume fraction are very important with regard to the calculation of radiative heat fluxes from luminous flames. The use of mean properties such as mean temperature and mean soot volume fraction may give erroneous answers in a turbulent diffusion flame because of the non-linear relationship between radiative heat flux and these properties. Therefore, it is instructive to examine the joint pdf of soot volume fraction and temperature and the correlation coefficients of mixture fraction, temperature and soot volume fraction. Fig. 31.12 presents the joint pdf of soot volume fraction and temperature at $x/d_{\mathrm{noz}} = 100$ and $r/r_{0.5} = 0.47$. The adiabatic situation would correspond to a pdf concentrated along the line of zero soot volume fraction. The variance of temperature would be greatest in the absence of soot. However, as more soot is present, the temperature variation decreases and the temperatures cluster around 1400 K. This finding bears a similarity to the measurements of *Sivathanu* et al. [31.19] in a turbulent ethylene air flame. They found a correlation betwen temperature and soot volume fraction that was more or less independent of soot loading and appeared to be clustered around 1500 K. Their results, in conjunction with the present calculations, suggest the possibility of a quasi-universal relationship between the soot layer and its temperature in hydrocarbon diffusion flames.

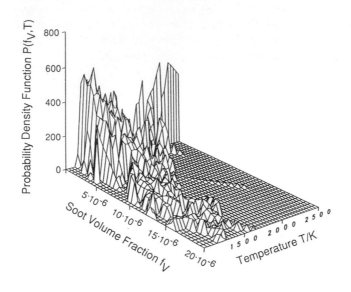

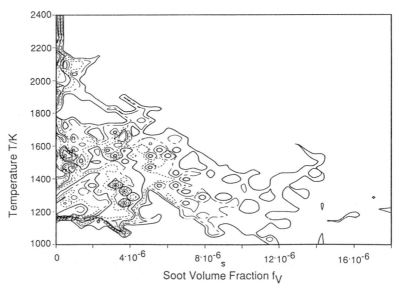

Fig. 31.12. Pdf of temperature T and soot volume fraction f_V at $r/r_{0.5} = 0.47$ and $x/d_{\mathrm{noz}} = 100$

The correlation coefficients between these parameters is shown at two axial stations $x/d_{\mathrm{noz}} = 100$ and $x/d_{\mathrm{noz}} = 150$ in Fig. 31.13 to Fig. 31.16. The solid line in each figure presents the results from restricting the mixing of the soot. The mean flame position at $x/d_{\mathrm{noz}} = 100$ is around $r/r_{0.5} = 1.0$.

It appears that soot and temperature are strongly negatively correlated on the fuel side of the reaction zone. This is a result of radiative heat loss. At $x/d_{\mathrm{noz}} = 150$ this correlation becomes even stronger. The implications for mean property modelling of radiation from this turbulent flame have not been analyzed as yet but some interesting insights should be gained from integration of a radiation model over the joint pdf.

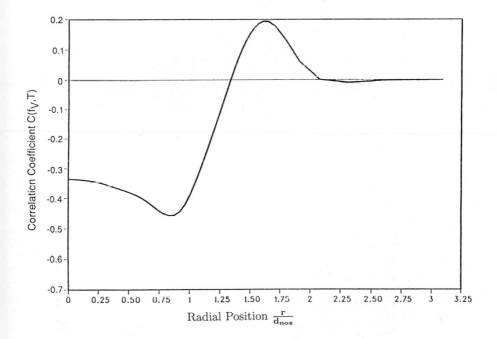

Fig. 31.13. Correlation coefficient between soot volume fraction f_V and temperature T at $x/d_{\mathrm{noz}} = 100$

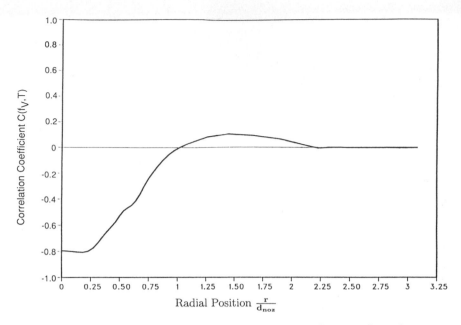

Fig. 31.14. Correlation coefficient between soot volume fraction f_V and temperature T at $x/d_{noz} = 150$

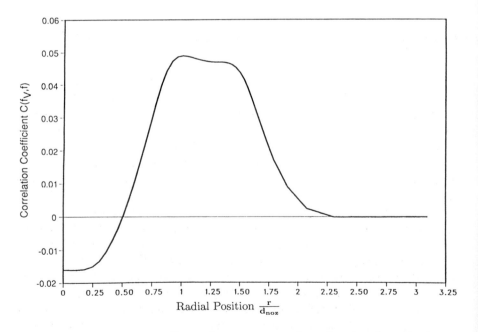

Fig. 31.15. Correlation coefficient between soot volume fraction f_V and mixture fraction f at $x/d_{noz} = 100$

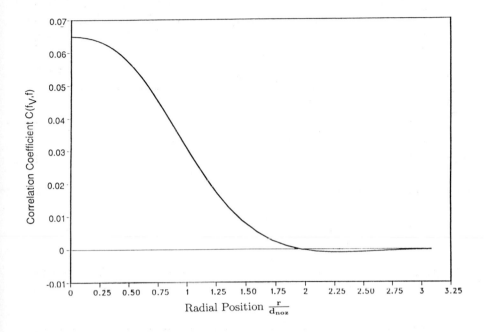

Fig. 31.16. Correlation coefficient between soot volume fraction f_V and mixture fraction f at $x/d_{\text{noz}} = 150$

Another important statistical aspect of the flow relates to the correlation of the joint pdf of mixture fraction and soot volume fraction. It would simplify turbulent flame modelling greatly if it were possible to assign a simple relationship (sometimes called a state relationship) between these variables. *Sivathanu* et al. [31.20] have found some indication of such state relationships in turbulent diffusion flames with a long residence time i.e., with a characteristic residence time greater than ten times the smoke point residence time. For ethylene the latter parameter is 41 ms [31.20]. Figure 31.15 shows the correlation coefficient of mixture fraction and soot volume fraction at $x/d_{\text{noz}} = 100$ and Fig. 31.16 presents the same statistics at $x/d_{\text{noz}} = 150$. An important observation is the very small correlation between the two variables at both axial locations. This suggests that the application of a state relationship between soot and mixture fraction is not likely to be correct for this flame. By way of comparison with the criterion of *Sivathanu* et al. [31.20], the mean axial velocity and flame length can be used to evaluate the residence time which in this case is of the order of 50 ms. Consequently, a good correlation between soot and mixture fraction may not be expected. The peak correlation increases with distance downstream; this may indicate that as residence time increases, the correlation between the two variables becomes stronger. The mechanism could be the self-limiting nature of the coupled soot formation and radiative heat loss processes. This model assumes Arrhenius kinetics for soot surface growth with a small activation energy for which there is some experimental evidence [31.2]; a similar description applies to the O_2 oxidation rate. As more soot is formed, the flame temperature drops with a concomitant drop in soot surface growth rates and soot oxidation rates. With the source terms turned off at low temperatures one would expect the correlation between the two scalars to increase with time as the soot volume fraction behaves increasingly like a conserved scalar although the large difference in diffusivities must always cause some difference to persist.

31.5 Conclusions

A model for soot formation in laminar, non-premixed flames has been incorporated into a pdf closure model for turbulent flames. In view of the nature of the measurements of the soot volume fraction, encouraging results have been obtained for a comparison of the predictions with experiments. The soot has a very substantial impact on the thermochemistry of the flame. Radiative heat loss reduces flame temperature by several hundred degrees with a concomitant effect on the flow field. The joint pdf of soot and temperature shows that most of the luminous radiation from the flame may be emitted at a temperature around 1400 K. The statistics of soot volume fraction and mixture fraction did not provide support for the notion of a state relationship under these flame conditions. The results indicate that it is important to conduct the calculation in the restricted scalar space; significant differences are observed in the soot volume fraction pdfs for the two mixing models that were used.

31.6 Acknowledgements

Financial support by a German DAAD-fellowship for M. M. (RWTH-Aachen) is gratefully acknowledged. I. M. Kennedy wishes to express his appreciation of the support of NSF through grant No. CBT-8857477. W. Kollmann and J.-Y. Chen are supported by the U.S. Department of Energy, Office of Basic Energy Sciences, Division of Chemical Sciences.

Nomenclature

$K_{\text{abs}}^{\text{Pl}}$	Planck mean absorption coefficient
C_p	Specific heat at constant pressure
P	Probability density function
$\dot{m}''$	Soot specific surface growth rate
h	Enthalpy
k	Turbulent kinetic energy
M	Molecular weight
N_s	Soot particle number density
Q	Source term in pdf equations
t	Time
T	Temperature
v	Velocity
R	Rate
W	Mass fraction
ε	Scalar dissipation rate
f	Mixture fraction
λ	Wavelength
ρ	Density
σ_{SB}	Stefan-Boltzmann constant
τ	Time scale
f_V	Soot volume fraction
ψ	Scalar variable

Superscripts

$^-$	molal
$^\sim$	mean
cd	convective-diffusive
r	radiative
th	thermophoretic

Subscripts

s	Surface growth
n	Nucleation
o	Oxidation, standard state

References

31.1   B.F. Magnussen, B.H. Hjertager, J.G. Olsen, D. Bhaduri: "Effects of Turbulent Structure and local Concentrations on Soot Formation and Combustion in C_2H_2 Diffusion Flames", in Seventeenth Symposium (International) on Combustion (The Combustion Institute, Pittsburgh 1978) p. 1383

31.2   P.A. Tesner, T.D. Snegiriova, V.G. Knorre: Combust. Flame **17**, 253 (1971)

31.3   J.B. Moss, C.D. Stewart, K.J. Syed: "Flowfield Modelling of Soot Formation at elevated Pressure", in Twenty-Second Symposium (International) on Combustion (The Combustion Institute, Pittsburgh 1989) p. 413

31.4   S.B. Pope: Progr. Energy Combust. Sci. **11**, 119 (1985)

31.5   W. Kollmann: Theor. Comp. Fluid Dyn., in press

31.6   I.M. Kennedy, W. Kollmann, J.-Y. Chen: Combust. Flame **81**, 73 (1991)

31.7   I.M. Kennedy, W. Kollmann, J.-Y. Chen: "Predictions of Soot in Laminar Diffusion Flames", AIAA Aerospace Sciences Meeting, Paper 90-0459, Reno, Nevada (1990)

31.8   J.-Y. Chen, W. Kollmann: "PDF Modeling of Chemical Nonequilibrium Effects in Turbulent Nonpremixed Hydrocarbon Flames", in Twenty-Second Symposium (International) on Combustion (The Combustion Institute, Pittsburgh 1989) p. 645

31.9   T. Neill, I.M. Kennedy: AIAA J. **29**, 931 (1990)

31.10   B.S. Haynes, H.Gg. Wagner: Prog. Energy Combust. Sci. **7**, 229 (1981)

31.11   K.G. Neoh, J.B. Howard, A.F. Sarofim: "Soot Oxidation in Flames", in Particulate Carbon: Formation During Combustion, ed. by D.C. Siegla, G.W. Smith, (Plenum Press, London 1981) p.261

31.12   J. Nagle, R.F. Strickland-Constable: "Oxidation of Carbon between 1000 − 2000°C", in Proceedings of the Fifth Conference on Carbon (Pergamon Press, New York 1962) p.154

31.13   C. Dopazo: Phys. Fluids **22**; 20 (1979)

31.14   J. Janicka, W. Kolbe, W. Kollmann: J. Non-Equilib. Thermodyn. **4**, 27 (1989)

31.15   J.-Y. Chen, Kollmann, R.W. Dibble: Combust. Sci. Technol. **64**, 315 (1989)

31.16   J.-Y. Chen, W. Kollmann: Combust. Flame **79**, 75 (1990)

31.17   R.W. Dibble, W. Kollmann, M. Farshchi, R.W. Schefer: "Second-Order Closure for Turbulent Nonpremixed Flames: Scalar Dissipation and Heat Release Effects", in Twenty-First Symposium (International) on Combustion (The Combustion Institute, Pittsburgh 1986) p. 1319

31.18   I.M. Kennedy: Combust. Sci. Technol. **59**, 107 (1988)

31.19   Y.R. Sivathanu, G.M. Faeth: Combust. Flame **81**, 150 (1990)

31.20   Y.R. Sivathanu, G.M. Faeth: Combust. Flame **81**, 133 (1990)

Discussion

Warnatz: The critical point in your laminar modeling seems to me this tabulation where you have soot yield or something like this as a function of temperature and mixture fraction. Where does it come from?

Kennedy: You are referring to the growth rate expression which is a function of mixture fraction and temperature. The input there is empirical. It comes from laser scattering measurements in a counterflow ethylene diffusion flame, where we can get the mixture fraction and we can get a correlation between that and the surface growth rate.

Warnatz: Is this empirical input from the same type of flames or even from the same flame?

Kennedy: No, it is from a counterflow diffusion flame, so it is a different flame. You can get this correlation which should be fairly universal for coflow, and counterflow diffusion flames and the results also indicate that this is true. We have a similar approach to what John Kent has done. We are trying to get an empirical correlation between mixture fraction and temperature as an input to the model which we can then integrate through the flow field. So we are using an empirical input because, for a particular fuel like a diesel fuel or an aviation fuel, it is very hard to do this from first principles.

Warnatz: How far can you transfere this to other flames?

Kennedy: This is only for one fuel; that is an other complication. For the effect of the temperature we have used an empirical Arrhenius expression. But we assumed the pressure to be 101.325 kPa. To do this calculation for a higher pressure situation we require measurements again in a counter flow flame, at higher pressure, to get the empirical input we need.

Frenklach: Why did you assume an optically thin model?

Kennedy: One of the obvious limitations would be increasing the pressure where you get a large soot volume fraction. The flame would be optically thick then. B. Moss reported about that with his kerosine flames. The optically thin assumption is reasonable for our flames. We measured in a turbulent flame radiated heat fluxes and that scales linearly with the measured soot volume fractions. This is a reasonable indication that we are in the optically thin regime. If you go up with pressure where we have a lot more soot then we can't make the optically thin assumption.

Kent: Your oxidation modeling is very temperature dependent. I don' t think you showed a lot of your temperature predictions in the coupling between the actual soot you have got. The radiation loss feeds back into the formation and oxidation.

Kennedy: I didn' t show you any temperature predictions. The temperature predictions are reasonable compared to Santoro's measurements. The feedback between radiation, temperature, oxidation and growth is essential. And if you don' t get the radiation right you are in trouble, nothing is going to be correct. There are still a lot of open questions about the oxidation. There are various parameters in calculating the oxidation which are uncertain. For example, the density of the soot which is commonly taken at 1.8 or 2 g / cc. I found that lower value actually worked better in the modeling. But that of course could be covering up the multitude of errors such as the fact that the surface area might be greater than I have calculated. The OH concentration is certainly greater than what we predict from equilibrium. It becomes rather difficult to get a lot of precision in that oxidation step.

Santoro: J. Kent showed some predictions from his modeling work using conserved scalars and comparing with flames. Both of you are not predicting a fast enough oxidation rate higher up in the flame. And in your modeling results, the rates are faster than what we have seen. I would like J. Kent and I. Kennedy to comment why they think the oxidation is slow.

Kennedy: In fact we are using different mechanisms for the oxidation. J. Kent is not using my approach for the OH.

Kent: The predicted values I showed were based on measured temperatures using the Nagle-Strickland-Constable constants with OH and O taken from equilibrium values. We were underpredicting the oxidation so, we knew we had the right temperature and we knew we were getting underpredictions in the oxidation rates.

Dobbins: We heard in previous sessions that there were two critical important temperatures. One temperature at which the soot has formed, and the other where soot burns out. I think your model must predict the latter but not the former.

Kennedy: To get the soot break through we obviously have to get the latter one. There is going to be a temperature we get below because of radiative heat loss and so we don' t get complete oxidation. I don't think that we can predict the first temperature at all. This is done from measurements and that correlation between mixture fraction and growth is more or less fixed. The really essential thing is not just the temperature but the whole history of the soot.

Glassman: I didn' t see the buoyancy term in your written equations and would you comment on the effect between buoyancy and the Froude number effect as you go from laminar to the turbulent flames.

Kennedy: The laminar flame is certainly buoyancy controlled. The velocity down stream has little to do with the velocity coming out of the nozzle. Even the turbulent flames still have Froude numbers where they are still not in the forced convection regime but in a mixed regime. So it is hard to say that they were in exactly the same regime as a gas turbine combuster. For example, buoyancy is important even in turbulent flames in the laboratory.

Model of Soot Formation:
Coupling of Turbulence and Soot
Chemistry

Annie Garo, Rachid Said, Roland Borghi

CORIA / URA CNRS 230,
76134 Mont-Saint-Aignan Cedex, France

Abstract: The aim of our study is to model soot formation in the extreme conditions of Diesel engines. The first concept has been to use a formulation for soot formation deduced from the most recent experimental work. However, in order to take into account the effects of turbulence and pressure, associated to soot production, we had to use very simplified assumptions. This ensemble of hypotheses is justified in the case of the Diesel engine only. The first step of this study consisted in obtaining a model coupling the turbulence of the flow and soot formation, by computing turbulent jet flames at atmospheric pressure. This is what we are presenting here. To model soot formation in Diesel engines, it will be necessary to distinguish the following three different processes, formation, coagulation and oxidation, then to take into account the thermal radiation of soot particles. Though the mechanisms for oxidation and coagulation have not been tested for Diesel conditions, they are reasonably understood. On the contrary, the formation mechanism must be analyzed to derive a very simplified relationship to fulfill the requirements of numerical fluid modelling. We propose to model soot formation by a method of an "extended flamelets approach" which includes high temperatures zones where pyrolysis and soot oxidation reactions may occur as well as thermal radiation. The second step will consist to introduce the expression for soot production at high pressures, in order to simulate soot inside the Diesel engine. This is in progress within the european research program CEC-JRC/IDEA. In this paper we will present first the characteristic properties of Diesel soot and the different time scales produced by an engine cycle. Then, we will see how these characteristics impose the expression of the coupling between turbulence and soot production. We will describe the semi-global model for soot production that has been tested in a laminar diffusion flame. After a detailed description of the turbulent model, a comparison between experiments and calculations will be presented in a turbulent ethylene-air jet. The inclusion of our model in a large computer code for simulating Diesel engine is actually under testing, and will not be presented here.

32.1 Characteristics of Diesel Soot

(I) The published experimental results are mainly for the emitted soot in the exhaust gases: the aggregates of particles are very similar to those obtained in fuel-rich flames and present a size distribution approximatively log-normal with a geometric mean diameter of about 25nm and a geometric standard deviation of 1.3-1.6, *Vuk* et al. [32.1].

(II) Using a sampling technique in a combustion chamber, *Kittelson* et al. [32.2] have shown that more than 93% of the soot that is formed, is oxidized before leaving the cylinder. This has been demonstrated recently by *Lepperhoff* et al. [32.3] using a time resolved technique.

(III) A yellow luminous radiant wavelength emission, which is characteristic for glowing soot, has often been observed in Diesel combustion. Although they are affected by large experimental uncertainties, instantaneous radiant heat flux measurements may help to understand Diesel engine combustion and to estimate characteristic times of soot presence in the combustion chamber.

Radiation peaks just after the end of premixed burning at 5-10 crank angle after the start of combustion (*Borman* [32.4]). About 30 to 40% of the total cylinder flux is due to radiation.

Using a flame luminosity detector in a D.I. Diesel, *Hiroyasu* [32.5] showed this yellow radiation to persist late into the expansion stroke (after the opening time of the exhaust valve), well past the period of heat release due to fuel combustion. Then, radiation will have to be considered in our model, but the problem of coupling radiation with the other thermal phenomena is very complex. The radiation exchange may cause the temperature within the chamber to be much more uniform than would be the case in the absence of radiation.

32.2 Characteristic Times

In order to discuss the influences of chemistry and turbulence on soot formation, it is of primary importance to compare first the time scales involved in each phenomenon.

(I) The time scales related to the soot formation have been considered by *Smith* [32.6], in the conditions of Diesel combustion. In flames at atmospheric pressure, these time scales are larger.

In premixed flames, formation occured in two separated steps: Nucleation, then surface growth, which last about 2-3 ms for nucleation, and 10-30 ms for total formation. (*Baumgärtner* et al. [32.7], *Bockhorn* et al. [32.8]).

In laminar diffusion flames, the representative time for the oxidation step is generally superior to 10 ms, and lasts about twice the time of the formation step (*Kent* et al. [32.9], *Garo* et al. [32.10]).

Table 32.1. Time scales related to the soot formation from *Smith* [32.6]

Processes	Time Constant or Duration
1. Formation of Precursors/Nucleation	$\sim$ few μs
2. Coalescent Coagulation ($\sim$ 10-fold)	$\sim$ 0.05ms after local nucleation
3. Spherule Identity Fixed	After coalescence ceases
4. Chain-forming Coagulation	$\sim$ few ms after coalescent
5. Depletion of Precursors	$\sim$ 0.2ms after nucleation
6. Non-sticking Collisions	$\sim$ few ms after nucleation
7. Oxidation of Particles	$\sim$ 4ms
8. Combustion Cycle Complete	3-4ms
9. Deposition of Hydrocarbons	During expansion and exhaust

Furthermore, the formation of soot seems more sensitive to pressure than the oxidation process.

(II) The turbulent mixing time has a predominant role in the production of soot particles. *Magnussen* et al. [32.11] deduced from experiments on free diffusion flames that soot is formed and contained in the turbulent eddies within the flame, and that the burn-up of soot is related to the dissipation of the turbulence. The characteristic times chosen by *Kiriakides* et al. [32.12] are based on the dissipation of the turbulent kinetic energy in the cylinder, during the post-injection period. In the D.I. Diesel engine with strong air swirl, the mixing rate depends on the fuel injection process and air swirl. The mixing process is controlled by the slower phenomenon which is the air swirl. Then, for angular velocities ranging from 1000 to 5000 radians per second, the mixing time is about 10^{-4} second.

In order to get realistic estimates for the turbulence time scales within a Diesel engine, we obtained results of a typical calculation of turbulence around the jet of droplets in a Diesel engine from *D. Gosman* [32.13]. The calculations have been done with a k-ε model; the result gives certainly the most exact values that can be expected at this time. If we define the integral time scale as usual ($\tau_L = 0.3\ k/\varepsilon$), the results of these calculations give a map of time scales. As examples, for three different points in the flow field, we find :

A : $\tau_t = 1.02$ ms close to the injector,
B : $\tau_t = 0.88$ ms in the fuel jet,
C : $\tau_t = 13.5$ ms in the middle of the bowl.

In fact, in the domain close to the jet, the turbulence time scale is almost constant. The mean engine speed here was 2.7 m/s for 1000 rpm.

By comparing of the time scales, we see that the oxidation of soot particles will be influenced by turbulent motions, the nucleation and coalescent coagulation will not. The formation of chain will be influenced to some extend. This estimation is only based on τ_t, which is the largest of the turbulence time scales. The Kolmogorov time scale being clearly smaller than τ_t the smallest eddies of the turbulence are likely to influence the coagulation.

(III) We can deduce then that the nucleation step and the beginning of the coagulation and surface growth step may be analysed in a pure flamelet approach ; the oxidation step likely not.

This conclusion is confirmed by the fact that the main reason for soot particles to be emitted is a non sufficient residence time in the flame zone. The emitted particles represent the part of the particles which have not been completely oxidized before the opening of the exhaust valve. To take into account incomplete oxidation, time must be considered, and this is not consistent with a pure flamelet approach. *Moss* et al. [32.14] have found that the lack of invariance of soot related quantities in a diffusion flame to the local mixture fraction, prevents the use of a classical flamelet approach.

We have used an "extended flamelet" approach like *Moss* et al. [32.14], but with the simplifying assumptions that soot formation may be relatively fast with respect to the turbulence time scales, and that soot oxidation has a predominant role.

32.3 Basic Equations

The 3-D calculations of the mean flow field, mean temperature and mean species concentration field require the solution of the coupled set of partial differential equations that comprises:the continuity equation, the mean momentum equation, the mean energy equation, the mean species mass fraction equations and the equations for the turbulence model.

The mean equations are classically written in terms of Favre or density weighted averages ; i.e. each variable g (except ρ and P) is written as $g = \tilde{g} + g'$ where $\tilde{g} = \bar{\rho}\bar{g} / \bar{\rho}$.

Eulerian transport equations are solved using the Genmix code (*Patankar* et al. [32.15]) and turbulent combustion is computed with the MIL model (*Gonzalez* et al. [32.16]). We focus our description on the turbulent simulation.

32.4 Basic Principles of the Model

Our "extended flamelet" model consists of three parts :
- a semi-global chemical model for the kinetics of soot formation and oxidation - a model of turbulent combustion, in order to know the local temperature, local equivalence ratio and local reaction progress statistics - a model able to take into account the turbulent fluctuations of temperature, equivalence ratio and reaction progress, the statistics of which are given by the turbulent combustion model, and able to deduce their effects on the

mean yield of soot. The structure of the model will, consequently, be as follows:

(I) A chemical semi-global model for soot formation is then needed. The nucleation, coagulation and surface growth phenomena can probably be represented empirically, only globally, taking advantage that these processes are fast. The oxidation process, at the contrary is to be described more exactly.

a) We consider that the soot is well known when the soot volume fraction f_V and the number density of soot particles N are known. From these quantities, we can derive a mean diameter if we consider that the soot particles are all spherical. Taking into account the agglomeration processes, we can in addition estimate a surface to volume ratio for chain-like particles.

The rate for f_V is modelled as :

$$\frac{\mathrm{d}f_V}{\mathrm{d}t} = \dot{W}_{f_V} = k_{\mathrm{sg}}\left(f_{V\infty} - f_V\right) - \dot{W}_{\mathrm{Ox}} . \qquad (32.1)$$

The formation term is deduced from the experimental results of $\dot{W}_{f_v}$ (*Baumgärtner* et al. [32.7], *Bockhorn* et al. [32.8]), in fuel rich premixed flames. k_{sg} is the inverse of a time scale for soot formation, including nucleation, coagulation and surface growth. $f_{V\infty}$ is the maximum value that the volume fraction may reach without oxidation ; it is supposed to be a function of the temperature, equivalence ratio (or C/O ratio) and pressure. Although the situation is different for diffusion flames, the function $f_{V\infty}(T, C/O)$ that is needed looks probably like the one established by *Böhm* et al. [32.17] and plotted in Fig. 32.1.

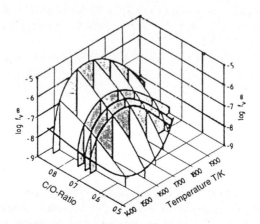

Fig. 32.1. $f_{V\infty}$ plotted as a function of the C/O ratio and the flame temperatures for C_2H_4 − air premixed flames at normal pressure (*Böhm* et al. [32.17]); T means flame temperature at 10mm height above the burner; the broad solid line is the threshold for soot formation

532   Annie Garo, Rachid Said, Roland Borghi

For the first tests of the model, we have adopted as empirical entry this type of plot. This envelope-curve (Fig. 32.1) has been fitted using polynomial expressions to give $f_{V\infty}$ versus C/O ratio (number of carbon atoms/oxygen atoms), when C/O is in the range 0.57-0.9, and versus the local temperature. When the value of C/O is in the range 0.9-3.0 (flammability limit), $f_{V\infty}$ is kept constant at its maximum value $f_{V\infty}(T, C/O = 0.9)$.

We define the threshold of soot particles when f_V is larger than 10^{-8} and the corresponding diameter d_0 larger than 1nm. The inferior boundary conditions are thus:

$$d_0 = 10^{-9}\text{m}, \ f_{V,0} = 10^{-8}, \ N_0 = 1.9 \cdot 10^{19} \text{ particles} \cdot \text{m}^{-3} \qquad (32.2)$$

by assuming that initial particles are spherical and present a monodisperse size distribution.

The minimal temperature allowing the formation of soot particles has been taken at 1450 K.

The time constant for this equivalent formation rate k_{sg}^{-1} has been measured (*Wieschnowsky* et al. [32.18]), and found almost constant $(\sim 10^{-2}\text{s}^{-1})$ in several cases; as a first value, we have considered it constant.

The recent experiments which have been done at high pressures indicate that $f_{V\infty}$ can be considered proportional to the pressure above 1 MPa (*Jander* et al. [32.19]).

The oxidation term is written as a function of the local oxidant concentration and the surface of soot particles. We have used the expression proposed by *Lee* et al. [32.20]:

$$\dot{W}_{\text{Ox}} = \frac{6.5}{\rho_s \cdot d_s} \ p_{O_2} \ T^{-1/2} \ f_V \ exp\left(-\frac{19750}{T}\right) \qquad (32.3)$$

where d_s is the soot particle diameter and ρ_s the soot density. The oxidation due to OH radicals is not taken into account, for sake of simplicity.

b) The total number density N is an other major characteristic of soot particles. Correlated to the particle diameter and surface, it indicates the status under which soot is emitted. This criterion must be considered in a precise model, because the oxidation depends on the surface area, which is larger for a cloud of small particles than for a few large particles of the same f_V.

At atmospheric pressure (molecular free regime), the evolution of the number density of spherical soot particles may be expressed by the following equation, based on Smoluchovski's theory (Wagner [32.21], Prado et al. [32.22]):

$$\frac{dN}{dt} = -1.1 \left(\frac{3}{4\pi}\right)^{1/6} \left(\frac{6k_B T}{\rho_s}\right)^{1/2} f_V^{1/6} N^{11/6} \qquad (32.4)$$

with k_B the Boltzmann's constant.

This expression has been linearized in the program. However, this equation cannot be applied for high pressure media, like Diesel engines, and has to be modified in this case.

c) In order to precise the empirical constants we had to use in our model, we have compared our simulation to experimental results in laminar diffusion flames, at atmospheric pressure. Since the experimental curve of $f_{V\infty}$ proposed by *Böhm* et al. [32.17] has been established for ethylene-air premixed flames, we have chosen to simulate such an hydrocarbon-air flame. The following experiment of *Santoro* et al. [32.23] has been selected: an ethylene-air laminar diffusion flame presenting a cylindrical symmetry has been simulated.

C_2H_4 flow rate : $\dot{v} = 3.85$ cm^3/s   velocity : $v = 3.98$ cm/s

Air flow rate : $\dot{v} = 713.3$ cm^3/s   velocity : $v = 8.90$ cm/s

The experimental and computed radial profiles of temperature are shown on figures 32.2 a, at several heights above the burner. We notice that they are overestimated higher in the flame; this can be due to radiation which has not been taken into account in this computation.

The computed and experimental profiles of soot volume fraction are given on figures 32.2 b; the mean diameter has been kept constant (1 nm) in these plots. The order of magnitude is correct, but the computed f_V values do increase like the experimental ones. The important role of the temperature is one of the reasons of this disagreement, but more likely, our semi-global model is not correct.

On figures 32.3 the computed radial profiles of f_V at 1 cm above the burner are shown, when the diameter is deduced from the number density (Fig. 32.3 a) and when the diameter is constant (Fig. 32.3 b).

However, this model allows us to see that the computation of the number density does not bring a very different result. Then, in turbulent flows, we did not consider N in our first tests. This conclusion has been obtained previously by *Kennedy* [32.24].

The conclusion is that the semi-global soot model is not sufficient to predict soot formation with accuracy, it is a first tool which gives reasonable orders of magnitude and it may be used to test the coupling with turbulence. The formation aspect will have to be improved using new experimental results. Oxidation is a predominant process which will be kept considered in the model. The constants which have been determined in simulating the laminar diffusion flame will be inserted then in the turbulent model, without any adjustment.

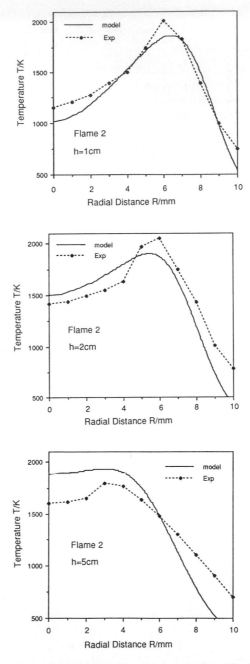

Fig. 32.2.a. Comparison of experimental and computed results in a laminar ethylene-air diffusion flame (Experiment 2, *Santoro* et al. [32.23]); computed (solid line) and experimental (dashed line) profiles of temperature versus the radial distance, at several heights (1cm, 2cm and 4cm) above the burner

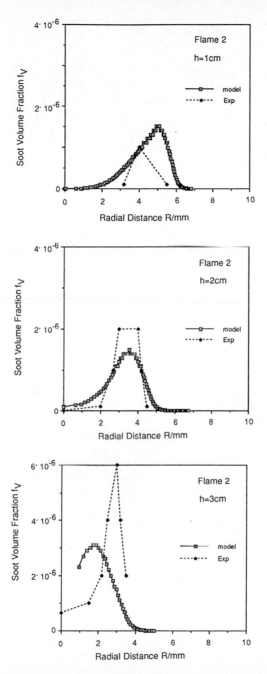

Fig. 32.2.b. Computed (solid line) and experimental (dashed line) profiles of soot volume fraction f_V versus the radial distance, at several heights (1cm, 2cm and 3cm) above the burner

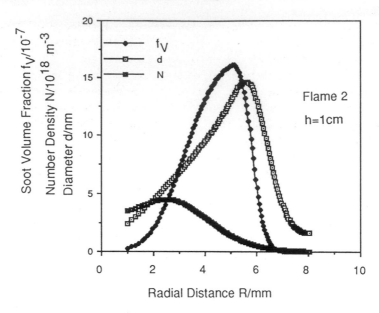

Fig. 32.3.a. Computed profiles of f_V (soot volume fraction), N (soot number density of spherical soot particles), and d (diameter of soot particles), versus the radial distance at 1cm above the burner; d is deduced from the local value of N; experimental conditions: flame 2 (*Santoro* et al. [32.23])

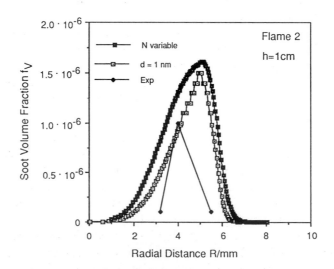

Fig. 32.3.b. Profiles of f_V (soot volume fraction) versus the radial distance at 1 cm above the burner; Exp: experimental f_V from *Santoro* [32.23]; d = 1nm: computed f_V using constant particle diameter; N variable: computed f_V using a particle diameter deduced from the calculation of the number density

(II) In turbulent combustion, the temperature, equivalence ratio and all other chemical and thermodynamical quantities are fluctuating. The "turbulent combustion model" takes into account these fluctuations in calculating the heat release; it provides us a statistical description of the flame brush. Not only the mean values of the temperature, the equivalence ratio and the reaction progress are needed but also a joint probability density function (pdf) of these three quantities. More informations, as oxygen atoms or other radical concentrations, could be of interest, but probably the need of these quantities would not give a definitely improved prediction, taking into account that they are clearly far more difficult to get than the former ones.

The turbulent ignition and combustion model we have used has been developed in Rouen, and is named M.I.L. (Modèle Intermittent Lagrangien, *Borghi* [32.25]). Another model of combustion would also be convenient. This model gives the joint probability density function versus the conserved scalar ϕ (equivalence ratio) and the reaction progress Y_O

$$\tilde{P}(\phi) = \int_0^1 \tilde{P}(\phi, Y_O)\, dY_O \qquad (32.5)$$

computed as a β-function. Some assumptions are made :

- ignition is very sudden (high activation energy). The chemical reaction begins very slowly; there is first a quasi negligible consumption of reactants and low heat release. After a delay, it terminates very rapidly, the major part of the consumption of reactants and heat release occuring in a short time interval compared to the "ignition" delay. The energetical chemical mechanism is reduced to a one-step global reaction.

- a distribution of turbulence time scales exists, to be compared with an "ignition chemical time scale". More details about this model can be found in a previous paper (*Gonzalez* et al. [32.16]).

(III) The coupling model which is able to compute the effect of fluctuations on soot formation is based on a lagrangian model where turbulent mixing interacts with the chemical processes. The joint pdf $\tilde{P}(\phi, Y_O)$, and $P(\tau)$ the distribution of time scales, given by the combustion model, allow us to compute the mean temperature, mean oxidizer mass fraction, and mean fuel mass fraction, as well as the moments and correlations.

In order to compute $\tilde{f}_V$ and $\tilde{N}$, the local mean characteristics for soot, we need more than $\tilde{P}(\phi, Y_O)$ since the production/oxidation rates of f_V and N depend not only on ϕ and Y_O but also on f_V and N. The consequence is that we have to take into account now $\tilde{P}(\phi, Y_O, f_V, N)$, the 4-dimensional joint pdf. This pdf is given by the lagrangian model.

The model I.E.M. (Interactions by exchange with the Mean) replaces the exact equation for Y at the point i:

$$\rho \frac{\mathrm{d}Y_i}{\mathrm{d}t} = \frac{\partial}{\partial t}(\rho Y_i) + \frac{\partial}{\partial x_j}(\rho u_j Y_i) = \frac{\partial}{\partial x_j}\left(\rho D_i \frac{\partial Y_i}{\partial x_j}\right) + \rho \dot{W}_i \qquad (32.6)$$

by $\frac{\mathrm{d}Y_i}{\mathrm{d}t} = \frac{\tilde{Y}_i - Y_i}{\tau_{\text{ex}}} + \dot{W}_i$ with $\tilde{Y}_i$ the local mean value and τ_{ex} the turbulent exchange time related to k/ε.

The first term is the small scale turbulent mixing, the second ($\dot{W}_i$), "chemical" rate, is purely laminar.

Then, the IEM model gives the set of equations:

$$\frac{\mathrm{d}\phi}{\mathrm{d}t} = \frac{\tilde{\phi} - \phi}{\tau_{\text{ex}}} \qquad (32.7)$$

$$\frac{\mathrm{d}h}{\mathrm{d}t} = \frac{\tilde{h} - h}{\tau_{\text{ex}}} \qquad (32.8)$$

$$\frac{\mathrm{d}f_V}{\mathrm{d}t} = \frac{\tilde{f}_V - f_V}{\tau_{\text{ex}}} + \dot{W}_{f_V} \qquad (32.9)$$

$$\frac{\mathrm{d}N}{\mathrm{d}t} = \frac{\tilde{N} - N}{\tau_{\text{ex}}} + \dot{W}_N \ . \qquad (32.10)$$

These equations (O.D.E.) have to be solved numerically, within each grid cell in the cylinder, (or the flame brush), for fluid particles coming from the fuel jet ($\phi = 1$) or from the cylinder air ($\phi = 0$), with $T(\phi)$ prescribed from the combustion model.

At each position in the cylinder ($\tilde{f}_V, \tilde{\phi}, \tilde{Y}_O, \tau_t, ...$known), $P(\phi, f_V)$ is supported by the "trajectories" solutions of the IEM lagrangian equations; an example of these trajectories can be :

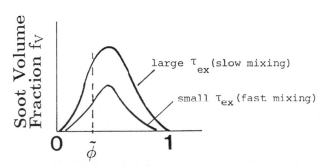

Graph 32.1.

We can then calculate the mean reaction rate along the trajectory for Y_O given, in the (f_V, ϕ) plane :

$$\tilde{\dot{W}}_{fv} = \int P(\phi)\dot{W}_{fv,\,\text{laminar}}\,[T(\phi), f_V(\phi), Y_O(\phi), ...]\,d\phi \ . \qquad (32.11)$$

This reaction rate is then used in the eulerian balance equation :

$$\frac{\partial}{\partial t}\left(\bar{\rho}\tilde{f}_V\right) + \frac{\partial}{\partial x_j}\left(\bar{\rho}\tilde{u}_j\tilde{f}_V\right) = \frac{\partial}{\partial x_j}\left(\bar{\rho}\,D_{turb}\frac{\partial \tilde{f}_V}{\partial x_j}\right) + \bar{\rho}\tilde{\dot{W}}_{fv} \qquad (32.12)$$

which is solved in each point of the grid.

To include the influence of radiation, we need $\tilde{P}(\phi, Y_O, T, f_V, N)$. The temperature is not only linked with ϕ and Y_O but also with f_V and N.

We use this extended lagrangian model with the assumption of radiation in the optically thin limit. Furthermore, in the case of Diesel engines, it is necessary to introduce the pressure variation effects ; the enthalpy equation which has to be integrated to give $T(\phi)$ is then :

$$\frac{dh}{dt} = \frac{\tilde{h} - h}{\tau_{ex}} - 4k\sigma_S B T^4 + \frac{1}{\rho}\frac{\partial \bar{P}}{\partial t} \qquad (32.13)$$

with $\sigma_S B$ the Stefan Boltzmann constant, k the Planck mean absorption coefficient for absorption. Emission from soot particles is deduced from the works of *Habib* et al. [32.26]:

$$k(T, f_V) = (aT + b)f_V \qquad (32.14)$$

with

$$a = 666.6 K^{-1} cm^{-1} \text{ and } b = 5.334 \cdot 10^5 cm^{-1} \ .$$

32.5 Results and Discussion

A turbulent jet diffusion flame, described by *Kent* et al. [32.27], has been modelled using the computation method described above. The experimental conditions are the following:

ethylene fuel (Flame A), nozzle diameter = 3mm, jet velocity (at 49°C) = 52 m/s.

The flow field has been computed with a parabolic numerical procedure including a standard $k - \varepsilon$ model with the classical constant for the turbulence. Let us notice that k and ε are not well known in the jet.

The eulerian equations have been solved for the following variables :

$$\tilde{u}, \tilde{h}, \tilde{Y}_{\text{fuel}}, \tilde{Y}_{\text{soot}}, k, \varepsilon, \tilde{\phi}, \tilde{\phi'^2} \ .$$

A global single reaction with an Arrhenius finite rate chemistry has been used for simulating combustion :

$$C_2H_4 + \nu\, O_2 \rightarrow Products + \Delta Q \qquad (32.15)$$

The contours of equivalence ratios are represented on Figure 32.4.a.

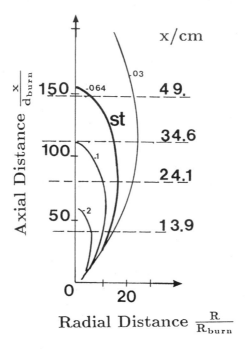

Fig. 32.4.a. Contours of equivalence ratios in the flame (*Kent* et al. [32.27]); Abscissa: the ratio of the radial distance to the radius of the burner; Ordinates: the ratio of the axial distance to the diameter of the burner

Some experimental and computed temperature profiles are compared on figures 32.4 b, c, d. There is generally good agreement over most of the flame length; this agreement is improved when radiation losses are taken into account (Fig. 32.4 d, e).

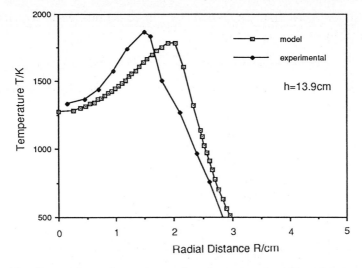

Fig. 32.4.b. Comparison of experimental and computed profiles of mean temperature in a turbulent ethylene-air diffusion flame (*Kent* et al. [32.27]) versus the radial distance (radiation not taken into account) at 13.9 cm above the burner

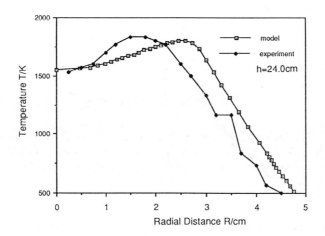

Fig. 32.4.c. Same as Fig. 32.4.b at 24.0 cm above the burner

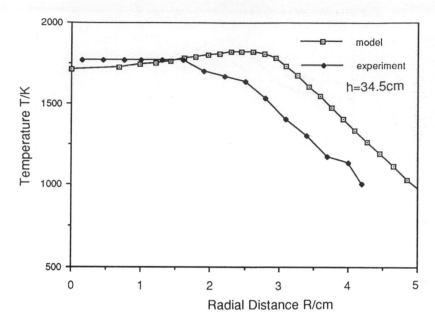

Fig. 32.4.d. Same as Fig. 32.4.b at 34.5 cm above the burner

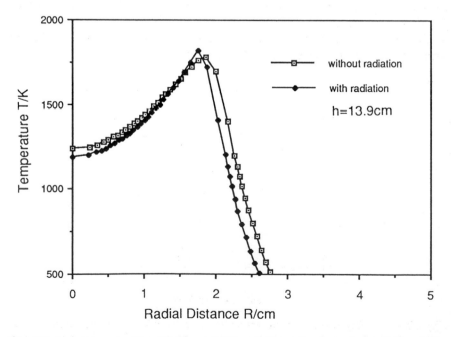

Fig. 32.4.e. Same as Fig. 32.4.b at 13.9 cm above the burner (radiation taken into account)

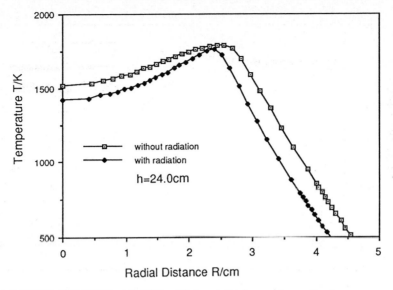

Fig. 32.4.f. Same as Fig. 32.4.e at 24.0 cm above the burner

On figures 32.5 the radial profiles of mean soot volume fraction, predicted without radiation, and the experimental values published by *Kent* et al. [32.27] are represented at several heights above the burner. The agreement is good at 34 cm (Fig. 32.5 c), but close and far from the burner exit, the agreement is unsatisfactory (Fig. 32.5 a, b, d).

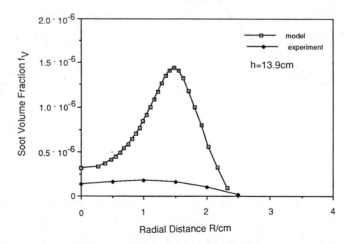

Fig. 32.5.a. Comparison of computed and experimental mean soot volume fraction profiles in a turbulent ethylene-air diffusion flame (*Kent* et al. [32.27]) versus the radial distance (radiation not taken into account) at 13 cm above the burner

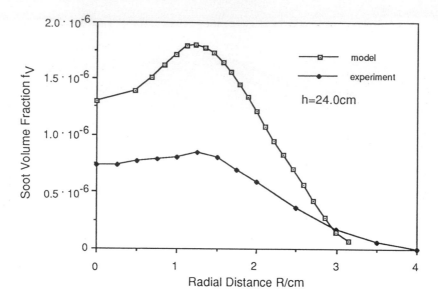

Fig. 32.5.b. Same as Fig. 32.5.a at 24 cm above the burner

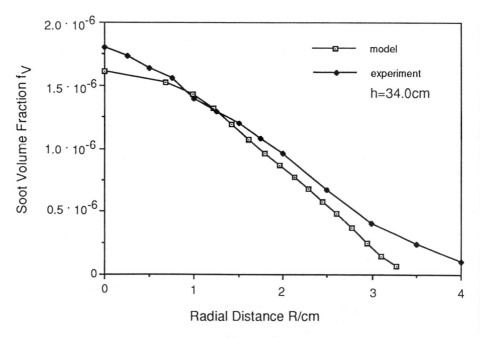

Fig. 32.5.c. Same as Fig. 32.5.a at 34 cm above the burner

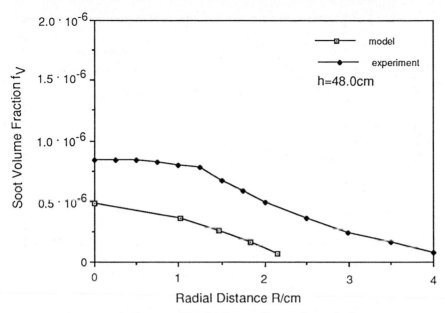

Fig. 32.5.d. Same as Fig. 32.5.a at 48 cm above the burner

When radiation losses are taken into account, the agreement is not very much improved, as can be seen on figures 32.6 a, b.

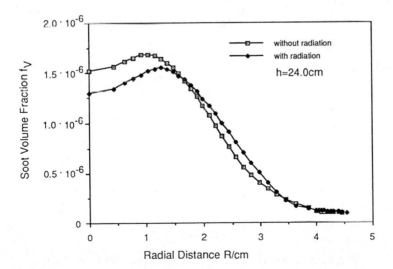

Fig. 32.6.a. Influence of radiation on mean soot volume fraction profiles versus the radial distance, in a turbulent diffusion flame (*Kent* et al. [32.27]) at 24 cm above the burner

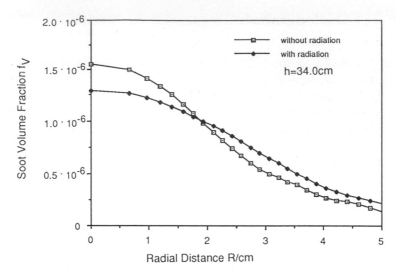

Fig. 32.6.b. Same as Fig. 32.6.a at 34 cm above the burner

32.6 Conclusions

The main characteristics of the model presented here are the following :

- it takes into account the soot oxidation in more details than the soot formation,

- it uses a semi-global chemical model, based on empirical data, some of them being improved with new experiments, or calculations.

- it uses an extended flamelet approach, because a pure flamelet approach is not suitable in order to treat, in particular, the soot oxidation.

However, we notice that some constants of the model which have been determined in the laminar case, before being inserted in the turbulent model, could be better adjusted.

The agreement between experimental and calculated soot concentrations, observed at several heights above the burner, could be improved by changing the laminar model of soot formation and by describing the oxidative process with more accuracy ; this work is in progress. The aim, which is to predict different conditions of flames and velocities, is still to realize.

32.7 Acknowledgments

Part of this work is supported by the Commission of the European Communities within the frame of the JOULE Programme, by the Swedish National Board for Technical Development, and by the Joint Research Committee of european automobile manufacturers (Fiat, Peugeot S.A., Renault, Volkswagen and Volvo) within the IDEA programme. C.N.R.S. is acknowledged too. Dr. A. Coppalle and Dr. M. Gonzalez (CNRS/CORIA) are acknowledged for helpful collaboration.

References

32.1 C.T. Vuk, M.A. Jones, J.H. Johnson: SAE Transactions, paper 760131 (1976)

32.2 D.B. Kittelson, M.J. Pipho, J.L. Ambs, D.C. Siegla: SAE Paper 56 1569 (1986)

32.3 G. Lepperhoff, M.Houben: IDEA/JRC, Periodical Report 11/1988 - 9/1989 (1989)

32.4 G. Borman, K. Nishiwaki: Prog. Energy Combust. Sci. **13**, 1 (1987)

32.5 H. Hiroyasu, M. Arai, K. Nakanishi: SAE paper 800252 (1980)

32.6 G.W. Smith: SAE paper 820466 (1982)

32.7 L. Baumgärtner, D. Hesse, H. Jander, H.Gg. Wagner: "Rate of Soot Growth in Atmospheric Premixed Laminar Flames", in Twentieth Symposium (International) on Combustion (The Combustion Institute, Pittsburgh 1984) p. 959

32.8 H. Bockhorn, F. Fetting, A. Heddrich, G. Wannemacher: "Investigation of the Surface Growth of Soot flat low Pressure Hydrocarbon Oxygen Flames", in Twentieth Symposium (International) on Combustion (The Combustion Institute, Pittsburgh 1984) p. 979

32.9 J.H. Kent, H. Jander, H.Gg. Wagner: "Soot Formation in a Laminar Diffusion Flame", in Eighteenth Symposium (International) on Combustion (The Combustion Institute, Pittsburgh 1981) p. 1117

32.10 A. Garo, J. Lahaye, G. Prado: "Mechanisms of Formation and Destruction of Soot Particles in a Laminar Methane-Air Diffusion Flame", in Twenty-First Symposium (International) on Combustion (The Combustion Institute, Pittsburgh 1986) p. 1023

32.11 B.F. Magnussen, B.H. Hjertager: "On Mathematical Modelling of Turbulent Combustion with special Emphasis on Soot Formation and Combustion", in Sixteenth Symposium (International) on Combustion (The Combustion Institute, Pittsburgh 1972) p. 719

32.12 S.C. Kiriakides, J.C. Dent, P.S. Mehta: SAE paper 860320 (1986)

32.13 D: Gosman: private communication

32.14 J.B. Moss, C.D. Stewart, K.J. Syed: "Flowfield Modelling of Soot Formation at Elevated Pressure", in Twenty-Second Symposium (International) on Combustion (The Combustion Institute, Pittsburgh 1988) p. 413

32.15 S.V. Patankar, D.B. Spalding: "Heat and mass transfer in boundary leyers", in A general calculation procedure (Intertext Books, London 1970)

32.16 M. Gonzalez, R. Borghi: in In Turbulent Shear Flows 7 (Springer-Verlag, Berlin 1991)

32.17 H. Böhm, D. Hesse, H. Jander, B. Lüers, J. Pietscher, H.Gg. Wagner, M. Weiss: "The Influence of Pressure and Temperature On Soot Formation in Premixed Flames", in Twenty-Second Symposium (International) on Combustion (The Combustion Institute, Pittsburgh 1988) p. 403

32.18 U. Wieschnowsky, H. Bockhorn, F. Fetting: "Some New Observations Concerning the Mass Growth of Soot in Premixed Hydrocarbon-Oxygen Flames" in Twenty-Second Symposium (International) on Combustion (The Combustion Institute, Pittsburgh 1988) p. 343

32.19 H. Böhm, M. Bönig, Ch. Feldermann, H. Jander, G. Rudolph, H.Gg. Wagner: "Pressure Dependence of Formation of Soot and PAH in Premixed Flames", this volume, sect. 9

32.20 K.B. Lee, M.W. Thring, J.M. Beer: Combust. Flame 6, 137 (1962)

32.21 H.Gg. Wagner: "Soot formation - an overview", in Particulate carbon: Formation during combustion, ed. by D.C. Siegla, G.W. Smith (Plenum Press, New York 1981), p. 1

32.22 G. Prado, J. Lahaye: "Morphology and Internal Structure of Soot and Black Carbon Blacks", in Particulate carbon: Formation during combustion, ed. by D.C. Siegla, G.W. Smith (Plenum Press, New York 1981), p. 33

32.23 R.J. Santoro, T.T. Yeh, J.J. Horvath, H.H. Semerjian: Combust. Sci. Technol. 53, 89 (1987)

32.24 I.M. Kennedy, W. Kollmann, J.Y. Chen: Combust.Flame 81, 73 (1990)

32.25 R. Borghi: Prog. Energy Combust. Sci. 14, 245 (1988)

32.26 Z. Habib, P. Vervish: Combust. Sci. Technol. 59, 4 (1988)

32.27 J.H. Kent, D.R. Honnery: Combust. Sci. Technol. 54, 383 (1987)

32.28 H. Mätzing, H.Gg. Wagner: "Measurements about the Influence of Pressure on Carbon Formation in Premixed Laminar C_2H_4-Air Flames", in Twenty-First Symposium (International) on Combustion (The Combustion Institute, Pittsburgh 1986) p. 1047

32.6 Discussion

Kent: Were you using a joint pdf between temperature and soot concentration to determine the oxidation rate.

Garo: For the oxidation rate we need to know the temperature and the fluctuation of temperature. So we have to know whether our temperature will be effected by radiation or not. We split our multidimensional pdf since there are parameters which are statistically independent. We obtain from these parameters a system of ODEs which gives us finally the mean reaction rate. With this information we can solve the balance equation for the soot volume fraction. In the oxidation, we take into account the fluctuations of the oxidant and soot concentration as well as temperature fluctuations; our model is able to give us an approximation of the joint pdf of the 3 variables.

Moss: Have you tried to switch the oxidation off low down in the Kent and Honnery flame.

Garo: Yes.

Moss: And does it not improve?

Garo: No. What we have is a very balanced system. If we suppress oxidation we will have too much soot.

Moss: I will show some predictions not of the Kent and Honnery flame but one of our own which is exactly like the Kent and Honnery flame. And I would want to advance the argument that in fact part of the problem is intermittency. The Beta-function pdf will suggest that you can have mean oxidation rate in circumstances where you don' t actually get coexistence of the states. In other words if the soot is mainly at rich mixtures and the oxidation can not occur at those mixtures then you have a classical intermittent source term problem which the Beta-function can not handle.

Garo: We may calculate several trajectories and use several exchange times. I think that we can weight each trajectory and give several possibilities like pyrolysed mixtures. We have then to simulate these different situations. This is more difficult to handle, but this allows us to avoid the problem of using a unique Beta-function pdf. Then the Beta-function and our trajectories $Y_{ox}(f)$ and $Y_{soot}(f)$ can indeed handle this situation.

Kent: Do you think that the discrepancies between the model and the experiments are due to the oxidation model or to the formation model?

Garo: The formation model is too simple. We are introducing an intermediate fuel from experimental data. This will push the soot formation a little bit further. But this work is still in progress.

Sarofim: Are you using your order of magnitude analyses to separate the formation from the oxidation based on time scales in an engine?

Garo: Yes.

Sarofim: But having done that you then went back and applied it to flames where the time scales do not apply.

Garo: We have to test the model. The first thing we had to do was to adjust our constants using a known laminar flame and then apply them to a turbulent flame. We know that there are some assumptions which are still not correct in the laminar atmospheric case but this seems us to be the only reasonable way to approach this problem. This work is in progress we can not extract conclusion from this approach now. The characteristic times are given to test the applicability of the flamelet theory. We mainly wanted to know how to couple the flamelet system for the ignition and turbulent combustion and the soot formation process which has a time history, because this is the major problem. This is the reason why we focus the study on characteristic times. But we must keep in mind that the approach we are proposing here for soot formation should be applied to high pressure systems only.

Modelling Soot Formation for Turbulent Flame Prediction

J. Barrie Moss

School of Mechanical Engineering,
Cranfield Institute of Technology,
Bedford, England

Abstract: A strategy is described for the development of simplified models of soot production suitable for incorporation in turbulent flame predictions. Detailed measurements in laminar diffusion flames are used to establish critical parameters in a reduced model of soot formation and burn-out based on two soot variables, the soot volume fraction f_V, and number density, N. This model incorporates representations of the key processes of nucleation, coagulation and surface growth.

The govering state properties are related to mixture fraction, a conserved scalar, which permits the ready extension of the model to turbulent flame prediction. The soot model is interpreted as a development of the laminar flamelet approach in which the source terms in the balance equations for f_V and N are closed over the pdf for mixture fraction.

Measurements of mean mixture fraction by microprobe sampling and mass spectrometric analysis, temperature by fine wire thermocouple and soot volume fraction by laser extinction and tomographic inversion are reported in a confined turbulent jet flame burning at atmospheric and elevated pressures. Comparisons are made between these measurements and model predictions. Whilst the experimental data is plausibly reproduced, uncertainties in the modelling are identified in relation to turbulence interaction, particle size and soot oxidation.

33.1 Introduction

Wide-ranging concerns for the environmental impact of exhaust emissions from diesel engines and, to a lesser extent, from gas turbines has recently added fresh impetus to research into the mechanisms of soot formation in flames and smoke production in practical combustion systems. The focus of such studies has also been modified to reflect the increasing role of numerical simulation in the process of design evaluation. Detailed flowfield prediction, employing sophisticated CFD techniques, affords the prospect of incorporating more comprehensive models of combustion chemistry, responsive to local flame properties, in such simulations. Even where smoke is not of direct concern, the influence of soot production on temperature within burning zones, through enhanced thermal radiation, may significantly affect both the formation of other pollutants and the durability of the combustor liner.

This paper will explore how insights into the fundamental chemical processes, largely derived from experimental observations of laminar flames, might be incorporated in turbulent flame predictions, more immediately relevant to practical applications. This latter emphasis will in turn impose constraints on the type of detailed laminar flame modelling which can be so incorporated and we shall therefore also seek to identify the most appropriate forms for those underlying data.

33.2 Flame Flowfield Prediction

Despite the continued expansion of computer capability, the prediction of three-dimensional elliptic combustion flows is constrained by computer storage and speed. Whilst models are under development which will admit the direct inclusion of complex chemistry into the turbulent flowfield calculation, based on balance equations for multi-dimensional probability density functions [33.1], the number of scalar variables which can be accommodated in this way will remain limited for the foreseeable future. In parallel, reduced chemical schemes are being developed for the simpler hydrocarbons [33.2], which restrict the number of reactive species computed to four (in the case of methane), a level which might be incorporated in a flowfield calculation. The extent to which the more detailed schemes under investigation for complex fuels [33.3] can in turn be simplified remains to be demonstrated. The further extension to soot formation is therefore limited presently by several factors. Systematically reduced chemical kinetic schemes for the gaseous species are not available for the hydrocarbon fuels of current interest in transport applications involving essentially non-premixed combustion. The dimension of such models, when developed, will clearly be greater than presently applying to methane, for example, and will be further enlarged by any soot formation mechanism. Laminar flame experiment will naturally provide the basis for validating such reduced chemical schemes but it seems appropriate to also examine their more direct role in soot model development, analogous to that of laminar flamelet modelling for the gaseous phase in turbulent combustion [33.4, 33.5].

The starting point for most current models of turbulent combustion is that the chemistry is fast and that mixing is the rate limiting process. As with the design of the combustion chamber, so the primary focus for model development has been the characterisation of turbulent mixing. The key scalar variable calculated in non-premixed flames is then the mixture fraction f which describes the local mixed state and, under the fast chemistry assumption, can be used to define the accompanying composition field [33.6]. This approach has proved successful in the prediction of combustion heat release and turbulent flowfield properties in a wide range of practical applications, whether coupled with the assumption of local chemical equilibrium or incorporating aspects of finite rate chemistry through flamelet modelling.

Although the chemistry of soot formation is clearly not fast by comparison with the heat release process, the central position of mixture fraction in modelling approaches must evidently be accommodated in attempts to incorporate soot models into turbulent flame predictions. Two particular constraints on soot model development are identified from this perspective; firstly, the number of additional variables introduced into the turbulent calculation must be small, there is no scope for the direct incorporation of complex schemes; secondly, models must capitalise where possible on the central role occupied by mixture fraction in characterising turbulent mixing, arguably the critical process in all turbulent flames of practical interest.

33.3 Laminar Flamelet Modelling of Soot Formation

Whilst very many studies of laminar diffusion flames have been reported over the years, leading to substantially improved understanding of sooting processes [33.7], there have been comparatively few systematic attempts to develop mathematical models of sooting flames. The complexity of the problems posed has tended to encourage parametric investigations, which focus on a limited range of measured properties, rather than more detailed mechanistic models.

It is clear that extensions of models for counter-flow and co-flowing laminar diffusion flames, incorporating quite comprehensive, if reduced, chemical schemes, will emerge for sooting hydrocarbon flames. One of the unresolved issues which will arise is that of the intimately coupled soot, thermal radiation and temperature fields and this does represent a significant departure from most existing models directed towards weakly sooting, near-adiabatically burning fuels like CH_4. How readily these formulations might be further simplified to meet the constraints of turbulent flame prediction set out earlier remains to be seen.

The most direct extension of flamelet modelling to incorporate sooting processes is that which simply introduces an additional state relationship for soot volume fraction f_V vs mixture fraction f based solely on laminar diffusion flame measurement. *Sivathanu* et al. [33.8, 33.9] have embarked on an extensive compilation of experimental data, embracing both gaseous species and soot, to generate such state relationships for a range of fuels.

These relationships can only be approximate however since they offer limited accommodation for such effects as residence time, radiative heat loss and local property gradients.

Soot modelling investigations which emphasise the ultimate goal of turbulent flame prediction, when establishing a model for the microscopic soot formation and oxidation processes, have been recently reported by *Moss* et al. [33.10] and *Kennedy* et al. [33.11]. Guided by laminar diffusion flame experiment, burning ethylene and pre-vaporised kerosene at atmospheric and elavated pressure, *Moss* et al. [33.10] and *Stewart* et al. [33.12] have developed a two-equation model for soot volume fraction f_V and number density N which reproduces key features of the observed sooting behaviour. Following the work of *Gilyazetdinov* [33.13], a simplified soot formation mechanism has been constructed which introduces modelled forms for the critical processes of nucleation, coagulation and surface growth. These source terms take the general form

$$\frac{d(N/N_A)}{dt} = \underset{\text{(nucleation)}}{\alpha} - \underset{\text{(coagulation)}}{\beta(N/N_A)^2} \tag{33.1}$$

$$\frac{d}{dt}(\rho_S f_V) = \underset{\text{(surface growth)}}{\gamma(\rho_S f_V)^{\frac{2}{3}} N^{\frac{1}{3}}} + \underset{\text{(nucleation)}}{\delta}$$

where

$$\alpha \equiv C_\alpha \rho^2 T^{\frac{1}{2}} X_{\text{fuel}}^{m\alpha} \exp(-\frac{T_\alpha}{T})$$

$$\beta \equiv C_\beta T^{\frac{1}{2}}$$

$$\gamma \equiv C_\gamma \rho T^{\frac{1}{2}} X_{\text{fuel}}^{m\gamma} \exp(-\frac{T_\gamma}{T})$$

$$\delta \equiv C_\delta \alpha \,.$$

N_A and ρ_S denote Avogadro's number and an assumed soot density (1800 $\text{kg}\,\text{m}^{-3}$) respectively, whilst ρ is the gas density, T is the temperature and X_{fuel} is the fuel mole fraction.

The model coefficients, $C_{\alpha\beta\gamma\delta}$, activation temperatures for nucleation and surface growth, T_α and T_γ, and fuel exponents, m_α and m_γ, which may each be fuel type dependent, are inferred from detailed comparisons between flame flowfield prediction and measurement. Computational predictions are achieved through the solution of balance equations for f_V and N which incorporate fluxes due to convection and thermophoresis in addition to the sooting mechanism outlined above. Flamelet relationships for the gaseous properties, notably density, temperature and fuel concentration, reduce the computed gasous scalars to simply that of mixture fraction. The temperature flamelet is prescribed from measurement and thereby circumvents the coupled radiation calculation which is itself soot property dependent.

Figures 33.1a. to 33.1c illustrate the streamwise development of measured and predicted properties in a kerosene-air flame, supported on a Wolfhard-Parker burner at a pressure of 2.26 atmospheres and with nitrogen dilution of the fuel (74% N_2 by mass). The soot volume fraction f_V was measured by laser extinction, temperature by fine wire thermocouple and mixture fraction by microprobe sampling and mass spectrometric analysis. Representation within the models of the processes of nucleation, coagualtion and surface growth appear to be the minimum necessary to reproduce the development shown.

33.4 Modelling the Sooting Turbulent Flame

This strategy has recently been extended to confined turbulent flames burning the same fuels *Young* et al. [33.14]. The experimental configuration is illustrated in Fig. 33.2. The turbulent jet flames are confined within a borosilicate tube, which is in turn mounted inside a cylindrical pressure vessel. Optical and probe access is provided at a fixed measurement station, relative to which the flame may be traversed vertically. The measurement techniques employed in the ealier laminar flame experiment are repeated. A mathematical inversion is now necessary to deduce the radial variation of time-averaged soot volume fraction from the path-integrated laser absorption measurements. Figures 33.3a. and 33.3b. show the centre line variation of mean mixture fraction, temperature and soot volume fraction for ethylene and pre-vaporised kerosene burning at atmospheric pressure. The kerosene data indicate a roughly four-fold increase in peak soot volume fraction in comparison with that for ethylene, but with an accompanying reduction in temperature. Illustrative data at elevated pressure [33.15] is presented in Fig. 33.4. Whilst differences between individual flames make simple inferences uncertain, relative to the pressure dependence, the roughly linear dependence illustrated is consistent with earlier observations in laminar diffusion flames [33.16, 33.10]. The peak levels of soot concentration are such that complete laser extinction is observed intermittently at the highest pressures investigated.

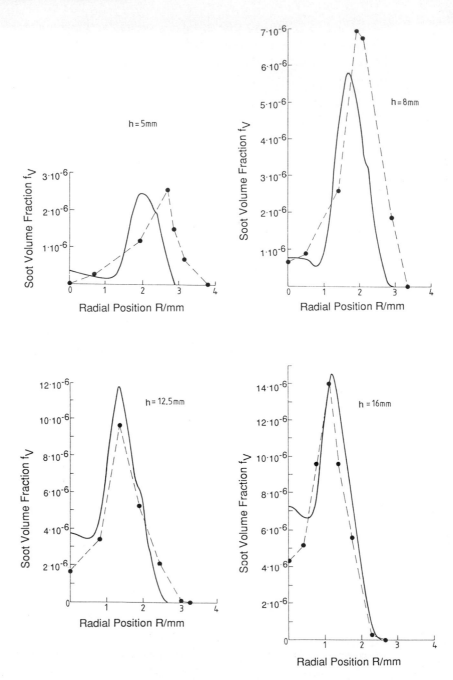

Fig. 33.1a. Comparison between prediction and experimental measurement of soot volume fraction at heights $h = 5$, 8, 12.5 and 16 mm in a prevaporised kerosene-nitrogen-air laminar diffusion flame supported on a Wolfhard-Parker burner

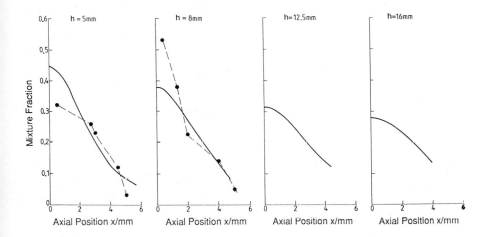

Fig. 33.1b. Comparison between prediction and experimental measurement of mixture fraction at heights $h = 5$, 8, 12.5 and 16 mm in a prevaporised kerosene-nitrogen-air laminar diffusion flame supported on a Wolfhard-Parker burner

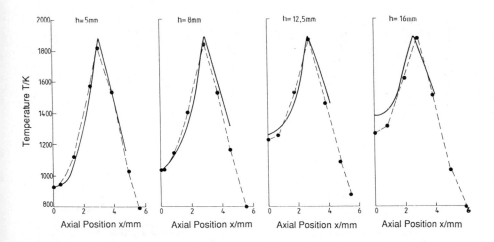

Fig. 33.1c. Comparison between prediction and experimental measurement of temperature at heights $h = 5$, 8, 12.5 and 16 mm in a prevaporised kerosene-nitrogen-air laminar diffusion flame supported on a Wolfhard-Parker burner

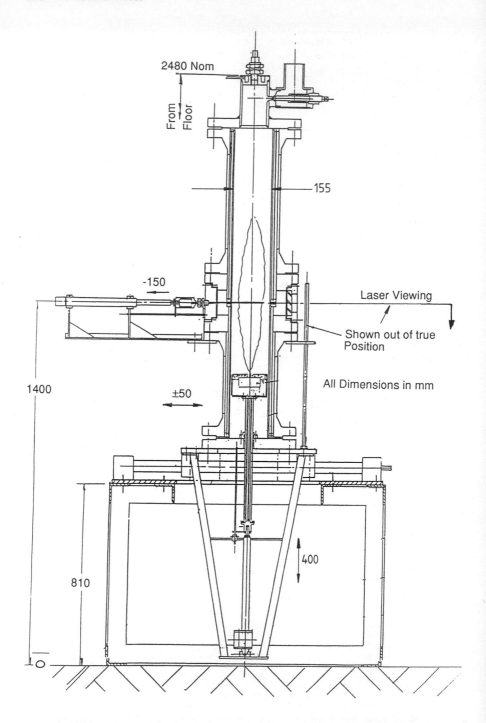

Fig. 33.2. Schematic of the confined turbulent jet flame rig

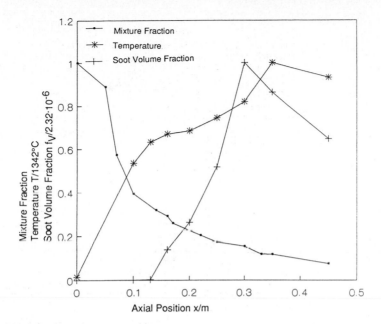

Fig. 33.3a. Centre-line measurements of mean soot volume fraction, mixture fraction and temperature in turbulent jet flames burning ethylene

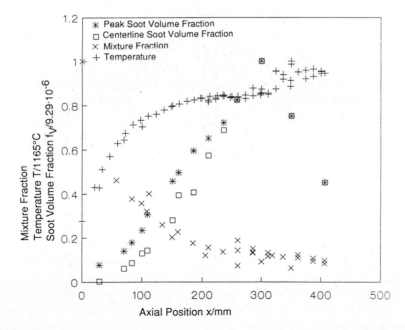

Fig. 33.3b. Centre-line measurements of mean soot volume fraction, mixture fraction and temperature in turbulent jet flames burning prevaporised kerosene

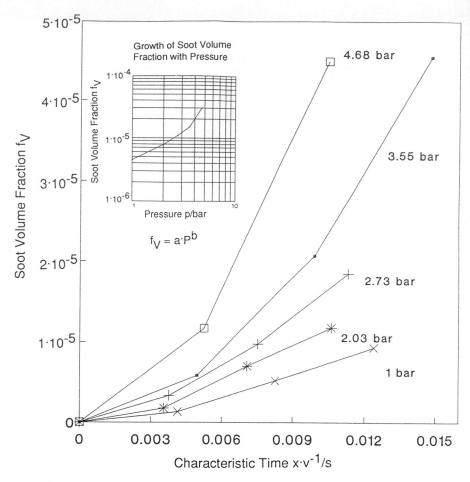

Fig. 33.4. Variation of peak soot volume fraction with characteristic time (axial position/jet exit velocity) for a range of operating pressures

These data provide a framework for testing the implementation of the laminar flame-based soot formation model. Figures 33.5a. and 33.5b. present a comparison of prediction with experiment for the ethylene flames. The complex coupling of temperature , soot formation and thermal radiation has been avoided by first computing the turbulent flowfield, using the conserved scalar approach, together with a laminar flamelet representation of density and temperature. The temperature-mixture fraction flamelet relationship is based on laminar flame measurement and incorporates radiative loss. It is then possible to effectively post-process the soot production onto the computed flowfield.

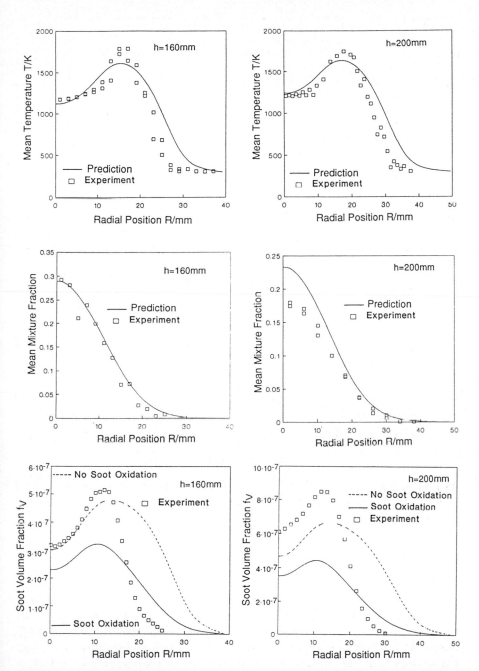

Fig. 33.5a. Comparison between prediction and experimetal measurement of mean temperature, mixture fraction and soot volume fraction at heights of 160 and 200 mm in a turbulent ethylene-air jet flame

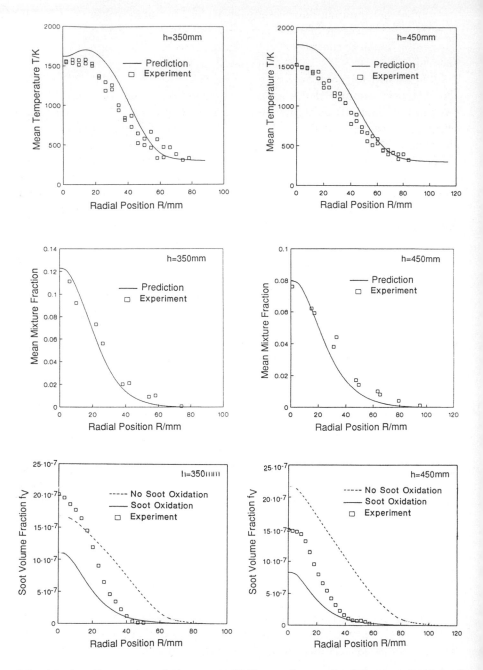

Fig. 33.5b. Comparison between prediction and experimetal measurement of mean temperature, mixture fraction and soot volume fraction at heights of 350 and 450 mm in a turbulent ethylene-air jet flame

Unlike the laminar flame study, in which it is practicable to separate in space the processes of soot formation and burn-out, the turbulent computations demand the modelling of both processes throughout the flame flowfield.

The model for soot oxidation is here taken to be that of *Nagle* et al. [33.17]. This assumes that the primary oxidising species is molecular oxygen, rather than OH as recent studies might suggest [33.18]. Turbulent averaging makes it particulary difficult to distinguish between the influence of alternative chemical kinetic mechanisms and local unmixedness.

Whilst the agreement between prediction and experiment is broadly encouraging, a number of issues specific to the turbulent flame emerge. The formation process, derived from the laminar flame studies, predicts plausible peak levels for soot volume fraction but significantly overpredicts the radial spread. The inclusion of oxidation in the source terms for soot properties does not then preferentially consume the "wings" of the distributions however. The centre-line values are observed to reduce substantially. Several possible explanations for this discrepancy suggests themselves:

(i) The flamelet approach assumes that the scalar variables, determining the source terms for soot formation and oxidation, are functions of mixture fraction. Surface growth and oxidation are heterogeneous processes however and therefore also particle surface area dependent. Whilst an average soot particle diameter d_p can be defined from soot volume fraction and number density, namely

$$d_p = \left(\frac{6f_V}{\pi N}\right)^{\frac{1}{3}}, \tag{33.2}$$

the laminar and turbulent averages each involve different ensembles and the source term scaling is uncertain. Discriminating data on soot particle size in turbulent flames is particularly sparse. Preliminary measurements in the Cranfield jet flame experiment, described earlier, by two-colour laser extinction and by differential mobility sizing imply diameters ≥ 300 nm.

Estimates of turbulent eddy lifetimes might be used to modify the effective particle surface area to account for the decrease in active sites which accompanies high temperature aging [33.19]. The discrimination of such a process in turbulent flames however is poorly substantiated at this time and likely to be masked by grosser uncertainties associated with turbulent averaging.

(ii) The more customary aspect of turbulent averaging introduces particular uncertainty in the oxidation term. The soot oxidation rate proposed by *Nagle* et al. [33.17] may be written in the form

$$\dot{W}_{ox} = M_C \left(\frac{xk_aP_{O_2}}{1 + k_zP_{O_2}} + (1-x)k_bP_{O_2}\right) \tag{33.3}$$

where

$$x = \left(1 + \frac{k_t}{k_b P_{O_2}}\right)^{-1}.$$

This rate expression, with its dependence on temperature and oxygen concentration, may be interpreted as a function of mixture fraction, soot number density and volume fraction. If we then incorporate the mean available particle surface area, $\pi^{1/3}(6f_V/N)^{2/3}$, the source term for soot mass consumption is given by

$$S_{ox} = \kappa \dot{W}_{ox}(f) N^{\frac{1}{3}} f_V^{\frac{2}{3}}. \tag{33.4}$$

Formal averaging of such a term for turbulent flowfield calculation requires a multi-dimensional scalar pdf embracing mixture fraction, soot number density and volume fraction. This is inaccessible in presently envisaged model closures. For the purposes of the calculations cited, the soot property correlations are negleted and the source term is written in the form

$$\overline{S_{ox}} = \kappa \overline{\dot{W}_{ox}} \overline{N}^{\frac{1}{3}} \overline{f_V}^{\frac{2}{3}} \tag{33.5}$$

where

$$\overline{\dot{W}_{ox}} = \overline{\rho} \int_0^1 \frac{\dot{W}_{ox}(f) \tilde{P}(f) df}{\rho}.$$

Whilst such a representation of the average accommodates the principal non-linearity in the source term, normally associated with the exponential temperature dependence of the rate expressions, it does not address satisfactorily the posssible unmixedness of particulate soot and the oxidizing species.

The soot oxidation term, in particular, is highly sensitive to mixture intermittency. In the flame regime in which the soot concentration is dominated by local formation rather than transport, for example, soot is confined to mixture fractions of order 0.1 from which oxidizing species are essentially absent. The *pdf* for mixture fraction, $P(f)$, will typically embrace leaner mixtures however and hence predict non-zero mean oxidation. It would appear from the predictions in Figs. 33.5a. and 33.5b. that it is arguably more plausible to neglect oxidation entirely at flame heights ≤ 450 mm. The nature of this argument would appear little influenced by the particular oxidizing species, whether O_2 or OH.

Much computationally-based turbulent combustion research, and in particular laminar flamelet modelling, has been largely prompted by the need to avoid explicit source term closures of this type. Sooting turbulent flames therefore invite an extremely careful re-appraisal of modelling strategies and a demanding experimental database.

References

33.1 S.B. Pope: Prog. Energy Combust. Sci. **11**, 119 (1985)

33.2 N. Peters, R.J. Kee: Combust. Flame **68**, 17 (1987)

33.3 C.K. Westbrook, J. Warnatz, K.J. Pitz: "A Detailed Chemical Kinetic Reaction Mechanism for the Oxidation of Iso-Octane and n-Heptane over an Extended Temperature Range and its Application to Analysis of Engine Knock", in Twenty-Second Symposium (International) on Combustion (The Combustion Institute, Pittsburgh 1988) p. 893

33.4 N. Peters: Prog. Energy Combust. Sci. **10**, 319 (1984)

33.5 S.K. Liew, K.N.C. Bray, J.B. Moss: Combust. Flame **56**, 199 (1984)

33.6 R.W. Bilger: Prog. Energy Combust. Sci. **1**, 87 (1976)

33.7 H. Jander, H.Gg. Wagner: "Soot Formation in Combustion" - an International Round Table Discussion, (Vandenhoeck and Ruprecht, Göttingen 1990)

33.8 Y.R. Sivathanu, G.M. Feath: Combust.Flame **81**, 133 (1990)

33.9 Y.R. Sivathanu, G.M. Faeth: Combust. Flame **82**, 211 (1990)

33.10 J.B. Moss, C.D. Stewart, K.J. Syed: "Flowfield Modelling of Soot Formation at Elevated Pressure", in Twenty-Second Symposium (International) on Combustion (The Combustion Institute, Pittsburgh 1988) p. 413

33.11 I.M. Kennedy, W. Kollmann, J.-Y. Chen: "Predictions of Soot in Laminar Diffusion Flames", AIAA Paper 90-0459 (1990)

33.12 C.D. Stewart, K.J. Syed, J.B. Moss: Combust. Sci. Technol. **75**, 211 (1991)

33.13 L.P. Gilyazetdinov: Khim. Tverd. Topl. (Moscow) **3**, 103 (1972)

33.14 K.J. Young, C.D. Stewart, K.J. Syed, J.B. Moss: "Soot Formation in Confined Turbulent Flames Fuelled by Pre-vaporised Kerosine and by Ethylene", in Proceeding on Tenth ISABE Meeting (AIAA, Nottingham, England 1991) p. 239

33.15 J.B. Moss, C.D. Stewart, K.J. Young, Q.-P. Zheng: "Smoke Production in Aircraft Engines: Flowfield Modelling and Experimental Validation", in Proceeding on Third European Propulsion Forum (ONERA, Paris 1991)

33.16 L.W. Flower, C.J. Bowman: "Soot Production in Axisymmetric Laminar Diffusion Flames at Pressure from one to Ten Atmospheres", in Twenty-First Symposium (International) on Combustion (The Combustion Institute, Pittsburgh 1986) p. 115

33.17 J. Nagle, R.F. Strickland-Constable: "Oxidation of Carbon between 1000 − 2000°C", in Proceedings of the Fifth Conference on Carbon (Pergamon Press, New York 1962) p. 154

33.18 A. Garo, G. Prado, J. Lahaye: Combust. Flame **79**, 226 (1990)

33.19 S.J. Harris: Combust. Sci. Technol. **72**, 67 (1990)

Discussion

Colket: Earlier in the session we saw the presentation by C.Vovelle who showed that even in laminar premixed flames the amount of aromatics substantially affect the soot that you will eventually produce. I am sure that this is even more true in the case of a diffusion flame. So, I am asking how do you characterise the fuel that you are using and how might that be different from the application in which you are interested.

Moss: At the moment we are working with one batch of standard aviation kerosine which has about 18% aromatic in it. This is higher than the levels that Dr. Vovelle discussed.

Bockhorn: You showed a large discrepancy between the calculated volume fractions of soot and the measured ones. Are these discrepancies due to a discrepancy in the flow field or temperature field?

Moss: In the sense that the only indications we have are the measurements that we have made, the distribution of mean mixture fraction and mean temperature look reasonable. They are as good as the kind of turbulent jet flame predictions that you normally expect. There are clearly some discrepancies as I suggested, on the center line for example the predicted mixture fractions look a little high. The flames are not very suitable for making probe measurements. They are equally unfriendly to making other optical measurements however. So until we have more information it will be difficult to tell. But within the framework of the mean mixture fraction and mean temperature that we have measured, the flow field predictions look reasonably good.

Wagner: You had intermittent periods when you had nearly complete absorption in your measurements. Could you tell us a little bit about the time variation of these absorptions? And did you try to insert light pipes so that you can get spatial resolution of these fluctuations?

Moss: Off the top of my head I'm not sure what the time traces look like. We have all this information but not here. At this point in time at 0.5 MPa they are very short. So they look like turbulent fluctuations during which the whole path is extinguished. We have a third experiment which is currently under way which is on a gas turbine combustor. So we have the whole spectrum running simultaneously. There we are using similar techniques and we are shilding the beam for a part of its path in the combustor for the reasons that a tomographic inversion is not possible because the flow field is not axisymmetric. We have tried this in the jet flame too but not in the high pressure case. At atmospheric pressure it works reasonably well and we get plausible looking predictions by backing out the sight tube and also doing the tomographic inversion. We will try to raise the pressure since that is one of the options available to us.

Sarofim: I noticed that you were predicting a fairly high oxidation rate along the axis. I think you are running into the same problem that you addressed to A. Garo. Could you not allow for intermittency along the centerline.

Moss: Yes we could. It's a slightly more complicated issue. If you look back at the genesis of the basic model one finds mixture fraction is an important characteristic that is colouring the whole of the flow-field. That is a simple carry over from the success that people have had with flamelet modeling in non premixed flames generally . There will be a region where soot volume fraction is determined essentially by the formation mechanism. This will be low down in the flame and I think under those circumstances oxidation actually is relatively unimportant. As you move further downstream then you would expect scrambling between mixture fraction and volume fraction because of different transport processes. In the oxidation region where soot breaks through then clearly, in terms of treating soot as a perturbation to mixture fraction, it now no longer correlates. Because you have zero mixture fraction but you have also got soot. So there is an evolution in the relative balance as you move along the flame. Another problem area is that we calculate mixture fraction as a gaseous conserved scalar in turbulent flows at this point in time but as more of the carbon is bound in the soot, in a solid phase, it must be increasingly tempting to start to distinguish the contribution to mixture fraction as solid carbon from the contribution to mixture fraction in the gas phase. We have made some predictions which do exactly that. They say that the mixture fraction is made of two parts, analogous to the treatment of droplets. It does make a significant difference even in ethylene. It tends to make the gasphase mixtures rather leaner than we thought they were on the basis of total mixture fraction and therefore it slides the reaction zone closer to the nozzle, not by a great amount in ethylene but clearly as the soot loadings go up that will become more and more of an issue. That unfortunately will add some more equations to the calculation domain if you actually have to partition between gas phase mixture fraction and a solid phase mixture fraction.

Williams: I'm surprised by the Nagle-Strickland-Constable expression used by yourself and a number of people. One principal objection is that in recent years the variation of the oxygen content has been found to be significantly different than what was used in the original expression. The current powers are something like 0.2 to 0.5 or 0.6 as opposed to the unity used in the original expression. This makes a very significant change, particulary if the pressure has been changed. On top of that there are recent measurements of intrinsic reaction rates and one can set up now the whole of the expression for soot oxidation in a fairly precise way. I'm rather surprised that you are not using this information.

Moss: I think it is a question of ones view of the priority of the issues. I didn't put the emphasis on using revised kinetic information of that type. What I'm not sure of is how plausibly I can reproduce the turbulence interaction. I think this is a question of ones background. I would be quite happy to adopt such revisions, but I don't think that will solve the problem. We have run some cases using OH mechanisms of the type that we have described earlier and looked at revisions to the Nagle-Strickland-Constable expression. It makes a difference but it doesn't close the argument that we have identified. I think there is another problem underlying it.

Santoro: You mentioned the problem of probing in these flames. And both ourself and, parallel to us, the Göttingen group are using microprobes with oscillating fibres in these high density soot flames. In our butene and butadiene flames we are able to get measurements. We didn't have such a difficult environment as you had but these probes are not very difficult to build. Secondly, you were talking about the applicability of your experiment to a gas turbine. Most gas turbine fuel nozzles now are air blast nozzles, so it seems to me to be quite appropriate for you to put a high velocity gas near your jet at the appropriate ratio that a gas turbine engine would have in the air blast part. This will give you some premixing early on which might reduce the levels of soot and yet still be comparable to what a gas turbine environment would develop. To get your turbulent fluctuating temperature measurements are you using thermocouples? I was wondering about time constants and soot deposition and how they would affect the those spectral measurements.

Moss: We are not reporting fluctuating temperatures; those are time averaged temperatures. Even making time averaged measurements involves a series of backward and forward maneuvers trying to clean the probe in between. But we are not trying to measure fluctuation levels, I think that would be unrealistic.

General Discussion on Models for Soot Formation in Turbulent Flames

chaired by

Henning Bockhorn, John Kent

Bockhorn: First I have to thank all the speakers and contributors for the dicussion on the papers this morning. Just to start the general discussion on the three papers I would like to recall the statement that B. Moss gave yesterday evening. We don't have any chance to predict a practical turbulent combustor like a gas turbine or a diesel engine if we expand the field of scalar by more than three or five scalars considering the soot chemistry. I think this point has been exactly worked out by the contribution of Ian Kennedy, Annie Garo, and B. Moss today. They presented models where all the soot chemistry had been described by two or three additional scalars. You presented as additional scalars the soot volume fraction and a kind of coupling between equilibrium chemistry and soot chemistry.

Kennedy: And a sort of energy equation, too.

Bockhorn: Barry Moss did it by introducing number density and volume fraction and the coupling between the chemistry which came from a flamelet approach has been done by semi-empirical rate equations which describe nucleation growth and coagulation. The same procedure has been applied by Annie Garo with a different type of turbulent calculation. So the situation is, that we end up with a model which has to be calibrated to different chemical systems. I think it has also to be calibrated to flow field situations since most of the turbulent models that are used need some improvements because of the anisotropy of turbulence in real combustion like gas turbines or diesel engines. So again these models of turbulence have to be calibrated or at least they should be improved from a theoretical point of view. But in any case the chemical system has to be calibrated to different fuels or fuel mixtures. I would like to ask the three contributors of this morning how many chances do they give this calibration methods to predict real engines or situations where we have a complicated chemical fuel and not a simple system as in acetylene diffusion flame or in cerosine flame.

Moss: I think for kerosine or for the diesel it is a major problem. I think from a scientific point of view it would be more satisfactory to work through the argument of constructing fuels that would be somewhat more accessible. But that is a relativly long term process. I don't want to go back over the ground of last night but I think the windows of opportunity for modeling and the design of combustors is not infinite. If I take a gas turbine as an illustration. Redesigning an engine is a great issue for a company. Take a small company like Rolls Royce when they redesign an engine they risk the net worth of the company. It is therefore something that you will not do very often. There is whole flow of activity in the combustion area currently because there are gas turbine combustor designs going through to meet the next round of expected legislation. That is a five year or less than ten year window it seems to me. And maybe the whole emission thing has to be resolved within that kind of time frame. So we have to make fairly rapid progress if we want to influence the design. If we don't then the designer will solve the problem for us without knowing any more about combustion than we know now. It is an expensive way of doing it and it is an empirical way of doing it. But a gas turbine combustor is 99.9 % thermally efficient. You don't really need to know a great deal about combustion to do that. I think for the first time there is really a requirement to put some chemistry in there to lower the emissions. Most devices have got away by just worrying about the distribution of heat release under the control of mixing. For emissions that is not true. But people are very clever and they will come up with schemes. They will influence mixture preparation and we shall be left short if we are not careful.

Kennedy: I think the question of calibration and testing these things is an important one. It is probably more a general question than just a plan to turbulent combustion. I think it probably applies to the kinetics modeling as well. We are at a point now in modeling where we can predict things that we can not measure and that is a problem. In terms of turbulent combustion we can measure a sort of rough integrated line of sight soot volume fraction. We have problems even measuring velocities in very sooty flames. Measuring the flow and getting LDV signals is not simple. We have problems measuring the temperature because of coating thermocouples. So in fact we are stuck with very gross overall parameters that we can test but we can not measure the detailed quantities which would tell us how well we are doing with the models. I would think that the same thing probably applies to some of the detailed chemistry that people have talked about. How do you measure some of those rate constants and things like that? You know we see some measurements now of PAH but how do you get the rate constants? The only thing we can do at the moment is to compare mean time averaged parameters like soot volume fraction. This is unsatisfactory.

Garo: We have to keep in mind that, now, we only want to find tendencies. If by taking a fuel with a given percentage of aromatics and we are able to determine the soot formation tendency, this will be a success. We need to know if we may consider the fuel as a precentage of aromatics and aliphatics. Then we could introduce two kinetics and look at the soot production rate depending on the percentage of both compounds. Our approach of the interaction between turbulence and chemistry would thus be completely tested. For the moment we have not enough practical simplification of the chemistry. We know the main theoretical paths of soot formation but not how to introduce these informations in a general and simple way. What we need is a simple reduced kinetic mechanism from any initial fuel to a first or a second intermediate species, then to soot, without complex chemistry.

Lepperhoff: If the experimentalist were able to measure exactly we would not need any model. We really need to know how sensitive our parameters are. I think our models have just to be able to give such results we can measure. We discussed the temperature profile measured by thermocouples or measured by CARS, that is quite a big difference. This is why it is very important to know the sensitivities of the input data.

Dobbins: I found that the experiment is alway horrendously more expensive compared to modeling. Furthermore that if you don' t make some attempt to formulate a theoretical framework then you will not understand the experiment.

Smyth: I have a specific question for the modellers. I havn' t seen a model that ignored surface growth and that' s no surprise. But John Kent put up a comparison between premixed and diffusion flames at the same stoichiometry in acetylene, the surface growth rate differed by a factor of five. Some models are extremely sensitive to the specific surface growth rate. The question is how well do they expect to capture the time-temperature history or residence time variation and the surface specific growth rate. You can put in activation energies and put in some pretty simple fixes, but if your models are extremly sensitive to it, then it is really a hopeless part of the problem.

Kent: I haven' t tackled the turbulent problem, but I can say how we use the parameters. For the laminar case we had a growth rate which was a function of mixture fraction and temperature. The question is whether that translates in the turbulent flame. The extra parameters which come in turbulence is flame stretch and scalar dissipation and it is probably an important factor that the flamelet model that you develop from laminar flames probably doesn' t really apply to the turbulent flame case.

Frenklach: We talk about of how theory interacts with experiments. First of all, we should stop talking about surface growth, we should talk about mass growth. There are two mechanisms for mass growth, one is a chemical reaction on the surface. That is the surface growth, but it's only one part of the story. The other part of the story is addition of the PAH which may not proceed through simple chemical means. There are serveral possibilities involved sticking to the surface, collision, Van der Waals forces, and some more. Now, the two mechanisms are operational to a different degree in different environment. I think that is the reason why we see different behavior in different flames. And the difference that you see may not come from wrong numbers, but mainly because you don't account for the PAH condensation on soot particles.

Glassman: I am not sure that my lack of physical fear for pdf's and Favre averageing, may show ignorance of the question I' m asking but Bob Santorro's comment a lot earlier revived certain things that always troubled me. I think everybody knows I shatter when people mix turbulence and chemistry. But yet in 1955 I had to built a turbulent flow reactor and we made temperature measurements in that particular flow reactor. One of the things that concern me at the time is how people handle the actual rates that they report. It becomes very apparent in a little paper that we puplished in 57 which is on this matter which the turbulence community completely ignored, of course. Basically is whether the rate at the averaged temperature can be significantly different from the average rate, particularly when you have a high activation energy process. Some of us convinced us that the actual particle formation is a high activation energy process. I am addressing myself and my good friend I. Kennedy whether his types of formulation handles that problem of the rate because otherwise an agreement between inceptive particle formation and experiment can't be reached.

Kennedy: I think you are referring to a classical problem with turbulent combustion which is that the rates are non linear functions of temperatures, so therefore the function of the average is not the same as the average of the function. Particularly these exponential terms like in chemical kinetics have always been one of the fundamental problems of turbulent combustion. You can deal with this problem on a k-ϵ basis for the flow equations and by formulating a pdf and integrating over pdf to get mean reaction rates. That is one way of doing it. My formulation directly calculates the pdf so I do not have to assume one. So we can handle those kind of approaches if you can get some handle on the distribution of temperatures. That is the pdf of temperature and I think that problem has been around along enough and I think we are not too troubled about that if we can make up a resonable pdf for temperature.

Glassman: What is the variance for the temperature?

Kennedy: It could be anything. It depends on the calculated pdf. The mean reaction rate can be a lot different if you base that on the mean temperature.

Glassman: My question was just what are the temperature fluctuations in the turbulent field that you are concerned with? How is this reflected on your averageing technics?

Bockhorn: If you take a presumed shape pdf, then you affect the procedure of averaging by the presumed shape of the pdf. So this is, I think what some people have pursued in the past. The approach that I. Kennedy persues is quite different from this, he calculated the shape of the pdf and the temperature variance is a result of this calculation. So there is no assumption on the shape of the pdf. I don't know exactly how you average and how the temperature fluctuations come in your model.

Moss: The difference is that I have closed source terms like formation by integrating over the pdf for mixture fraction. But this is an assumed shape pdf so it's much more sensitive to the shape of the beta function than the process that I. Kennedy describes. The trade off that you are making is that

I'm computing equations for the mean and variance of the mixture fraction and I. Kennedy is computing a pdf transport equation which is a much more complicated animal. But they are both attempting to accomodate scalar fluctuations of the type that Prof. Glassman describes.

Williams: My comment goes back perhaps to I. Kennedy's paper. One of the problems that I can see is that one assumes the combustion reaction to be fuel and oxygen to give CO_2. Do you have a detailed chemistry in your problem.

Kennedy: Since I'm not a chemist I like to get the chemistry out of it. What we often do in this case is to assume something like a laminar flamelet model. You take actual measurements in laminar flames and some sort of strain rate and then you can apply that as library for the chemistry. So you calculate the mixture fraction, you go into your library of flamelets and look up what the composition and temperature would be for that mixture fraction. The other approach to what we have done is what we call the constrained equilibrium. That is equilibrium as a function of two scalars mixture fraction and enthalpy. You have then a look up table where you go in with mixture fraction and enthalpy and get your thermo-chemical properties that way. So we don't have any rate equation in the model for fuel consumption or fuel oxidation. In my case it's an assumption that the kinetics are fast and that they are in equilibrium.

Williams: There are relationships not only between CO and OH which could be coupled by putting in OH reactions for soot burn out. But there are relationships between CO and PAH which could be coupled although these are very lose assumptions.

Santoro: I think you are asking about species that have concentrations that are determined by kinetic processes like CO or soot. Intermediates are not well treated with a conserved scalar approach in my opinion. We have looked recently at CO in a range of flames that have different soot loadings and we do not find that the conserved scalar approach, of having a library of flames, will predict the CO level because it's too sensitive to the local conditions in terms of that CO to OH rate. I think those kind of models can be added into a conserved scalar library. So that you can get a lot of species like CO, O_2, water, fuel, remaining fuel from that type of calculation. But in my opinion you have to couple in some limited kinetics to handle things. And this is just what I. Kennedy does in his soot model. He doesn't try to use a conserved scalar to get the soot oxidation process. He puts a kinetic rate in to do that.

Bockhorn: I think this was a kind of concluding remark covering the problems that are inherent in the flamelet approaches or in cutting down the soot chemistry to some few scalars which are linked by equilibrium chemistry or flamelet chemistry.

Part VII

Final Discussion
and Perspectives

The last part of this book resumes the final discussion to all items that have been presented during the workshop and reveals some perspectives for future work on the solution of unsolved problems or new approaches.

Mechanisms and Models of Soot Formation: Final Discussion and Perspectives

chaired by

Heinz Georg Wagner

Wagner: Now, we should try to determine the development into the future and I would like to take the chance to ask those of you who had certain prospectives concerning soot formation to take part and to contribute to this discussion. You have seen here different approaches to the soot problem. One is to determine the conditions under which you definitely do not get soot. That is information which is already accepted by the engineers and that is what they need. Some other people tried to look deep into the chemistry. There were also people who tried to build bridges across those parts which are unknown in chemistry and feed them into the models to describe everything. As you may remember, mankind got the fire about 10,000 generations ago and I sometimes imagine how God was thinking about which kind of fire should he give to these"animals" here on earth. If you had to make the decision you would decide to give them the diffusion flame. God did the same thing and he was of course absolutely aware of what he did. He preferred a complicated system with a high degree of intrinsic stability and therefor a small risk. So this is the reason why we have to face all these problems and have to make complicated models in order to understand what is going on in a diffusion flame. That was a little introduction and now I would like to start the dicussion. Firstly, I would like to hear about your prospectives, where are the gaps of knowledge, and where do we need information urgently? Secondly, I would like to ask those of you who are willing to tell us about their further plans in their research so that we can get an overview of what is going on. I know that the last question is probably not very easy to answer. I see Dr. Sarofim is willing to start the discussion.

Sarofim: I would like to address a question to B. Moss. He gave us his rule of how many variables you can add to a turbulence model. In his own models of sooting he uncoupled and post-processed the kinetics from the turbulence and the flow field calculations. When dealing with pollutants which in many cases don't influence the flow and temperature field why not have a complex pollutant model which is post-processed.

Moss: The straight forward answer is, for some pollutants I'm sure that this is true. The simplest case is to take the thermal NO_x. Something which is thermally insignificant you can post process increasingly elaborate models. The issue would clearly be that you design your device ever better and the NO_x concentration falls and the balance shifts from being thermal NO_x to prompt NO_x or to fuel bound contribution or what ever. Then your simple model ceases to be simple. I'm not sure that it is possible really in the context of soot because of the strong radiative coupling. That is an issue that I. Kennedy has partially addressed and about which I have serious doubts. We have probably all flirted with trying to put detailed radiative calculations into even atmospheric turbulent flames and that rapidly becomes a very complicated issue. I believe that if we do offer more then two or three additional scalar variables whether you treated them in a very superficial way as I have done or whether you adapt the Kennedy, Kollmann, Chen formulation and go to a multidimensional scalar pdf. I suppose the greatest exponent in this area is S. Pope. Even there, even for the gas phase scalars, we are not prepared to go beyond four scalars at this point in time. The complexity of doubling that by adding the soot model is unacceptable at this point in time. We may be able to compute some very simple flame geometries but I can't see that there is a practical solution in which the soot problem has added more than two or three variables to the scalar dimension.

Santoro: I want to try to get things going by giving a summary of what I thought came out from this meeting that was an advance on the last meeting and where I see some controversies. If you give me a few minutes, I'll run through some points. I think at the last meeting we were just seeing in the preparticle chemistry area, the emergence of this idea of what is now called reactive coagulation. And at that time we were arguing about how do these things that are 1000 or 2000u stick together. At this meeting people are proposing reactive coagulation of 100 and 200u species and saying that this route may be, under certain flame conditions, the dominante way to soot as opposed to the acetylenic build up mechanism. So I see a challenge to further look at the kinetics of those types of isomerisations and dimerisation versus the Frenklach type of build up mechanism. I see that as an issue in my mind since, as much as I like the data, I haven't seen the physics that makes me say I'm comfortable with reactive coagulation of such smaller species. The second change has come in a subtler way and I think it deals with our focusing on the surface area question from a different perspective. We had a debate also in Göttingen about the Bockhorn and Harris flame results and what they meant. Now the active site argument has arisen, it is not a new concept, but what I think it does is make us focus much more on the nature of the soot particle surface. And that is something that I haven't seen in a while. There are other groups of people who presently do work in this area such as people who investigate additives for fuels or those in the carbon black industry. They know a lot about how to treat surfaces of particles. There are measurement techniques that we haven't taken advantage of yet. I'm trying to collaborate with workers in this area. The third point I'd like to make is, I firmly believe that the work that is going on in the measurements of the fractal aspects of soot particles is changing the way we are going to have to do light scattering measurements. The final topic we always talk about in soot formation is oxidation. It is clear that there are only three or four names that come up when you talk about measurements that have been done in this area and what they mean. And I think much more consideration of this topic is needed, given the wide diversity of results that we are seeing. My own work in diffusion flames is going along the lines similar to that of Prof. D'Alessio. With the help of K. Smyth from NIST, we are using UV pulse laser techniques in these flames to measure not only small radicals like OH and may be CH, but also to start to look at the evolution of these large fluorescing species in the flames. Because as gratified as I am to see people who have been able to use the measurements that we made and the excellent laminar diffusion flame measurements of Kent and Wagner in their model evaluation, they are really missing information on the chemistry in those flames to do the same kind of comparison. I think the models are getting to the point where they are asking questions dealing with these chemical mechanisms. M. Frenklach always is urging me to give him that kind of information because he would like to model diffusion flames too. So we have to address their problem as well as move on in the surface active site area. Furthermore, I agree with Prof. Wagner that we need to examine diffusion flames at very high pressures, similar to what he has done with the premixed flames. I think from what we heard about the diesel problem that this point is clear. I am not studying any flames such over 10 atm and may be I should rethink that because of this meeting. On the modeling area I'm afraid I can't comment. I'm personally very pleased with what I see. People are now taking very complicated fluid mechanic situations and trying to put soot mechanisms into the model, that is a true advance in the field since our last workshop.

Colket: I just wanted to respond to the first item that you brought up about the differences between coagulation processes of 1000u and greater and the use of models which use small molecular weight species. First of all, in our model we are allowing very small molecular weight species to coagulate; these species are smaller than what Prof. Frenklach is using. And that is really an artificial simplification due to a lack of understanding of how to model the higher molecuar weight species. This also reduces the cost of those calculations. That is not necessarily an indication that those processes are important in any flame but rather a demonstration of the ability to use such simplifying assumption to try to make predictions. In terms of experimental evidence indicating that these coagulation processes or recombination processes occur for these small species you have to look at the time scales that Prof. Sarofim presented for example. These times were on the order of seconds which are three orders of magnitude longer than what we were expecting in time scales for the flame process. This difference in time scales seems quite reasonable because recombination processes are radical additions with loss of an H to form a very strong bond which doesn't fall apart. First of all the concentration of the radical itself is probably an order magnitude lower than the parent. Secondly the rate constant for that process is about two orders of magnitude lower than a sticking process might be with Van der Waals forces. So there are about three orders of magnitude difference in the rates of those processes. The importance of them in a flame, is still yet to be determined.

Wagner: You didn't tell us something about your future plans but may be we can come back to you. I would like to ask M. Frenklach, he always has a bunch of ideas and he also can express them in a short convincing way.

Frenklach: First I would like to make a comment about this diagram. I personally don't like separation between modelers and experimentalists because I am myself an experimentalist who is modeling to understand my experiments. I want to give a simple example. If someone is doing theory and predicting a rate coefficients and he doesn't compare it to the experiments no one will accept the number. The same is in experimental kinetics: if you measure a number but don't back it up by a theory it is also not accepted. I would like to propose the same attitude in our research. So far experimental data are taken as a kind of bible whereas modeling predictions are not taken that seriously. I think the coupling between modeling and experiment should come forth. The biggest difference that I see from prior meetings to this one is that many more people doing modeling and trying to link observation to some basic ideas. Some of the new developments that R. Santoro pointed out come exactly from that. We can't explain new experimental results with the same old models. That is where new development will come in. The second point, I think we desperately need an experiment against which we can calibrate our models. For example, such things like calibrating PAH, soot particles and so on. So far, in my opinion, we have may be one flame, Bockhorn's flame which has both the determination of PAH and soot at the same conditions. But we need similar type of information with much higher accuracy for temperature and major species concentration profiles.

Wagner: Would you recommend to have standard flame conditions?

Frenklach: We need to agree on some standard flames but it shouldn't be one because for example surface deposition of mass may be different under different flame conditions. So I would like to see at least two premixed flames, one high-temperature one low-temperature flame. I would like to have a standard flame that everyone can contribute to and we can test our model on. Then we need some standard diffusion flames and some turbulent flames. I think we need to have standard flames and we have to complete them with high quality information because this essentially will determine the success in modeling. In the near future we will try to apply our model to more practical conditions as far as we can stretch it. We are working now on explaining the nature of sooting limits for laminar premixed flames. We are incorporating new radiative models so we do not need to rely on the thin optical model approximation. That is done. In the future I would like to go back and look more at the details of chemistry.

Pfefferle: One thing is missing up till now. That is testing of submodels and conditions where species and transport parameters can be characterized. Obviously in a system with many unknowns each with large error bars it is critical to define subsystems where a mathematical well-posed problem can be formulated. If I was doing an experiment to study kinetics, I am not going to do it in a situation were I don't know the temperature or the temperature variations and the flow field and whether things are coupled or not. In the case of soot formation, the field of large hydrocarbon chemistry under combustion conditions is in it's infancy. We have blind zones where it is critical to obtain detail chemical information. One modeling approach on the chemistry side is to use a model such as that Huston Miller is now working with, where you input experimental data for the meassured species field and temperature field and then test your soot growth model with a partial parameter field that you know. Experimental and theoretical submodel testing aspect, that is missing from the current background. Also time and length scales are very important. It is important to understand which processes are coupled and which are not and when. This is what helps us to decide how to make these submodels. Since incineration is going to become a very important field we should look at surface processes which can be very important in both soot and hazardous PAH formation. Although this community will have a lot to offer we have done very little with surface interactions up till now. My future plans are to do more to investigate the role of PAH reaction in soot formation and try to determine rational theoreticaly-based correlation for pyrolysis and oxidative pyrolysis reactivity as a function of fuel structure in terms of small ring and soot formation. We would also like to explore the role of the surfaces in effecting soot formation processes in incineration applications.

Wagner: I had the impression that we are going into a Renaissance of the PAH. Concerning the particles I would like to remind you that Tesner went through that very laborious business of measuring the different kinds of surfaces of growing soot particles and these data should not be forgotten.

Smyth: I think the prediction of flame structure in premixed flames is still dissappointingly poor. I remember the Westmoreland paper in the 1986 Symposium in Munich where he took an extensive data base with much more information (174 species) and compared it with five published mechanisms. The mechanisms of J. Warnatz were clearly superior but they were far from perfect, particularly getting into hydrocarbon radicals, and I think since I have seen no follow-up and no such comparison made in the last five years that there is a general reluctance to really pin down even the early steps to benzene formation in premixed flames. It gets to be a very personal situation. Somebody is right and somebody is calling somebody else wrong. It also gets into nasty boundary condition problems with H-atoms diffusing back into the burner. This is a very difficult model to start with. But I think we don't knock heads together very successfully in terms of ironing out where the problems are. I think you take your best estimates of thermodynamics, your best rate data, you construct the model that is the best you can do and then it disagrees with the experiment and there it sits. A millimeter is a mile on an H-atom profile at atmospheric pressure from the experimental point of view. Those kinds of shifts are significant disagreements on the basic radicals and the more complicated hydrocarbon radicals which are not being ironed out.

Wagner: I was expecting you to bring up that problem. Firstly, the detailed mechansims for the single species are getting examined. Quite a number of people are looking at the elementary reaction of the consecutive products of the ring systems and secondly the benzene formation and destruction problem has now its hundred anniversary. It was Berthelot who spent a lot of time and work and did a lot of very nice experiments in order to try to find out how benzene is being formed and what are the products of benzene destruction. This is also one of those problems which will accompany us for quite some time and now we are asking Prof. Glassman to give his views of the future.

Glassman: Not quite of the future, I have to pay some compliments to the chairman he always impresses me with his total intellect in approaching a problem. This intellect today started with nature and God as he put it. But unfortunately Prof. Wagner you didn't carry it one step further, now not only did God give us the diffusion flame as you put it but he basicly gave us condensed phase diffusion flames. About my own plans, I have no future plans I could not get a graduate student to follow up. I am no longer working in soot except that I am supervising and doing some simple testing with my equipment. But I received a call about six month ago from somebody who was doing droptower experiments, microgravity experiments. If you take a droplet in a drop tower you see no soot formation and if you take the same sized droplet at the earth condition you see a lot of soot. This obviously has to do with the convective profile which comes around the drop. We have heard absolutely nothing at this Symposium and the previous workshop on the creation of soot in liquid fuels which is probably the greatest number of fuels we use. People touched them and then throw them away. None of us had looked at the condition if you take a droplet in a normal field that is a non microgravity field and want to get soot formed if it is bigger than a certain size. If it is smaller than a certain size then the convective motions give you a completely enclosed spherically symmetric flame. Therefore, the soot particles don't appear. I think it is time that someone looks carefully at the problem of soot formation arround liquid droplets in a convective atmosphere.

Wagner: I know a few people who already stepped into that problem. I completely agree with you at these things as usual and I mean it even goes a little bit further. You also should look at solid material. Now we have one of our top modelers here, J. Warnatz

Warnatz: The present state in my opinion is that we have a rather good picture on the chemistry of aliphatic hydrocarbons, the aromatics are just starting with benzene. We have presented a first attempt to that in New York last month. The problem is there are no real measurements on benzene flames. There are a lot of measurements from the MIT, the Bittner dissertation, but the temperature profiles are missing which is a very weak part in that work. The next thing is to use this chemistry to model soot formation. The problem will be to build these models into flame codes and in a droplet code; we shall do both in the near future. The second point is how to combine this work with turbulent modeling. We are doing flamelet modeling with the KIVA2 code using large libraries and we are doing pdf work together with S. Pope. My concern is the small number of variables of scalars which is allowed. There is the magic number of four. What would be much more interesting is to talk how this number will go up in the next years. I can imagine that with ten variables we could make a good soot model. We can for instance model combustion with two variables. We need four or five for soot modeling, and, if you consider the diesel engine we need another four or five more for ignition. Then we end up with 10 or 12 and can expect good results. Finally, I would be very interested to hear the people's opinion who have worked on soot formation whether it is possible to have a tabulation of surface growth rates as function of the mixture fraction and temperature (in the diesel engine the pressure would be a third independant variable).

Wagner: We shouldn't forget one point and I wanted to come back to M. Frenklach but I didn't want to do that immediately because it may sound a little bit unpolite. In the past the experimentalist had a thermocouple and a pencil and modelers had a pencil and that was fair. Then the experimentalists got a spectroscope and the modelers still had just their pencil, that was unfair. Today it is the other way round. The experimentalists have the spectroscope and the lasers and the modelers get the growing computers and of course they will be very soon in a situation where they can process much more informations than the experimetalists can deliver. That puts the experimentalist under pressure and if science is done under pressure it usually has a tendency to get kind of a little bit rushing across the surface. Now we are coming to the next contribution, Prof. Homann. He is one of those who are going to look in the very details trying to figure out the chemistry as closely as possible.

Homann: I think the big success of the modeling was due to the use of very carefully measured elementry rate constants for small particles. That started with the hydrogen-oxygen-flame which is very well understood. I believe that there is a bad need for the same information about growing hydrocarbon species. We almost know nothing about the reactions of medium sized radicals. We know nothing about the thermal decomposition of PAH with and without side chains. This is why something has to be done on that field. I know it is a very difficult problem to investigate the thermal decomposition of PAH in a shock tube for instance but may be there are some other ideas. I have a question to the modelers, may be to M. Frenklach. With this lumping technique that bridges the gap between particle let say benzene and the beginning of the small soot particles you are calculating the concentration of these lumped species. Could you tell us what concentrations do we have to expect for a lumped species with a mass of 600 or 800u. Can you tell us a reasonable concentration from your calculations where we have to look. So that we can choose better experimental methods.

Frenklach: We don't lump species by masses, they are lumped by chemical nature. We are calculating the total of all different masses but of the same chemical character. So that doesn't answer your question because we don't have this answer. But if you ask me what is the concentration level of a 20 membered ring, I think the answer was provided yesterday by Mr. Mauss. He had this graph calculated and I think that is exactly the answer to your question.

Mauss: This is calculated with a lumping procedure. So if I have here 10^{-6} you will be three orders of magnitude lower.

Wagner: But that is not the way how it really happens.

Mauss: Yes the peak concentration of very stable species are not taken into account.

Homann: If it is an order of magnitude lower then you would expect a mole fraction of 10^{-7}.

Mauss: Yes.

Homann: So I don't think this would be a problem but I guess that we have to look at lower concentrations.

Mauss: I think the most stable one will have this mole fraction.

Wagner: That is at least a hint. But I know there are a few people here who think that not only the PAH are responsible for the soot growth. There is something in between soot and the PAH. I know Prof. Dobbins has that opinion and D' Alessio is not here in order not to be forced to fight for that concept and I myself also think that it is true. We usually have the habit to look at things we can catch easily and look away from things which we don't understand. That is a kind of human behaviour which has been introduced during evolution of course. Are there any further remarks on that topic before Mr. Bockhorn will finish the session. Shall we get some statements from the other side of the earth?

The other side of the earth: No.

Wagner: That means you agree completely with us.

The other side of the earth: Yes.

Wagner: We had various essential points which were being considered here and one of those points is to get more quantitative and reliable informations.

In the past there was this habit that somebody came new into the area, made one or two experiments, hopefully not more, and then he came up with a theory and that is the reason why we have such a large number of theories in soot formation. Now we are in a state where we can collect reliable information, we were talking about the possibility to introduce standard flames. I would like to follow that point. We must think about appropriate conditions for the modelers as well as for the experimentalists i.e. we have to use standard flames. But we must be able to measure particles, the chemical species, temperature, radicals, ions and we should have conditions where the sytems are as low-dimensional as possible.

Lindstedt: I would like to agree with that. There has been a lot of discussions about coflowing diffusion flames. I would like to add the counterflow geometry as a possible standard geometry to do studies of this kind. There has not been many flames measured and it is a very variable geometry. At the moment there is group at Aachen which is setting up such a flame and we extensively used measurements produced by Prof. Glassman's group and we have found them extremely helpful because you can vary conditions in these flames easily.

Wagner: I would like to go a little bit further. I would not only like to say, use a counter flow diffusion flame, I would like to get the information, for a counter flow diffusion flame with a particular geometry and a particular fuel so that you really can get comparable experimental results.

Lindstedt: I agree.

Colket: I agree exactly with what you are saying that we should not just fix the flame conditions but the burner type, size, and geometry. As an example (this is something I don' t know how many people are aware of this) but M.Zabielski back at UTRC has done a lot of work on different types of burners and finds that burners with different pressure drops across the burner face can have a dramatically different flame structure high above the burner which leads to a diffferent amount of cooling which goes back into the burner and causes dramatic differences in the production of species above the flame. Whatever one expect of unifying these flame conditions, the burner configuration should be also well defined. At this point shall I say my future plans?

Wagner: It would be very kind of you if you would mention your future plans.

Colket: Firstly, I shall be looking for a while at PAH formation because for the modeling we need some more information on PAH production rates so that we can compare them with models such as the ones which have been put forward in the literature. It is necessary to try to find out how we can do a better job with them. Hopefully I'll have something at the Symposium if I'm lucky and if I have the time to do it. Secondly, another area is the modeling work together with B.Hall and M.Smooke and trying to integrate the soot models that we have been building into an opposed jet diffusion flame. Perhaps at some point we would have the flexibility of calculating for a given pressure, temperature, fuel conditions what the soot growth rates might be in such a flame. Those results could be potentially fit into flame models such as I. Kennedy has been putting together.

Wagner: May I repeat, the next Symposium should be the location where we should meet and come to a firm conclusion. I would really like to do that and I think the President would accept that proposal that we take the next Symposium as a location to agree upon a certain standard flame set.

Calcote: I have always been a very strong advocate of a standard flame and a standard burner. When I got into this field there was already such a burner, the Homann / Wagner burner with acetylene/oxygen at 20 Torr and that turned out to be a pretty poor burner as we all know because of heat transfer. Nevertheless it has served as a standard flame on which many people have made measurements. A standard burner of that nature is extremely important. I think we should also pick a fuel other than acetylene as a standard fuel. May be benzene would be a good choice, we should have both an aromatic and an aliphatic fuel. As far as doing something at the Symposium I think that is a good idea. We can arrange sometime to get a group together if everyone will give some thought to it before that and we shall see what we can do.

Sarofim: I find myself very rarely in disagreement with Prof. Wagner but I think this is one time. My feeling is that if you standardize on a burner and set up conditions then you will stifle creativity. I was very excited by the Peters counter flow burner. It was a new idea and that is because we didn't have a standard burner. Let the thousand flowers bloom philosophy is a good way to go. I'd like to come back and say that your basic premise was that the modelers had computers and they could work so much faster. Then if they can work so much faster why not let them struggle to match every single burner configuration and that will at least make sure that the models are robust, that they can be applied to wide ranges of conditions. But this seems to be a minority opinion.

Wagner: No, that is not a minority opinion. You know that young human beings are guided for some time and may be with the modelers we are in a comparable situation.

Santoro: I guess I want to say a couple of things about the burner debate. M. Frenklach had mentioned the co-flow burner because we have seen quite a few references to results in that burner. In the diffusion flame area we have three burners that are used: the coflow, counterflow and Wolfhard-Parker burners. The problem I see with the counter flow diffusion flame burner is that the stagnation point location is very critical and you can vary its location by the choice of the flow conditions. If you recall the W. Flower et al. experiments, they were very careful about maintaining the stagnation point to flame front location constant, so that they could compare results when the flow conditions were varied. So I think if you are going to standardize on that type of burner you have to be very careful about the flow conditions that are being used or it starts to get difficult, in my opinion. I do see some problems with the co-flow burners because they require a lot of measurements. May be more effort should be put into the Wolfhard-Parker burner which has some stability problems. That is why the co-flow burner is being used so much, because everybody who operates that burner finds it is very stable. One quick comment on A. Sarofim's suggestion. The other thing that a standard burner allows, which I found when I did some work with K. Smyth, is to do different measurements in different labs. You can also do one or two similar measurements to check each other and make sure that everybody is getting the same results which adds confidence to the results. I think to the people who are trying to reproduce those results in a calculation, they are more confident when more than a single individual obtained the needed results. Whether there should be a single burner, I don't know? But there is a lot of benefit by having exactly the same conditions.

Wagner: We are here in order to get different opinions and I'm very pleased to see that development to the counter flow diffusion flames because I remember when Felix Weinberg started again, investigated that burner extensively he solved most of his problems and nobody continued this type of work. Now we are coming towards the end of our discussion. M. Colket wants to make a statement and then you come to a final conclusion about the model.

Colket: I want to vote against the opposed jet diffusion flame as a standard diffusion flame because it is impossible to put a mass-spectrometer in it and a mass-spectrometer is still the most powerful tool we have got to get the profiles that are needed.

Moss: Over the last 10 years it seems to me one of the great unifying burner themes has been the laminar flamelet concept, in the sense of it's relationship both to premixed and non-premixed flames, because it's provided a flame forum for chemists to talk to computational specialists who look at flow field. Soot is not in that position yet. There is this feeling that actually flamelet modeling is inappropriate in the context of soot. We are into the business of genuine finite rate chemistry; if the chemistry is not fast, the method is not going to work. I'm happy with the idea of consolidating on a few laminar flame burner geometries and I think one would want to include both the Wolfhard-Parker and the counter-flow in that. To find out about soot in the turbulent flame field we must actually have a much better database in the turbulent flames than we have and I do not think it will be as easy to carry over the laminar flame concept to the turbulent flame because we will not have an equivalent of the laminar flamelet concept. What I hope to see is the development of diagnostics in sooting flames. It seems to be in a primitive state and it is a very difficult regime, may be it is a limit problem and may be there are no simple ways through this. I have gone away from the two day at this meeting being much more sceptical about most of the measurements that I was likely to make before I came here. And I would want to encourage people who have been interested in making measurements to take the turbulent flame more seriously. Unless we actually have some combination of simultaneous measuremets we are really not going to build and validate any of those joint pdfs. I spent 10 very happy years working in non-premixed turbulent flames, my future plans will be to move back to turbulent premixed flames that are non-luminous.

Wagner: Ladies and gentlemen, I would like to thank those who contributed to this discussion. I think there is some very valuable information coming out here. I would also like to thank those who showed us that they can keep their secrets. I think we can now finish this outlook into the future. Apparently we are somewhere in between and that's a good state for science. You know there is a lot of work to be done, many things are needed. Now I would like to give the microphone to Prof. Bockhorn and before that I would like to thank him very much for organizing this workshop and I would also like to thank everybody who has been involved in the organisation. I hope that he will not be suffering to much when he has to work up through all these manuscripts, through all these discussion remarks and to all that trouble with the publishers!

Index of Contributors

Subject Index

Springer Series in Chemical Physics
Editors: Vitalii I. Goldanskii   Fritz P. Schäfer   J. Peter Toennies

Printing: Mercedesdruck, Berlin
Binding: Buchbinderei Lüderitz & Bauer, Berlin